Lecture Notes in Electrical Engineering

Volume 991

The book series *Lecture Notes in Electrical Engineering* (LNEE) publishes the latest developments in Electrical Engineering—quickly, informally and in high quality. While original research reported in proceedings and monographs has traditionally formed the core of LNEE, we also encourage authors to submit books devoted to supporting student education and professional training in the various fields and applications areas of electrical engineering. The series cover classical and emerging topics concerning:

- Communication Engineering, Information Theory and Networks
- Electronics Engineering and Microelectronics
- Signal, Image and Speech Processing
- Wireless and Mobile Communication
- Circuits and Systems
- Energy Systems, Power Electronics and Electrical Machines
- Electro-optical Engineering
- Instrumentation Engineering
- Avionics Engineering
- Control Systems
- Internet-of-Things and Cybersecurity
- Biomedical Devices, MEMS and NEMS

For general information about this book series, comments or suggestions, please contact leontina.dicecco@springer.com.

To submit a proposal or request further information, please contact the Publishing Editor in your country:

China

Jasmine Dou, Editor (jasmine.dou@springer.com)

India, Japan, Rest of Asia

Swati Meherishi, Editorial Director (Swati.Meherishi@springer.com)

Southeast Asia, Australia, New Zealand

Ramesh Nath Premnath, Editor (ramesh.premnath@springernature.com)

USA, Canada

Michael Luby, Senior Editor (michael.luby@springer.com)

All other Countries

Leontina Di Cecco, Senior Editor (leontina.dicecco@springer.com)

**** This series is indexed by EI Compendex and Scopus databases. ****

Min Xu · Li Yang · Linghao Zhang · Shu Yan
Editors

Innovative Technologies for Printing and Packaging

Springer

Editors
Min Xu
China Academy of Printing Technology
Beijing, China

Li Yang
China Academy of Printing Technology
Beijing, China

Linghao Zhang
China Academy of Printing Technology
Beijing, China

Shu Yan
China Academy of Printing Technology
Beijing, China

ISSN 1876-1100 ISSN 1876-1119 (electronic)
Lecture Notes in Electrical Engineering
ISBN 978-981-19-9023-6 ISBN 978-981-19-9024-3 (eBook)
https://doi.org/10.1007/978-981-19-9024-3

This Springer imprint is published by the registered company Springer Nature Singapore Pte Ltd.
The registered company address is: 152 Beach Road, #21-01/04 Gateway East, Singapore 189721, Singapore

Committees

Supervisors

The Printing Technology Association of China
Chinese Society for Imaging Science and Technology

Sponsors

China Academy of Printing Technology
Qilu University of Technology, Shandong Academy of Science
Printing Technology Journals Press

Organizers

Editorial Department of *Digital Printing*
Beijing Key Laboratory of New Technology of Packaging and Printing
Key Laboratory of Environmental Protection and Intelligence Technology for Printing Industry
Faculty of Light Industry, Qilu University of Technology, Shandong Academy of Science
State Key Laboratory of Biobased Material and Green Papermaking
Printing Technology Professional Committee, Chinese Society for Imaging Science and Technology

Support

Shandong Printing Association

Co-sponsors

School of Printing and Packaging Engineering, Beijing Institute of Graphic Communication
School of Media and Design, Hangzhou Dianzi University
School of Mechanical and Electrical Engineering, Beijing Institute of Graphic Communication
College of Light Industry Science and Engineering, Tianjin University of Science and Technology
Faculty of Printing, Packaging Engineering and Digital Media Technology, Xi'an University of Technology
College of Light Industry and Engineering, South China University of Technology
Research Center of Graphic Communication, Printing and Packaging, Wuhan University
State Key Laboratory of Modern Optical Instrumentation, Zhejiang University
College of Bioresources Chemical and Materials Engineering, Shaanxi University of Science and Technology
School of Light Industry and Chemical Engineering, Dalian Polytechnic University
Light Industry College, Harbin University of Commerce
School of Packaging and Material Engineering, Hunan University of Technology
College of Materials Science and Engineering, Beijing University of Chemical Technology
College of Material Science and Engineering, Zhengzhou University
School of Art and Design, Henan University of Engineering
School of Light Industry, Beijing Technology and Business University
College of Communication and Art Design, University of Shanghai for Science and Technology
School of Mechanical Engineering, Tianjin University of Commerce
School of Mechanical Engineering, North University of China
Shanghai Publishing and Printing College
School of Mechanical Engineering, Jiangnan University
College of Light Industry and Food Engineering, Nanjing Forestry University
College of Mechanical Engineering, Quzhou University
School of Media and Communication, Shenzhen Polytechnic
Tianjin Vocational Institute
College of Communications, Taiwan University of Arts

Conference Executive Committee

Chairmen

Yu Liu	Vice President of Qilu University of Technology, Shandong Academy of Science
Xiaoxia Chang	Vice-General Manager of China Culture Industry Development Corporation, President and Manager of Beijing Keyin Media Culture Co., LTD, President of *Printing Technology* Journals Press
Pengfei Zhao	President of China Academy of Printing Technology

Vice Chairmen

Jinhong Li	Director of the Academic Affairs Office of Qilu University of Technology, Shandong Academy of Science
Yiping Liu	Vice-General Manager of Beijing Keyin Media Culture Co., LTD
Zhuangzhi Ye	Vice President of China Academy of Printing Technology

Honorary Chairmen

Desen Qu	Former President of Beijing Institute of Graphic Communication
Tingliang Chu	Former President of China Academy of Printing Technology, Executive Vice Chairman of the Printing Technology Association of China
Wencai Xu	Former Vice President of Beijing Institute of Graphic Communication, Vice Chairman of the Printing Technology Association of China
Jialing Pu	Former Vice President of Beijing Institute of Graphic Communication, Honorary Chairman of Chinese Society for Imaging Science and Technology

Secretary-Generals

Maohai Lin	Vice Director of Faculty of Light Industry of Qilu University of Technology, Shandong Academy of Science, Vice Director of State Key Laboratory of Biobased Material and Green Papermaking
Min Xu	Chief Editor of *Digital Printing*

Conference Academic Committee

Chairman

Deren Li Academician of Chinese Academy of Sciences, Academician of China Engineering Academy, Academician of International Eurasian Academy of Sciences, Academician of New York Academy of Sciences, Academician of International Academy of Astronautics, Professor of Wuhan University, Director of State Key Laboratory of Information Engineering in Surveying, Mapping and Remote

Vice Chairmen

Kefu Chen	Academician of China Engineering Academy, Professor of South China University of Technology, Director of Academic Committee of State Key Laboratory of Pulp and Paper Engineering
Jinping Qu	Academician of Chinese Academy of Engineering, Professor of South China University of Technology, Director of National Engineering Research Center of Novel Equipment for Polymer Processing, Director of the Key Laboratory for Polymer Processing Engineering of Ministry of Education
Guangnan Ni	Academician of China Engineering Academy, Researcher of Institute of Computing Technology Chinese Academy of Sciences, Board Chairperson of Chinese Information Processing Society of China
Songlin Zhuang	Academician of China Engineering Academy, Director and Professor of School of Optical-Electrical and Computer Engineering of University of Shanghai for Science and Technology, Optical Expert

Commissioners

Yuri Andreev	Head of Moscow State University of Printing Arts
Stephen W. Bigger	Doctor, Professor, Vice President of Faculty of Engineering and Science of Victoria University
Jinda Cai	Doctor, Professor of University of Shanghai for Science and Technology
Congjun Cao	Doctor, Professor of Xi'an University of Technology
Guorong Cao	Doctor, Professor of Beijing Institute of Graphic Communication

Shaozhong Cao	Doctor, Professor of Beijing Institute of Graphic Communication
Guangxue Chen	Doctor, Professor of South China University of Technology
Jinzhou Chen	Professor of Zhengzhou University
Yunzhi Chen	Doctor, Professor of Tianjin University of Science and Technology
Kamal Chopra	Professor, President of All India Federation of Master Printers
Fuqiang Chu	Doctor, Professor of Qilu University of Technology, Shandong Academy of Science
Tingliang Chu	Professorial Senior Engineer of China Academy of Printing Technology
Martin Dreher	Doctor, Director and General Manager of DFTA Technology Center, Professor of the Hochschule der Medien (HdM)
Changqing Fang	Doctor, Professor of Xi'an University of Technology
Patrick Gane	Doctor, Professor of Printing Technology at the School of Chemical Technology, Aalto University
Phil Green	Professor of Colour Imaging, London College of Communication
Weibing Gu	Doctor, Senior Engineer of Suzhou Institute of Nano-Tech and Nano-Bionics, Chinese Academy of Science
Jeroen Guinée	Doctor, Associate Professor of Leiden University
Aran Hansuebsai	Doctor, Associate Professor of Chulalongkorn University
Jon Yngve Hardeberg	Doctor, Professor of Norwegian University of Science and Technology
Songhua He	Doctor, Professor of Shenzhen Polytechnic
Roger D. Hersch	Doctor, Professor of Computer Science and Head of the Peripheral Systems Laboratory at the Ecole Polytechnique Fédérale de Lausanne (EPFL)
Mathias Hinkelmann	Doctor, Professor of Stuttgart Media University
Thomas Hoffmann-Walbeck	Professor of the Faculty of Print and Media Technology, Stuttgart University of Media
Yungcheng Hsieh	Doctor, Professor of National Taiwan University of Arts
Kun Hu	Doctor, Senior Engineer of Beijing Institute of Graphic Communication
Min Huang	Doctor, Professor of Beijing Institute of Graphic Communication
Takashi Kitamura	Doctor, Professor of Graduate School of Advanced Integration Science, Chiba University
Shijun Kuang	Consultant Engineer of China National Pulp and Paper Research Institute

Yanlin Song	Doctor, Professor of Institute of Chemistry, Chinese Academy of Sciences
Zhicheng Sun	Doctor, Professor of Beijing Institute of Graphic Communication
Zhihui Sun	Professor of Harbin University of Commerce
Zhi Tang	Doctor, Researcher of Wangxuan Institute of Computer Technology, Peking University
Anita Teleman	Doctor, Research Manager of Printing Solutions at the Research Institute Innventia, Sweden
Yuemin Teng	Professor of Shanghai Publishing and Printing College
Junfei Tian	Doctor, Professor of South China University of Technology
Martti Toivakka	Doctor, Professor and Head of the Laboratory of Paper Coating and Converting at Åbo Akademi University
Alexander Tsyganenko	Professor of Media Industry Academy
Philipp Urban	Head of Emmy Noether Research Group, Institute of Printing Science and Technology, Technische Universität Darmstadt
Howard E. Vogl	Visiting Professor of Rochester Institute of Technology
Xiaoxia Wan	Doctor, Professor of Wuhan University
Haiqiao Wang	Doctor, Professor of Beijing University of Chemical Technology
Lijie Wang	Professor of Shenzhen Polytechnic
Mengmeng Wang	Doctor, Associate Professor of Jiangnan University
Qiang Wang	Doctor, Professor of Hangzhou Dianzi University
Wei Wang	Doctor, Professor of Beijing Institute of Graphic Communication
Xiaohui Wang	Doctor, Professor of South China University of Technology
Yufeng Wang	Senior Engineer of Tianjin University of Science and Technology
Minchen Wei	Doctor, Associate Professor of the Hong Kong Polytechnic University
Xianfu Wei	Doctor, Professor of Beijing Institute of Graphic Communication
Yan Wei	Doctor, Professor of Tsinghua University, Beijing Institute of Graphic Communication
Jimei Wu	Doctor, Professor of Xi'an University of Technology
Wei Wu	Doctor, Professor of Wuhan University
Kaida Xiao	Doctor, University Academic Fellow of University of Leeds
Punan Xie	Professor of Beijing Institute of Graphic Communication

Xiulan Xin	Doctor, Professor of Beijing Technology and Business University
Jinlin Xu	Professor of Xi'an University of Technology
Wencai Xu	Professor of Beijing Institute of Graphic Communication
Bin Yang	Doctor, Researcher of Peking University
Haigen Yao	Professor of Shanghai Publishing and Printing College
Yadong Yin	Doctor, Professor of University of California, Riverside
Ruien Yu	Doctor, Associate Professor of North University of China
Haiyan Zhang	Professor of Xi'an University of Technology
Pengfei Zhao	Senior Engineer of China Academy of Printing Technology
Yunfei Zhong	Professor of Hunan University of Technology
Haihua Zhou	Doctor, Associate Researcher of Institute of Chemistry, Chinese Academy of Sciences
Shisheng Zhou	Doctor, Professor of Xi'an University of Technology
Yingquan Zou	Doctor, Professor of Beijing Normal University

Reviewers

Congjun Cao	Doctor, Professor of Xi'an University of Technology
Shaozhong Cao	Doctor, Professor of Beijing Institute of Graphic Communication
Guangxue Chen	Doctor, Professor of South China University of Technology
Liangzhe Chen	Doctor, Lecturer of Jingchu University of Technology
Fuqiang Chu	Doctor, Professor of Qilu University of Technology, Shandong Academy of Science
Guirong Dong	Doctor, Associate Professor of Xi'an University of Technology
Yi Fang	Doctor, Lecturer of Beijing Institute of Graphic Communication
Xue Gong	Doctor, Associate Professor of Harbin University of Commerce
Weibing Gu	Doctor, Senior Engineer of Suzhou Institute of Nano-Tech and Nano-Bionics, Chinese Academy of Science
Linghua Guo	Doctor, Professor of Shaanxi University of Science and Technology
Minghui He	Doctor, Vice Researcher of South China University of Technology
Heping Hou	Doctor, Associate Professor of Xi'an University of Technology
Min Huang	Doctor, Professor of Beijing Institute of Graphic Communication
Lijiang Huo	Professor of Dalian Polytechnic University
Xiaoshan Jiang	Doctor, Senior Engineer of Beijing Institute of Graphic Communication

Fazhong Zhang	Doctor, Senior Engineer of Beijing Linjiang Technology Co., LTD
Gaimei Zhang	Doctor, Professor of Beijing Institute of Graphic Communication
Guowei Zhang	Doctor, Lecturer of Tianjin University of Science and Technology
Lizheng Zhang	Lecturer of Beijing Institute of Graphic Communication
Zhengjian Zhang	Doctor, Associate Professor of Tianjin University of Science and Technology
Haihua Zhou	Doctor, Associate Researcher of Institute of Chemistry, Chinese Academy of Sciences
Jinda Zhu	Doctor, Associate Professor of Hebei University of Science and Technology
Lei Zhu	Doctor, Senior Engineer of Beijing Institute of Graphic Communication
Wei Zi	Doctor, Lecturer of Zhengzhou University

Preface

"2022 13th China Academic Conference on Printing and Packaging and Forum of Collaborative Innovation of Industry-University-Research", one of the series "China Academic Conference on Printing and Packaging" which is mainly hosted by China Academy of Printing Technology, was held on November 10–12, 2022, in Jinan City, China. Under the guiding of the Printing Technology Association of China and Chinese Society for Imaging Science and Technology, the conference was co-hosted by China Academy of Printing Technology, Qilu University of Technology (Shandong Academy of Sciences), and *Printing Technology* Journals Press, was organized by Key Laboratory of Environmental Protection and Intelligence Technology for Printing Industry in CAPT, Faculty of Light Industry of Qilu University of Technology (Shandong Academy of Sciences), State Key Laboratory of Biobased Material and Green Papermaking Printing Technology, Editorial Department of *Digital Printing*, and Printing Technology Professional Committee of Chinese Society for Imaging Science and Technology.

By far, "China Academic Conference on Printing and Packaging (CACPP)" and its series of events have been held for thirteen sessions and have already become the most influential academic exchange activities in printing and packaging fields in China.

In recent five years, China's printing industry has been expanding in scale, improving in quality and efficiency, optimized in industrial layout, and obvious in structural adjustment. The proportion of output value of key printing enterprises above designated size has exceeded 60%. Traditional drivers being replaced by new ones, the compound annual growth rate of digital printing has exceeded 30%; effectively responding to the COVID-19 outbreak and other risks and challenges, printing support work is strong and powerful. In 2021, there were 6010 enterprises above designated size in the printing industry, with a total operating income of 744.2 billion yuan, a growth rate of 10.3% over the previous year. Universities, research institutions, and enterprises have been paying more attention to scientific research, actively introducing computer technology, Internet, artificial intelligence, material science, and other basic research results as support. Many scientific research achievements have emerged in the intelligentization of printing equipment, the greening and functionalization of printing materials, digitization of printing processing. These results will continue to be reported and communicated at the 2022 13th China Academic Conference on Printing and Packaging and Forum of Collaborative Innovation of Industry-University-Research.

For this academic conference, we specially invited Prof. Zhicheng Ji, the former Vice President of Jiangnan University, Prof. Yao Li of Harbin Institute of Technology, Prof. Yunhai Wang of Shandong University, and Xingxiang Ji of Qilu University of Technology (Shandong Academy of Sciences) to make keynote speeches on topics of Intelligent Optimization Manufacturing Based on Industrial Internet, Research of Photothermal Materials, Research of Replacing Plastic with Paper, Automatic Visualization of Combining Visual Perception and Aesthetic Art. At the same time, four parallel symposiums on Printing and Digital Media Technology, Innovative Packaging Technology, Printing Electronics Technology, and Intelligent Technology were held, in which the professors, the young scholars, and the authors of papers from related universities participated and conducted reports.

At the same time, in order to promote the industry-university-research collaborative innovation, the conference organized the relevant scientific researchers to publish their scientific findings that can be industrially applied and invite the enterprises to disclose their requirements for the scientific collaboration, which will strongly promote the combination of scientific research and industry.

The conference received 142 papers this year, among which about 84 papers were selected to be published on Lecture Notes in Electrical Engineering (LNEE) (ISSN: 1876-1100) by Springer.

Here we greatly acknowledge all the co-sponsors that offered great support for the conference, and they are: Beijing Institute of Graphic Communication, School of Light Industry and Engineering of South China University of Technology, Research Center of Graphic Communication, Printing and Packaging of Wuhan University, Faculty of Printing, Packaging Engineering and Digital Media Technology of Xi'an University of Technology, School of Media and Design of Hangzhou Dianzi University, College of Light Industry Science and Engineering of Tianjin University of Science and Technology, State Key Laboratory of Modern Optical Instrumentation of Zhejiang University, College of Bioresources Chemical and Materials Engineering of Shaanxi University of Science and Technology, College of Communication and Art Design of University of Shanghai for Science and Technology, Light Industry College of Harbin University of Commerce, School of Mechanical Engineering of Jiangnan University, School of Light Industry and Chemical Engineering of Dalian Polytechnic University, School of Packaging and Material Engineering of Hunan University of Technology, College of Engineering of Qufu Normal University, College of Light Industry and Food Engineering of Nanjing Forestry University, College of Materials Science and Engineering of Beijing University of Chemical Technology, School of Material Science and Engineering of Zhengzhou University, School of Light Industry of Beijing Technology and Business University, School of Mechanical Engineering of Tianjin University of Commerce, School of Mechanical Engineering of North University of China, College of Mechanical Engineering of Quzhou University, School of

Art and Design of Henan University of Engineering, School of Media and Communication of Shenzhen Polytechnic, Tianjin Vocational Institute, Shanghai Publishing and Printing College, College of Communications of Taiwan University of Arts.

We would like to express our gratitude to the 42 experts from the fields of color, image, computer, material, machinery, and information engineering for their strict reviewing and recommending papers for the conference with strict standards.

We also thank Springer for offering us an international platform for publishing.

We look forward to our reunion at 2023 14th China Academic Conference on Printing and Packaging.

Beijing, China
November 2022

Edited by
China Academy of Printing Technology
Printing Technology Journals Press

Contents

Color Science and Technology

Testing the Performance of the CIECAM16 for Cross-Media Colour Reproduction Using Real Scene Experiment

Yuechen Zhu and Ming Ronnier Luo(✉)

State Key Laboratory of Modern Optical Instrumentation, Zhejiang University, Hangzhou, China
m.r.luo@zju.edu.cn

Abstract. The existing colour appearance models are mainly based on the experiments using colour patches and simple images in a cabinet or on a display, which may have problems in predicting images in real scene. In this study, a real scene experiment was conducted to restore real-world scenario, and study the impact of CCTs and illuminance levels of the lighting conditions on the cross-media colour reproduction. The experiment was conducted in a spectrally tuneable LED lighting room. A total of 48 groups of real scenes were selected, including four CCTs, four illuminance levels and three types of objects. The objects included oil paintings, fruits and vegetables, natural skin tone test charts. The results showed that the colour appearance of real scene and the image on the display were relatively different, especially for low CCT and illuminance level, and the illuminance affected the chromaticity in the case of low illuminance conditions. The contents of scene didn't show a significate impact. In addition, the prediction of CCT in the CIECAM16 should be revised especially under low illuminance conditions, and there was a big deviation in the prediction of luminance.

Keywords: Cross-media colour reproduction · Colour appearance model · Real scene experiment

1 Introduction

The colour appearance model, CIECAM16, was recommended by the International Commission on Illumination (CIE) to predict the colour appearance under different viewing conditions, such as media, surround and background [1]. An approximate cross media colour reproduction (CMCR) is obtained by inputting different parameters. The accuracy of the model is mostly affected by the dataset collected by psychophysical experiments. The most widely-used dataset is the LUTCHI dataset, which was obtained by a large-scale experiment using colour patches against various viewing conditions [2, 3]. Braun and Fairchild performed experiments to compare the colour appearance between printed images and CRT reproductions [4]. Huang et al. studied the impact of different viewing light illuminance on CMCR between display and printed samples using colour matching method [5]. But these studies only included colour patches and simple images, which may have problems in predicting images in real scene. Xu et al.

M. Xu et al. (Eds.): CACPP 2022, LNEE 991, pp. 3–8, 2023.
https://doi.org/10.1007/978-981-19-9024-3_1

reproduced a real lit environment as an image on a self-luminous display and studied equal perceived brightness between the room and the display [6]. But the chromaticity was not taken into consideration in the study, which was important in colour appearance reproduction.

In order to reproduce the colour appearance between real scene and images on self-luminous display. This study conducted a series of psychophysical experiments using threshold method. The goal is to find the trend and test the performance of in the CIECAM16 for cross-media colour reproduction.

2 Methods

The experiment was conducted in a lighting room controlled by ten LED light sources with 11 channels. Apple Pro Display XDR was used as a standard display (size: 32 inch, luminance level: 324.4 cd/m^2) in this study. A Konica Minolta CS2000A spectroradiometer was used to calibrate the illumination conditions and the display. CIE1931 standard colorimetric observer was used to calculate the XYZ values.

Sixteen adapting conditions were designed including four CCTs (3000 K, 4500 K, 6500 K, 8000 K) and 4 illuminance levels (10 lx, 100 lx, 500 lx, 1000 lx).

Three types of real scenes were built up, containing oil paintings, fruits and vegetables, natural skin tone test charts, as shown in Fig. 1. Note that all the scenes contained a Macbeth ColorChecker Chart (MCCC) for subsequent analysis.

a.oil paintings b.fruits and vegetables c.natural skin tone test charts

Fig. 1. Three types of real scenes

Original images of the real scenes were first captured using a Nikon Z6 digital camera under a standard condition (6500 K, 500 lx). A camera characterization model was implemented using a polynomial regression technique with a precision of 1.5 ΔE^*_{ab} averaged from the 24 test colours of the MCCC. Each original image transformed from camera image in RGB format to image in CIE XYZ tristimulus values on a pixel-by-pixel basis. An image rendering database was built up using the CIECAM16 model. Each original image was then processed to give 513 images with different white points, including 27 CCTs ($X_wY_wZ_w$) and 19 luminance levels (k), shown in Table 1, (X'_w, Y'_w, Z'_w) = (X_w*k, Y_w*k, Z_w*k). The XYZ data were processed via the forward CIECAM16 model to predict the perceptual attributes under current standard condition (6500 K, 500 lx, surround = average). Then, using the reverse CIECAM16 model, the attributes were transformed to XYZ tristimulus data for target viewing conditions. Finally, the XYZ values were transformed to RGB values on a display via the Gain-Offset-Gamma model.

Table 1. White points of 513 image rendering database, including 27 CCTs and 19 luminance levels (expressed by k).

27 CCT (K)	2900	3100	3300	3500	3700	3900	4100	4300	4500
	4700	4900	5100	5300	5500	5700	5900	6100	6300
	6500	6700	6900	7100	7300	7500	7700	7900	8100
19 k	0.2	0.3	0.4	0.5	0.6	0.7	0.8	0.9	1.0
	1.1	1.2	1.3	1.4	1.5	1.6	1.7	1.8	1.9
	2.0	/	/	/	/	/	/	/	/

Observers were asked to do adaptation for two minutes by looking around the neutral wall and the objects in the real scene. The next step was 'Rough selection', 20 images differed in CCT (CCT = 3500 K, 4500 K, 5500 K, 6500 K, 7500 K) and luminance (k = 0.5, 0.9, 1.3, 1.7) were presented simultaneously, as shown in Fig. 2a. Observers were asked to choose one initial image (labelled as 'CCT_i, k_i') which was most similar to the real scene on the display. When they finished the selection, the program automatically found 49 images which were close to the initial image in the image rendering database (7 CCT: $CCT_i \pm 600, \pm 400, \pm 200, +0$; 7 k: $k_i \pm 0.3, \pm 0.2, \pm 0.1, +0$). Each image was shown in the centre of the display in a random order, the remaining part of the display was set to a grey which was as same as the white wall of the real scene to stabilize the adaptation, as shown in Fig. 2b. This step was 'Fine selection', observer compared the images on the display with the real scene, and a threshold method was used for observers to judge each image as 'matched or not matched'. The sequence of test images, lighting conditions and real scenes was randomized for each observer.

a.Rough selection b.Fine selection

Fig. 2. Experimental interface

Twenty normal colour vision observers (10 males and 10 females) between 22 and 28 years of age took part in the experiment. In total, 49,980 evaluations were accumulated, i.e., (4 CCTs $\times$ 4 illuminance levels + 1 repeat) $\times$ 3 real scenes $\times$ 49 evaluations $\times$ 20 observers.

3 Results

Mean Colour Difference from the Mean (MCDM) were calculated to represent the observer variation of the result. All colour differences were calculated using CIELAB colour space [3]. The overall MCDM values of inter-observer variations were found to be 8.2, 7.6, 5.9 of ΔE^{*}_{ab} units respectively. The results of natural skin tone test charts showed the highest stability. And the observer variations were typical for psychophysical experiments using threshold method.

Because the images were generated on a pixel-by-pixel basis and each image contained a MCCC, the CIE XYZ tristimulus values of No. 21 patch (natural grey) of each image were measured to represent its colorimetric values. Thus, for each experimental setting, observer evaluated 49 images corresponding to different XYZ values. To average the XYZ values of the images judged by 'matched', the results of each experimental setting were obtained.

Figure 3 shows the results of three types of real scenes individually. The crosses are the chromaticity coordinates of the adapting conditions of the real scene, and the squares are the chromaticity coordinates of the visual results on the display. The distances between them represented the appearance shift between display and real scene. It can be found that there is a consistent trend, which means that the contents of real scene didn't impact the results. Thus, the averaged results were plotted in CIE1931 $u'v'$ plane, as shown in Fig. 4. The results showed that the visual results of the display under low CCT are quite different from the real scene, especially for low illuminance.

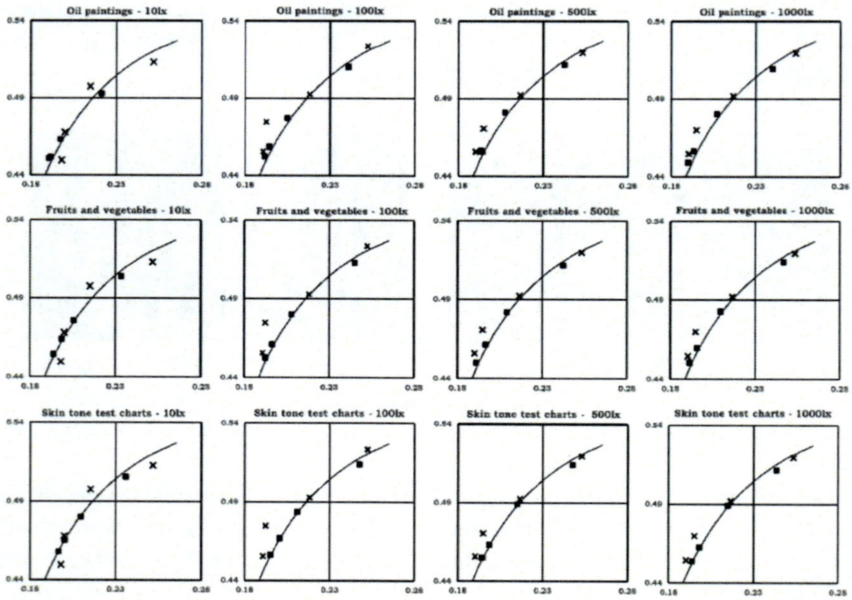

Fig. 3. Results of three types of real scenes plotted in CIE1931 $u'v'$ plane individually.

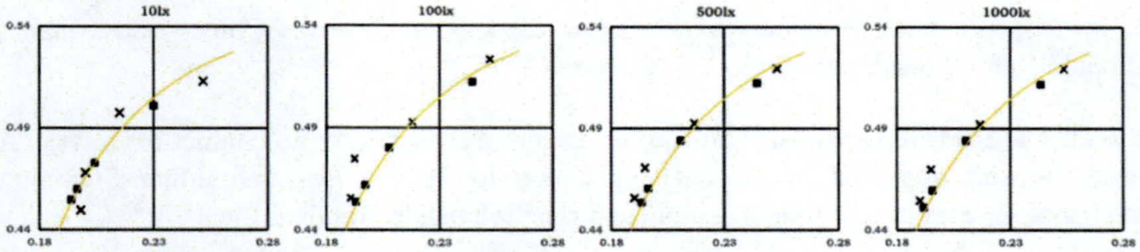

Fig. 4. Averaged results were plotted in CIE1931 $u'v'$ plane.

Then, the visual results were used to test the performance of the CIECAM16. Figure 5 shows the CCT and luminance of No. 21 patch predicted by the CIECAM16 comparing with the visual results. It can be found that there is a big deviation in the prediction of luminance. And the prediction of CCT in the CIECAM16 should be revised especially under low illuminance conditions.

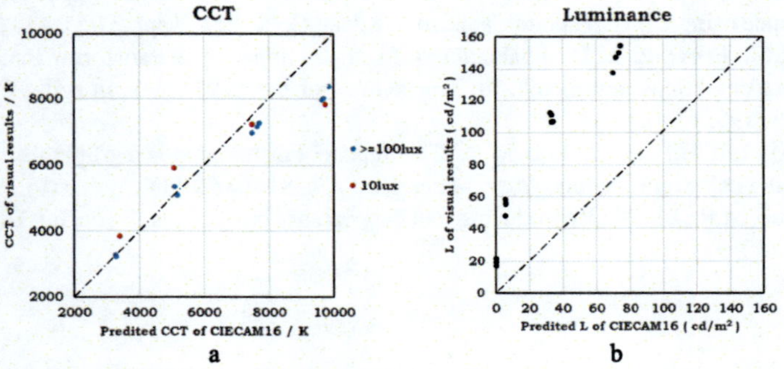

Fig. 5. CCT and luminance predicted by the CIECAM16 comparing with the visual results.

4 Conclusions

A psychophysical experiment was conducted to restore real-world scenario, and study the impact of CCTs and illuminance levels of the lighting conditions on the cross-media colour reproduction. The experiment was conducted in a spectrally tuneable LED lighting room. A total of 48 groups of real scenes were selected, including four CCTs, four illuminance levels and three types of objects. The contents of real scenes included oil paintings, fruits and vegetables, natural skin tone test charts.

The results showed that the colour appearance of real scene and the image on the display were relatively different, especially for low CCT and illuminance level, and the illuminance affected the chromaticity in the case of low illuminance conditions. The contents of scene didn't show a significate impact. In addition, the performance of the CIECAM16 colour appearance model was tested. It can be found that the prediction of CCT in the CIECAM16 should be revised especially under low illuminance conditions, and there is a big deviation in the prediction of luminance. Thus, the CIECAM16 should be modified when applied to cross-media colour reproduction.

Acknowledgements. The authors like to thank the support from the Chinese Government's National Science Foundation (Project Number on 61775190).

Statement of Ethical Approval The authors declare that they have no conflict of interest. All procedures were approved by the Zhejiang University Human Research Ethics Committee. Participants gave written informed consent and were reimbursed for their time.

References

1. CIE 248:2022: The CIE 2016 Colour Appearance Model for Colour Management Systems: CIECAM16 (2022)
2. Luo, M.R., Clarke, A.A., Rhodes, P.A., Schappo, A., Scrivener, S.A.R., Tait, C.J.: Quantifying colour appearance: part I – LUTCHI colour appearance data. Color Res. Appl. **16**(3), 166–180 (1991)
3. Luo, M.R.: Quantifying colour appearance. Part II. Testing colour appearance models performance using LUTCHI colour appearance data. Color Res. Appl. **31**(5), 438 (2006)
4. Braun, K.M., Fairchild, M.D.: Evaluation of five color-appearance transforms across changes in viewing conditions and media. In: Proceedings of the 1995 Color Imaging Conference, pp. 93–96 (1995)
5. Huang, M., Liu, H., Cui, G., Luo, M.R.: The impact of different viewing light illuminance on cross-media color reproduction. Adv. Mater. Res. **174**, 81–84 (2011)
6. Xu, L., Luo, M.R., Liu, X.: Reproducing room appearance on displays. J. Soc. Inf. Disp. **22**(12), 623–630 (2015)

Colour Matching Results from Two Distinct Observers via a Visual Trichromator System

Keyu Shi[1], Ming Ronnier Luo[1](✉), Tingwei Huang[2], and Jianlong Zhang[2]

[1] State Key Laboratory of Modern Optical Instrumentation, Zhejiang University, Hangzhou, China
m.r.luo@zju.edu.cn

[2] Thousand Lights Lighting (Changzhou) Limited, Changzhou, China

Abstract. Human colour vision is represented by Colour Matching Function (CMF). A colour matching experiment based upon Maxwell's matching method was performed. A visual trichromator based on spectrum tunable LED system consisting of control panels was built. The fixed "standard" half-field was white, while the "test" half-field was illuminated by three standard primaries one of which was replaced by a spectrally offset primary. The results from two distinct observers were reported in terms of inter- and intra-observer variations, 95% perception ellipses and individual CMFs.

Keywords: Colour matching function · Spectrum tunable LED system

1 Introduction

Human color vision varies between observers because of individual differences in macular, lens, and photopigment optical densities, and spectral shifts in the cone sensitivity spectra [1]. The most commonly used standard colorimetric observers are the CIE 1931 and CIE 1964 standard colorimetric observers, i.e. the 2° and 10° observers [2]. The CIE 1931 standard colorimetric observer was based upon the data accumulated by Guild (1931) [3] and Wright (1929) [4, 5], while the CIE 1964 standard colorimetric observer was based on experimental investigations made by Stiles and Burch (1959) [6], and Speranskaya (1959) [7]. However, it was found recently that the 1931 observer could lead to errors in colour reproduction [8, 9].

In 2015, the CIE proposed a physiological observer model (CIEPO06) [10, 11], which includes eight additional physiological parameters with corresponding population variability defined by standard deviations [6, 12].

This paper describes a multi-primary visual trichromator (named LEDMax supplied by the Thouslite®, China) to perform colour matching experiment.

It includes three parts: control panel, two identical left and right LED Cube illuminators and a viewing compartment. Figure 1 is the schematic diagram of letter two parts of the system. Each illuminator contains 18 LEDs with different centre wavelengths, ranging from 400 to 700 nm. The light emitted by the left and right LED Cube units illuminate a white reflective surface in the viewing compartment, and then uniformly

M. Xu et al. (Eds.): CACPP 2022, LNEE 991, pp. 9–14, 2023.
https://doi.org/10.1007/978-981-19-9024-3_2

fill the two side-by-side semi-circular apertures. Due to the use of a diffuse reflector, the spatial uniformity of the device reached a very high level. The field of view seen by the observer is a circle divided into two halves. The size of the field of view can be changed by exchanging the aperture.

Fig. 1. Internal structure and appearance of the visual trichromator.

Before conducting the experiment, a series of tests were conducted to evaluate system performance. Their spectral power distributions (SPDs) were measured using a Konica-Minolta CS2000 tele-spectroradiometer, denoted as TSR. The long-term repeatability over six months was represented by RMSE% (normalized) was 0.52% for left cube and 0.73% for right cube. The short-term repeatability within 24 h in RMSE% was 0.40% (left cube) and 0.42% (right cube). The difference between the left and right was 0.44%. Also, the color difference calculated by CIEDE2000 formula was used to judge the items mentioned above, the value did not exceed 0.5 for both repeatability and consistency. The above test results showed that the device is quite stable and the right and left cubes agree well.

2 Experimental

Subsequently, the visual trichromator was used to perform colour matching experiment using a 2° field of view. Maxwell's matching method [13] was chosen as the experimental method. The method is illustrated in Fig. 2.

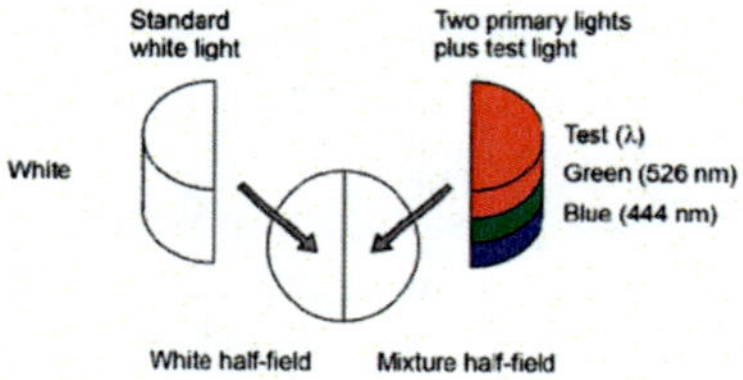

Fig. 2. Maxwell's colour matching method.

Table 1 shows different primary sets used to match the reference stimulus, which was always in the White half-field using 3 fixed LEDs at 640, 530 and 445 nm, to match a Correlated Colour Temperature (CCT) of 6500 K at luminance of 100 cd/m^2. The field of view was fixed to 2°. Figure 3 showed the SPDs of primary.

Table 1. 11 different primary sets for mixture half-field.

Primary sets	1	2	3	4	5	6	7	8	9	10	11
R (nm)	640	640	640	640	640	640	640	595	605	660	675
G (nm)	530	530	530	530	505	545	560	530	530	530	530
B (nm)	430	445	460	475	445	445	445	445	445	445	445

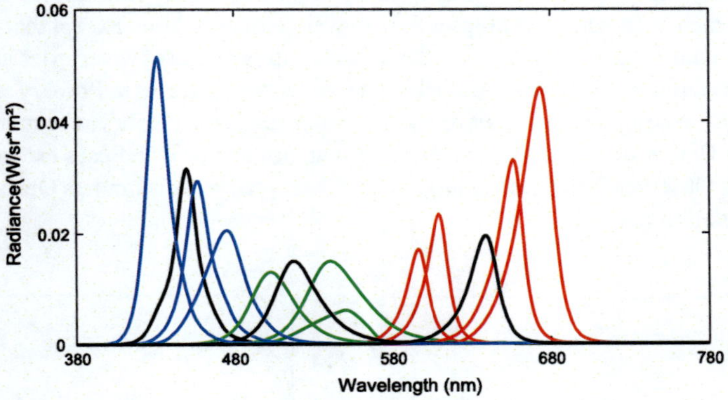

Fig. 3. SPDs of primary LEDs (black lines for white half-field).

In total, 25 normal vision observers took part in the experiment. In this paper, the results from the two distinct observers, who performed the most and the least precise respectively, were reported. They are male observers with the ages of 24 and 22, respectively. The goals of this paper are to systematically report the data analysis of the colour matching results from individual performance until the final colour matching function.

The experiment was conducted in a dark room. Prior to the real experiment, observers received a training section to adjust the channel intensity of the three primaries. They were adapted in the testing environment for two minutes, after which they began to make colour matches until 11 matches were completed. The observers could Each observer repeated the experiment five times. In total, 55 separate matches were made. They came to the lab twice, in total about two hours.

3 Results

The SPDs of all the stimuli that the observers matched to the white standard were measured at the completion of the experiment with the TSR placed at the observer's eye position. The color difference between observer's matching results and reference stimuli could be used to estimate the matching error. Also, Inter-, and Intra-observer variation could be represented by ΔE_{00} calculated using the CIEDE2000 colour difference formula [14] with CIE 1931 2° CMF.

However, each observer's performance had very large difference, i.e. for Observer 1, the average matching error, Intra-observer were 5.38 and 1.11 ΔE_{00} units, but 9.01 and 2.31 for Observer 2, respectively. For calculating intra-observer variation, the mean colour difference from the mean (MCDM) was used. The results showed Observer 1 performed much more consistently than Observer 2. Two of the matches were found to have large observer variations (see later). After removing the two, the results became 3.88 and 1.03 for Observer 1, and 5.57, 2.38 for Observer 2. Finally, the average colour difference between two observers was 4.02.

Figure 4 shows 11 ellipses fitted from observer's matching results with 11 primary sets. For each observer, most ellipses agreed well with each other except that Ellipses 4 and 5 were larger than those of the others. The centres of these two ellipses are also far from the reference color. This is probably due to the two primaries in Primary sets 4 and 5 being close to complementary wavelengths, which made the colours to be more difficult to match. Both sets agreed well on the shape and orientation of ellipses except the size, for which Observer 2 ellipses are larger than Observer 1's. This implies Observer 2 had larger variation than Observer 1.

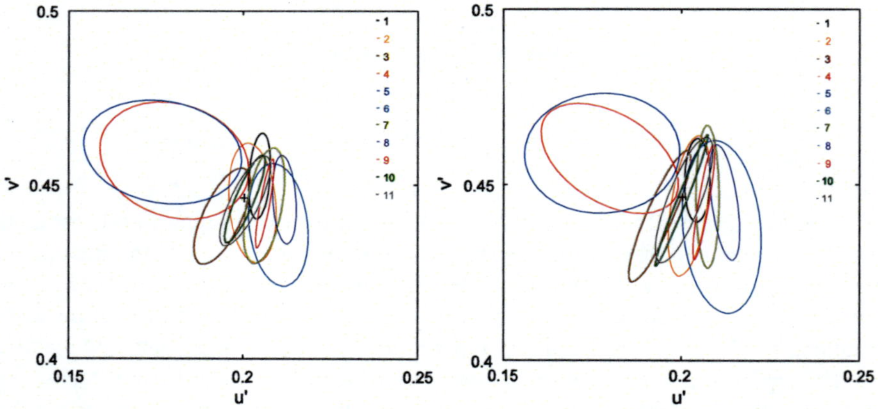

Fig. 4. 95% confidence perceptibility ellipses in u′v′ space for Observer 1 (left) and Observer 2 (right). (The colour centre of reference colour was drawn as the plus sign.)

The present observer color matching data were used to derive individual estimates of the cone fundamental CMFs based on a computational method introduced by Stockman and Rider [15]. The strategy was first to derive mathematical functions for the logarithmic L-, M- and S-cone absorbance spectra and for the standard lens and macular optical density spectra. Figure 5(a)–(c) plot the two individual observer's CMFs together with CIE 2006 2° CMF for all three, S-, M- and L-functions respectively.

Everyone's CMF was compared with the CIE 2006 2° in terms of Root Mean Square percentage (RMSE%). Their mean values are reported in Table 2.

The results in Table 2 show the RMSE% values between everyone's CMF and CIE 2006 2° observer, and between the two observers. Their mean values for L-, M-, S-functions were 2.0%, 7.7% and 3.71 respectively. The average difference between the

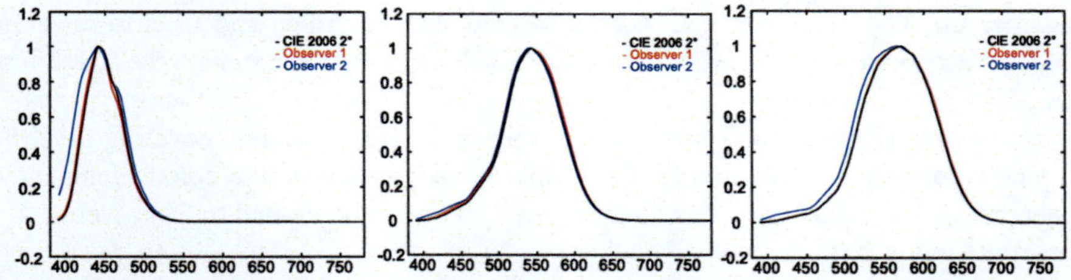

Fig. 5. Cone fundamentals from CIE 2006 2° standard colorimetric observer, and cone fundamentals from the two observer's results: S- (left), M- (mid) and L-cone (right).

Table 2. *RMSE%* and between the fitted L-, M-, S-curve for the observers and those from the CIE 2006 2° observers.

	L-	M-	S-	Mean *RMSE%*	Mean ΔE_{00}
OBS 1 vs CIE 2006 2°	0.58	1.54	3.88	2.00	3.35
OBS 2 vs CIE 2006 2°	5.39	2.99	14.77	7.72	7.09
OBS 1 vs OBS 2	4.49	2.31	8.06	3.71	4.24

two observers was 3.7%. The results supported Fig. 4 that Observer 1 is closer to CIE 2006 observer than Observer 2.

In addition to the RMSE% error between the L-, M-, S-curves, the 24 colors from X-Rite ColorChecker chart were also used to compare the difference between calculation results using the three CMFs. Their mean ΔE_{00} values are also reported.

Similar to the RMSE% results, CMF from Observer 1 gave smaller prediction error with 2006 2° CMF, while the prediction error between CMF from Observer 2 and 2006 2° CMF was much larger. And the difference between CMFs from Observer 1 and 2 was not so obvious.

4 Conclusion

A visual trichromator was constructed to perform a colour matching experiment. Its side-by-side semi-circular matching fields were illuminated by two spectrally tunable LED systems each with 18 different types of LED with center wavelengths ranging from 400 nm to 700 nm.

A series of tests were conducted to evaluate system performance. The calibrations were found to be stable and repeatable, and matched well across the two matching fields. The Maxwell colour matching experiment was performed by five observers with normal colour vision. Observers adjusted the three primaries to match the white standard. Each observer performed 11 colour matches in the experiment, which was repeated five times.

The colour matching data were used to fit confidence ellipses and individual color matching functions. The results were reported in terms of inter- and intra-observer variations together with a plot of their perceptibility ellipses to represent observer metamerism. The Stockman and Rider's method was described and used to develop CMF for each individual observer. The mean CMF were used to compare the CIE 2006 2° CMF.

In the next phase of the study, more observers will be recruited, covering the 10–80 years-old range. In addition to the normal vision observers, the colour deficiency observers will also take part. The filed size will no longer be limited to 2° as well.

References

1. Stockman, A., Sharpe, L.T.: Cone spectral sensitivities and color matching. In: Gegenfurtner, K., Sharpe, L.T. (eds.) Color Vision: From Genes to Perception, pp. 53–87. Cambridge University Press, Cambridge (2000)
2. CIE 015:2018: Colourimetry, 4th edn. Commission Internationale de l'Eclairage, Vienna, Austria (2018)
3. Guild, J.: The colorimetric properties of the spectrum. Philos. Trans. R. Soc. Lond. Ser. A **230**, 149–187 (1929)
4. Wright, W.D.: A re-determination of the trichromatic coefficients of the spectral colours. Trans. Opt. Soc. Lond. **30**(4), 141–164 (1929)
5. Wright, W.D.: A re-determination of the mixture curves of the spectrum. Trans. Opt. Soc. Lond. **31**(4), 201–218 (1930)
6. Stiles, W.S., Burch, J.M.: N.P.L. colour-matching investigation: final report. Opt. Acta **6**(1), 1–26 (1959)
7. Speranskaya, N.I.: Determination of spectral colour co-ordinates for twenty-seven normal observers. Opt. Spectrosc. **7**, 424–428 (1959)
8. Wu, J., Wei, M.: Colour mismatch and observer metamerism between conventional liquid crystal displays and organic light emitting diode displays. Opt. Express **29**(8), 12292–12306 (2021)
9. Huang, M., Li, Y., Wang, Y.: Effect of primary peak wavelength on color matching and color matching function performance. Opt. Express **29**(24), 40447–40461 (2021)
10. CIE: Fundamental chromaticity diagram with physiological axes – part 1. CIE 170-1:2006 (2006)
11. CIE: Fundamental chromaticity diagram with physiological axes – part 2: spectral luminous efficiency functions and chromaticity diagrams. CIE 170-2:2015 (2015)
12. Stockman, A., Sharpe, L.T.: The spectral sensitivities of the middle- and long-wavelength sensitive cones derived from measurements in observers of known genotype. Vis. Res. **40**, 1711–1737 (2000)
13. Maxwell, J.C.: On the theory of compound colours and the relations of the colours of the spectrum. Philos. Trans. R. Soc. Lond. **150**, 57–84
14. Johnson, G.M., Fairchild, M.D.: A top down description of S-CIELAB and CIEDE2000. Color Res. Appl. **28**(6), 425–435 (2003)
15. Stockman, A., Rider, A.T.: Practical formulae for generating human cone fundamentals with both normal and shifted lmax values as well as with variable macular, lens and photopigment optical densities consistent with CIE 2006 and 2015 standards. In preparation

Analysis of Consumers' Emotional Preference for Color Reproduction on Mobile Phone Screens

Xuejie Song, Jing Liang(✉), Nianyu Zou, Tiantian Li, Aibo Wang, and Caiyin Wang

Dalian Polytechnic University, Dalian, Liaoning, China
ljlove426@163.com

Abstract. This paper studied the effects of color temperature and clothing color in mobile phones with LCD screens and OLED screens under LED light sources. This paper analyzed the emotional differences of consumers on mobile phone color reproduction from two aspects of atmosphere perception and psychological perception. The experiment was based on the adjustable color temperature LED light source, which used the lighting environment with three different color temperatures. 11 observers were invited to evaluate the lighting environment, atmosphere perception, and psychological perception, respectively. The stability of the data was analyzed and the data was verified to be valid by the difference coefficient (CV). Through the analysis of atmosphere perception, it showed that the observer's comfort value reached the peak value at CCT of 4500 K, that was the most comfortable. Through the analysis of psychological perception, it could be seen that when the lighting environment was different, observers preferred to use mobile phones with LCD screen. The results of psychological perception analysis of each color of clothing showed that the change of color temperature had a higher effect on yellow and gray. It was better to go offline shopping for colored clothes, and the experience of an appropriate lighting environment could promote the consumption behavior of shoppers extremely.

Keywords: LCD mobile phone screen · OLED mobile phone screen · Color temperature · Clothing color · Atmosphere perception · Psychological perception

1 Introduction

With the development of smart phones, manufacturers pay more attention to the screen. The screen is getting bigger, the color performance is getting better. The current research shows that the emergence of OLED solves some of the shortcomings of LCD performance, and it also has some drawbacks. Therefore, OLED will not completely replace LCD in the future [1, 2].

Scholars have studied the impact of lighting parameters on the change of emotional and psychological. Zhai et al. studied the impact on visual comfort of changes in illumination and color temperature in the art museum [3]. Zhai et al. studied the atmospheric perception of dynamic lighting in the range of warm and cool colors [4–7]. In addition,

M. Xu et al. (Eds.): CACPP 2022, LNEE 991, pp. 15–23, 2023.
https://doi.org/10.1007/978-981-19-9024-3_3

the color-based feature has gradually become an important factor in the design and production of mobile phones. Ding et al. analyzed the emotional image with color theory, which provided a theoretical basis for the smart phone design [8]. Li et al. compared chromatic aberrations to analyze color reproduction on mobile phone screens, which was conducive to design immersive mobile phones [9].

In recent years, with the spread of COVID-19 throughout the country, our life is increasingly inseparable from online shopping. However, different mobile phone screens have different degrees of color display for clothing, and the color display degree of mobile phones in different light environments is also different [10, 11]. Previous studies have shown that light environments with different color temperatures can affect consumer psychology and consumer behavior [12]. Based on the above conclusions, this paper explores consumers' emotional preferences for online shopping with mobile phones with different screens.

2 Visual Psychophysical Experiments

2.1 Experimental Light Environment

The experiment was carried out in room B405 of the laboratory in Dalian Polytechnic University. The laboratory space was 5.5 m*2 m*3 m (L*W*H). In this experiment, three color temperature levels were selected: I, II and III. Specific parameters were measured by a spectral color illuminometer, as shown in Table 1.

Table 1. Lighting parameters of ambient lighting environment

Number	CCT/K	Illuminance/lx	Ra
I	3150	302.7	81.6
II	4339	305.1	84.4
III	5897	302.5	82.7

2.2 Observers

In this experiment, a total of 11 college students (6 males, 5 females) were invited to simulate consumers, with an average age of 21 years old. All of them were undergraduate and graduate students studying at Dalian Polytechnic University. 11 observers had normal vision, all of them passed the color blindness test of Ishihara, meeting the visual needs of the visual psychophysical experiment.

2.3 Mobile Phone and Clothing Colors

This experiment includes two mobile phones. The screen of mobile phone A is OLED, and the screen of mobile phone B is LCD. OLED has the characteristic of self-illumination, while LCD does not [2]. Before the experiment, the two mobile phones were set at 1500 lx and placed on the working surface of 0.75 m. There were six colors in total, including: yellow, blue, gray, green, pink, and purple. The semantic difference scale method [13] was adopted to allow consumers to conduct the subjective evaluation of the atmosphere perception and the subjective evaluation in experiment. Psychological perception evaluation words include: liking, attraction, ugliness, desire to buy, dazzling and color distortion.

The display structure of mobile phone screen is complex, and there is no obvious difference in the parameters of different colors of clothing measured by contact spectrophotometer, so we used a non-contact spectral radiometric luminance meter (CS-2000) to measure the color parameters (A *, B *) of six clothing images of two mobile phones under three different color temperatures under a 2° perspective. The contact spectrophotometer can measure data more accurately and make the measurement results more reliable. Therefore, we chose to measure the color parameters of real clothing (A *, B *) under the standard light source D_{65} with A field of view of 2°. The comparison is shown in Fig. 1.

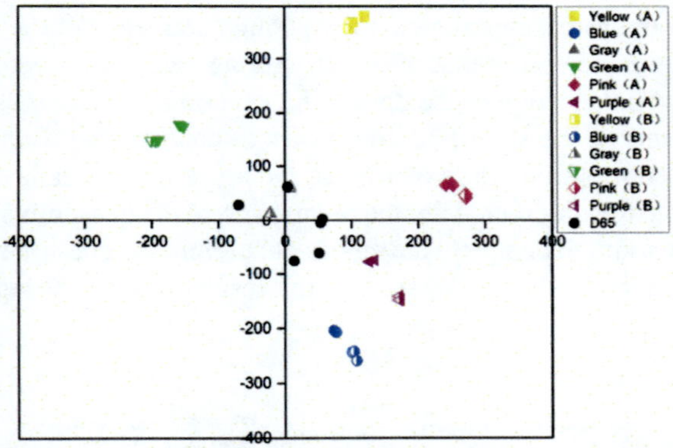

Fig. 1. Comparison of mobile phone reproduced clothing color parameters and clothing color parameters

CIE standard illuminant D_{65} is the most commonly used artificial daylight with a color temperature of 6500 K. As can be seen from Fig. 1, the coordinate point spacing of each color is very small for mobile phone A and mobile phone B at different color temperatures, so the color temperature has no significant influence on the screen observation effect. The color of the gray clothing in the two mobile phones is similar to that in the standard illumination volume D_{65}, indicating that the color of the gray clothing in the mobile phone is close to the natural color.

2.4 Questionnaire

This experiment included two subjective evaluation questionnaires, one was the atmosphere perception questionnaire, the other was the psychological perception questionnaire. The 20 quantifiers used a positive and negative rating scale, as shown in the second row of Table 2. The psychological perception questionnaire included five evaluation items for each color, which were liking, attraction, ugliness, desire to buy, dazzling and color distortion. A six-point Likert scale [10] was adopted, as shown in the third row of Table 2.

Table 2. Subjective questionnaire scoring standard

Degree	Very negative	Negative	A little negative	A little positive	Positive	Very positive
Score	−3	−2	−1	1	2	3
Score	1	2	3	4	5	6

2.5 Experimental Process

Before the psychophysical experiment, the lighting parameters of the lighting environment were measured to ensure that the other lighting parameters were constant except for the change of color temperature during the experiment. Ishihara's color blindness test was conducted on the simulated consumers to confirm the visual accessibility of the observers. Before starting the experiment, the process and details of the experiment were introduced to the observer. After the organizer sent the questionnaire, the observers were led into the room to complete the emotional preference questionnaire. Finally, the experimental data was collected to analyze the results of ratings. Figure 2 shows the detailed process.

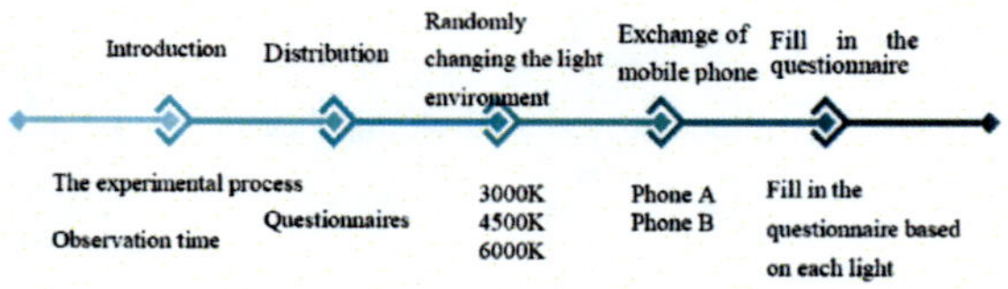

Fig. 2. Experimental process

3 Results and Discussion

3.1 Data Stability Analysis

In this paper, the stability of subjective evaluation data was judged by using the coefficient of variation (CV). After careful calculation, the CV values of 11 observers were shown in Fig. 3, with an average value of 31.11%. Although there were two data that were over 40%, they were still within the range of the required CV values to conduct the experiment [10]. The visual experiment data of 11 observers was reliable and could be used for subsequent analysis.

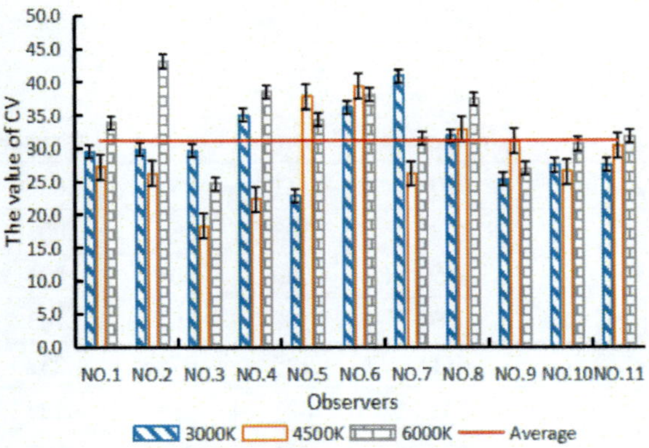

Fig. 3. Stability of 11 observer evaluation data

3.2 Analysis of Emotional Preference for Atmosphere Perception in Lighting Environment

20 quantifier pairs were reduced in dimension through principal component analysis and maximum variance rotation method. Table 4 shows the results.

Six factors were extracted by principal component analysis, and the total variance was 75.521%. The bold values was the quantifier with high factor load proportion in Table 4. Therefore, each factor was named according to the common characteristics of the quantifier pair, which were: factor 1: integration, factor 2: beauty, factor 3: comfort, factor 4: advanced, and factor 5: simplicity.

In this paper, after the five dimensions of the lighting atmosphere perception were obtained, the influences of color temperature changes on the five dimensions of perception were analyzed by univariate analysis (ANOVA). The experiment results showed that color temperature had a significant effect on factor 3 (comfort), as shown in Table 5.

The analysis of color temperature influences factor 3, as shown in Fig. 4. Color temperature had a significant effect on comfort. The higher the color temperature was, the comfort first increased and then decreased. Around 4500 K, the comfort reached the peak value, which was the most comfortable.

3.3 Analysis of Emotional Preference of Psychological Perception

In this experiment, the positive evaluation quantifiers were liking, attraction, and desire to buy, while the negative evaluation quantifiers were ugliness, dazzling, and color distortion. Therefore, evaluating the popularity of a color depended on the difference between the forward and reverse values. The positive and negative evaluations of mobile phone A are shown in Fig. 5(a) and (c), while the positive and negative evaluations of mobile phone B are shown in Fig. 5(b) and (d) The mean value of negative evaluation was larger than the mean value of positive evaluation. This showed that the effect observed was not ideal on the two mobile phones. Therefore, it was suggested that offline sales can win consumers' satisfaction and desire to buy products more than online sales.

Table 4. Simplified quantifier pairs by actor analysis

Quantification	Composition				
	Integration	Beauty	Comfort	Advanced	Simplicity
Colorful-uncolorful	**.907**	−.161	−.056	.007	−.161
Lifeless-lifely	**.750**	.393	.085	.003	.021
Dim-bright	**.669**	−.080	−.305	−.095	.319
Boring-interesting	**.639**	.366	.265	.243	.216
Unpopular-popular	**.611**	−.125	−.075	−.232	.363
Tacky-vulgar	−.185	**.808**	−.036	.223	−.173
Glamorous-unglamorous	−.074	**−.741**	−.246	.307	.327
Old-fashion	.406	**.721**	−.204	.070	.036
Ugly-beautiful	−.084	**.603**	.470	.177	.213
Uncomfortable-comfortable	.114	−.062	**.909**	−.145	−.041
Dazzling-soft	−.471	.171	**.738**	.194	−.114
Dislike-like	.137	.225	**.601**	.315	.419
Ordinary-elegant	.202	.030	.056	**.767**	−.308
Frivolos-solemn	−.319	−.103	.142	**.678**	.223
Plain-luxury	−.038	.202	−.009	**.654**	.017
Casual-formal	.087	−.055	−.079	.574	.566
Gorgeous-simple	−.117	−.027	.502	−.043	**.732**
Uneven-even	.172	−.090	−.071	.002	.702
Vague-clarity	.412	.072	−.039	−.072	−.050
Low grade-high grade	.236	.535	.128	.070	.150

Table 5. Univariate analysis results of color temperature on atmosphere perception

Factor	F	Significance
Factor 3 (comfort)	10.772	**.000**

We compared the mean value of emotional preferences for psychological perception of two mobile phones under three color temperatures, as shown in Fig. 6. In addition to yellow clothing color perception preference effect on mobile phone B was worse than that on mobile phone A, other clothing color perception effects on mobile phone B were better than that on mobile phone A. Thus observers were more likely to use a phone B with an LCD screen.

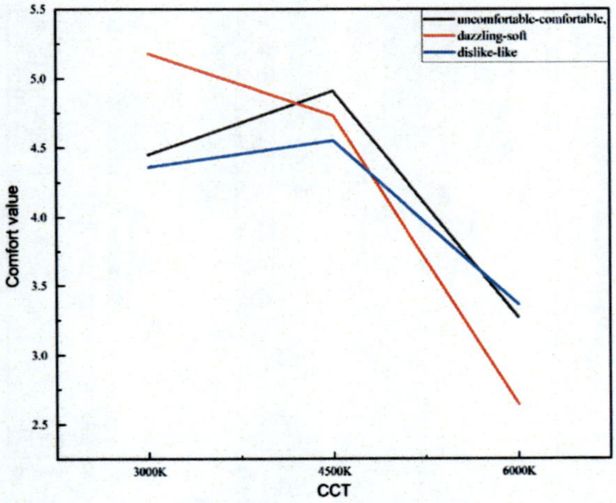

Fig. 4. Effect of color temperature on factor 3 (comfort)

Fig. 5. **a** Mean comparison of positive evaluation of mobile phone A. **b** Mean comparison of positive evaluation of mobile phone B. **c** Mean comparison of negative evaluation of mobile phone A. **d** Mean comparison of negative evaluation of mobile phone B

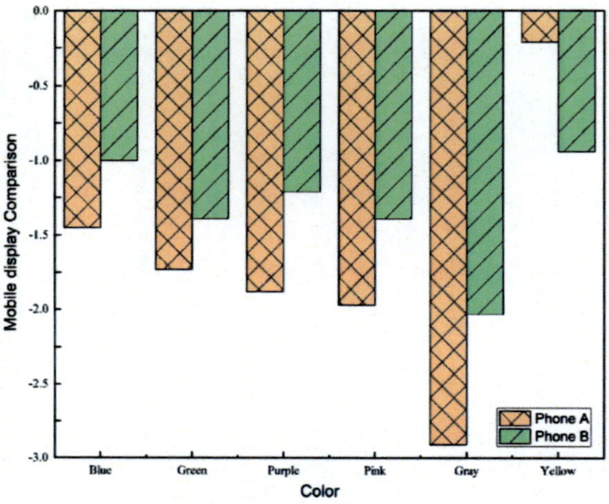

Fig. 6. Overall comparison positive and negative evaluation of two mobile phones

At the same time, it could be seen from Fig. 7 that each color presentation was affected by the change of color temperature, overall, yellow was the most popular and gray was the least. Around 3000 K, blue was the most popular color, gray was the least popular color, and all other colors were in the order of yellow-green-pink-purple. Around 4500 K, yellow was the most popular, gray was the least popular, blue and purple were equally popular, followed by pink and green. Around 6000 K, yellow was the most popular color, gray was the least popular color, and purple, green and pink were in the middle.

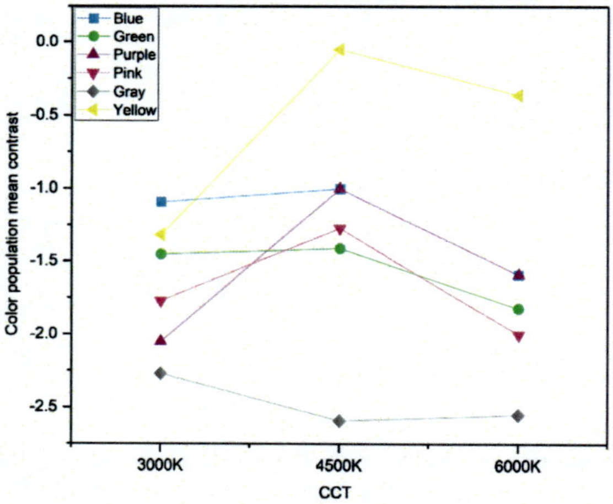

Fig. 7. Overall color summary line chart

4 Conclusion

This paper studied the results of consumers' atmosphere perception and psychological perception under light environments with different color temperatures, and analyzed the emotional preference of consumers for color reproduction of mobile phones. Through the analysis of atmosphere perception, it showed that the observer's comfort value reached the peak value at CCT of 4500 K, that was the most comfortable. Through the analysis of psychological perception, it could be seen observers preferred to use mobile phones with LCD screen. The results of psychological perception analysis showed that yellow was the most popular and gray was the least.

It was found that the change of external lighting environment had no significant impact on consumers' online shopping. It is suggested that clothing managers adopt offline sales to win consumers' satisfaction and desire.

Acknowledgments. National Natural Science Young Foundation of China (No. 61802041); QianBaiHui Research Fund Project (2017-228195) and Dalian Estar Intelligent Technology Co. Ltd Research Fund Project (2121062).

References

1. Zhang, L.W.: Development status and technical analysis of transparent OLED display. Technol. Innov. Appl. **8**, 135–136+139 (2020)
2. Wei, S.W., Jiao, L.T., Han, L.: Mobile phone screen display technology analysis and comparison. Heilongjiang Sci. **18**, 80–81 (2019)
3. Zhai, Q., Luo, M., Liu, X.: The impact of illuminance and colour temperature on viewing fine art paintings under LED lighting. Light. Res. Technol. **47**, 795–809 (2014)
4. Stokkermans, M., Vogels, I., de Kort, Y., Heynderickx, I.: Relation between the perceived atmosphere of a lit environment and perceptual attributes of light. Light. Res. Technol. **50**, 1164–1178 (2018)
5. Ming, Z.Y., et al.: The visual influence of LED light source with different illumination and color temperature on calligraphy works. Chin. J. Light. Eng. **30**, 55–60 (2019)
6. Li, B., Zhai, Q.Y., Hutchings, J.B., Luo, M.R., Ying, F.T.: Atmosphere perception of dynamic LED lighting over different hue ranges. Light. Res. Technol. **51**, 682–703 (2019)
7. Bi, K.: Research on Smartphone Color of Young Users Based on Emotional Image. Hunan University (2019)
8. Jun, Z., Ying, W., Lu, Z.: Research on color reproduction ability of mainstream brand mobile phones. In: 2015 Fourth China Printing and Packaging Academic Conference Abstracts 18 (2015)
9. Li, H., Luo, M.R., Liu, X.Y., Wang, B.Y., Liu, H.Y.: Evaluation of colour appearance in a real lit room. Light. Res. Technol. **48**, 412–432 (2016)
10. Huang, Z., Liu, Q., Luo, M.R., Pointer, M.R., Wu, B., Liu, A.: The whiteness of lighting and colour preference, part 2: a meta-analysis of psychophysical data. Light. Res. Technol. **52**, 23–35 (2020)
11. Li, W.X., Li, L.S., Ying, Z.X.: A study on college students' leisure view—a survey based on semantic difference scale. J. Adult Educ. Hebei Univ. **17** (2015)
12. Bartoszek, D., Fiutowski, J., Dohnalik, T., Kawalec, T.: Optical surface devices for atomic and atom physics. Opt. Appl. **40**(3) (2010)
13. Nazir, T., Irtaza, A., Starovoitov, V., Harun, S.W.: Optic disc and optic cup segmentation for glaucoma detection from blur retinal images using improved mask-RCNN. Int. J. Opt. (2021)

Camera Spectral Sensitivity Estimation Based on a Display

Hui Fan and Ming Ronnier Luo(✉)

State Key Laboratory of Modern Optical Instrumentation, Zhejiang University, Hangzhou, China
m.r.luo@zju.edu.cn

Abstract. In this study, the spectral sensitivities of a camera were estimated by capturing the colour patches presented on a display. Two different patterns (RGBW and CMYW) were tested. Principal component analysis based on a prior art spectral sensitivities database was implemented. The accuracy of the estimated spectral sensitivities was evaluated in terms of four metrics, spectral error, RGB error, colour difference and Vora value. It was found that the accuracy of the proposed method based on a display was similar to that of the same algorithm applied to a reflective colour chart. The accuracy in terms of spectral error was 12.6%, 12.1%, and 12.4% for RGBW, CMYW patterns, and reflective colour chart respectively.

Keywords: Camera · Spectral sensitivity · Display

1 Introduction

Spectral sensitivities represent the spectral response of a camera at different wavelength. It can be applied in multispectral imaging, colour constancy, spectral reflectance recovery et al. [1–3]. However, this information is generally not provided by the manufacturer. The standard method of camera spectral calibration is to use a monochromator [4, 5]. This device can generate monochromatic lights at different wavelength, and the camera is used to capture a series of images over the desired range of wavelength. But it is a costly device and has not been widely applied. Various algorithms have been applied to estimate the camera spectral sensitivities by capturing colour samples. The colour samples mainly included reflective colour charts [6, 7], LED-based samples [8], fluorescence [9] and a display [10]. Among different media of colour samples, a display was the most common device in our daily life. In the study of Zhu et al. [10], 81 colour patches designed from orthogonal test were presented on a LCD display, and the camera spectral sensitivities were estimated by window filters on frequency and spatial domains.

In this study, the principal component analysis (PCA) algorithm was implemented with colour samples presented on a display. Our hypothesis is that accurate estimation of camera spectral sensitivities can be achieved by employing the three primaries (red, green and blue) of a display and their combinations (white, cyan, magenta and yellow) as training samples.

M. Xu et al. (Eds.): CACPP 2022, LNEE 991, pp. 24–30, 2023.
https://doi.org/10.1007/978-981-19-9024-3_4

2 Method

The linear responses of a camera can be formulated by Eq. (1), where X_i is the camera response, $r(\lambda)$ is the spectral reflectance of object, $e(\lambda)$ is the spectral power distribution (SPD) of the light, $s_i(\lambda)$ is the spectral sensitivity of the camera, and $[\lambda_{min}\lambda_{max}]$ is the range of visible spectrum. For a common trichromatic camera, there are three channels corresponding to R, G and B.

$$X_i = \int_{\lambda_{min}}^{\lambda_{max}} r(\lambda)e(\lambda)s_i(\lambda)d\lambda, \quad i = R, G, B \tag{1}$$

After sampling the spectrum from 400 nm to 700 nm with 10 nm interval, the above equation can be written in matrix donation as Eq. (2), where X is the N by 3 matrix of camera responses, r is the N by 31 spectral reflectance of objects, E is the 31 by 31 diagonal matrix indicating the SPD of light, and S is the 31 by 3 camera spectral sensitivities. N is the number of the objects.

$$X = rES \tag{2}$$

Suppose L to be the spectral signals received by the camera. It is the product of the spectral reflectance of objects and the SPD of light. Then Eq. (2) can be written as Eq. (3),

$$X = LS \tag{3}$$

In general, the estimation of camera spectral sensitivities requires the known spectral signals L and the corresponding camera responses X.

In the study of Jiang et al. [11], spectral sensitivities database including 28 cameras was collected. They performed PCA on the spectral sensitivities and found the space was two-dimensional. In other words, the spectral sensitivities could be expressed by the first two eigenvectors as Eq. (4), where B_i is the eigenvector matrix and a_i is the coefficients.

$$S_i = B_i a_i, \quad i = R, G, B \tag{4}$$

Substituting Eq. (4) into Eq. (3), then the least squares solution of S_i can be found by Eq. (5). The superscripts ^{T} and $^{-1}$ denote respectively the matrix transpose and inverse.

$$min_{a_i} \|LB_i a_i - X_i\|^2$$

$$S_i = B_i\Big[(LB_i)^T(LB_i)\Big]^{-1}(LB_i)^T X_i, \quad i = R, G, B \tag{5}$$

Considering that the dimensionality of spectral sensitivities has been decreased to only two by means of PCA, the colour samples with more than two dimensionalities can be used to recover the spectral sensitivities theoretically. As a result, the traditional trichromatic display that contains RGB primaries could be used as the target. In this study, the colour samples were presented on a display, including the primaries (red, green and blue) of a display and their combinations (white, cyan, magenta and yellow). These colours were divided into two groups. One includes red, green, blue and white, marked as RGBW, while the other is cyan, magenta, yellow and white, marked as CMYW. Figure 1 shows the patterns of colour samples presented on the display. By capturing the colour samples using a given camera, and measuring the corresponding SPD, it is expected that the spectral sensitivities of the camera can be estimated by PCA algorithm following Eq. (5).

a.RGBW b.CMYW

Fig. 1. Patterns of colour samples presented on a display

3 Experiment

The experiment was conducted in a dark room. Figure 2 shows the experimental situation. The patterns of colour samples as shown in Fig. 1 were presented on the mobile display of OPPO FindX3 Pro. It was a 6.7-inch OLED display with 3216 × 1440 pixels. A Canon 650D digital camera was used to capture the colour patterns. It should be noted that the camera of this model was not included in the spectral sensitivities database used for PCA. The setup of the camera (ISO, shutter speed, F-number) was adjusted to obtain appropriate exposure. The responses of the colour patches were extracted from the RAW images. An image under dark condition was also captured to subtract the dark noise from camera responses.

Fig. 2. Experimental situation of capturing the colour samples on the display

The SPD of each colour patch was measured using a JETI-Specbos 1211 spectroradiometer. Figure 3 shows the SPD of the colour samples. Figure 4 shows the chromaticity in CIE 1976 u′v′ plane.

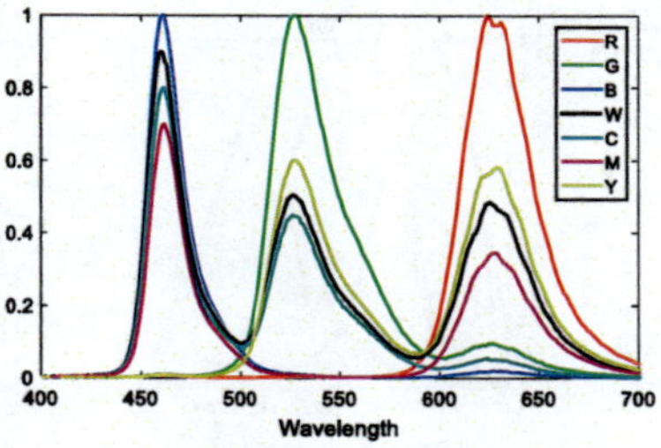

Fig. 3. Spectral power distribution (SPD) of the colour samples

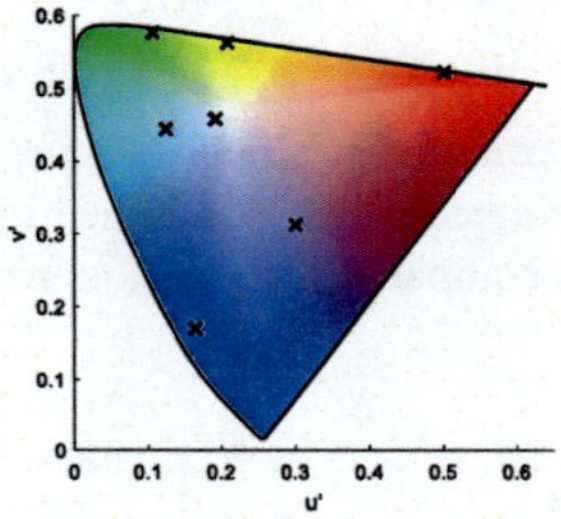

Fig. 4. Chromaticity of the colour samples in CIE 1976 u′v′ plane

In addition, a reflective colour chart was captured to test the accuracy of the estimated spectral sensitivities. An Xrite Macbeth ColorChecker chart (MCCC) was placed in a Thouslite® LED viewing cabinet. The test lights were simulated D50, D65 and A. The camera was used to capture the chart with a viewing/illumination geometry of 0/45°. The SPD of the light was measured using the JETI-Specbos 1211 spectroradiometer. The spectral reflectance of the colour patches was measured using a Datacolor SF600 spectrophotometer. The spectral signals reflected by the colour chart were calculated as the product of the SPD of light and the spectral reflectance of the colour patches.

4 Results and Discussion

The spectral sensitivities of the camera were estimated from the two groups of display-based colour samples, RGBW and CMYW. And the same PCA algorithm was also applied to the 24 colour samples on MCCC under D65 to obtain the estimated spectral sensitivities. The ground truth spectral sensitivities were calibrated by a Labsphere QES1000 monochromator. Figure 5 plotted the comparison of the estimated spectral sensitivities from different colour samples, and the results were compared with the ground truth results by monochromator. It can be seen that all the three estimated spectral sensitivities had a close match with those calibrated by the monochromator. Differences rarely exist between the shapes of the three estimated results.

Four different metrics were used to test the accuracy of the estimated spectral sensitivities, including spectral error (*SE*), RGB error, colour difference ΔE^*_{ab} and Vora value. These metrics were also used in the study of Finlayson et al. [6]. Let us denote

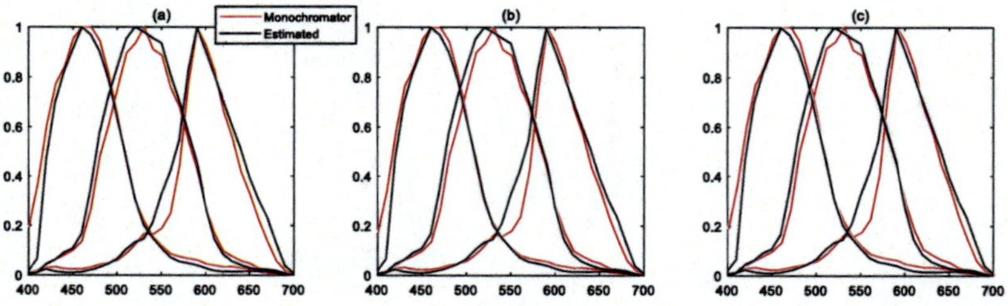

a.RGBW b.CMYW c. MCCC

Fig. 5. Comparison of the spectral sensitivities calibrated by monochromator and those estimated from (a) RGBW samples on display, (b) CMYW samples on display, (c) MCCC under D65.

the estimated and the ground truth spectral sensitivities as $\hat{S}$ and S, respectively. The spectral error (*SE*) can be calculated as Eq. (6), indicating the spectral accuracy of the estimated spectral sensitivities compared with monochromator.

$$SE = \frac{\left\|\hat{S} - S\right\|}{\|S\|} \times 100\% \tag{6}$$

The percentage RGB error was calculated between the measured RGB and the predicted RGB by spectral sensitivities. The predicted RGB is calculated by Eq. (2). Let $\widehat{X_i}$ and X_i donate the predicted and measured camera responses of the sample *i*. The percentage RGB error can be calculated by Eq. (7), where N is the number of the testing samples. In this study, the testing samples were the 24 colours on MCCC under D50 and A.

$$\Delta X\% = \frac{1}{N}\sum_{i=1}^{N}\frac{\left|\widehat{X_i} - X_i\right|}{|X_i|} \times 100\%,\quad \mathrm{X = R, G, B} \tag{7}$$

The colour difference ΔE^*_{ab} is calculated as the Euclidean distance in CIE 1976 $L^*a^*b^*$ colour space. Firstly, the linear color correction matrix (CCM) between camera responses and CIE tristimulus XYZ values under CIE D65 is developed. The camera responses are calculated via ground truth spectral sensitivities. The testing samples are the Leeds 100876 reflectance dataset [12]. Then the RGB values predicted by both the estimated and the ground truth spectral sensitivities are transformed to XYZ values and then to $L^*a^*b^*$. The ΔE^*_{ab} between them is calculated to indicate the accuracy of estimated spectral sensitivities.

Vora value [13] is defined by Eq. (8). Its value ranges between 0 and 1. Larger *Vora* value means a closer match between $\hat{S}$ and S.

$$Vora = \frac{1}{3}trace\left(S\left(S^TS\right)^{-1}S^T\hat{S}\left(\hat{S}^T\hat{S}\right)^{-1}\hat{S}^T\right) \tag{8}$$

Table 1 lists the accuracy of the estimated spectral sensitivities in terms of different metrics. The best result of each metric was marked in bold. It can be found that the spectral sensitivities estimated from CMYW samples based on display had the highest accuracy in terms of *SE*, RGB error and Vora value, and slightly larger ΔE^*_{ab} value. However, the variations between different samples were not significant. Overall, it was concluded that using display-based samples to estimate the spectral sensitivities of a camera by PCA method could result in similar accuracy compared with that using a reflective colour chart MCCC.

Table 1. Accuracy of the estimated spectral sensitivities from different colour samples

Samples	Display		MCCC
	RGBW	CMYW	
SE	12.6%	**12.1%**	12.4%
ΔRGB%	1.92%	**1.65%**	1.79%
$\Delta E_{ab}{}^*$	1.76	1.82	**1.66**
Vora	0.985	**0.987**	0.985

The advantage of using a display to present colour samples is that it is quite convenient and simple to use. It only needs a single shoot, so that it has higher efficiency compared with the method using a monochromator or multi-channel LED. The method includes only a spectroradiometer and a display. A display, especially a mobile display, is a common device in our daily life. In the future work, it will verify whether displays of different primaries will give similar spectral sensitivity functions and compare with their performance.

Nevertheless, using colour samples based on a display had its limitations. Due to the low dimensionality of display samples (usually only three for a trichromatic display with RGB primaries), only PCA algorithm could be applied because it considered the spectral sensitivities to be two-dimensional. So its accuracy was restricted by the PCA algorithm and the present spectral sensitivities database, and thus this method could only be applied to a trichromatic camera. In addition, the setup of the camera should be carefully adjusted to prevent the Moire fringe in the image when capturing the display. With the demand of higher estimated accuracy, a reflective colour chart or LED-based samples could be used together with other algorithms such as regularization and quadratic programming [5].

5 Conclusion

In this study, the spectral sensitivities of a camera were estimated by capturing colour samples presented on a display using PCA algorithm. The colour samples were arranged as RGBW and CMYW patterns respectively. The results showed the estimated spectral sensitivities to give a close match with those calibrated by a monochromator. The accuracy was evaluated in terms of spectral error, RGB error, colour difference and Vora value metrics. The results showed that the display-based method could have similar accuracy compared with the method based on a reflective colour chart MCCC.

References

1. Park, J.-I., Lee, M.-H., Grossberg, M.D., Nayar, S.K.: Multispectral imaging using multiplexed illumination. In: Proceedings of IEEE International Conference on Computer Vision (2007)
2. Finlayson, G.D., Hordley, S.D.: Color constancy at a pixel. J Opt. Soc. Am. A Opt. Image Sci. **18**, 253–264 (2001)
3. Liang, J.X., Xiao, K.D., Hu, X.R.: Investigation of light source effects on digital camera-based spectral estimation. Opt. Express **29**, 43899–43916 (2021)
4. Vora, P.L., Farrell, J.E., Tietz, J.D., Brainard, D.H.: Digital color cameras - 2 - spectral response (1997)
5. Darrodi, M.M., Finlayson, G., Goodman, T., Mackiewicz, M.: Reference data set for camera spectral sensitivity estimation. J. Opt. Soc. Am. A **32**, 381–391 (2015)
6. Finlayson, G., Darrodi, M.M., Mackiewicz, M.: Rank-based camera spectral sensitivity estimation. J. Opt. Soc. Am. A Opt. Image Sci. Vis. **33**, 589–599 (2016)
7. Zhou, M., Chen, W., He, T., Zhang, Q., Shen, J.: Scan-free end-to-end new approach for snapshot camera spectral sensitivity estimation. Opt. Lett. **46**, 5806–5809 (2021)
8. DiCarlo, J.M., Montgomery, G.E., Trovinger, S.W.: Emissive chart for imager calibration. In: 12th Color and Imaging Conference, pp. 295–301 (2004)
9. Han, S., Matsushita, Y., Sato, I., Okabe, T., Sato, Y.: Camera spectral sensitivity estimation from a single image under unknown illumination by using fluorescence. In: Proceedings of IEEE Conference on Computer Vision and Pattern Recognition (CVPR), pp. 805–812 (2012)
10. Zhu, J., Xie, X., Liao, N., Zhang, Z., Wu, W., Lv, L.: Spectral sensitivity estimation of trichromatic camera based on orthogonal test and window filtering. Opt. Express **28**, 28085–28100 (2020)
11. Jiang, J., Liu, D., Gu, J., Susstrunk, S.: What is the space of spectral sensitivity functions for digital color cameras? In: IEEE Workshop on the Applications of Computer Vision, pp. 168–179 (2013)
12. Li, C.J., Luo, M.R., Pointer, M.R., Green, P.: Comparison of real colour gamuts using a new reflectance database. Color Res. Appl. **39**, 442–451 (2014)
13. Vora, P.L., Trussell, H.J.: Mathematical methods for the design of color scanning filters. IEEE Trans. Image Process. **6**, 312–320 (1997)

Prints Clarity Evaluation Indexes Spatial Frequency Response

Zimo Yan[1(✉)], Yuxia Yuan[1], Xiao Yang[2], Xiaofang Wang[2], and Yanfang Xu[1]

[1] School of Printing and Packaging Engineering, Beijing Institute of Graphic Communication, Beijing, China
yanzimo12@163.com

[2] Beijing Academy of Printing and Packaging Industrial Technology, Beijing, China

Abstract. The definition quality of prints has been concerned. According to relevant international standards, this paper realized the spatial frequency response (SFR) evaluation technology to characterize the sharpness of printed images. The slant edge method was used to solve the SFR curve of the prints based on the scanning edge image. In addition, the characteristic frequency and effective characteristic frequency quality indexes of printing SFR were established. The measurement of digital printing samples shows that the quality index characteristic frequency and effective characteristic frequency is able to reflect and represent the clarity quality of prints effectively.

Keywords: Characteristic frequency · Digital printing · Sharpness · SFR

1 Introduction

Digital printing is becoming more and more popular with its variability, timeliness and shorter plate, whose market share is increasing. The quality of digital printing has the same commonalities with that of traditional printing, as well as the unique aspects of its technical characteristics. In the field of quality evaluation, international standards, such as ISO/IEC 24790 and ISO/TS 15311-2, have been established specifically for digital prints quality [1, 2].

Sharpness is an important quality attribute of image. The sharpness of printed image stems from its detail resolution. The detail resolution of image is a function of spatial frequency. Therefore, in order to describe the correlation between spatial frequency and the resolution of output system, the relationship should be reflected by the curve which called spatial frequency response (SFR) curve [3]. ISO/IEC TS 29112:2018 stipulates that SFR is one of the quality attributes to describe the sharpness of digital printed images. Meanwhile, the slant edge method is used as the first method to measure the prints SFR [4].

The image quality of digital printing mainly determined by the printing equipment and printing process, but also affected by the paper performance. For example, in the ink-jet printing output, the paper performance has a significant impact on the raggedness

M. Xu et al. (Eds.): CACPP 2022, LNEE 991, pp. 31–37, 2023.
https://doi.org/10.1007/978-981-19-9024-3_5

and blurriness of the printing line image [5]. And affect the clarity of the image further to a great extent.

For this purpose, the SFR measurement technology based on the slant edge has been implemented in this paper, which realized the SFR quality evaluation of prints according to the output experiment carried out by different digital printing equipment, paper and printing conditions.

2 Printing SFR and Measurement Techniques

2.1 Spatial Frequency Response

The spatial frequency response (SFR), which represents the contrast ratio varies with the spatial frequency between the output image and the original image, reflects the ability to retain the details contrast characteristics of the original image.

According to ISO/IEC TS 29112:2018, the SFR of prints is defined as the brightness contrast varies with spatial frequency, which could be measured by the slant edge method. Specifically, the edge image is output to the substrate with a white background. The edge image is a square block with different inclination. The inclination in the vertical direction is usually 0°, 8° and 24°. The printed image is transformed to digital image by scanner, and the spatial frequency response is solved by digital image processing.

2.2 SFR Measurement Method

According to ISO/IEC TS 29112:2018, the measurement of SFR can be achieved depending on the professional scanners, which acquire the printed digital images, and SFR quality would be measured based on image processing technology [4]. The specific process and steps are as Fig. 1 follows.

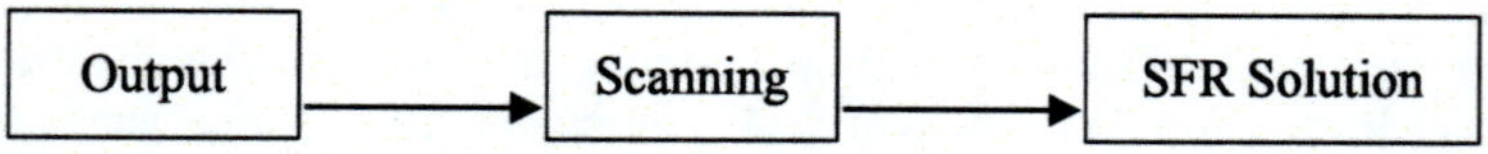

Fig. 1. Flowchart of SFR measurement processes and steps

(1) The output of the test target
The square blocks for SFR measurement in the above standard are output using the printing equipment to be evaluated.
(2) The scanning of the printing images
First, the scanner is calibrated according the method in ISO/IEC TS 29112:2018, that is, to establish the opto-electronic conversion function (OECF).
Subsequently, the output printed products in step (1) is scanned to digital images. The scanning resolution is required to be no less than 1200 dpi and no image optimization is required. The resulting digital image must be stored as *.tif or *.bmp format.

(3) Solution of SFR curve
For the image quality attributes to be evaluated, the SFR curve is solved from the prints described in step (2). The solution steps are as follows:

a. Extract the sub-pixel edge position according to the light reflection coefficient data of the printed edge;
b. The edge spread curve is fit according to the sampling data extracted from sub-pixel edge position;
c. The line spread function curve is obtained by first derivative of the edge spread curve;
d. The spatial frequency response curve is obtained by Fourier transform to the line spread function curve.

3 SFR Measurement Practice

A total of 6 test samples were prepared by combining different digital printers and paper in different printing conditions. Samples and output condition parameters are shown in Table 1.

Table 1. Sample output parameters

No.	Printing equipment	Printing conditions	Paper
A1	Inkjet devices 1	UV printing 1200*1200 dpi 50 m/min	300 g white cardboard
A2	Inkjet devices 1	UV printing 1200*1200 dpi 20 m/min	300 g white cardboard
A3	Inkjet devices 1	UV printing 1200*1200 dpi 50 m/min	Coated paper
B1	Inkjet devices 2	Non-UV printing 1200*1200 dpi	Offset paper
B2	Inkjet devices 2	Non-UV printing 1200*1200 dpi	Digital paper
C1	Inkjet devices 3	Non-UV high quality desktop printing 1440*1440 dpi	Ordinary printing paper

In addition, the square with an inclination of 8° was selected, and the printing SFR quality was characterized by the average measurement results of the top, bottom, left and right edges.

Epson Expression 12000XL scanner was selected for digital imaging the prints, and the resolution was 3200 dpi.

The OECF of the scanner is established as shown in Fig. 2. According to the OECF, the scanned edge image can be transformed from the gray value of the image to the light reflection coefficient of the print itself, which represents the brightness value of the printed image. Furthermore, the OECF curve is fitted, and the standard error is 0.001.

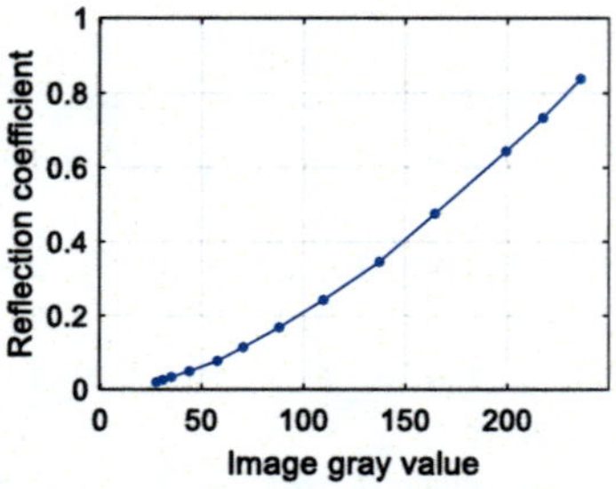

Fig. 2. Scanner OECF relationship curve

3.1 SFR Curve

According to the edge light reflection coefficient image calculated by OECF conversion, the SFR curve of edge can be solved according to the solution algorithm. The edge diagram of a test sample and its four edge SFR curves are shown as result in Figs. 3 and 4. The abscissa is the spatial frequency (cy/mm), and the ordinate is the SFR value.

Fig. 3. Edge scan image of sample A1

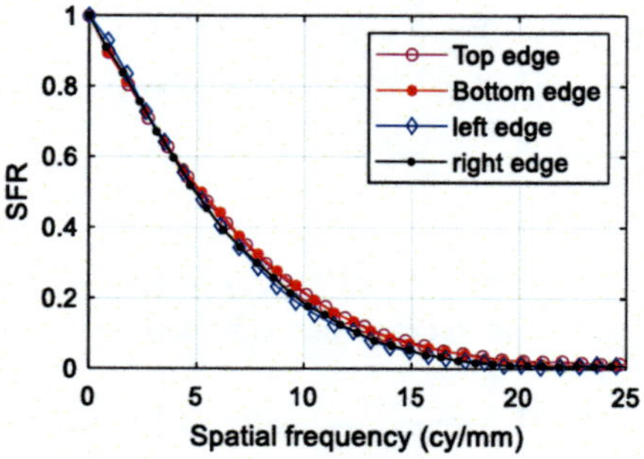

Fig. 4. SFR curve of sample A1

The remarkable feature of the curve shown in Fig. 4 is that the SFR value is the variation of the normalized brightness contrast following spatial frequency, that is, the brightness contrast of 0 cy/mm spatial frequency (equivalent to the change between the infinite field color and the infinite substrate background color) is regarded as 1. In fact, as shown in Fig. 3, the brightness contrast between the field color inside the edge and the color of the background substrate is not necessarily 1, and different prints have different values.

3.2 SFR Quality Index

ISO/IEC TS 29112:2018 standard only specifies the measuring method of the prints SFR curve, that is, how to resolve this curve. However, SFR curve is a variational relationship represented by multiple data, which is not conducive to compare the quality differences represented by SFR curves simply and clearly. For this purpose, this paper tries to establish a simple quality index to characterize the SFR quality grade of prints easily.

Firstly, the spatial frequency with value of 0.5 is defined as a characteristic frequency by means of the characterization method of MTF quality characteristics of optical system, which represents the decrease rate of SFR as spatial frequency variation. It is recorded as Fth in this paper. Then, considering that the surrounding area under the SFR curve (the cut-off frequency of area solution was selected as 25 cy/mm) was the sum of all effective spatial frequencies corresponding to brightness contrast in the prints, which reflected the complete clarity information of the prints, it was defined as the clarity quality index, which is recorded as Q0.

In addition, the brightness contrast between edge field color and background substrate was defined as edge contrast, which is recorded as C. Because of the different color density of printing samples, the edge contrast is different. Considering that the edge contrast has a direct impact on the visual sharpness of prints, multiply the edge contrast C by the SFR value, so the original characteristic frequency Fth and the area under the curve Q0 are also multiplied by the edge contrast C. The corresponding values are named as the effective characteristic frequency and the effective quality index, which is recorded as eFth and Q respectively.

After the experiment, the above quality indexes of each test sample are shown as result in Table 2.

Table 2. SFR quality indexes

No.	Fth	Q0	C	eFth	Q
A1	5.07	5.88	0.98	4.97	5.76
A2	4.59	5.27	0.97	4.45	5.12
A3	5.08	5.92	0.97	4.92	5.74
B1	3.33	3.91	0.72	2.40	2.80
B2	3.93	4.54	0.78	3.07	3.55
C1	3.27	3.77	0.95	3.10	3.58

Further, the correlation between Q0 and Fth, and the correlation between Q and eFth is investigated. The correlation curves are shown in Fig. 5a and b, and the linear correlation coefficients are 0.9991.

a. Correlation between Q0 value and Fth b. Correlation between Q value and eFth

Fig. 5. Correlation diagram

Figure 5(a) and (b) show that there is a good linear correlation between Q0 and Fth, and the same between Q and eFth. This result shows that Fth is equivalent to Q0, and eFth is equivalent to Q, that is, one of the defined (effective) characteristic frequency and (effective) quality index is able to characterize the SFR quality. In this paper, it is suggested that the characteristic frequency Fth and the effective characteristic frequency eFth should be used to characterize the normalized SFR curve characteristics after multiplying the edge contrast. Figure 6 shows the comparison of Fth and eFth of the experimental sample.

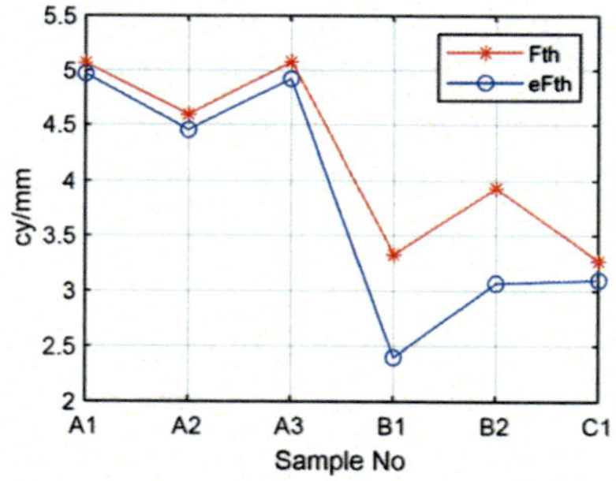

Fig. 6. Comparison of samples' SFR quality characteristics

The comparison in Fig. 6 shows that the image sharpness quality of samples is generally divided into two groups by means of the image sharpness characterized by normalized SFR curve, which represented by Fth. The sharpness of samples A1, A2 and A3 is relatively higher, while that of samples B1, B2 and C1 is relatively lower. According to the effective image sharpness represented by the SFR curve including the edge contrast represented by eFth, the relationship between the image sharpness of the two samples has not changed, but the effective sharpness of B1 and B2 samples relatively declined much more. According to the edge contrast C in Table 2, it is the relatively lower edge contrast of B1 and B2 samples that reduces their effective sharpness. Edge contrast is actually the maximum contrast of image brightness, so it is not difficult to understand that the comparison which based on eFth can evaluate the visual image sharpness with a better result.

Comparing the differences between the SFR quality indexes Fth and eFth of these samples, it can be concluded that due to the UV ink, the rapid curing characteristics of samples A1, A2 and A3 can significantly reduce the lateral infiltration of ink on the paper surface which creates an obvious advantage of sharpness. Accordingly, the B1 and B2 samples printed by non-UV ink do not have the unique advantage (fast curing speed) of UV ink, which showed that the faster attenuation of image contrast. Meanwhile, compared with B1, the higher eFth value of B2 showed that digital paper was superior to offset paper. Samples C1 and B2 of the photo grade inkjet have equivalent effective sharpness quality.

4 Conclusions

In this paper, the spatial frequency response (SFR) measurement on prints sharpness is practiced. It effectively evaluated the influence of different printing conditions on prints clarity. The quality indexes eFth and Q can reflect and characterize the sharpness quality of printed matter with a better result. The effective characteristic frequency and quality index proposed in this paper can reflect the impact of contrast, which further improve the characterization of prints sharpness.

References

1. ISO/IEC TS 24790: Information Technology-Office Equipment-Measurement of Image Quality Attributes for Hardcopy Output-Monochrome Text and Graphic Images. ISO, Geneva (2017)
2. ISO/TS 15311-1: Graphic Technology-Requirements for Printed Matter for Commercial and Industrial Production-Part 1: Measurement Methods and Reporting Schema. ISO, Geneva (2016)
3. Xu, Y.: Printing Image Analysis and Measurement. Cultural Development Press, Beijing (2021)
4. ISO/IEC 29112:2018: Information Technology—Office Equipment—Test Pages and Methods for Measuring Monochrome Printer Resolution. ISO, Geneva (2018)
5. Gao, C., Xu, Y., Xu, Y., Sone, Y.: Analysis of the effect of paper type on output line's quality on inkjet printer. Inf. Record. Mater. **13**(06), 4–8 (2012)

Investigation of Parametric Colour Difference on Physical Size Effect for Sample Pairs with Separation

Qiang Xu[1], Changjun Li[2], and Ming Ronnier Luo[1](✉)

[1] State Key Laboratory of Modern Optical Instrumentation, Zhejiang University, Hangzhou, China
m.r.Luo@zju.edu.cn

[2] School of Electronics and Information Engineering, University of Science and Technology Liaoning, Anshan, China

Abstract. Present study investigates the use of 1931 and 1964 standard colorimetric observers in colorimetry. It was recommended by the CIE to apply them below and above 4° field of view (FoV), respectively. An experiment was carried out to evaluate parametric colour difference on different physical sizes using a grey-scale method on a display. 280 sample pairs of display colours were assessed 20 times by 46 normal colour-vision observers (mean age 24 years). Sample pairs were selected surrounding the 5 CIE colour centres, having mean colour difference of 4 CIELAB units, with separation between two colours on a pair, having 4 physical sizes, 2°, 4°, 10° and 20°. The visual results were used to test the effects of different colour matching functions (CMFs) and colour difference equations. It was found that CIEDE2000 performed the best, followed by CAM16-UCS and CIELAB the worst. Finally, CIE 1931, CIE 1964, CIE 2006-2°, CIE 2006-10° and 2006-4° CMFs had similar performance in calculating colour differences.

Keywords: Parametric colour difference · Physical size effect · Colour difference equation · Colour matching function

1 Introduction

CIE 1931 colour matching functions (CMFs) define 2° standard colorimetric observers, applied for physical size or field of view (FoV) less than 4°. If the physical size is increased, colour discrimination becomes more pronounced, and 2° matches tend to break down. For this reason, in 1964, the CIE recommended a different set of CMFs, CIE 1964 standard colorimetric observers, to define 10° standard colorimetric observers for samples having physical sizes greater than 4°. Taking into account variation of the lens, macula pigment and cone response as a function of age, CIE 2006 CMFs [1] for FoV from 1° to 10°, for observers of different ages and physical sizes, have been recommended.

To study the effect of physical size on colour-difference evaluation and compare the performance of different CMFs and colour-difference equations, the present experiment was conducted using square sample pairs in 4 different sizes.

M. Xu et al. (Eds.): CACPP 2022, LNEE 991, pp. 38–42, 2023.
https://doi.org/10.1007/978-981-19-9024-3_6

2 Methods

2.1 Apparatus

The experiment was carried out using a 31-inch NEC PA311D liquid crystal display, with a resolution of 4096 × 2160 pixels. The display peak white was set to 6500 *K* with a luminance of 300 cd/m^2, characterized using the Gain-Offset-Gamma (GOG) [2] model. The GOG model's predictive accuracy had an average of 0.35 CIEDE2000 [3] (ΔE_{00}) units over 70 sample colours in the present experiment, with a standard deviation of 0.20 ΔE_{00} units. All the measurements were conducted using a Konica Minolta CS2000A tele-spectroradiometer in terms of spectral power distribution (SPD) for each stimulus. Their colorimetric values were calculated by multiplying SPD to CIE 1964 standard colorimetric observer, unless stated otherwise.

2.2 Sample Preparation

Five CIE recommended colour centres were studied, e.g., grey, red, yellow, green, and blue [4]. 14 sample pairs were selected around each colour centre, including 7 pairs with only lightness difference and 7 pairs without lightness difference. Sample colours were distributed uniformly from 0 to 180° in $\Delta a^* \Delta b^*$ plane, from 0 to 90° in $\Delta L^* \Delta a^*$ or $\Delta L^* \Delta b^*$ planes, respectively. All sample pairs had a colour-difference of about 4 CIELAB (ΔE^*_{ab}) units (from 3.45 to 4.94, with a mean value of 4.09 and standard deviation of 0.29). There was a hairline between two colours on a pair, produced by adding a black pixel across the edge between two colours. Sample pairs were shown on the display in 4 sizes, having 2°, 4°, 10°, and 20° field of view angle. Totally, 280 sample pairs (5 centres × 14 samples × 4 FoVs) were studied. 80 of them were repeated to evaluate intra-observer variation.

2.3 Visual Assessment

The grey-scale method [5] was used to assess colour-differences of sample pairs. The grey-scale samples consisted of 9 ISO 105 A02 [5] samples (GS-1 to GS-5 with an interval of 0.5) and 1 additional sample (GS-0.5). The grey-scale pairs were constructed between the standard (GS-5) and each of GS-0.5 to GS-5 samples. Equation (1) was used to scale the visual judgments in terms of grey-scale values (GS) to visual colour-difference values (ΔV).

$$\Delta V = 0.1172GS^4 - 1.7394GS^3 + 9.6987GS^2 - 26.0010GS + 31.8068 \quad (1)$$

Forty-six normal colour vision observers participated in the experiment, including 21 males and 25 females, with ages from 19 to 30 years (mean 24 years, standard deviation 2.9). The experiment consisted of 4 parts according to FoV of 2°, 4°, 10° and 20°. Each observer was involved in 1–4 parts of the experiment. Each part was assessed by 20 observers.

The experiment was carried out in a darkened room. Observers seated 60 cm in front the display. After the 1-min adaptation, observers were asked to assessed the colour-difference of sample pairs using grey-scale with one decimal. Sample pairs were shown in a random order.

3 Results and Discussion

3.1 Observer Variation

The standard residual sum of squares (STRESS) [6] value calculated from Eq. (2) was used to indicate the disagreement between two sets of data compared.

$$STRESS = 100\sqrt{\frac{\sum(F\Delta E_i - \Delta V_i)^2}{\sum \Delta V_i^2}} \tag{2}$$

where $F = \left(\sum \Delta E_i \Delta V_i\right)/\left(\sum \Delta E_i^2\right)$, is a scaling factor to adjust ΔE and ΔV on to the same scale. The percent STRESS values are always between 0 and 100. Values of STRESS near to zero indicate better agreement between two sets of data.

For the inter-observer variation, the STRESS values were 46 (2°), 41 (4°), 44 (10°) and 41 (20°), with an average of 43. For the intra-observer variation, the STRESS values were 24 (2°), 25 (4°), 23 (10°) and 22 (20°), with an average of 23. The inter-observer variation was larger than the intra-observer variation by about 200%.

3.2 Physical Size Effect

The STRESS values between all possible combinations of ΔV values between two FoVs were calculated. Table 1 lists the STRESS values between ΔV values from different FoVs and the mean ΔV. The largest discrepancy was found between 2° and 10° FoVs to have 25 STRESS units. The mean STRESS between ΔV values of individual FoVs compared is 19. However, comparing the inter-observer variation of 43 units, this difference is considered to be small.

Table 1. STRESS values between ΔV values from different FoVs and the mean ΔV

	4°	10°	20°	Mean ΔV
2°	20	25	21	16
4°	–	14	17	9
10°	–	–	19	12
20°	–	–	–	11

3.3 Comparison of CMFs

The spectral power distribution (SPD) of each sample was used to calculate tristimulus values XYZ using 5 sets of CMFs, named CIE 1931, CIE 1964, 2006-4°, CIE 2006-2°, CIE 2006-10°, for which the latter two were acquired from the website of Colour & Vision Research Laboratory [7] and 2006-4° CMF was calculated based on CIE publication

[1] for the 24 years old observer, the average age of the present observers. And then, colour-differences values (ΔE) of sample pairs from CIELAB [8], CIEDE2000 [9] and CAM16-UCS [10] were obtained. And their values were inter-compared between all 10 combinations of CMFs. It was found the results in terms of the STRESS value were very similar between the three colour models. The results from CIELAB were taken as an example. They can be divided into three groups: (1) 2° CMFs (CIE 1931 and CIE 2006-2°), (2) 10° CMFs (CIE 1964, and CIE 2006-10°), and (3) CIE 2006-4° respectively. The STRESS values were 2.9 and 0.6 for the two CMFs in Groups 1 and 2 respectively. This means the two 10° CMFs agreed better than the two 2° CMFs. Comparing to CIE 2006-4° CMF, the closest was CIE 2006-2°, followed by CIE 1931, CIE 1964, CIE 2006-10° with 1.2, 3.5, 4.3 and 4.4 STRESS units respectively. The largest difference was found between CIE 1931 and 1964 CMFs (5.0 units). This STRESS value is still considered to be extremely small comparing to the inter-observer variation (43 units) and physical size effect (19 units), respectively.

3.4 Model Performance

The visual data were used to test the performance of three colour models, CIELAB [8], CIEDE2000 [9] and CAM16-UCS [10]. Again, the performance is reported using the STRESS value calculated between the visual differences ΔV and the predicted differences ΔE. Figure 1 shows models' performance in STRESS units under each CMF and FoV. CIEDE2000 and CAM16-UCS outperformed CIELAB significantly as previously reported by many researchers. The former two gave quite good performance, i.e., much smaller STRESS values than that of inter-observer variation. Furthermore, with FoV increasing, model's performance in STRESS units increased first and then deceased. The best performance appeared in 20° FoV stimuli.

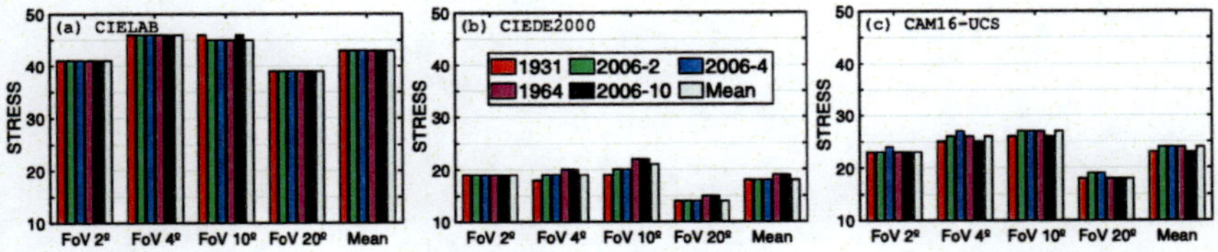

Fig. 1. Models' performance in STRESS units under each CMF and FoV

It can also find that the ranges of 1, 3, 2 STRESS values between the best and the worst CMFs were CIELAB, CIEDE2000 and CAM16-UCS, respectively. This implies that there is hardly any difference between CMFs in calculating colour differences.

4 Conclusion

The experiment was conducted using display colours to reveal the FoVs of test stimulus (2°, 4°, 10° and 20°) affected by physical sizes. The results clearly showed that different FoVs had little impact on colour differences. We can conclude the parametric effect due to stimulus sizes to be small. Different CMFs used in colour difference formulae had little difference. The largest difference was found between CIE 1931 and 1964 CMFs (5.0 STRESS units).

Acknowledgements. The work is supported by National Natural Science Foundation of China (Grant number: 61775190).

References

1. CIE 170-1:2006: Fundamental Chromaticity Diagram with Physiological Axes - Part 1. Commission Internationale de l'Eclairage, Vienna, Austria (2006)
2. Berns, R.S.: Methods for characterizing CRT displays. Displays **16**, 173–182 (1996)
3. ISO/CIE: Colorimetry – part 6: CIEDE2000 colour-difference formula. In: ISO/CIE 11664-6:2014(E) (ISO/CIE, 2014), pp. 11664–11666 (2014)
4. Robertson, A.: CIE guidelines for coordinated research on color-difference evaluation. Color Res. Appl. **3**, 33–39 (1978)
5. ISO 105-A02: Textiles - Tests for Colour Fastness - Part A02: Grey Scale for Assessing Change in Colour. ISO, Geneva (1993)
6. CIE 217:2016: Recommended Method for Evaluating the Performance of Colour-Difference Formulae. Commission Internationale de l'Eclairage, Vienna, Austria (2016)
7. Colour & Vision Research Laboratory. http://www.cvrl.org
8. CIE 015: 2018: Colourimetry. Commission Internationale de l'Eclairage, Vienna, Austria (2018)
9. Luo, M.R., Cui, G., Rigg, B.: The development of the CIE 2000 colour-difference formula: CIEDE2000. Color Res. Appl. **26**, 340–350 (2001)
10. Li, C., et al.: Comprehensive color solutions: CAM16, CAT16, and CAM16-UCS. Color Res. Appl. **42**, 703–718 (2017)

A New Black Generation Algorithm for Color Printing

Hao Qin[1,2] and Ming Zhu[2](✉)

[1] Faculty of Light Industry, Qilu University of Technology (Shandong Academy of Sciences), Jinan, China
[2] School of Art and Design, Henan University of Engineering, Zhengzhou, China
115457719@qq.com

Abstract. To obtain a unique solution of the reverse equation of the cellular Neugebauer model for color printing, the method of gamut range grouping is proposed. First, the gamut sets based on different black plate values are obtained by experiment. Then, the range of black plate is determined by deciding which the gamut sets an input color belongs. On this basis, the optimal black value is determined according to the tone and color characteristics of the image, then the unique solution is calculated by inputting the optimal black value into the reverse Neugebauer equation. The color difference evaluation results show that the color reproduction accuracy of the new algorithm is better than the grey balance algorithm and is close to the i1Profiler software. The image quality evaluation results show that the new algorithm can reproduce the dark details of the image well. In addition, the SMGBD algorithm can speed up gamut boundary calculation. Therefore, the new algorithm can meet the requirements of the printing industry.

Keywords: Printing black generation · Color gamut range · Neugebauer equation · Color reproduction

1 Introduction

Black generation is an important technology in the color space conversion for printing reproduction. To solve the defect of three primary color reproduction only used cyan (C), magenta (M) and yellow (Y), it is necessary to add additional black in color printing [1]. However, in both traditional four-color printing and multicolor high-fidelity printing [2], the complexity of the color separation model is increased because of the addition of black. At present, most of the research works focus on the colorspace conversion and color gamut mapping algorithms [3], while there are few reports on the black generation algorithm for color separation model [4]. The traditional black generation algorithm is based on the printing grey balance curve [5], but it is usually not satisfactory when it is used to reproduce the darktone area of printed images.

M. Xu et al. (Eds.): CACPP 2022, LNEE 991, pp. 43–47, 2023.
https://doi.org/10.1007/978-981-19-9024-3_7

2 Algorithm Principle

2.1 Algorithm Framework

The new algorithm is proposed based on the cellular Neugebauer model, and the overall flow chart is shown in Fig. 1. First, the input color is mapped to the printer color gamut. Then, there will be different methods to calculate K when the input color is located in different tones. Finally, input the K into the reverse cellular Neugebauer equation for calculating C, M, and Y. The ICC color conversion experiment is designed on the Matlab platform, and the conversion accuracy of the new algorithm is verified. Meanwhile, the practical application effect of the new algorithm is analyzed in printing image color separation.

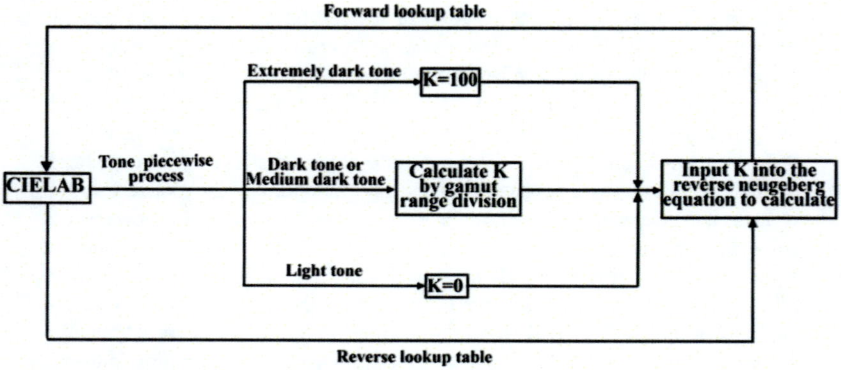

Fig. 1. Algorithm framework

2.2 Tone Piecewise Strategy

In order to do different treatments for different color based on its tone characteristics, the tone piecewise strategy is designed. The tone range is divided into three pieces based on the L channel of printing greyscale: extreme darktone, middletone, and highlight. For an input color, after the color separation by the reverse Neugebauer equation, the black will be added if there are at least one result values (C, M, or Y) greater than 1.0, because, in this situation, the input color can not be reproduced accurately by only C, M, Y. The input color can be considered within the darktone area, and the new algorithm is used to determine the K value with a large black replacement coefficient. The point K = 1.0 of the printing greyscale is regarded as the boundary of the extreme darktone. An input color will be considered within the extreme darktone area if its L channel value is below that of the boundary point (K = 1.0); in this situation, the black can be set to 1.0 directly.

2.3 Principle of Black Generation Algorithm

To make the experimental model more accurate, the five-level cell is selected to divide the Neugebauer model [6], and the correction coefficient of the Neugebauer model is set to 1.2. Besides, the bilateral filter is used to smooth lookup table curves. SMGBD algorithm is used for gamut mapping [7].

The new algorithm is shown in Fig. 2. In the first step, K is divided into five grades (with 20 for the interval). Then, the color patches are designed and printed for each k-value level, moreover, the CIELAB value of the color patches are measured. According to the CIELAB value, each K value level's print gamut is calculated and generated gamut sets corresponding to different K.

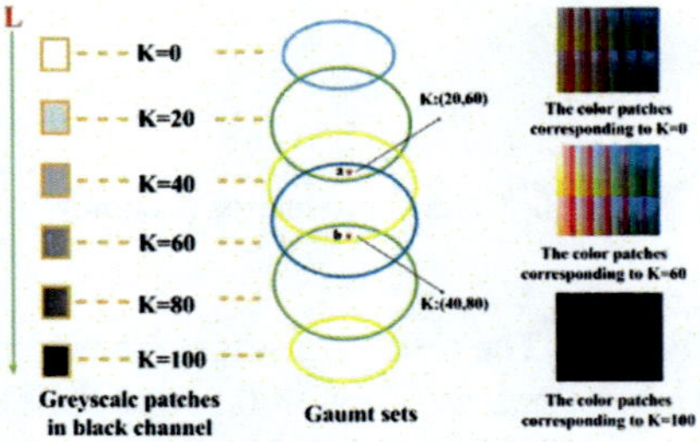

Fig. 2. Black generation algorithm flow chart

Based on the input color's CIELAB, judge the colorspace in gamut sets, then the range of K is determined according to the colorspace. For example, point "a" is located in gamut sets corresponding to the K of 20%, 40%, and 60%, so the K range is from 20% to 60%. Similarly, the K range of point "b" is from 40% to 80%.

The third step is to set the black plate replacement rate "t". It will set different replacement rates based on different color tones of images. According to the maximum K (Kmax) and minimum K (Kmin) in the K range obtained in the previous steps, the optimal K is calculated by using the formula (1). Finally, input the K into the cellular Neugebauer equation to solve the unique solution of the CMY.

$$K = Kmin + (Kmax - Kmin) * t \tag{1}$$

3 Experimental Results and Analysis

3.1 Evaluation of Color Reproduction Accuracy

Experiment 1: The standard ECI2002 color patches and color difference theory were used to evaluate the color reproduction accuracy. The grey balance algorithm and the i1Profiler algorithm were also participating in the performance comparison. The results are shown in Fig. 3.

As shown in Fig. 3, the red line was obtained using the new algorithm when the black replacement rate of the extremely darktone remained at 0.9 and the black replacement rate of the middletone changed from 0 to 1 with 0.2 intervals. It can be concluded that when the black plate replacement rate of middletone was set to 0.5, the mean color difference and maximum mean color difference of the color patches were the smallest

Fig. 3. Color reproduction accuracy

and were close to the i1Profiler. The result was shown in the green line when the black replacement rate of the middletone was kept at 0.5, and the black replacement rate of the extremely darktone area varied from 0.1 to 0.9. Compared with the last experiment, the mean color difference and maximum mean color difference of color patches were decreased.

The blue line represents the grey balance algorithm when the black plate replacement rate changed between 0.1 and 0.9. It can be seen that the color difference decreased with the decrease of the black plate replacement rate, and the performance of the grey balance algorithm was inferior to the new algorithm.

3.2 Image Reproduction Effect Evaluation

Experiment 2: First, the black generation algorithm was used to calculate the reverse lookup table, and then the lookup table was written into ICC color profiles. Second, collected test images, which should contain dark, medium, and light tones. Lastly, Photoshop was used to load ICC color profiles, converted the image into CMYK mode, and observed it through screen preview. The results are shown in Fig. 4.

In Fig. 4, the areas where the blue box, yellow box, and red box respectively represent the highlight, middletone, and darktone. There was little difference among different methods in the highlight area. In the middletone, the image processed by grey balance will enlarge saturation. In the darktone, both the new algorithm and i1Profiler software can reproduce the details of the image well; however, the grey balance algorithm had the phenomenon of the loss of image details.

4 Conclusions

A new black generation algorithm was proposed in this paper to solve the problem of color printing separation of the reverse cellular Neugebauer model. The test results show that the color reproduction accuracy of the new algorithm is better than the grey balance algorithm and is close to the i1Profiler algorithm. The image test results show that the

Fig. 4. Color separation results of the different algorithms

new algorithm can reproduce the details of the dark tone area well. Therefore, the new algorithm can meet the requirements of the printing industry.

Acknowledgement. This research is supported by Scientific and Technological Projects of Henan Province (No. 212102210397), The Key Scientific Research Project of Education Department of Henan Province in 2022 (No. 22A510012), Lab of Green Platemaking and Standardization for Flexographic Printing (No. ZBKT202005), The Key Laboratory of Intelligence and Green Flexographic (No. ZBKT202104), Research Development Projects of Henan University of Engineering in 2020 (No. PYXM202010).

References

1. Zhu, M., Liu, Z., Chen, G.X.: Research on six-color separation model based on subarea Neugebauer equations. Acta Opt. Sin. **31**(07), 288–297 (2011)
2. Wu, G.Y.: Charamer mismatch-based spectral gamut mapping. Laser Phys. Lett. **16**(9), 095206 (2019)
3. Zhu, M., Tian, Z.: A spatial gamut mapping algorithm based on adaptive detail preservation. J. Imaging Sci. Technol. **62**(1), 1–12 (2018)
4. Zhang, Z.J., Liu, Z., Wu, M.G.: Polynomial regression multi-color separation method based on subspace partition. Packag. Eng. **34**(7), 65–76 (2013)
5. Cholewo, J.: Conversion between CMYK spaces preserving black separation. In: Color and Imaging Conference, 2000, no. 1, pp. 257–261 (2000)
6. Jiao, H.M., Zhu, M.: Influencing factors of color prediction of cellular Neugebauer model. In: Lecture Notes in Electrical Engineering (2021)
7. Zhong, Y.F., Fu, L.J., Wang, X.J.: Research of black plate generation method based on Neugebauer equation. Packag. Eng. **034**(11), 98–101 (2013)

Optimization of Standard Object Color Spectrum Database Based on Human Eye Color Discrimination Threshold Cycle Algorithm

Qian Cao(✉)

Department of Printing and Packaging Engineering, Shanghai Publishing and Printing College, Shanghai, China
caoqian8889@163.com

Abstract. The Standard Object Color Spectrum (SOCS) database is a database in the Technical Report (TR) of the Japanese Organization for Standardization (JIS) and International Standardization (ISO) and is widely used in color reproduction evaluation and color science research. This paper proposes an optimized loop algorithm recommended by Qian Cao et al. in 2022 to remove the data redundancy points of SOCS and improve the problem of uneven distribution of SOCS in LAB color space. Compared with SOCS, the color gamut volume of the optimized SOCS is almost unchanged, and the data is reduced by 33.35%, which overcomes the shortcomings of accumulating too many sample points in the local space, and the uniformity distribution of optimized SOCS in the LAB color space is greatly improved.

Keywords: Standard object color spectra (SOCS) database · Loop algorithm · Human eye color discrimination threshold

1 Introduction

The Standard Object Color Spectrum (SOCS) database for color reproduction evaluation is a database in the Technical Report (TR) of the Japanese Organization for Standardization (JIS) and International Standardization (ISO). Since its release, SOCS has been widely used in color science research. Kato et al. [1] used offset printing data in SOCS to test a new printing color prediction model; Inui et al. [2] studied the color gamut of SOCS in 2004, and compared it with the color gamut of other spectral databases; Li et al. [3] studied the color gamut of SOCS and other spectrum databases in 2013. Park et al. [4] used the spectral data of flowers and leaves in SOCS to study the three primary colors of laser displays in 2006. Sun [5] used SOCS to optimize the color rendering index. Cao et al. [6] used SOCS as a testing sample to study the multi-spectral data compression algorithm in 2020. However, when SOCS was established, it was considered to collect as many spectral data of different types of real objects as possible, without considering the problem of data redundancy, that is, a lot of spectral data is transformed into colorimetric data that cannot be distinguished by the human eye. After SOCS is transformed

M. Xu et al. (Eds.): CACPP 2022, LNEE 991, pp. 48–53, 2023.
https://doi.org/10.1007/978-981-19-9024-3_8

into LAB color space, the spatial distribution of SOCS is uneven, and too many colors are accumulated in some areas, as shown in Fig. 1. From a colorimetric point of view, there is a lot of data redundancy.

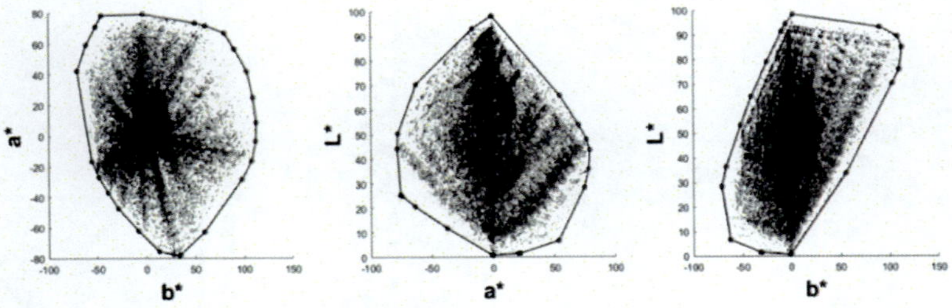

Fig. 1. Distribution of SOCS database in LAB color space

2 Method

Cao et al. [7] proposed a loop algorithm to remove redundant points of multispectral datasets in 2022. In order to remove the data redundancy points of SOCS and improve the problem of uneven distribution of SOCS in LAB color space, this paper improves and optimizes the loop algorithm proposed by Qian Cao et al. in 2022. The specific steps are as follows:

First, the spectral database is converted into a chromaticity database under the CIE1931 standard colorimetric observer and the CIE standard illuminant D_{65}.

Loop algorithm for deleting redundant points is shown in Fig. 2. At the beginning of the loop, judge whether the chromaticity dataset is empty. If so, the loop ends. If the chromaticity dataset is not empty, in the chromaticity dataset, look for the farthest target point away from the center point (50, 0, 0) of the LAB color space and save the target point to the target points dataset. Then, delete target point and points in the chromaticity dataset whose color difference from the target point is less than the human eye color discrimination threshold. Finally, update the chromaticity dataset, start the next loop, … until the end of the cycle.

The target points dataset after the end of the loop is converted to the corresponding spectral dataset, that is, the optimized spectral dataset.

3 Experiment and Process

3.1 Data Sources

The SOCS database contains the spectra of 53486 objects color, such as photographic materials, offset printing, gravure printing, computer color prints, painting (not for art), painting (oil painting, watercolor), textiles, flowers and leaves, outdoor scenes, human skin and calibration data.

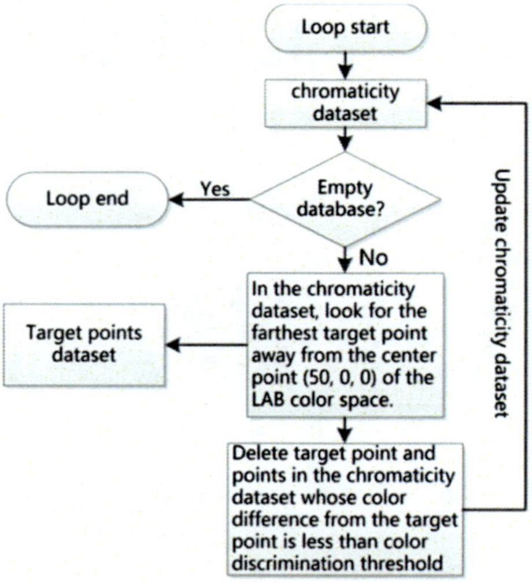

Fig. 2. Flow chart of loop algorithm for deleting redundant points

3.2 Human Eye Color Discrimination Threshold

The ability of human eyes to distinguish color is different in different color regions and in different colored substances such as displays and printed samples [8–10]. Two colors less than the human eye color discrimination threshold can be considered as the same color visually, and the color point close to the center point (50, 0, 0) of LAB color space is considered as redundant point. Since SOCS is mainly composed of spectral data such as offset printing, gravure printing and color printing, this paper adopts the research results of human eye color discrimination threshold of printed samples by Min [8], and the human eye color discrimination threshold is 0.76 ΔE^*_{ab}. According to the method in the second part of this paper, the color discrimination threshold of human eye is 0.76 ΔE^*_{ab}.

4 Results and Discussion

When the human eye color discrimination threshold is 0.76 ΔE^*_{ab}, SOCS is optimized according to the experimental method, and the optimized spectral dataset is named Optimized SOCS. The comparison between SOCS and Optimized SOCS is shown in Table 1.

Color gamut volume is an important attribute of color gamut, which refers to the amount of color information. In this paper, the spectral dataset is transformed into CIELAB chromaticity value under the conditions of illuminant D_{65} and CIE1931 standard colorimetric observer, and then the convex hull algorithm [11] is used to calculate the color gamut volume. It can be seen from Table 1 that compared with SOCS, the color gamut volume of the optimized SOCS remains almost unchanged.

It can also be seen from Table 1 that the data volume of Optimized SOCS is reduced by 33.35% compared with SOCS. Three chromaticity diagrams L^* a^*, L^* b^* and a^*

Table 1. Comparison between SOCS and optimized SOCS

	SOCS	Optimized SOCS
The number of samples	53486	35649
Redundant point	17837	
Proportion of redundant points	33.35%	
Color gamut volume	9.3707×10^5	9.3701×10^5
File size	6.15 MB	3.76 MB

b^* are used to compare the colorimetric spatial distribution of Optimized SOCS and SOCS, as shown in Figs. 3, 4 and 5. It can be clearly seen from Figs. 3, 4 and 5 that compared with SOCS, the Optimized SOCS overcomes the disadvantage of stacking too many sample points in local space, and greatly improves the uniformity distribution in LAB color space. Since the optimized SOCS dataset has no redundant data, it is more reasonable to use it in other color science studies, and the amount of computation will be reduced.

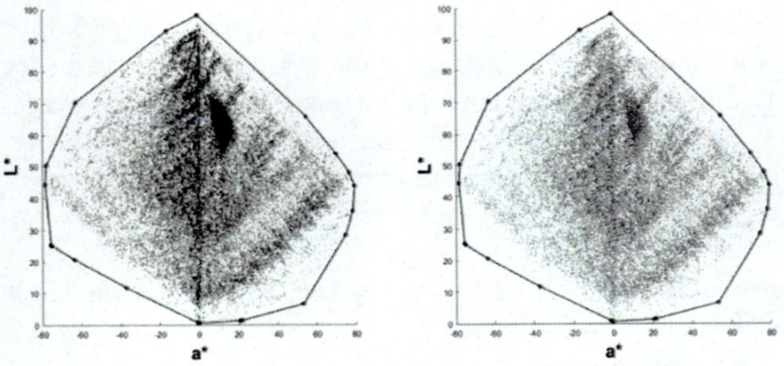

(a) SOCS (b) Optimized SOCS

Fig. 3. L^* a^* chromaticity diagram (D_{65}/2°)

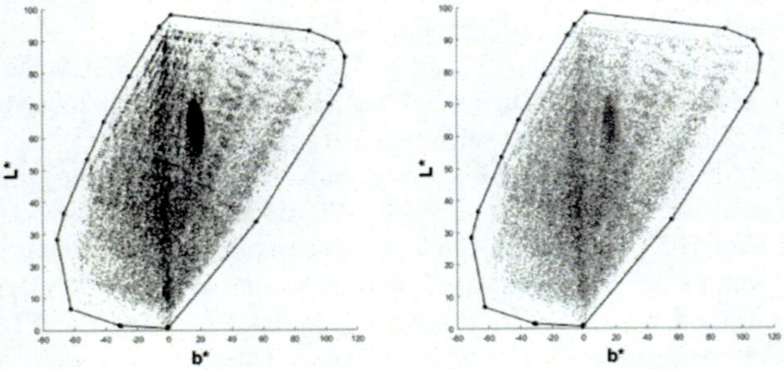

(a) SOCS (b) Optimized SOCS

Fig. 4. L^* b^* chromaticity diagram (D_{65}/2°)

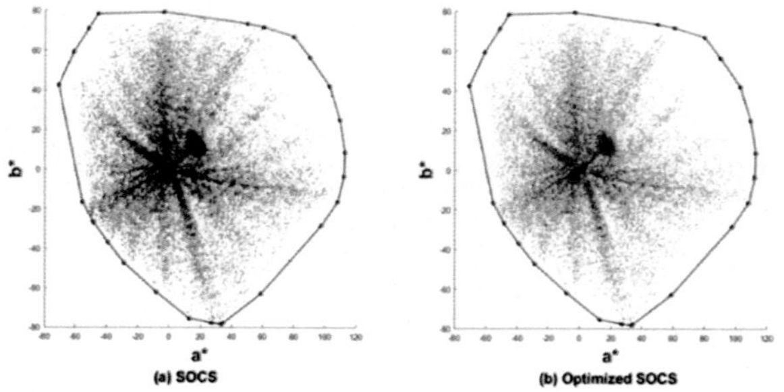

Fig. 5. a^* b^* chromaticity diagram (D_{65}/2°)

5 Conclusion

In view of the existence of data redundancy and uneven distribution of the SOCS database in the LAB color space, this paper proposes a loop algorithm based on the human eye color discrimination threshold to delete redundant spectral data in the spectral database. This paper adopts the human eye color discrimination threshold is 0.76 ΔE^*_{ab} according to the research results of human eye color discrimination threshold of printed samples by Huang Min. Compared with the SOCS database, the color gamut volume of the optimized SOCS is almost unchanged, and the data is reduced by 33.35%, which overcomes the shortcomings of accumulating too many sample points in the local space, and the uniformity distribution in the LAB color space is greatly improved. The research method in this paper is also used to optimize other spectral databases or chromaticity databases, to delete redundant points and improve the uniformity of spectral databases or chromaticity databases in LAB color space.

Acknowledgments. This study is funded by Key Lab of Intelligent and Green Flexographic Printing (KLIGFP-01, No ZBKT201905).

References

1. Kato, M., Inui, M., Azuma, Y.: Colorimetric prediction of halftone prints with pale-ink model. Acad. Rep. Fac. Eng. Tokyo Polytech. Univ. **34**(1), 71–77
2. Inui, M., Hirokawa, T., Azuma, Y., Tajima, J.: Color gamut of SOCS and its comparison to pointer's gamut. In: NIP & Digital Fabrication Conference, Jan 2004, vol. 2004, no. 1, pp. 410–415. Society for Imaging Science and Technology
3. Li, C., Luo, M.R., Pointer, M.R., Green, P.: Comparison of real colour gamuts using a new reflectance database. Color Res. Appl. **39**(5), 442–451 (2014)
4. Park, S.O., Kim, H.S., Kwon, Y.J.: Guideline on designing laser display primary for reproducing real world object colors. In: Conference on Colour in Graphics, Imaging, and Vision, Jan 2006, vol. 2006, no. 1, pp. 216–219. Society for Imaging Science and Technology
5. Sun, P.L.: Optimizing Color Rendering Index Using Standard Object Color Spectra Database and CIECAM02
6. Cao, Q., Li, X., Li, J.: Effects of cone response function on multispectral data compression. In: Advances in Graphic Communication, Printing and Packaging Technology and Materials, pp. 82–88. Springer, Singapore (2021)

7. Cao, Q., Li, J., Li, X.: Multispectral data optimization using loop algorithm to remove redundant point. In: Interdisciplinary Research for Printing and Packaging, pp. 82–88. Springer, Singapore (2022)
8. Min, H.: Study on just-noticeable color difference discrimination threshold by using printed samples I: analysis of visual evaluation experimental data. Acta Opt. Sin. **29**(5), 1431–1436 (2009)
9. Luo, M.R., Minchew, C., Kenyon, P., Cui, G.: Verification of CIEDE2000 using industrial data. In: AIC Color and Paints, Interim Meeting of the International Colour Association, Proceedings, pp. 97–102 (2004)
10. Zhehong, W., Haisong, X.: Study on color discrimination threshold using CRT display part I: analysis of experimental data and human color vision characteristics. Acta Opt. Sin. **27**(6), 1139 (2007)
11. Barber, C.B., Dobkin, D.P., Huhdanpaa, H.: The quickhull algorithm for convex hulls. ACM Trans. Math. Softw. (TOMS) **22**(4), 469–483 (1996)

Solution to the Reproduction of Multiple Pantone Colors on Six-Color Offset Press

Chunyan Bai(✉) and Yan Liu

Department of Printing and Packaging Engineering, Shanghai Publishing and Printing College, Shanghai, China
chunyanbai1984@163.com

Abstract. Printing is the technique for reproducing many copies of a graphic message. Accurate reproduction of color is an important parameter to evaluate the quality of printing. In the current offset printing industry, six-color press still occupies a large market share. For six-color offset press, when the Pantone colors are more than six it is difficult to accurately reproduce all colors. In this paper the solution of printing C, M, Y, K four primary colors and two spot colors is proposed. When this method is applied to an eleven colors artwork, color differences for main colors are below 3, for the auxiliary colors are below 15, especially some difficult auxiliary colors the average color differences is reduced by 59%. It is demonstrated the method can effectively control the color difference and make all Pantone colors reach the industry standard or customer's requirement. Therefore, it can achieve accurate color reproduction of multiple pantone color by the six-color press.

Keywords: Offset · Pantone · Spot · Artwork · Color difference

1 Introduction

Color reproduction in the offset printing is based on the principle of subtractive mixture of pigments [1, 2]. The subtractive primary colors are C (Cyan), M (Magenta), and Y (Yellow) [2, 3]. In additionally, in order to increase the saturation of black color generated by subtractive primary colors, the black ink is introduced. That is, the variety of printing colors presented in the eyes of observers are produced through overprinting the four primary colors of C, M, Y, and K according to different dot area proportion.

The six-color offset press has six printing units, it means in addition to the four primary colors of C, M, Y, K, two extra spot colors can be printed in one printing process. The six-color press can solve most of the printing tasks, however, when it encounters artworks with multiple Pantone colors, the problem will become tricky. Here, multiple Pantone color artworks has two categories: one is designed with C, M, Y, K colors and more than two Pantone colors, and the other is designed with more than six Pantone colors. Some foreign brands and customers, such as the Hello Kitty brand, Target and Walmart supermarkets, their package artworks belong to the second category. All the colors involved in the artwork are Pantone colors, the quantity of colors is usually more than six, moreover, customers have high requirements for all colors. This is a challenge for many factories and operators that lack experience. This paper will analyze the technical difficulties of printing such artworks and propose practical solution.

M. Xu et al. (Eds.): CACPP 2022, LNEE 991, pp. 54–61, 2023.
https://doi.org/10.1007/978-981-19-9024-3_9

2 Technical Difficulties for Printing Multiple Pantone Colors on the Six-Color Press

2.1 Defects of Printing All Pantone Color as Spot Color

At present, the most widely used presses in offset industry are still four-color and six-color presses. Only some large enterprises have more than six-color press. A six-color press can reproduce up to six colors in one printing process, so when the color in the artwork exceeds six colors, it will bring some technical difficulties to operators. Take the artwork in Fig. 1 as an example. This artwork contains eleven Pantone colors. If all Pantone colors are printed as spot color, the substrate needs to run twice on a six-color press. Although this printing process can achieve accurate reproduction of multiple Pantone colors, there are many defects and potential quality problems. Firstly, many times of overprint in the process will induce inaccurate overprint. Secondly, the distortion and expansion of the dot will cause the distortion of the pattern. In additionally, the serious shrinkage of the paper due to the twice printing process will also affect the accuracy of overprint. Finally, the mixing and printing of eleven spot colors itself will consume a lot of manpower and materials, thus reducing the production profit. Therefore, it is obvious that print all multiple Pantone color as spot colors are not feasible.

Fig. 1. Multiple Pantone colors artwork

2.2 Defects of Printing All Pantone Color as C, M, Y, K Colors

A common way for factories is directly converting all Pantone colors into C, M, Y and K colors to print. However, this method can cause difficulty to color match. In order to match the Pantone colors, it is necessary to increase or decrease the amount of one or two certain inks. However, because C, M, Y, K plates are fixed, the amount of ink in the entire plate will increase or reduce synchronously. As a result, when some of the Pantone colors are accurately reproduced, the others will have serious color difference. In turn, if you focus on the other colors, the previously adjusted color will change again. Finally, you will find that no matter how you adjust the ink amount, you will lose some colors, and the printing product cannot meet the customer's requirements. No matter how many times of proofing or printing, it is a waste of manpower and materials.

3 Solution for Reproduction Multiple Pantone Colors on the Six-Colors Press

Through above analysis, we know that it is not feasible to convert all Pantone colors into four-color to print or to print all colors as spot colors. Therefore, in the six-color printing, we put forward a solution of four primary color (C, M, Y, K) add two spot colors printing. The prerequisite to this solution is to divide all Pantone colors in the artwork into two parts, main colors and auxiliary colors. The main colors are the two Pantone colors with the largest area proportion and are the most important. The auxiliary colors refer to all colors other than the main colors. After the division, two categories of colors will be processed separately.

3.1 Reproduction of Auxiliary Colors

Auxiliary colors can be reproduced through three steps: color separation, black processing, and reverse setting.

Color Separation. The color separation here is different from the process of converting RGB images into CMYK images. Here it refers to the process of converting all auxiliary colors in the artworks to C, M, Y, K colors. Take Fig. 1 as an example. After observation, two colors (PMS 224 and PMS 225) have the largest area and are very important, so they are selected as the main colors, while all other colors are auxiliary colors. Then the color separation process, converting all the auxiliary colors to C, M, Y, K, is automatically executed by pre-press software. The result of color separation is shown in Table 1.

As can be seen from Table 1, color separation can convert eleven Pantone colors in the artwork to six printed colors. It can effectively reduce the quantity of printed colors. Color separation not only make the colors suitable for the six-color press, but also make color match become simple.

Black Processing. Pantone black is a color that needs extra attention. If it is directly converted to C, M, Y, K to print, there will be many problems [4]. Such as, when Pantone Black C is converts to four colors, the proportion is 13% M, 49% Y, and 98% K. On the one hand, two times overprinting will have the failure of overprinting, especial for small black text and lines. It will lead to the defect of false font, double shadow and so on. On the other hand, the overprinted black is often not saturated enough, and dries slowly. Therefore, in order to avoid these problems, pantone black can be treated as 100% K and 30–35% C. The processed data are shown in Table 2. This treatment has the advantage of eliminating the four-color overprint, making printing simple and color more saturated. However, the disadvantage is that pre-press operators need to have a lot of experience.

Table 1. Multiple Pantone colors were converted to four primary colors and two spot colors

Pantone colors in the artwork		Convert to (C, M, Y, K) + (two spot colors)					
		C (%)	M (%)	Y (%)	K (%)	PMS 224C (%)	PMS 225C (%)
1	PMS 1205	0	5	31	0	0	0
2	PMS 1375	0	40	90	0	0	0
3	PMS 165	0	59	96	0	0	0
4	PMS 4655	0	26	45	18	0	0
5	PMS 186	0	100	81	4	0	0
6	PMS 376	50	0	100	0	0	0
7	PMS 355	94	0	100	0	0	0
8	PMS 293	100	57	0	2	0	0
9	PMS black C black BBbLACK	0	13	49	98	0	0
10	PMS 224	0	0	0	0	100	0
11	PMS 225	0	0	0	0	0	100

Reverse Setting. Reverse setting is based on pre-press proofing and swatch lookup [5]. After the above steps, all auxiliary colors have been converted to four colors in different proportions. At this point, the first pre-press proofing is needed. The printed colors will be compared with the Pantone Swatch. If any color is deviates from the original, it can be corrected by reverse setting. Take Fig. 1 as example, it was found that the printed green color is deviated seriously from the Pantone color PMS 376. Firstly, find PMS 376 in the Pantone Swatch and the most similar color in the Complete Process Color Chart (CMYK swatch) through comparing, then use the C, M, Y, K proportion (C: 55%, M: 0, Y: 100%, K: 0) to reverse set the PMS 376 by the pre-press software. It is the same for PMS 1205 and PMS 4655. Data processing result is shown in Table 3.

All above three steps can be used to solve the reproduction of auxiliary color. Next, we will study the reproduction of the main colors.

3.2 Reproduction of Main Colors

The main colors can be reproduced using spot color inks. The key step is the mixing of spot color ink.

Mixing the Dark Spot Color Ink. In printing industry, a variety of spot color ink are often mixed by different proportion of primary color ink. The proportion of primary color ink is determined according to the operators' experience, swatch and calculation formula [6]. Except for C, M, Y, K ink, the configuration of spot colors also needs special color ink and diluent [7–9]. According to whether need diluent, spot colors can be divided

Table 2. Setting the Pantone black

Pantone colors in the artwork		Convert to (C, M, Y, K) + (two spot colors)					
		C (%)	M (%)	Y (%)	K (%)	PMS 224C (%)	PMS 225C (%)
1	PMS 1205	0	5	31	0	0	0
2	PMS 1375	0	40	90	0	0	0
3	PMS 165	0	59	96	0	0	0
4	PMS 4655	0	26	45	18	0	0
5	PMS 186	0	100	81	4	0	0
6	PMS 376	50	0	100	0	0	0
7	PMS 355	94	0	100	0	0	0
8	PMS 293	100	57	0	2	0	0
9	PMS black C black BBbLACK	*40*	*0*	*0*	*100*	0	0
10	PMS 224	0	0	0	0	100	0
11	PMS 225	0	0	0	0	0	100

into dark colors and light colors. The mixture of dark colors only needs C, M, Y, K ink and special color ink, do not need diluent. Suck as PMS 2593C in Fig. 2(a), the mixture requires only Pantone Ruby Red and Pantone Violet. For experienced operator, it can be directly matched. But for most workers it is recommended to mix the colors step by step. First find the required spot color from the Pantone swatch, and then mix the ink according to the type and proportion of primary inks in the swatch. Due to the influence of ink, paper, printing equipment and other factors, the proportion on the swatch is just for reference. In the actual mixing process, a color with the largest proportion is used as quantitative, and the other colors are used as variables and added gradually. After a set of mixing and testing, the optimum proportion was determined by densitometer measurement.

Mixing the Light Spot Color Ink. To mix the light color, it needs to add some diluents. Commonly used diluents in the printing industry include white ink, white lake and so on. When using white lake as a diluent, the mixed ink has a certain transparency, but insufficient hiding power, and is not bright enough. On the contrary, when white ink is used as a diluent, the ink is bright and has strong hiding power, but the pigment content is high, so the fluidity is poor, and there will be slow drying and other phenomena. Therefore, it is recommended to use both white ink and white lake as diluent when mixing light ink. According to the Neugebauer equation, the amount of white ink used depends on the proportion of primary colors, it is expressed as [6]

$$W = (1 - Ink_1)(1 - Ink_2) \cdots (1 - Ink_n) \quad (1)$$

Table 3. Reverse setting

Pantone colors in the artwork		Convert to (C, M, Y, K) + (two spot colors)					
		C (%)	M (%)	Y (%)	K (%)	PMS 224C (%)	PMS 225C (%)
1	PMS 1205	*0*	*5*	*60*	*0*	0	0
2	PMS 1375	0	40	90	0	0	0
3	PMS 165	0	59	96	0	0	0
4	PMS 4655	*10*	*40*	*50*	20	0	0
5	PMS 186	0	100	81	4	0	0
6	PMS 376	*55*	*0*	*100*	*0*	0	0
7	PMS 355	94	0	100	0	0	0
8	PMS 293	100	57	0	2	0	0
9	PMS black C black BBbLACK	*40*	*0*	*0*	*100*	0	0
10	PMS 224	0	0	0	0	100	0
11	PMS 225	0	0	0	0	0	100

Fig. 2. (a) PMS 2593C, (b) PMS 224C

where Ink_1, Ink_2 … Ink_n are the amount of different primary colors, W is the amount of white ink. Taking PMS 224C in Fig. 2(b) as an example, according to the swatch: Pantone Rose Red 15.6, Pantone Ruby Red 9.4, so the white ink calculated based on Eq. (1) is 77, the reference proportion is 16:9:77, namely 16 parts Pantone Rose Red ink, 9 parts Pantone Ruby Red, and 77 parts white ink.

3.3 Color Test

Color test is used to check the printed colors. The tools needed are high-precision electronic balance, color measuring instrument and ink proofer. All spot color inks are tested by the ink proofer, and the color differences are measured and recorded in each experiment [10]. The observation condition is D65 standard light source. The color measuring instrument is densitometer.

The three attributes of color include hue, lightness, saturation, the difference of any kind of attribute can cause the color difference on the vision [11]. Thus, the color difference is a comprehensive evaluate obtained by combining three attributes [3]. Due to the randomness, emotionality, and other unstable factors in the judgment of color by human eyes [12], the color differences measured by densitometers are used as the common and objective judgment in the industry. When the color difference between the printed color and the original is controlled within the range of industry standard or customer required, it indicate the color is accurately reproduced.

In the test, the color difference for spot color inks is below 3. Except for PMS 4655, PMS 1205 and PMS 376, all auxiliary colors are printed by four color inks directly and their color difference is below 15. For difficult auxiliary colors PMS 4655, PMS 1205 and PMS 376, through reverse setting, the average color difference is reduced from 30.14 to 12.23, about 59%. It is obvious that spot color ink printing is easy to control, although the color differences for auxiliary colors have exceeded the industry standard, considering they are reproduced with four color inks, so they are confirmed by the customer. It is demonstrated the four primary color (C, M, Y, K) add two spot colors printing method can effectively solve the problem of printing multiple Pantone colors on the six-color offset press.

4 Conclusion

It is a challenge to print multiple Pantone color artwork on a six-color offset press. This paper presents an effective method for accurately reproducing multiple Pantone colors. Distinguishing main colors and auxiliary colors in artwork is a premise to this solution. Then they can be solved effectively by different steps respectively. On the one hand, put aside the main color, it is relatively simple to ensure the accurate reproduction of the auxiliary color using C, M, Y, K four colors. On the other hand, printing with mixed spot color inks can ensure the accuracy reproduction of the main colors. The color test results show that the color differences for main colors are below 3, for auxiliary colors are below 15. For difficult colors PMS 4655, PMS 1205 and PMS 376, through reverse setting, the average color difference is reduced by 59%. All colors belong to the range of customer required. Therefore, this solution can realize the accurately color reproduction of multiple pantone color artwork on the six-color offset press. Of course, other new problems may be encountered in the production, they need to be solved according to the actual situation. It is hoped the solution proposed here can bring some enlightenment to the peer and help someone.

Acknowledgments. The study is supported by Key Lab of Intelligent and Green Flexographic Printing, and Shanghai Publishing and Printing College High-level Talents Start-up Fund (2022RCKY03).

References

1. Weijie, H., Shunqing, T., Zhengfang, Z.: Modern Color Science and Application. Beijing Institute of Technology Press, Beijing (2007)
2. Changxian, C.: Professional English for Printing. Cultural Development Press, Beijing (2013)
3. Liang, Z., Yan, L.: Color Principle and Application. Cultural Development Press, Beijing (2013)
4. Xiaohui, H.: Printing Quality Inspection and Control. China Light Industry Press, Beijing (2018)
5. Xiaowen, Z., Wenshun, S.: Spot color ink deployment method. Print. J. **06**, 43–44 (2008)
6. Quanzhong, L., Guo, L.: Discussion on the calculation of white ink proportion when mixing spot color ink. Net Friend World **12**, 69–71 (2013)
7. Learn how to quickly prepare spot color ink. Screen Print. Ind. **09**, 47–49 (2021)
8. Cuiqing, C., Zhengning, T.: Study of computer color matching model for spot color ink. Packag. Eng. **29**(3), 75–77 (2008)
9. Chang, X., Qi, W., Lin, Z.: Preliminary study of computer color matching model for spot color ink. China Print. Packag. Study **3**(6), 9–13 (2011)
10. Brainard, D.H.: Color appearance and color difference specification. Sci. Color 191–216 (2003)
11. Katoh, N., Ito, M., Ohno, S.: Three-dimensional gamut mapping using various color difference formulae and color spaces. J. Electron. Imaging **8**(4), 365–379 (1999)
12. Brainard, D.H., Wandell, B.A.: Asymmetric color-matching: how color appearance depends on the illuminant. J. Opt. Soc. Am. A **9**, 1433–1448 (1992)

Application Research for the Printing Process of Spot Color on the Product Packaging

Enyin Fang, Chuan Zhang(✉), and Shengwei Yang

Department of Printing and Packaging Engineering, Shanghai Publishing and Printing College, Shanghai, China
zhangchuanxs@126.com

Abstract. With the upgrading of consumption and brand awareness in the commercial market, Spot colors are used more and more widely in modern advertising design and product packaging design to attract consumers' attention and enhance brand recognition. Spot color printing can not only solve the problems of overprint and ink-water balance in four-color printing, but also has the characteristics of ink saving, stable color and wide color gamut. If the spot color is not well controlled in the printing process, the grade of the product will be greatly reduced, which will affect the competitiveness of the product in the market. Therefore, it is very important that how to select different processes according to the properties of spot color to realize the color consistency and accuracy of spot color printing. In this paper, different printing processes are used to print spot colors based on the solid and screening. Experiments are carried out to verify the feasibility and accuracy of different printing processes, which can provide a method for the printing of spot colors on product packaging.

Keywords: Spot color · Printing process · Spectral reflectance · Lab value

1 Introduction

In packaging printing, four-color printing is generally used, that is, cyan, magenta, yellow, and black inks, which are overprinted to reproduce color originals. However, with the upgrading of consumption in the commercial market and the stronger brand awareness, modern advertising design and product packaging design are increasingly using spot colors to attract consumers' attention and enhance brand recognition. To accurately realize the exclusive color of the terminal brand, realize the color consistency of different batches of products, and meet customer requirements, the application of spot color printing on packaging will be very important.

Spot color printing is widely used in product packaging, the saturation of spot color is high, and the uniform spot color is usually printed using the solid. Moreover the amount of ink should be appropriately increased. When the thickness of the ink layer on the plate is large, the changes of ink layer thickness are less sensitive to color, so it is easier to get uniform and thick prints. In addition, the gamut of the spot color is larger than the four-color. The color that cannot be printed by the four-color printing can be achieved

M. Xu et al. (Eds.): CACPP 2022, LNEE 991, pp. 62–69, 2023.
https://doi.org/10.1007/978-981-19-9024-3_10

by using the spot color printing, especially some high-purity colors can only be printed with the spot color, metallic colors and fluorescent colors are all printed in spot color. Since spot color printing does not require register, the printing can be completed at one time, avoiding other problems caused by inaccurate register. The effect of spot color printing is obviously better than that of four-color printing.

There are two common methods of spot color control, including density and chromaticity control. In the printing industry, density can be defined as the proportion of incident light absorbed by the surface of an object, and can indirectly represent the property of the amount of light absorbed by an object. An object absorbs a large amount of light, indicating that its density is high; an object absorbs a small amount of light, and its surface density is low. Densitometry was developed by the American National Standards Committee, which specifies that the optical density of a color is the base 10 logarithm of the opacity. Colorimetric measurement is a method to obtain color chromaticity information according to the principle of chromaticity. The measurement of chromaticity is currently the most accurate method for spot color control.

2 Ink Color Control Method for Spot Color

2.1 Solid Ink Control for Spot Color

(1) Determine the first sample of spot color. The operator of the printing press measures the density of the standard color sample provided by the customer and ink-adjuster in turn, than makes a judgment and can control the amount of ink through the ink key of the printing press. When the color density is also relatively close, record the Lab value at this time, which called its wet chromaticity.

(2) Determine the dry chromaticity of the spot color. Since spot colors have two states, wet and dry, some spot colors have little change in wet and dry states, and some spot colors have great changes. In order to understand this change of spot color, it is necessary to record the time-sharing of spot color, establishing the database of factory's own spot color, so people must record the chromaticity values of 2 h, 4 h, 8 h, 16 h, etc. If it is a new brand spot color, this operation will definitely cause a waste of time, so many printing houses evaluate the dry chromaticity Lab value of the spot color based on experience or drying the printing samples with a hair dryer.

(3) Control during the printing process. The printing press takes samples every 300 sheets to measure the chromaticity of the spot colors. Guarantees spot color consistency.

2.2 Dot Ink Control for Spot Color

The Lab value of spot dot ink is small, which is greatly affected by the whiteness of the paper, which must be confirmed before the printing. So the Lab of the spot dot is usually calculated. For example, GMG, CGS often use the Lab value that are the paper white and solid spot, to do linear calculation to obtain the Lab value of the spot dot. The calculation method can be shown in formula (1). Through the above method, the Lab value of the spot dot can be obtained for reference in subsequent printing.

$$
\begin{aligned}
L &= L_{paper\ white} - \left(L_{paper\ white} - L_{solid}\right) \times spot\ dot\ area \\
a &= a_{paper\ white} - \left(a_{solid} - a_{paper\ white}\right) \times spot\ dot\ area \\
b &= b_{paper\ white} - \left(b_{solid} - b_{paper\ white}\right) \times spot\ dot\ area
\end{aligned} \tag{1}
$$

3 Ink Color Control Process for Spot Color

3.1 Solid Ink Process for Spot Color

Modify the Four Colors that are not Easy to Control in the Digital Original to be Spot Colors For each digital original sent by the customer, the color must be carefully reviewed. When there are large-area color blocks produced by overprinting and light-colored dots composed of two colors in the digital original, they must be converted form four colors to spot color. In the digital original used in the experiment, the background color in the file is a large-area color block, its four-color value is C52M26Y57K21, and no spot color is used as shown in Fig. 1.

Fig. 1. Four-color original file

Because such a large area with four-color is inconvenient for color control. First, there are some problems such as uneven color from left to right, easy color jumping, etc. Second, the size of the box is also relatively large; the size from left to right is 1100 mm, if the color of the docking part is inaccurate, trapping will be appearance. In order to avoid these problems, the mattress, background image and barcode use four-color printing process, as shown in Fig. 2, the color of background is changed to a spot color, as shown in Fig. 3. Since the background is changed to a spot color, there will also be more room to adjust the color of the mattress after adjusting.

Set Spot Color. Use the illustrator software to convert the background color composed of the four colors into a spot color, and then define the spot color name as Tesco Cyan Green C, use an inkjet proofing system that conforms to the ISO12647-2 standard for outputting, and use eXact to measure the printed color. The chromaticity data of the background color are shown in Table 1.

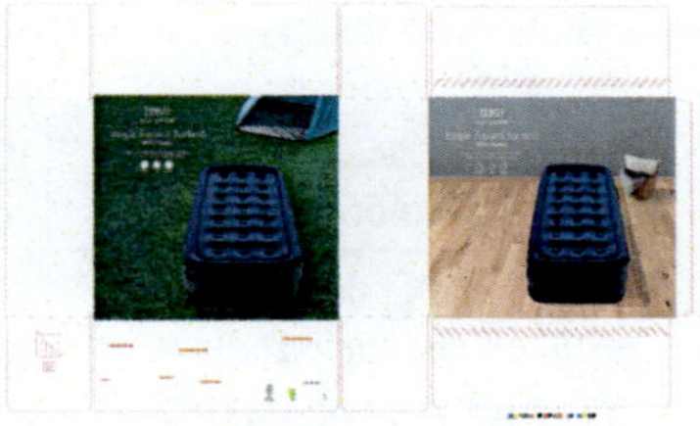

Fig. 2. Four-color part in original

Fig. 3. Spot part converted form four-color

Table 1. Chromaticity data of ground color spot color

Spot name	Chromaticity data				
	L	*a*	*b*	*c*	*h*
Tesco Cyan Green C	53.80	−11.90	12.98	17.54	132.74

Since the gamut of the inkjet proofing system is normally larger than that of traditional offset printing, the color defects caused by digital proofing can be avoided through color management by simulating outputting of offset printing. However, if the color of quality control is high, which needs to be accurately presented, it needs further verification. Therefore, after the spot color is set, it needs to be delivered to the printing factory for IGT proofing, and then machine proofing.

IGT Proofing. IGT can completely imitate the working principle of the printing machine, with adjustable pressure and a wide range of substrates. It is suitable for printing, packaging, and ink enterprises. It accurately predicts the hue of the ink, and it can show color in 2 min. It is also very convenient to clean, and save paper, ink, water and electricity. At the same time, it avoids the excessively long occupation time of the printing machine due to color matching, and effectively improves the production efficiency. Table 2 shows the comparison between IGT and the traditional method for dealing with spot color.

After the printing factory receives the color card proofing request, it immediately arranges IGT to show the color, and makes a color card to judge whether it can meet the requirements through subsequent printing.

Table 2. Comparison between IGT and traditional method for dealing with spot color

Content	Traditional way for dealing with spot colors	IGT for dealing with spot colors
Check hue of the spot color	Manual color development is difficult and the error is large. After the printing machine is installed, it needs to be repeatedly tested, which consumes much time and many materials. Some materials cannot be hand-painted	The color card is made by simulating the printing machine, which is fast, accurate and easy to check. It does not need to occupy the time of the printing machine. Only small amount of materials are needed, and the trial printing cost is low
The effect of wet and dry ink changes on hue	The drying time of the ink is too long on the printing press, which is difficult to control and solve	The color card can be printed before printing, which can be compared after natural drying, and excluded in advance
Order spot colors from a spot color ink formulation company	There are different paper qualities, and the hue varies greatly; when the quantity is small, the order is difficult, the time is long, and the price is expensive	Color card can be printed on the paper to be produced to exclude discrepancies

In the experiment, we measured the data of 5 points on the IGT color card, as shown in Table 3. After inspection, it was found that the color difference fully met the requirements.

Table 3. Spot color chromaticity data

Spot name	Chromaticity data				
	L	a	b	ΔE	Δh
Tesco Cyan Green C	53.80	−11.90	12.98		
Measurement 1	52.11	−11.20	11.99	2.03	0.01
Measurement 2	53.11	−12.20	11.02	2.01	1.17
Measurement 3	51.50	−11.92	12.60	2.32	0.04
Measurement 4	52.45	−10.20	12.65	2.18	2.02
Measurement 5	53.50	−11.20	12.35	0.93	0.8

Printing and Product Inspection. After confirming the spot color, carry out pre-press file processing, modify the four-color to spot color, the spot color name is Tesco Cyan Green C, and notify the process modification, change the printing color from four-color printing to four-color plus spot color, and sent the file to the printing house for proofing. One week later, the proofs from the printing factory were received. After passing the preliminary test, the inspection was arranged by third party. The test passed as shown in Fig. 4.

Fig. 4. Product inspection passed

3.2 Spot Color Dot Ink

Obtaining the Spectral Reflectance of Spot Color Dots. In the actual production process, compared with the spot color solid, we will find that Lab value of spot color dot calculated by formula (1) deviates from the actual measured Lab value. The main reason is that the spot color solid usually has a standard Lab value or a Pantone card as the sample-tracking standard. We can know the precise Lab value. Through the color management system, the color of the spot color solid can be accurately reproduced; while the spot color dot some of them do not have references such as Pantone cards. The Lab value is calculated by software. If the calculated result is deviated, the printed color will naturally be deviated.

The CIELAB color space is proposed by the International Commission on Illumination and is mainly used for color evaluation. According to Grassmann's law, the spectral reflectance of an area unit on a printed matter is the sum of the spectral reflectance of each color. Since there are only paper white and spot color in each area unit of spot color dots, the formula for calculating the spectral reflectance of spot color dots is shown in formula (2).

$$\rho_{spot\ color\ dot} = (1 - \alpha) \times \rho_{paper\ white} + \alpha \times \rho_{spot\ color\ solid} \tag{2}$$

In the formula, $\rho_{spot\ color\ dot}$ is the spectral reflectance of the spot color dot, and α is the dot area of spot color dot. $\rho_{paper\ white}$ is the spectral reflectance of paper white,

$\rho_{spot\ color\ solid}$ is the spot color solid spectral reflectance. Since the Lab chromaticity value of a color is calculated from the XYZ tristimulus value, and the XYZ tristimulus value is calculated from the reflectance of each band of the spectrum, the Lab chromaticity value of the spot color dot can be obtained through formula (2).

Acquisition of Experimental Measurement Data. The measuring instrument is X-Rite's eXact. Taking brown solid and brown spot color dots of 90%, 70%, 60%, 45%, 20% as examples.

Use an inkjet proofing system that complies with ISO12647-2 standard for spot color dot output, then use eXact to measure the color patch, and use formulas (1) and (2) to calculate the data for the Lab, use formula (3) to calculate the data for ΔE as shown in Table 4.

$$\Delta E = \sqrt[2]{\Delta L^2 + \Delta a^2 + \Delta b^2} \tag{3}$$

Table 4. Comparison of the measured Lab value and the calculated Lab value

Destination point value		Paper white	100	90	70	60	45	20
Measured value	L	95.4	52.2	60.3	70.8	75.9	81.3	89.6
	a	−0.2	26.8	20.4	13.2	10.7	7.7	3.1
	b	4.6	51.2	34.4	21.1	17.5	13.5	8.0
Formula (1) calculated value	L_1	95.4	52.2	56.5	65.2	69.5	76.0	86.8
	a_1	−0.2	26.8	24.1	18.7	16.0	12.0	5.2
	b_1	4.6	51.2	46.6	37.2	32.6	25.6	13.9
Formula (2) calculates the value	L_2	95.4	52.2	58.5	67.8	72.8	77.6	87.6
	a_2	−0.2	26.8	22.8	16.5	14.1	10.0	4.6
	b_2	4.6	51.2	38.5	27.9	24.1	17.5	12.0
Chromatic aberration	ΔE (1)			13.3	17.9	17.2	13.8	6.8
	ΔE (2)			6.1	8.1	8.1	5.9	4.7

Analysis of Experimental Measurement Data. According to Table 4, the comparison chart of color difference analysis is shown in Fig. 5. It can be seen that the color difference value calculated by formula (1) is quite different from the measured data. Among them, the color difference ΔE of 70% spot color is the largest, which is 17.91. The overall average color difference ΔE is 13.83. Compared with the measured value, the ΔE color difference of the dots calculated by the formula (2) is smaller, which is based on the spectral reflectance of paper white and spot solid. The average color difference ΔE is 6.57, although it is different from the measured data. There is still a certain gap in comparison, but because of the better data stability and convergence, it can be used as the target value for printing. Therefore, when packaging printing involves spot colors, the customer cannot only provide the Lab value of the solid spot color, but also the spectral reflectance of the solid spot color. Only in this way can the simulation be accurate when printing spot color dots.

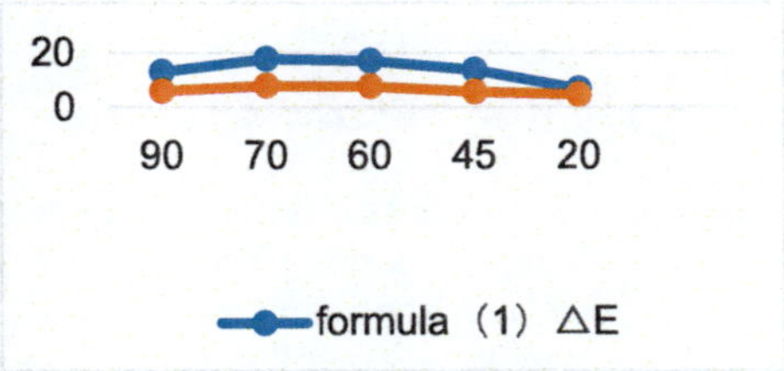

Fig. 5. Formula (1), formula (2) color difference comparison chart

4 Conclusion

With the development of science and technology and the continuous improvement of human aesthetic awareness, people have more and more requirements for the quality of printed matter, at the same time, higher requirements are put forward for printing color. Spot color printing has good effect in color presentation and relatively stable color, and is widely used in packaging products such as cigarette labels, wine labels, and cosmetics. In this paper, solid and screen, two kinds of spot colors, are printed by different printing processes. The feasibility and accuracy of different printing processes are verified through experiments, meanwhile, a method for printing spot colors in product packaging is provided.

Acknowledgements. This research is supported by Lab of Intelligence and Green Platemaking for Flexographic Printing (KLGFP-01).

References

1. Liang, Y.: Skillful use of spot color ink printing to improve the quality of printed matter. Print. World **11**, 42–45 (2011)
2. Zhang, L., Zhang, M., Yang, X.: Research on color difference in spot color printing. Packag. Eng. **19**, 28–31 (2010)
3. Yang, Y.: Solutions to the problem of spot color cast in digital printing. Digit. Print. **6**, 9–13 (2017)
4. Tian, Z.: Matters needing attention in spot color printing. Print. Mag. **12**, 54–57 (2007)
5. Zhao, X.: Spot color processing technology in color box processing. Print. Mag. **12**, 40–43 (2011)
6. Liu, J.: Some problems that should be paid attention to in the prepress processing of spot color printing. Guangdong Print. **2**, 23–29 (2005)
7. Nan, J.: How to print spot colors. Print. Qual. Stand. **4**, 26–28 (2008)
8. Sun, Q.: Research on density detection and characterization of spot colors. Print. Qual. Stand. **3**, 12–18 (2016)

Multi-output Least-Squares SVR Spectral Reflectance Reconstruction Model Based on Differential Evolution Optimization

Dongwen Tian[1,2](✉) and Jinghuan Ge[1]

[1] Department of Printing and Packaging Engineering, Shanghai Publishing and Printing College, Shanghai, China
dwtian@sppc.edu.cn
[2] University of Shanghai for Science and Technology, Shanghai, China

Abstract. To study the colorimetric characterization method of CCD digital camera based on multi-output least squares support vector machine regression (MOLSSVR). The differential evolution (DE) algorithm was used to optimize the hyperparameters and kernel functions of the MOLSSVR algorithm in order to avoid the influence of the artificial factors and improve the accuracy of the prediction. In accordance with the ColorChecker SG standard target and matt Munsell color chips, a nonlinear mapping model of RGB three-channel value to spectral reflectance was established by using DE-MOLSSVR. The R-squared of the model in the training set is 0.998, the R-squared of the validation set is 0.997, and the average color difference of CIEDE2000 between the training set and the verification set are 0.157 and 0.183, respectively. The spectral reflectance reconstruction model can better realize the spectral characterization of CCD digital camera and can be applied to print workshop for chromaticity detection of printed images.

Keywords: Digital camera · Differential evolution · Least squares support vector · Spectral characterization · Spectral reflectance reconstruction

1 Introduction

Spectral characterization of image color can be performed by obtaining the spectral reflectance characteristics of an object through ambient environmental information, lighting conditions, spectral characteristics of the camera, as well as the output multi-channel image, and this process is called Spectral Reconstruction. In the field of color measurement, the reconstructed spectral reflectance can be used to analyze the ink color in printed materials, and can be used for quality inspection of printed materials. The reconstructed spectral reflectance can be used for high-fidelity printing, which can accurately reproduce the color and avoid the phenomenon of metamerism, solving the problem of color consistency in different environments and different scenes [1–4]. For example, in the field of textile, color printing reproduction. It can be seen that the reconstruction of spectral reflectance is the core technology for the above applications,

M. Xu et al. (Eds.): CACPP 2022, LNEE 991, pp. 70–76, 2023.
https://doi.org/10.1007/978-981-19-9024-3_11

and its research has important significance and practical application value. The spectral reflectance reconstruction technology has become one of the hot spots of current research. In 2011, Chong et al. [5] proposed the MDST method, which uses discrete sine transform to gradually approximate the original spectral reflectance to improve the similarity of the two spectral curve values as much as possible. Alvarez-Gila et al. [6] proposed to use a convolutional neural network structure for end-to-end mapping between input RGB image pairs and hyperspectral image pairs. This method is based on a individual pixel, and when applied to spectral data, cannot effectively utilize the local context, so the spectral accuracy is low, the running speed is slow, and the running cost is high. In 2019, Wang et al. [7] obtained the complete spectral data and reconstructed all the spectra using the cubic spline interpolation method. The interpolation reconstruction method is affected by narrow-band filters. If the reconstruction accuracy of spectral reflectance is to be guaranteed, this method needs to use more narrow-band filters.

In the least squares support vector machine algorithm, all training sets are support vectors and the error is one of the optimization targets. This is an important difference between least squares support vector machine (LSSVM) and support vector machine (SVM). LSSVM is also faster because it solves a linear system of equations to yield the optimal solution. In this paper, we address the characteristics of spectral reconstruction and the problems of existing spectral reconstruction algorithms. The least squares support vector machine is combined with the difference evolution algorithm. An optimized least-squares support vector machine spectral reconstruction model based on the difference evolution algorithm is proposed. The fast and efficient acquisition of visible spectral information is achieved under the conditions of ensuring the use of conventional CCD digital cameras for spectral reconstruction and optimization for small sample data sets.

2 Spectral Reflectance Reconstruction Method Based on Differential Evolution LSSVM

2.1 Principle of Least Squares Support Vector Machine

The least squares support vector machine follows the kernel function machine learning with the least structural risk as the target, which is an extension of SVM. Compared to standard SVM, LS-SVM transforms convex quadratic programming into a linear equation solving problem. The solution requires less resources, simpler parameter selection, and faster computation, but the global optimal solution cannot be guaranteed. This method solves the practical problems of nonlinearity and small samples.

The low-dimensional input space samples are mapped to the high-dimensional feature space. Through the nonlinear mapping, the low-dimensional nonlinear regression problem can be transformed into the linear regression problem in the high-dimensional feature space, which is the core of the nonlinear support vector machine [8].

In the regression model, the training set $D = \{(x_i, y_i), i = 1, 2, \ldots, m\}$ of m sample points has n dimensional input and t dimensional output, which is transformed into a high-dimensional feature space through a nonlinear relationship, which is

$$f(x_i)_j = \omega_j^{\mathrm{T}} \varphi(x_j) + b_j \quad (i = 1, 2, \ldots, m;\ j = 1, 2, \ldots, t) \tag{1}$$

According to the principle of structural risk minimization, combined with the regularized loss function, the formula (1) can be equivalent to

$$\begin{cases} \min J\left(\omega_j, e_{ij}\right) = \dfrac{1}{2}\left(\omega_j^T \omega_j + \gamma \displaystyle\sum_{i=1}^{m}\sum_{j=1}^{t} e_{ij}^2\right) \\ s.t.\ \ y_{ij} = \omega_j^T \varphi(x) + e_{ij} + b_j,\ \ (i = 1, \ldots, t) \end{cases} \tag{2}$$

In the formula, γ is the penalty factor, e_{ij} is the prediction error, and $J\left(\omega_j, e_{ij}\right)$ is the structural risk function.

The LSSVM function model can be obtained as

$$f(x_i)_j = \sum_{i=1}^{m}\sum_{j=1}^{t} \alpha_{ij} K\left(x_i, x_j\right) + b_j \tag{3}$$

The kernel function $K(x_i, x_j)$ in Eq. (3) is the inner product of the high-dimensional feature space. According to the relevant theory of functionals, the kernel function needs to satisfy the mercer condition. The kernel function in this article chooses radial basis function (RBF):

$$K(x_i, x_i) = \exp\left(-\frac{\left\|x_i - x_j\right\|^2}{2\sigma^2}\right) \tag{4}$$

2.2 Differential Evolution Algorithm

The differential evolution algorithm is used to find the optimization of the kernel function and model parameters. Differential evolution algorithm [9] is an evolutionary algorithm proposed by Storn and Price in 1997, which is a greedy genetic algorithm based on real number encoding with the idea of optimality preservation. The computational steps are as follows.

Step 1: Population initialization. The DE population is initialized with *NP* (Number of Populations) individuals using uniform randomization.

$$x_i^j = x_{\min}^j + rand(0, 1) \cdot \left(x_{\max}^j - x_{\min}^j\right),\ \ \{i = 1, 2, 3, \ldots, NP\} \tag{5}$$

where x_i^j is the i-th vector in the j-th generation of vectors, $x_{\min}^j$ is the minimum value in the j-th generation of vectors, $x_{\max}^j$ is the maximum value in the j-th generation of vectors, and rand(0, 1) is a random number between 0 and 1.

Step 2: Select the target vector. In this paper, the first target vector X_i is generated randomly, and the subsequent target vectors are generated by the algorithm.

Step 3: Mutation operation, choose the scaling factor F and randomly select different individuals from the population to adopt the corresponding mutation strategy. For the variation vectors generated by the variation strategy, in order to ensure the usefulness of the variation vectors, the upper and lower bounds of the variation vectors need to be checked. For the variance vectors that do not satisfy the boundary conditions, this paper performs the corresponding processing to make them satisfy.

Step 4: Hybridization operation, select the hybridization probability CR, and hybridize the variance vector with the target vector to obtain the experimental vector U.

Step 5: Selection operation, the selection operator for the optimization problem can be described as:

$$X_i = \begin{cases} U_i & \textit{if } (E(U_i) \le E(X_i)) \\ X_i & \textit{otherwise} \end{cases} \quad (6)$$

The differential evolution algorithm is to continuously perform the second to fifth step operations until the maximum evolutionary algebra or the threshold of the objective function is reached.

The mean square error:

$$E(x_i) = \frac{1}{m}\sum_{i=1}^{M}(f(x_i) - y_i)^2 \quad (7)$$

where m is the number of samples, $f(x_i)$ is the predicted value, and y_i is the actual measured value.

2.3 Characterization Model of Image Input Device Based on Differential Evolution Least Squares Support Vector Regression

The red (R), green (G), and blue (B) feature quantities are used as the input of the LSSVM model as $x_i = [R\,G\,B]_i$, and the spectral reflectance are the output of the model as $y_i = [R_{400}, R_{410}, \ldots, R_{700}]_i$, so the set of training samples is $\{(x_i,\ y_i),\ i = 1, 2, \ldots, n\}$. The support vector machine maps the sample x of the input space to the high-dimensional feature space H through the nonlinear mapping function $\varphi(x)$, and establishes the linear regression function by using the principle of minimizing structural risk in the high-dimensional feature space H.

The image input device is characterized by the least-squares support vector machine regression, the steps are as follows.

(1) Normalization of experimental sample data. The sample set data is normalized to [0, 1] before the regression model is established. The normalization method can be written as $x_i' = (x_i - x_{\min})/(x_{\max} - x_{\min})$.
Where x_i is the data to be normalized, x_i' is the data after normalization, $x_{\min}$ is the minimum value, and $x_{\max}$ is the maximum value.
(2) Optimization of DE-LSSVM parameters. Taking sets of training sample data as input and the corresponding spectral reflectance as model output, it is judged whether the set termination conditions are met, and the optimal parameters σ and γ that meet the conditions are obtained.
(3) Establish the DE-LSSVM regression model and use the model to make predictions. Obtain the best parameters σ and γ from step (2), use the RBF kernel function to train DE-LSSVR regression modeling, and use the trained model to perform regression prediction on the validation set samples.

(4) Evaluate the performance indicators of the prediction model RMSE, R-squared and CIEDE2000. If it does not meet the requirements, go to step (2) and continue to find the optimal parameters through differential evolution algorithm.
(5) Compare the measured value and predicted value of the training and verification set, respectively. Obtained the RMSE and R-square of the model.
(6) DE-LSSVR is applied to spectral reconstruction model. The results are inverse normalized to obtain the CIEDE2000 color difference for the prediction and training sets respectively. The performance of the spectral reconstruction model is finally evaluated.

3 Experimental Verification and Analysis

In this paper, two different sets of samples were selected for experimentation, Munsell color chips and Xrite ColorChecker SG. Several studies have shown that the Munsell color system, which contains a wide range of colors, is a relatively perfect color system [10–12]. The Munsell semi-gloss color cards containing 1269 colors can depict most of the colors seen by the human eye in nature and has been widely used in the design of printed packaging products. Therefore, Munsell color cards were selected as training samples for the experiments. The spectral reflectance of Munsell color cards and ColorChecker SG was measured by Xrite eXact advanced spectrophotometer. The collected spectra range from 400 to 700 nm, and the experiments were conducted in the interval of 10 nm, so that the spectral reflectance values with a spectral dimension of 31 were obtained.

According to the acquired image RGB value and the spectral reflectance value of the color target measured by the spectrophotometer, a transformation DE-LSSVM model is established. The 1269 data samples of Munsell color chips were used as the training set, and the 140 data samples of X-Rite ColorChecker Digital SG standard card were used as the validation set. The difference between the predicted value and the measured value of the training set and the validation set is evaluated by three parameters: Root Mean Squared Error (RMSE), correlation coefficient R-squared and (CIEDE2000). The prediction results of the training set and the validation set are shown in Table 1.

Table 1. Color difference CIEDE2000 and RMSE of different methods under D50

Methods	RMSE		ΔE_{00} (test set)		
	Train set	Test set	max	min	Mean
BP	0.033 0.002	0.039	4.106 0.024	0.024 0.024	0.914
LSSVM	0.021	0.027	3.307	0.014	0.445
Ours	0.011	0.015	1.959	0.010	0.183

The overall color difference CIEDE2000 distribution between the training set and the verifying set is shown in Fig. 1.

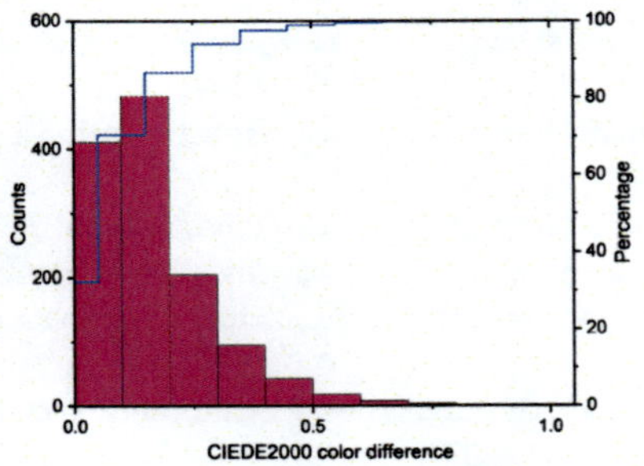

Fig. 1. Histogram of CIEDE2000 color differences for training and testing datasets

It can be seen from Table 1 that the average color difference of the test set is 0.183, the minimum color difference is 0.01, and the maximum color difference is 1.959. Compared with the results of traditional BP neural networks and LSSVM, the nonlinear mapping model obtained by DE-LSSVR has greatly improved the accuracy of fitting and the overall distribution of color difference, and requires fewer training samples. This spectral reconstruction model is in compliance with the current requirements of the printing industry for color reproduction of fine products.

4 Conclusion

In this paper, based on the least squares support vector machine regression, a spectral reconstruction model of image acquisition device is established, and the optimal parameters of the model are obtained by using the difference evolution algorithm, which can quickly and accurately obtain the nonlinear mapping relationship between the RGB values of the image acquisition device and the spectral reflectance. The model has good results and generalization ability for the problem of small samples and nonlinear mapping. The experimental results show that the prediction accuracy of the support vector regression model based on difference least squares is more excellent compared with the traditional support vector regression, polynomial fitting and neural network regression. The color difference has obvious advantages and can be used as a spectral reconstruction method for common image acquisition devices.

Acknowledgment. This study is supported by the Shanghai Green Packaging Professional Technical Service Center and the Lab of Green Platemaking and Standardization for Flexographic Printing.

References

1. Amiri, M.M., Fairchild, M.D.: A strategy toward spectral and colorimetric color reproduction using ordinary digital cameras. Color Res. Appl. **43**(5), 675–684 (2018)
2. Liang, J., Wan, X.: Spectral reconstruction from single RGB image of trichromatic digital camera. Acta Opt. Sin. **37**(09), 370–377 (2017)

3. Chen, S., et al.: Multi-spectral inversion of wetland soil organic matter content by soil spectral reconstruction. Spectrosc. Spect. Anal. **38**(3), 912–917 (2018)
4. Choi, I., et al.: High-quality hyperspectral reconstruction using a spectral prior. ACM Trans. Graph. **36**(6), 1–13 (2017)
5. Chong, E., Zak, S.: An Introduction to Optimization, 3rd edn. Wiley (2011)
6. Hajipour, A., Shams-Nateri, A.: Effect of classification by competitive neural network on reconstruction of reflectance spectra using principal component analysis. Color Res. Appl. **42**(2), 182–188 (2017)
7. Wang, W., Wang, J.: Research progress of spectral reflectance reconstruction technology. Packag. Eng. **41**(11), 254–261 (2020)
8. Suykens, J., Vandewalle, J., et al.: Optimal control by least squares support vector machines. Neural Netw. **14**(1), 23–35 (2001)
9. Storn, R., Price, K.: Differential evolution—a simple and efficient heuristic for global optimization over continuous spaces. J. Glob. Optim. **11**(4), 341–359 (1997)
10. Liang, J., Wan, X.: Optimized method for spectral reflectance reconstruction from camera responses. Opt. Express **25**(23), 28273–28287 (2017)
11. Mansouri, A., Sliwa, T., Hardeberg, J.Y., Voisin, Y.: Representation and estimation of spectral reflectances using projection on PCA and wavelet bases. Color Res. Appl. **33**(6), 485–493 (2008)
12. Amiri, M.M., Amirshahi, S.H.: A step by step recovery of spectral data from colorimetric information. J. Opt. **44**(4), 373–383 (2015). https://doi.org/10.1007/s12596-015-0299-9

Calibration of Gray Balance for Fluorescent Inkjet Image Based on Spectral Calculation

Wan Zhang, Linhong Huang, Yongjian Wu, Yingjie Xu, Hui Wang, and Beiqing Huang(✉)

School of Printing and Packaging Engineering, Beijing Institute of Graphic Communication, Beijing, China
22722234@qq.com

Abstract. The color reproduction of fluorescent image is related to the gray balance of color-light. The single channel control of printer was developed to realize the output of fluorescent ink-jet image. The curve of gray balance of color light based on integral area superposition of emission spectra was obtained and evaluated. The results showed the same tendency of the fluorescence intensity on these four substrates, blue > green > red. The color gamut of fluorescent ink-jet ink on Mengken and Daolin papers were close to sRGB. The calculation method of gray balance for color light based on the spectral curve integral area of superposition principle was suitable for fluorescent ink-jet ink, and the obtained gray balance curves revealed it was necessary to increase the dot area of red ink on Mengken paper in order to compensate the missing red color of color gamut. The level of dark tone was much clearer after using gray balance curve on Mengken paper.

Keywords: Fluorescent ink-jet image · Photophysical properties · Color light-gray balance · Color reproduction

1 Introduction

The fluorescent image is obtained through printing the colourless fluorescent ink on the substrate, which is colourless under the sunlight, but the colorized fluorescent image can be reproduced under the fluorescent light, because the fluorescent ink can absorbed the ultraviolet photon. Using this feature, it can be applied to anti-counterfeiting products to achieve high anti-counterfeiting function requirements, such as passport, securities, graduation certificates, stamps and other products [1–3]. In addition, the application scope is further expanded according to the market demand, for example, in the field of interior decoration, fluorescent images can present a special artistic effect. At present, fluorescent images are mainly printed through Offset printing [4], Screen printing [5], Gravure printing [6] and other traditional printing methods, while inkjet printing is a kind of digital printing method, which has the characteristics of variable data, personalized and instant printing [7–9]. The combination of fluorescent image and inkjet printing [10–12] can further expand its application fields, such as anti-counterfeit bar codes of packaging products, anti-counterfeit information on various commercial documents and tickets, personalized artistic effects, etc.

M. Xu et al. (Eds.): CACPP 2022, LNEE 991, pp. 77–86, 2023.
https://doi.org/10.1007/978-981-19-9024-3_12

At present, fluorescent images are mainly simple color, and rarely reproduce rich colors, which limit its application scope. The reason is that mature color reproduction control methods, such as color measurement, gray balance and color management, as well as prediction models, are established for ordinary ink prints and cannot be applied to color control and prediction of images [13]. Hersch et al. [14] proposed a prediction model for establishing fluorescence images, but the appropriate value of correction coefficient N was not obtained. Rogers [15] proposed a fluorescence image prediction model based on Neugebauer prediction model. Zhang and Du [16] studied and established a new spectral reflectance model of halftone fluorescence images according to the long and short interval rule of multiple reflection process of light in halftone images. Domestic and overseas researchers pay attention to the model prediction of fluorescence image, but have not established a practical and operable color reproduction control method of fluorescence image. In order to improve the printing quality of fluorescent images and reproduce rich colors, it is necessary to establish a color control system suitable for fluorescent images.

The color reproduction effect of fluorescent color images is closely related to the gray balance control of color light. If the control is appropriate, the color will be more ideal [16]. The ideal gray balance of color light should be obtained by the same amount of red, green, blue fluorescent ink printing, but the actual red, green, blue fluorescent ink has partial color, so it is necessary to determine the red, green, blue monochrome ink dot area rate which can obtain the ideal gray balance of color light. This paper implements the output of fluorescent color images based on ordinary printer, establishes the gray balance calculation method of color light according to the integral area superposition principle of fluorescence image emission spectrum, obtains the gray balance curve of color light based on different substrates, and evaluates the accuracy of curve. The results are very important for color reproduction control of fluorescence inkjet image.

2 Experimental

2.1 Materials and Instruments

Materials: Mengken paper and Daolin paper (100 g, A4, 100 g, A4, Jiangsu Good paper Industry), offset paper (80 g, B3, Shandong Sun Paper Industry), coated paper (100 g, Epson), fluorescence inkjet inks (500 g, red, green, blue, SO-KEN).

Instruments: Inkjet printer (EpsonStylusPro7600), Fluorescence Spectrometer (RF-5301PC, Shimadzu Japan), Lamp box (JudgeII, Guangzhou Libao Laboratory Testing Instrument Co., LTD), Spectral radiometer (PR-655, PhotoResearch Inc).

Softwares: Photoshop (CS3, Adobe), The single channel control software of printer (Applicable to Epson Style Pro series, independently developed).

2.2 Experimental Methods

Single Channel Control Software of Printer (Independently Developed). Since there is no commercially available printer for fluorescent inkjet ink, our research group independently developed the single-channel control software epson7600test.exe suitable for

EpsonStylusPro series printers, which is installed on EpsonStylusPro7600 printer. Three of the seven channels of the printer were selected as the red, green and blue channels of the fluorescent inkjet ink, as shown in Fig. 1. The red, green and blue fluorescent inkjet inks are respectively poured into clean ink cartridges and installed in three channels selected by EpsonStylusPro7600 printer. The single-channel control software controls the channel and outputs corresponding red, green and blue color separation files.

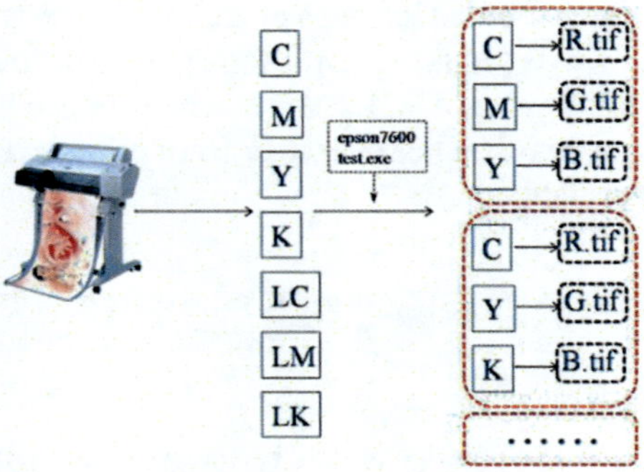

Fig. 1. Single channel control of printer

The specific steps are as follows: First, the output files were changed to RGB mode in Photoshop and saved in TIF format by channel, including R.tif, G.tif and B.tif files. The red, green and blue fluorescent inkjet inks were filled into cyan, magenta and yellow ink cartridges respectively. Using single-channel control software epson7600test.exe to merge the three TIF files into output file ttt.dat, as shown in Fig. 1. This file is in the DAT format that can be recognized by single-channel control software.

Calculation Method of Color Light-Gray Balance. The ideal color light-gray balance of fluorescent ink should be obtained by the same amount of red, green, blue fluorescent ink printing, but the actual red, green, blue fluorescent ink has partial color. Therefore, it is necessary to determine the desired red, green, blue monochrome ink dot area rate for the ideal color light-gray balance. According to the principle of integral area superposition of fluorescence inkjet ink emission spectrum, the spectral area of printing color block with three colors is equal to the sum of the integral area of each monochromatic block. It is assumed that under normal printing conditions, the proportion of each monochrome ink forming the color light-gray balance is red f_R, green f_G and blue f_B, and the spectral integral area of color block based on color light-gray balance is A_e, and the included red, green and blue spectral integral area is A_{er}, A_{eg}, A_{eb} respectively. According to the superposition relationship of spectral area, the integral area of the red, green and blue spectrum contained in the colorblock based on the color light-gray balance should be equal to the sum of the integral area of the red, green and blue spectrum contained in each monochrome ink, as shown in Eq. (1):

$$\begin{aligned} A_{er} &= f_R A_{Rr} + f_G A_{Gr} + f_B A_{Br} \\ A_{eg} &= f_R A_{Rg} + f_G A_{Gg} + f_B A_{Bg} \\ A_{eb} &= f_R A_{Rb} + f_G A_{Gb} + f_B A_{Bb} \end{aligned} \tag{1}$$

$$\begin{bmatrix} A_{\mathrm{er}} \\ A_{\mathrm{eg}} \\ A_{\mathrm{eb}} \end{bmatrix}_i = \begin{bmatrix} A_{\mathrm{Rr}} & A_{\mathrm{Gr}} & A_{\mathrm{Br}} \\ A_{\mathrm{Rg}} & A_{\mathrm{Gg}} & A_{\mathrm{Bg}} \\ A_{\mathrm{Rb}} & A_{\mathrm{Gb}} & A_{\mathrm{Bb}} \end{bmatrix}_i \begin{bmatrix} f_{\mathrm{R}} \\ f_{\mathrm{G}} \\ f_{\mathrm{B}} \end{bmatrix}_i \tag{2}$$

In Eq. (1): $A_{\mathrm{Rr}}, A_{\mathrm{Gr}}, A_{\mathrm{Br}}, A_{\mathrm{Rg}}, A_{\mathrm{Gg}}, A_{\mathrm{Bg}}, A_{\mathrm{Rb}}, A_{\mathrm{Gb}}, A_{\mathrm{Bb}}$ are respectively the integral areas of red, green and blue spectra contained in the separate printing of red, green and blue inks, which are written as matrix Eq. (2).

In Eq. (2): i is the number of ladder steps of the color light-gray balance;

$A_{\mathrm{Rr}}, A_{\mathrm{Gr}}, A_{\mathrm{Br}}, A_{\mathrm{Rg}}, A_{\mathrm{Gg}}, A_{\mathrm{Bg}}, A_{\mathrm{Rb}}, A_{\mathrm{Gb}}, A_{\mathrm{Bb}}$ form coefficient matrix, can be measured to obtain the corresponding value; $A_{\mathrm{er}}, A_{\mathrm{eg}}, A_{\mathrm{eb}}$ can be obtained by measuring the color block based on color light-gray balance of each order of printing, Fig. 2 is the step wedge of gray balance for color light.

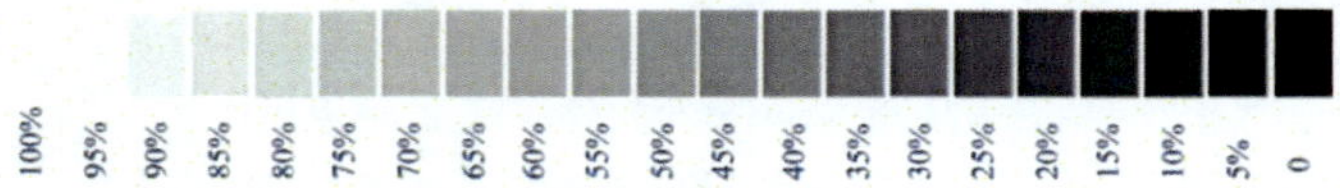

Fig. 2. Step wedge of gray balance for color light

Therefore, only $f_{\mathrm{R}}, f_{\mathrm{G}}, f_{\mathrm{B}}$ are unknowns in Eqs. (1)–(2). By using inverse matrix method, the proportion coefficients $f_{\mathrm{R}i}, f_{\mathrm{G}i}, f_{\mathrm{B}i}$ of the three primary colors can be calculated for the spectral area of the ideal color light-gray balance, as shown in Eq. (3).

$$\begin{bmatrix} f_{\mathrm{R}i} \\ f_{\mathrm{G}i} \\ f_{\mathrm{B}i} \end{bmatrix} = \begin{bmatrix} A_{\mathrm{Rr}} & A_{\mathrm{Gr}} & A_{\mathrm{Br}} \\ A_{\mathrm{Rg}} & A_{\mathrm{Gg}} & A_{\mathrm{Bg}} \\ A_{\mathrm{Rb}} & A_{\mathrm{Gb}} & A_{\mathrm{Bb}} \end{bmatrix}_i^{-1} \begin{bmatrix} A_{\mathrm{e}i} \\ A_{\mathrm{e}i} \\ A_{\mathrm{e}i} \end{bmatrix} \tag{3}$$

$$\begin{cases} A_{\mathrm{er}i} = f_{\mathrm{R}i} A_{\mathrm{Rr}i} \\ A_{\mathrm{eg}i} = f_{\mathrm{G}i} A_{\mathrm{Gg}i} \\ A_{\mathrm{eb}i} = f_{\mathrm{B}i} A_{\mathrm{Bb}i} \end{cases} \tag{4}$$

The main spectral integral area of the actual red, green and blue ink required for the color light-gray balance can be obtained, as shown in Eq. (4).

According to Murray-Davies formula, the formula of dot area rate can be obtained by the spectral integral area, such as Eq. (5).

$$\begin{cases} a_{\mathrm{er}i} = \dfrac{1 - 10^{-A_{\mathrm{er}i}}}{1 - 10^{-A_{\mathrm{SR}}}} \\ a_{\mathrm{eg}i} = \dfrac{1 - 10^{-A_{\mathrm{eg}i}}}{1 - 10^{-A_{\mathrm{SG}}}} \\ a_{\mathrm{eb}i} = \dfrac{1 - 10^{-A_{\mathrm{eb}i}}}{1 - 10^{-A_{\mathrm{SB}}}} \end{cases} \tag{5}$$

In Eq. (5): $a_{\mathrm{er}i}, a_{\mathrm{eg}i}, a_{\mathrm{eb}i}$ are the area rates of red, green and blue dot corresponding to the equivalent color light-gray balance;

$A_{eri}, A_{egi}, A_{ebi}$ are the spectral integral areas of red, green and blue ink required for equivalent color light-gray balance obtained by Eq. (4);

A_{SR}, A_{SG}, A_{SB} are 100% field spectral integral area.

Equation (5) can calculate the equivalent color gray balance 1 − i each tone corresponding to the red, green, blue three primary color fluorescent ink dot area rate, establish the gray balance curve for output.

3 Result and Discussion

3.1 Photophysical Properties of Fluorescent Inkjet Inks on Different Substrates

The test file containing red, green and blue color blocks is designed, as shown in Fig. 3a. The single-channel control software EPson7600Test.exe is used to output on four kinds of paper: Mengken paper, Daolin paper, offset paper and coated paper. The color of the output sample under UV lamp is shown in Fig. 3b. Fluorescence spectrometer (RF-5301PC, Shimadzu, Japan) was used to measure the spectral data of the color blocks of red, green and blue with 100% dot area rate on four kinds of paper, as shown in Fig. 4 and Table 1.

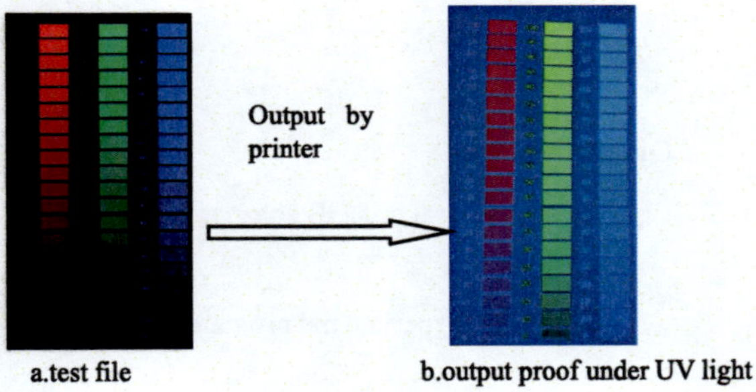

Fig. 3. Test file of step wedge

Figure 4 shows red, green and blue fluorescent light emitting spectrum on four substrates, obviously, the blue on the Offset paper and coated paper is unimodal, red and green are for bimodal. All of them contain a blue spectrum with the big range 380–500 nm, which can influence the impact of fluorescent image. Table 1 shows that the maximum emission wavelengths on Mengken paper are red 614 nm, green 522 nm and blue 443 nm respectively. The maximum emission wavelengths on Daolin paper are: red 616 nm, green 524 nm, blue 444 nm. The maximum emission wavelength on offset paper is red 615 nm, green 521 nm, blue 439 nm. The maximum emission wavelengths on coated paper are red 615 nm, green 522 nm, blue 440 nm.

The blue fluorescence intensity on the four substrates is Offset paper (44.20) > coated paper (35.98) > Mengken paper (11.06) > Daolin paper (10.59), which is largely related to the fact that both Offset paper and coated paper contain a certain amount of fluorescent whitening agent [17, 18]. In addition, it can be seen from Fig. 4(a) and (b) that the crest of red is longer and narrower than that of the other two colors, and the saturation of the a.Mengken b.Daolin

c.Offset d.Coated

Fig. 4. FL spectral of RGB color on proofs

Table 1. Experimental programs

Paper	Red ink		Green ink		Blue ink	
	$\lambda_{max}FL^a$ (nm)	Fluorescence intensity	$\lambda_{max}FL^a$ (nm)	Fluorescence intensity	$\lambda_{max}FL^a$ (nm)	Fluorescence intensity
Mengken	614	3.61	522	5.41	443	11.06
Daolin	616	3.90	524	5.82	444	10.59
Offset	615	2.92	521	6.59	439	44.20
Coated	615	2.79	522	5.57	440	35.98

Note: [a] Maximum absorption wavelength, test conditions: solid sample, room temperature

reproduced color is the highest. The crest of the other two colors is wider than that of red, but the fluorescence intensity is mainly concentrated near the peak, which can meet the needs of color reproduction.

3.2 Calculation of Color Light-Gray Balance of Fluorescent Inkjet Ink and Curve Verification

Calculation of Color Light-Gray Balance and Curve Establishment. According to Fig. 4, the integral range of red is 580–650 nm, green is 480–580 nm, and blue is 395–480 nm. Spectral data of color blocks in Fig. 3 were measured by using a Spectral radiometer. Then, according to the calculation method of 1.3.3, Finally, the red, green and blue dot area rates corresponding to the 21 color blocks with different dot area rates on 0–100% graded tone were calculated, and as shown in Table 2 and Fig. 5.

Table 2. Dot area of RGB inks required in equivalent gray balance

i	Dot area rate (%)	a_{eri} (%)	a_{egi} (%)	a_{ebi} (%)
1	100	100.00	100.00	100.00
2	95	100.00	94.25	99.55
3	90	100.00	85.26	91.08
4	85	100.00	80.58	86.56
5	80	100.00	73.16	78.58
6	75	100.00	66.54	70.76
7	70	94.00	59.43	62.25
8	65	87.26	53.60	55.50
9	60	80.35	47.26	48.82
10	55	72.58	41.16	42.23
11	50	65.44	35.63	36.44
12	45	58.89	31.09	31.65
13	40	50.53	25.81	25.77
14	35	43.78	21.44	20.88
15	30	38.19	17.49	16.63
16	25	31.25	13.56	12.49
17	20	24.34	9.95	8.72
18	15	15.75	0.61	5.01
19	10	7.84	2.86	2.05
20	5	2.28	0.73	0.41
21	0	0.93	0.22	0.09

It can be seen from Table 2 and Fig. 5 that, in order to obtain the ideal color light-gray balance, it is necessary to adjust the dot area rate of red, green and blue components. The red ink needs to be adjusted more to compensate for the loss of the red part of the color gamut, while the green and blue components need smaller values than the

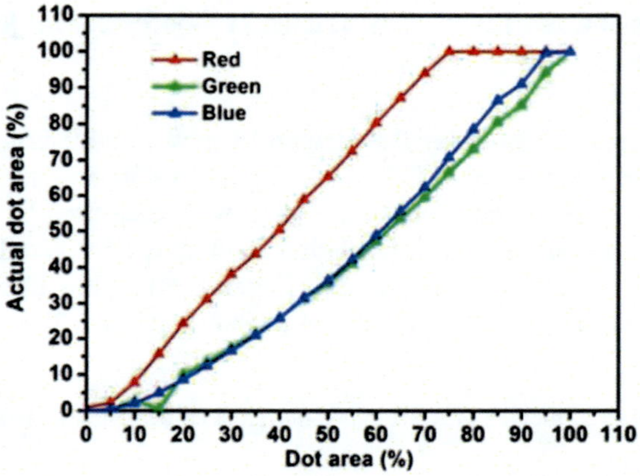

Fig. 5. Gray balance curve for color light

ideal dot area rate, in order to adjust the phenomenon of more spectrum content of green and blue ink, especially blue. Because the UV light source contains more blue spectrum, when the human eye observe the sample, it will feel more blue components. The decrease of the amount of blue ink will reduce the human eye's perception of the blue component of the sample, so as to improve the overall color effect [19, 20]. However, the gray balance calculation is based on the monitor standard, and the blue component is significantly higher than the red and green component. In order to keep consistent with it, the adjustment of the blue dot area rate in the obtained color light-gray balance curve is not too low, and even slightly higher than the green dot area rate in the bright part.

Curve Verification. According to the data in Table 2, the color blocks were draw in Photoshop (replacing the dot area rate of CMY), and then save them as R.tif, G.tif, B.tif. The self-developed single-channel control software EPson7600Test.exe and color separation software rgbhalftone.exe were used to control the output of RGB fluorescent inkjet ink in the CMY channel of the printer, and the obtained sample was the color light-gray balance color block, as shown in Fig. 6.

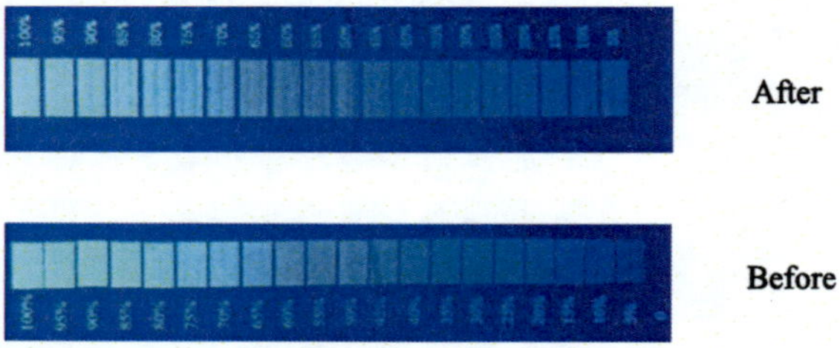

Fig. 6. Adjustment of gray balance

It can be seen that on Mengken paper, 0–100% of the overall tone is white before the adjustment of the color light-gray balance, and the hierarchy distinction is not obvious. However, after the use of the gray balance curve, the levels of the dark part are clearer, especially the 50–85% of the tone level is richer. It shows that the calculating method of color light-gray balance based on spectrum is suitable for color reproduction of fluorescent image.

4 Conclusions

The output of fluorescent image was realized by independently developed single-channel printer control software. The color reproduction of fluorescent images on different printing materials were analyzed, and the color light-gray balance calculation method was suitable for fluorescent images. The gray balance curve obtained by Mengken paper suggests that the red ink needs to be adjusted more to compensate for the loss of the red part of the color gamut, while the green and blue components need smaller values than the ideal dot area rate, in order to adjust the phenomenon of more spectrum content of green and blue ink, especially blue. After using the gray balance curve, the level of the dark part of the color block is clearer.

Acknowledgements. This study is funded by BIGC Project (Nos. Ea202203, Ee202205).

References

1. Hersch, R.D., Donzé, P., Chosson, S.: ACM Trans. Graph. **26**(3), Article 75 (2007)
2. Rossier, R., Hersch, R.D.: In: 19th Color Imaging Conference, Switzerland (2011)
3. van Renesse, R.L.: Printing Inks and Printing Techniques. Optical Document Security, London (2005)
4. Yang, Y., Xu, W., Sun, J., et al.: In: Proceedings of CACPP2010, no. 2, pp. 348–352 (2010)
5. Coudray, M.A.: Screen Print. **94**(6), 28–32 (2004)
6. Wei, J., Sun, C., Huang, L.: J. Tianjin Univ. Sci. Technol. **27**, 36–39 (2012)
7. Liu, H.M., et al.: J. Colloid Interface Sci. **465**, 106–111 (2016)
8. Jafarifard, S., Bastani, S., Atasheh, S.G., Morteza, G.S.: Prog. Org. Coat. **90**, 399–406 (2016)
9. Stempien, Z., Rybicki, E., Rybicki, T., Lesnikowski, J.: Sens. Actuators B Chem. **224**, 714–725 (2016)
10. Narita, S., Eto, K.: Method for fluorescent image formation: print produced thereby and thermal transfer sheet thereof. Patent Number: USA27346979, 3 Feb 2009
11. Angelo, P.D., Kronfli, R., Farnood, R.R.: J. Lumin. **136**, 100–108 (2013)
12. Coyle, W.J., Smith, J.C.: Methods and ink compositions for invisibly printed security images having multiple authentication features. Patent Number: USA33159785, 5 Apr 2004
13. Meng, Q.: Color Prediction Model of a Fluorescent Ink Print Under Ink Penetration Condition. Jiangnan University, Jiangsu (2008)
14. Hersch, R.D., Douzé, P., Chosson, S.: Color images visible under UV light. In: International Conference on Computer Graphics and Interactive Techniques, United States, 5 Aug 2007
15. Rogers, G.L.: Spectral model of a fluorescent ink halftone. J. Opt. Soc. Am. A **17**(11), 1975–1981 (2000)

16. Zhang, Y., Du, Y.: Clapper-yule spectral reflection and transmission of halftone Color fluorescent image. Acta Opt. Sin. **27**(2), 365–370 (2007)
17. Zhang, A.N.S., Nee, A.Y.C., Kamal, Y.T., Lu, W.F., Ma, B., Lan, W.: The new challenge for color management in digital printing. In: International Conference on Digital Printing Technologies (2017)
18. Yi, R., Liu, Z., Hui, L.: Classification of fluorescent brighteners and their affecting factors in papermaking. China Pulp Paper Ind. **38**(6), 52–55 (2017)
19. Li, Y., Chen, G.: Influence of optical brightening agents on color management. Packag. Eng. **37**(3), 160–164 (2016)
20. Gonome, H., Ishikawa, Y.K., Takahiro, K., Yamada, J.: Radiative transfer analysis of the effect of ink dot area on color phase in inkjet printing. J. Quant. Spectrosc. Radiat. Transf. **194**, 17–23 (2017)

Spectral Reflectance Reconstruction of Organic Tissue Based on Camera Responses

Yang Chen, Siyuan Zhang, and Lihao Xu(✉)

College of Media and Design, Hangzhou Dianzi University, Hangzhou, China
lihao.xu@hdu.edu.cn

Abstract. Reflectance information of a tissue is desired in the field of medical imaging, especially for disease diagnosis. In this paper, a new reflectance restoration algorithm is proposed to recover the reflectance information using a commercial camera. Initially, a color clustering method was applied to obtain the representative colors of tissue samples. These colors were then used to construct a look-up-table (LUT) using a lattice regression model. Interpolation methods can then be applied to the newly built LUT to acquire its matching reflectance information for any RGB input images. Present results showed that the proposed method further improve the accuracy in spectral reconstruction for tissue samples.

Keywords: Reflectance reconstruction · Lookup table · Lattice regression

1 Introduction

An accurate reflectance estimation algorithm is highly preferred in the field biomedical imaging, especially for disease diagnosis and image-guided surgery. The traditional way to obtain the reflectance data is to apply a multi-spectral imaging system (MSIS). However, it is usually bulky and time-consuming. Therefore, commercial digital cameras have sparked a lot of interest in developing an accurate reflectance estimation algorithm.

Accurate reflectance estimation using a digital camera has been a hot topic for years, resulting in a tremendous amount of methods. Xiao et al. [1] extended the traditional PCA method by embedding a polynomial model and applied it to skin colors, yielding a significant improvement. Arad et al. [2] described a method that uses sparse encoding and dictionary learning and works well on a variety of datasets. However, those methods are either lacking in accuracy or too complicated to be applied to real-time applications, such as image-guided surgery.

In this paper, a new lattice regression based algorithm is proposed to recover the reflectance information of tissue samples and its performance was compared with two other commonly adopted methods. Present results showed that, the proposed algorithm can further improve accuracy of reflectance reconstruction.

M. Xu et al. (Eds.): CACPP 2022, LNEE 991, pp. 87–91, 2023.
https://doi.org/10.1007/978-981-19-9024-3_13

2 Proposed Method

Image generation is an injection process that maps high-dimensional reflectance onto three-dimensional RGB space, implying that two independent reflectance data sets could have the same RGB values. This problem is known as *metamerism.* But lucky enough, the frequency of metameric pairs in natural scenes was found to be as low as 10^{-6} to 10^{-4} [3], meaning that two samples with different reflectance data are likely to produce two distinct colors. This is especially true for the organic tissue, where colors are in low effective dimension. Therefore, each RGB triplet corresponds to a unique reflectance. With these concepts in mind, a new reflectance recovering algorithm is developed and is given in Fig. 1.

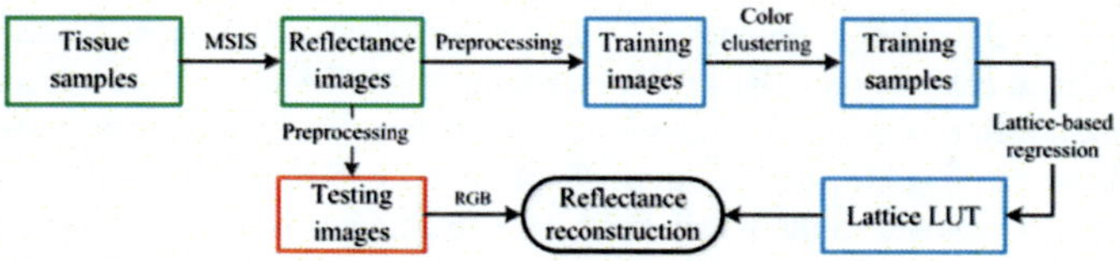

Fig. 1. Workflow of the proposed method

Initially, a series of hyperspectral images of organ samples were captured using an MSIS [4], and they were projected onto the spectral sensitivity functions (SSFs) of a commercial camera to make their corresponding RGB images. A pre-treatment was performed to remove the high-lights on the surface. Afterwards, a k-means color clustering algorithm was applied to extract their representative colors. Those colors were adopted to form a lattice-based regression LUT. On the other hand, this freshly developed LUT was utilized to conduct interpolation for input RGB images, and its corresponding reflectance image can be easily retrieved by interpolation.

2.1 Lattice-Based Regression LUT

The proposed method is demonstrated in Fig. 2. The training data is shown by red points. Each has a different RGB value that corresponds to a reflectance. The testing data point is in yellow, and its related reflectance must be calculated using the training dataset. However, since those training data points are irregularly distributed, it is difficult to perform this estimation. In this study, the lattice regression-based model [5] was adopted and further extended to calculate the regularized vertex and to build a new LUT called *lattice LUT*.

Assume that all of these training points are contained within a lattice, with its vertex points (m nodes) evenly spread along all R, G, and B axes. Let $\{\boldsymbol{r}_j \in \boldsymbol{R}^d\}_{j=1:m}$ be the corresponding reflectance for each node $\{\boldsymbol{s}_j \in \boldsymbol{R}^3\}_{j=1:m}$. $\boldsymbol{d}$ is the dimension of the reflectance spectrum. Given a set of $\boldsymbol{n}$ training RGB inputs $\{\boldsymbol{x}_i \in \boldsymbol{R}^3\}_{i=1:n}$, they corresponds to $\boldsymbol{n}$ reflectance $\{\boldsymbol{y}_i \in \boldsymbol{R}^d\}_{i=1:n}$. Therefore, each training data point $\boldsymbol{x}_i$ can be represented by a linear combination (interpolation) using the lattice nodes.

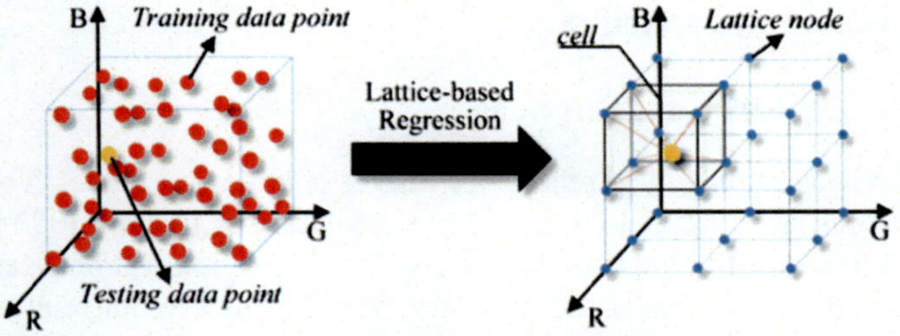

Fig. 2. An example of the proposed method. The unevenly distributed training data points were regularized using the lattice-based regression model.

The goal of lattice regression is to find a set of $\{s_j \in R^3\}_{i=1:m}$ that can accurately interpolate the training data using a certain interpolation method. Thus, a set of weights $\{w_{ij}\}_{i=1:n, j=1:m}$ to interpolate x_i can be described as:

$$\sum\nolimits_{j=1}^{m} w_{ij}s_j = x_i; \quad \sum\nolimits_{j=1}^{m} w_{ij} = 1, \ (i = 1, \ldots, n) \tag{1}$$

A trilinear interpolation method is used in this study due to its high efficiency. Given the linear interpolation weights $\{w_{ij}\}$, the output corresponding to x_i is calculated as $\hat{y} = \sum_j w_{ij}r_j$. Thus, the interpolated reflectance $\hat{r}$ that minimizes the interpolation error on the training data is given by

$$\hat{r} = arg \min_r \sum\nolimits_{i=1}^{n} (\hat{y}_i - y_i)^2 = arg \min_r \sum\nolimits_{i=1}^{n} \left(\left(\sum\nolimits_{j=1}^{m} w_{ij}r_j \right) - y_i \right)^2 \tag{2}$$

Generally, the above equation is underdetermined in most cases where there is a cell that contains none of the training data point. In this condition, the solution is not unique. In this study, a smoothness term is added using the second-order difference in each dimension, called *Hessian regularizer*. It can be written as Eq. (3).

$$\sum\nolimits_{k=1}^{d} \sum \left(\left(r_{h,k} - r_{j,k} \right) - \left(r_{j,k} - r_{l,k} \right) \right)^2 = \sum\nolimits_{k=1}^{d} \sum \left(r_{h,k} - 2r_{j,k} + r_{l,k} \right)^2 \tag{3}$$

where $r_{h,k}$, $r_{j,k}$, and $r_{l,k}$ correspond to the three adjacent nodes $s_{h,k}$, $s_{j,k}$, and $s_{l,k}$; h and l respectively represent the sample number before and after sample j; k denotes the spectrum dimension at each wavelength. Therefore, the mathematical model of the lattice-based regression is as follows:

$$\hat{r} = arg \min_r \| Wr - y \|_2^2 + \lambda r^T K_s r \tag{4}$$

where the training reflectance $y = [y_1, \ldots, y_n]^T$, $W \in [0,1]^{m*d}$ denotes the set of weights to interpolate the training data and K_s is an $d \times d$ matrix. The role of the regularization parameter $\lambda (> 0)$ is to tradeoff between solving accuracy and smoothness. Equation (4) has a closed-form solution,

$$\hat{r} = \left(W^T W + \lambda K_S \right)^{-1} W^T y \tag{5}$$

Eventually, the *Lattice LUT* was established and any input color can be transformed into reflectance by simply adopting this newly built LUT.

3 Experiment and Discussion

3.1 Sample Acquisition and Screening

This study collected a total of 34 reflectance images of biological tissues. Those images were captured using an MSIS [4], which has a resolution of 1040 by 1392 pixels. The filters' spectra extend from 400 to 700 nm, with a 10 nm interval between them. To convert these reflectance images into RGB images, SSFs of a commercial camera, i.e., Canon 60D, were obtained from Jiang et al. [6], and they are illustrated in Fig. 3(a). In this conversion, the illuminant is set as D_{65}. As a result, thirty-four RGB images were constructed.

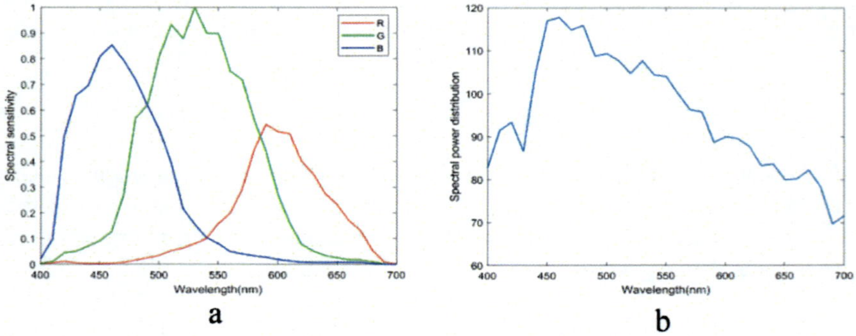

Fig. 3. (a) Spectral sensitivity of the camera, and (b) spectral power distribution of the light source.

To build the LUT, one has to have a set of training samples. In order to have a set of training colors that are evenly distributed in the RGB, the k-means algorithm was first applied to 33 of these reflectance images. For each image, a total of 100 clusters were obtained and the center of each cluster was adopted as the representative color. In total, 3300 (33*100) representative colors were extracted and they were hoped to be evenly distributed and cover the reflectance space well. One the organ heart image was left as the testing image. This is because it is representative for the organ dataset and has the most complex colors.

3.2 Results and Analysis

The performance of the proposed method is illustrated in Table 1, together with two other commonly adopted methods, i.e., Pseudo-Inverse (PINV) and Sparse encoding recovery (SR). They are evaluated using three metrics, i.e., CIE DE2000 color difference (ΔE_{00}), root mean square error (*RMSE*), and goodness-of-fit coefficient (*GFC*). ΔE_{00} is used to reflect the perceptual color difference between two color stimuli, while the other two metrics represents their spectral difference. The minimal errors (min), mean errors (mean), and maximum errors (max) for all three metrics are summarized in Table 1. The best result is marked bold.

Table 1. Performance of the investigated methods

Method	*RMSE* (%)			ΔE_{00}			*GFC* (%)		
	Min	Mean	Max	Min	Mean	Max	Min	Mean	Max
PINV	0.600	1.723	16.611	0.030	0.770	8.499	81.993	99.044	99.920
SR	0.178	1.467	16.892	0.003	1.089	12.751	77.470	99.092	99.981
Ours	**0.145**	**0.709**	**15.782**	**0.000**	**0.709**	**8.361**	**83.769**	**99.832**	**99.999**

As is shown in Table 1, all these methods performed well in terms of reflectance recovery on organ sample. In comparison to the present technique, the proposed method has the smallest RMSE and CIE DE2000 color difference, as well as the biggest GFC, demonstrating its superiority over others. An interesting finding is that, the PINV method gave a comparable performance to the SR method, which is likely due to the small effective dimension of the reflectance space formed by organ samples.

4 Conclusions

In this study, a new lattice regression based algorithm is proposed to recover the reflectance information of organ samples, and its performance is compared with two other commonly adopted methods, i.e. the Pseudo-Inverse method and the Sparse encoding recovery method. Present results show that the proposed method outperformed the other methods.

Acknowledgements. A Project Supported by Scientific Research Fund of Zhejiang Provincial Education Department; Supported by the Fundamental Research Funds for the Provincial Universities of Zhejiang (GK219909299001-019).

References

1. Xiao, K., Zhu, Y., et al.: Improved method for skin reflectance reconstruction from camera images. Opt. Express **24**(13), 14934 (2016)
2. Arad, B., Ben-Shahar, O.: Sparse recovery of hyperspectral signal from natural RGB images. In: Lecture Notes in Computer Science, pp. 19–34 (2016)
3. Foster, D.H., Amano, K., Nascimento, S.M., Foster, M.J.: Frequency of metamerism in natural scenes. J. Opt. Soc. Am. A Opt. Image Sci. Vis. **23**(10), 2359–2372 (2006)
4. Shen, H., Cai, P., et al.: Reflectance reconstruction for multispectral imaging by adaptive Wiener estimation. Opt. Express **15**(23), 15545–15545 (2007)
5. Garcia, E., Arora, R., et al.: Optimized regression for efficient function evaluation. IEEE Trans. Image Process. **21**(9), 4128–4140 (2012)
6. Jiang, J., Liu, D., Gu, J., Süsstrunk, S.: What is the space of spectral sensitivity functions for digital color cameras? In: IEEE Workshop on Applications of Computer Vision (WACV), 2013, pp. 168–179 (2013)

Digital Media Technology

Research and Application of Multi-dimensional Virtual Simulation Packaging Based on AR Technology

Changmei Ren[1], Zhanjun Si[1](✉), Zhiqiang Zhou[2], and Miao Yan[2]

[1] School of Light Industry Science and Engineering, Tianjin University of Science and Technology, Tianjin, China
szj@tust.edu.cn

[2] College of Artificial Intelligence, Tianjin University of Science and Technology, Tianjin, China

Abstract. In this study, AR technology was combined with product packaging to create a multi-objective recognition database and embed it into the community group buying mini program to improve the problem of single product packaging recognition and low functionality. Using Unity 3D engine and Vuforia to build AR scenes, realize the interconnection between virtual scenes and the real world, increase particle effects to enhance the fun of interaction, and intuitively understand the details of required products. The study combines the augmented reality technology and product packaging, from dimensional proceed with, put forward a new digital and recycled packaging new direction, to identify the personalized AR for specific population development, improve the functional sex of the packaging design, innovation of product packaging information display, realize the consumers and the depth of fusion products.

Keywords: Unity 3D · AR technology · Multi-target recognition · Model interaction

1 Introduction

The development of science and technology has promoted the informatization and digitization of The Times. Product packaging is the main source of product information, and consumers will not see the product until they have access to the product packaging [1]. Product packaging is carrying more and more information, and it is difficult for users to quickly understand these information in the process of purchase [2]. Therefore, in recent years, there has been a paradigm shift, with product packaging turning to more sensory attraction and interactive services [3]. Therefore, information content integration and product display design in the process of packaging design will become a new development trend of packaging design in the future.

Augmented Reality (AR) refers to the real-time registration of computer-generated virtual objects or information into the three-dimensional world [4]. The experiencer can not only see the real world, but also interact with virtual objects through devices [5]. The engineering research community has long focused on the application of AR

M. Xu et al. (Eds.): CACPP 2022, LNEE 991, pp. 95–105, 2023.
https://doi.org/10.1007/978-981-19-9024-3_14

technology in assembly operations [6]. AR technology has a wide range of applications in many fields, such as teaching and training, games and entertainment, military and digital reproduction of cultural relics and other fields [7]. With the development of computer technology and information network and the continuous innovation of smart devices, some AR devices have entered the market, and the rendering quality and comfort have been significantly improved. The user's experience in the design of various industries is increasingly important [8]. Due to the limitations of AR devices and the innovation of mobile devices such as smartphones, AR technology on mobile devices such as mobile phones is an important research direction.

It is no longer enough to meet social development only through conventional display methods. We need to always be at the forefront of technology and seek innovative ways to attract people's attention [9]. Mobile AR technology has changed user behavior habits, and also has a certain impact on the marketing environment and marketing methods [10]. Traditional packaging is very limited in terms of product information transmission, and packaging is less interactive, and many designs are difficult to convey to users. The research on the combination of AR technology and product packaging has also become a trend. In 2017, Guo Juan et al. [11] explored the specific implementation of packaging information design under AR technology through the design of rational information in AR technology; in 2018, Li Ying et al. [12] Interactive design of AR technology in food packaging; in 2019, Li Shiyao et al. [13] applied AR technology to Longjing tea packaging to change the display of product information from flat to three-dimensional. This research intends to combine a variety of products, use the newly created database, identify product packaging through AR, and accurately achieve interesting interactions with the product, and then view more information about the product.

In this study, the augmented reality technology is combined with product packaging. Starting from multiple dimensions, the brochure and product packaging are transformed into three-dimensional models with a more realistic sense, so as to increase users' dynamic visual experience and realize three-dimensional display of packaging information and diversification of communication forms. The interaction between users and products is promoted so that users can get a better sense of immersion. The detailed parameter information of products can be understood from various angles through rotation and zooming [14], providing more possibilities for displaying product information.

2 System Overall Framework Design

2.1 Development Environment

The operating system is Windows 10 Ultimate, the processor is Intel (R) Core (TM) i5-8250U CPU @ 1.60 GHz 1.80 GHz, the installed memory is 8.00 GB, and the Android smartphone model is HUAWEI nova3.

2.2 Design Ideas

This research is based on AR technology, product packaging combined with AR technology, to provide users with more product information in an interesting way. It consists of four parts: brochure design and production, multimedia resource design and production, UI interface design and Unity 3D interactive function design.

Product information display page, by clicking the button to view different products, has a simple introduction to the products, mainly including text and video to display the basic information of the products. The AR identification page is divided into two forms. The first one is supported by AR paper brochures, which are used to provide image identification. By making different brochures and adding different products, different products can be displayed. At the same time, it can rotate and zoom with the products to display product information in all directions. The second method takes the product packaging as the identification object, and displays the 3D model of the corresponding product by identifying the product packaging. You can click the button to watch the product video and display the information of the product, and at the same time, add creative special effects to interact with users. The design framework is shown in Fig. 1.

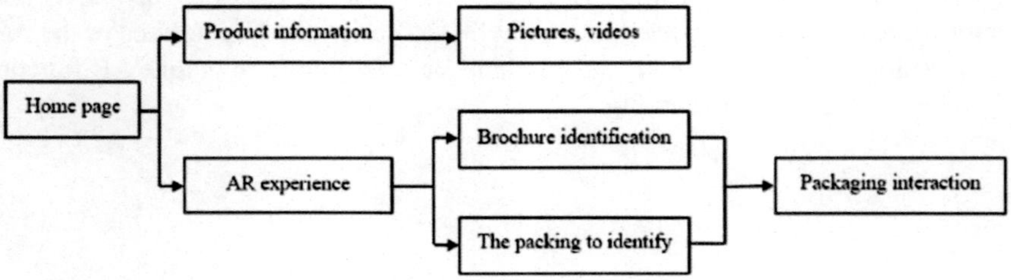

Fig. 1. Design framework

2.3 Key Technology

According to the technical means and manifestations, AR can be divided into AR based on computer vision and AR based on geographical location information.

In this study, the former is adopted. Firstly, the mapping relationship between the real world and the screen is established, then the screen coordinates are obtained, and then the video or 3D model associated with the AR identifier is displayed to the corresponding screen coordinates. At this time, the virtual scene superimposed on the screen can achieve the effect of attaching to the identifier. The matrix transformation formula for mapping the template coordinate system to the screen coordinate system is as follows:

$$s\begin{bmatrix} Xc \\ yc \\ 1 \end{bmatrix} = \begin{bmatrix} A11 & A12 & A12 \\ 0 & A22 & A23 \\ 0 & 0 & 1 \end{bmatrix} = \begin{bmatrix} R11 & R12 & R13 & T1 \\ R21 & R22 & R23 & T2 \\ R31 & R32 & R33 & T3 \end{bmatrix} \begin{bmatrix} Xm \\ Ym \\ Zm \\ 1 \end{bmatrix} = AWm \begin{bmatrix} Xm \\ Ym \\ Zm \\ 1 \end{bmatrix}, \tag{1}$$

where: s is the scaling factor; Matrix a is the camera internal reference matrix; The matrix Wm is the camera external reference matrix. The internal reference matrix is obtained by camera calibration, while the external reference matrix is unknown. It is necessary to estimate Wm according to screen coordinates (xc, yc), the previously defined identifier coordinate system and the internal reference matrix, and draw the graph according to Wm [15].

2.4 System Framework Design

The design and application of AR technology in product packaging is based on product packaging and brochures, and serves as a prominent function of product information and features. This application mainly includes the design of product information display and AR identification.

Firstly, the UI interface icons, buttons and AR paper publication materials are designed and made with AI and PS, and the video materials are edited and output with PR. Then, the product packaging model is made in 3Ds Max, and the materials and maps are added to the model with V-Ray renderer to further optimize the information. Finally, all materials are imported into Unity 3D to design the UI interface of the APP, design the interaction with users, and combine Vuforia plug-in to realize AR function. The technical route is shown in Fig. 2.

3 Development and Implementation of System

3.1 AR Paper Brochure Design

Product information design is a process of understanding and combing product content information, and organizing effective information by combining design elements such as words, graphics, colors and materials [16]. When designing a product brochure, it is necessary to have recognition degree. While highlighting the nature of the product, it is also necessary to consider the recognition degree of the brochure, so as to facilitate the combination with AR recognition. Collect materials as needed, then use AI and PS to draw and layout, and finally export the materials to PNG format. The model mapping should be combined with the model itself, not only to restore the product itself, but also to add more information so that users can know more about the product during use. When designing materials such as UI interface and button icons, combining with the style of the software itself, it conforms to the user's usage habits. The product promotion page renderings are shown in Fig. 3.

Fig. 3. Rendering of product promotion page

3.2 Design and Production of Multimedia Resources

Production of Video. AR technology is not only the interactive creation of pictures, 3D models, animation special effects and other media, but also can make deep-level interaction by using audio and video. The collected video materials are imported into PR for editing and integration, and various operations such as material splicing, creative production and special effects addition are adopted, and video subtitles are added, and finally saved in mp4 format.

Manufacture of Three-Dimensional Model. Product model is an important part of information display and interaction. At the same time, we can also use model animation and model mapping to convey more product hidden information. Select the appropriate products to model the external packaging and internal products, and build the packaging model in 3Ds Max. When modeling, consider the model parts that need to be given different maps in advance.

After modeling with the basic tools in 3Ds Max, set the corresponding material parameters through the plug-in of V-Ray renderer, add the corresponding bitmap to the texture ball to be mapped, drag the set texture ball onto the corresponding model and add UVW mapping modifier, adjust the direction and size of each model map in the modifier, and finally export the FBX format file, as shown in Fig. 4.

Fig. 4. Part of the product packaging modeling renderings

3.3 Build a Database

Vuforia provides the function of online database creation, log in to Vuforia official website, and obtain the product key for subsequent database and software development [17]. Vulia provides many kinds of recognition methods such as image recognition, multi-target recognition and column recognition on the developer's page. The AR recognition part of this study is divided into brochure recognition and package recognition, which use Vulia's image recognition function and multi-target recognition function respectively.

In the part of brochure identification, add picture identification and upload the designed brochure pictures one by one. In the part of packaging recognition, multi-target recognition is adopted. The implementation idea is as follows: build a project, import the installation package of easyAR, and fill in the product key obtained from the official website. Then create a new script to manage MultiTarget, hide all models before tracing, and display the specified model when tracing to the specified target.

Secondly, determine the ratio of length, width and height of the cube, adjust the position, proportion and size of each picture of the cube package and upload it, as shown in Fig. 5. Vuforia will analyze and identify the data uploaded by users, generate a database, and finally download all the processed databases (Unity 3D packages) to the local area.

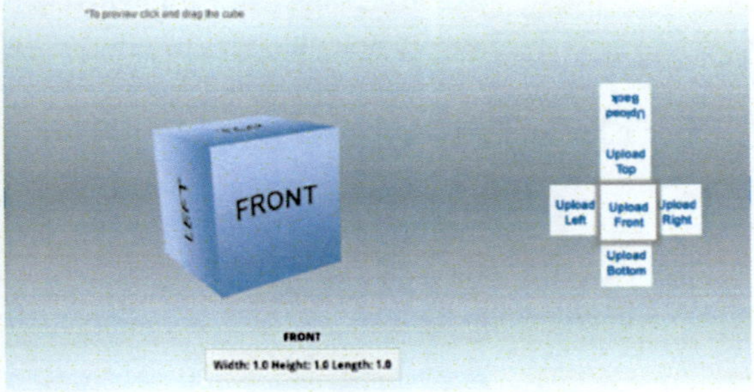

Fig. 5. Multi-target recognition

3.4 Design and Implementation of Each Function of the System

Create a Unity 3D project, and import the already made resources such as videos, pictures, models and databases into Unity 3D. UI interaction design needs the UGUI system in Unity 3D, and the main UGUI controls are Canvas, Image and Button, etc. According to the designed materials, different scenes are created, including the cover page, home page, product information page and AR experience page. The positions of icons and buttons are designed according to the content and functions, and the interactive functions are controlled by writing scripts with Visual Studio.

System UI Interface Design. The cover page is the first page that users see after opening the software. Unity 3D has a rich and complex Animation system, which not only supports imported animation, but also can make various animations in Unity 3D through the animation system. The interaction of the cover design is to click the "Click to enter" button, and the cover image opens from the middle to the left and right sides, turning to the front page like a door opening, animating the cover to move left and right, controlling the animation playing and page jumping. You need to write code for the scene page jumping. Select the button right under the script of the project folder to create two C# scripts, which are named "UIPlaneAll" and "UiManager" and drag the built UIPlaneAll scripts onto the Canvas. Using script control, realize the jump of scene page.

AR Identification and Production. This is an important part of the interactive experience that users can get through augmented reality APP. When they see real product packaging and brochures in the real world, the product model will be vividly displayed on the user's mobile client at the same time. Users can zoom in and out of the product model, rotate and drag it, and watch the product introduction video through gesture manipulation, and get a deep interactive experience.

In Unity 3D, Vuforia plug-in provides a series of tools to create and manage databases. The image recognition function is used in the production of brochure recognition. Vuforia supports the simultaneous recognition of a specified number of multiple images, and creates ImageTarget prefabricated parts according to the number of recognition images. Each ImageTarget represents one recognition image, and the recognition image is changed into the processed publicity image in the database. The product model, video and other resources are in one-to-one correspondence with the recognition image, and the model size and position are adjusted by moving, rotating and zooming. Pay attention to the position of each object, so as to avoid the distance between the model and the identifier, as shown in Fig. 6.

Fig. 6. Create image recognition

Multi-target recognition is needed in product packaging recognition. Multi-target preform is created in the scene, and its database is changed into the imported MultiTarget recognition database. The resources such as video and model are corresponding to the location of the model. After product packaging recognition is successful, you can view every part of the package in all directions, and each part has corresponding information.

In order to increase the interest and beauty of interaction, this study added particle special effects to product packaging. It is a two-dimensional image rendered in the three-dimensional space of particles, which can show smoke, fire, water droplets and other effects. Particle emitter is combined with particle renderer to create static particle effect. Particle animator can move particles in different directions and change their colors, and can also control each individual particle in the particle system through scripts, thus creating a unique particle effect, as shown in Fig. 7.

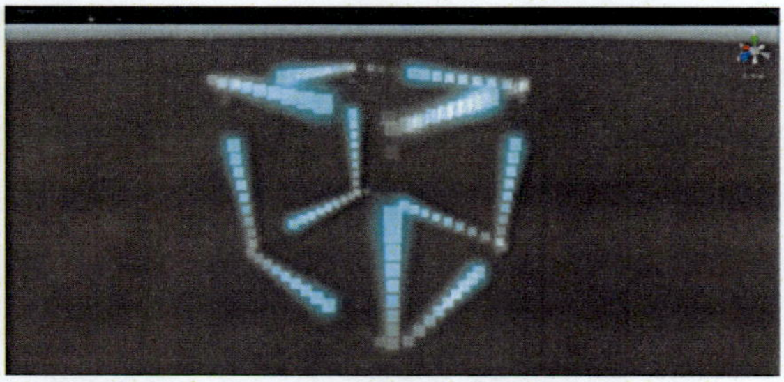

Fig. 7. Particle effects

By writing scripts, interactive actions such as zooming in and out, rotating and dragging are added to the model, and the functions of zooming and rotating the model are realized, so that users can view the product information more clearly. A music playing event is added to the recognition behavior, and the background music will be played at the same time after the model is recognized. At the same time, a click event is added for interaction, which can control the particle special effects, check the packaging information, and play the background music and video.

Interactive Function of Animation. The homepage classifies products. Select the products you want to know to jump to the product page, or click "AR Purchase" to enter the AR scene, where you can identify the product pictures in the brochure. After identifying the pictures, the corresponding product models will appear. You can view the detailed information of the products by rotating and zooming the products, as shown in Fig. 8 on the left.

The product page mainly displays the price, specifications, delivery information and other contents of the product, and then introduces the product in an all-round way through pictures and videos. Swipe your finger left to switch the product pictures, click the video play button to play the video, and click the "AR Packaging Experience" button to enter the AR scene for object recognition. When the product packaging is recognized, you can see product descriptions, interesting special effects, etc., and you can directly interact with the product. Watching each side has different effects, as shown in the second and third left of Fig. 8.

4 Test and Release

After the overall function of the software is developed, it needs to be debugged and optimized continuously. Test whether the internal functions of the software can be realized and whether the content is complete. After debugging, open Unity 3D, select Edit-Preferences-External Tools, and add the path of the installed Java JDK and Android SDK to Unity 3D. Set the output in Build Settings, name the application, click Build to package and generate Android apk file, and then install it in Android phone, which realizes the software release.

Fig. 8. Packaging identification

5 Conclusion

This study explores the combination of AR technology and product packaging, and adds model interaction. Users can view products by scanning brochures and product packaging, and all information of products will be presented to users, and some derivative information can also be added, which provides a new model of product information. This research can not only be combined with products, but also design AR product manuals, which can not only increase brand influence, but also bring unique shopping experience to consumers. Secondly, we can also improve the economic benefits of shopping malls by designing leaflets together with them. This research optimizes product information by updating product packaging and product model, which is a new direction of digitalization and reduction of packaging. AR technology will definitely provide great power for the development of packaging industry.

References

1. Liu, Y., Wang, D.: Application of computer software photoshop to tea packaging design. Fujian Tea **42**(04), 168 (2020)
2. Yang, Y.: Research on the application of modern information technology in packaging design. J. Hunan Univ. Sci. Technol. **245**(02), 154–156 (2018)
3. Krishna, A., Cian, L., Aydınoğlu, N.Z.: Sensory aspects of package design. J. Retail. **93**(1), 43–54 (2017)
4. Azuma, R.T.: A survey of augmented reality. Presence Teleoperators Virtual Environ. **6**(4), 355–385 (1997)
5. Sanna, A., Manuri, F.: A survey on applications of augmented reality. Adv. Comput. Sci. Int. J. **5**(1), 18–27 (2016)
6. Wang, X., Ong, S.K., Nee, A.Y.C.: A comprehensive survey of augmented reality assembly research. Adv. Manuf. **4**, 1–22 (2016)

7. Zhang, Y.: Application and analysis of augmented reality technology in modern packaging design. Western Leather 128 (2020)
8. Cascini, G., et al.: Exploring the use of AR technology for co-creative product and packaging design. Comput. Ind. **123** (2020)
9. Zhao, P.F., Chen, T.E., Wang, W., et al.: Research on the agricultural skills training based on the motion-sensing technology of the leap motion (2015)
10. Sabir, M., Prakash, J.: Virtual reality: a review. In: 2nd International Conference on Advance Trends in Engineering and Technology (2014)
11. Guo, J., Du, W.: Research on packaging information design based on AR technology. Packag. Eng. **38**(06), 26–29 (2017)
12. Li, Y., Su, Y.: Interactive design of food packaging based on AR technology. Packag. Eng. **40**(02), 60–64 (2019)
13. Li, S., Si, Z., Li, H.: Design and development of Longjing tea packaging APP based on AR technology. Packag. Eng. **41**(15), 176–180 (2020)
14. Tan, P., Wang, M., Fu, T., Ji, Y.: Cultural heritage AR interaction design based on experiential learning theory. Packag. Eng. 1–7 (2020)
15. Jianshu: Principles of AR Technology [EB/OL]
16. https://www.jianshu.com/p/3aaclec6ea2f. Accessed 18 Mar 2021
17. Ji, W., Li, J.: Research on the barrier-free design of packaging visual information based on the concept of universal design. Ind. Des. **11**, 61–63 (2020)
18. El Habbak, H., Cushnan, D.: Developing AR Games for iOS and Android, p. 9. Packt Publishing (2013)

App Design and Research Based on Traditional Art Intangible Cultural Heritage

Li Ma[1](✉) and Kehuan Cao[2](✉)

[1] Department of Printing and Packaging Engineering, Shanghai Publishing and Printing College, Shanghai, China
865798692@qq.com
[2] Guangdong ATV Professional Academy for Performing Arts, Guangdong Baiyun University, Guangdong, China
377082169@qq.com

Abstract. For the development of traditional art intangible cultural heritage, we use digital media technology and internet platforms to create applications, thus promoting product innovation and cultural heritage. Starting from the origins of traditional arts intangible cultural heritage, by sorting and classifying projects according to their use and characteristics, we analyze the main reasons that constrain their heritage and development, and propose design strategies. Finally, a new approach is provided for the transmission and promotion of the intangible cultural heritage of folk art.

Keywords: Intangible cultural heritage · Inheritance and promotion · Traditional art · Application program design

1 Role of APP in the Protection of Intangible Cultural Heritage

In terms of intangible cultural heritage dissemination, the design of relevant applications in China is still in its infancy, but there are already some applications designed to disseminate China's intangible cultural heritage. While some applications are relatively complete in terms of design functionality, most applications still have many problems, namely the uneven level of design in terms of form and function of each application, little content information, incomplete information and outdated news time. There are various types of intangible cultural heritage of traditional arts. They include the art of paper-cutting, the art of new year painting, the art of shadow painting, the art of batik, the art of embroidery, the art of lacquer, the art of weaving, etc. However, in art-themed non-fiction applications, most of them just list knowledge one by one. In a layered manner, graphic material is placed directly into the application, lacking effective information structure and user engagement [1]. In summary, it was decided to use the mobile terminal application as a display platform for the design practice of digital transmission of artistic intangible cultural heritage. The research idea is shown in Fig. 1.

M. Xu et al. (Eds.): CACPP 2022, LNEE 991, pp. 106–111, 2023.
https://doi.org/10.1007/978-981-19-9024-3_15

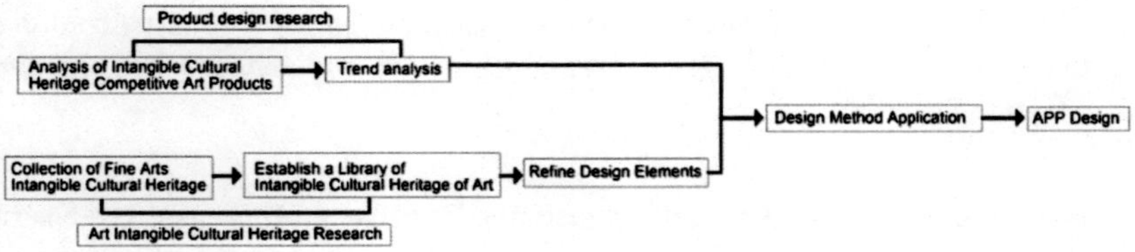

Fig. 1. Research approach

2 APP Design

2.1 Conception of "Intangible Cultural Heritage" APP Interface Interaction

Through the research on the four stores that mainly provide APP in the market, the author learned that there are a few kinds of "intangible cultural heritage" APP in the stores, and there are three types of problems

(1) Think that mobile terminal APP can do what computers can do, and simply design APP as a scaled-down version of computer applications;
(2) The interactive structure of the APP is single, and most of them simply put the digitized graphic data in the APP in a flat way, lacking effective information architecture and user participation;
(3) Some APPs are so elaborate that their functions are complex, which affects users' browsing of content.

The author found several excellent works from the APP boutique recommendation, which mainly introduced traditional cultural resources, and analyzed and summarized the interactive concept of APP interface design for the "intangible cultural heritage" category.

2.2 Realistic Reproduction

The live reproduction APP interface takes the simulation of the production or application scenario of "intangible cultural heritage" resources as the interface interaction elements or means. The presentation and browsing methods of page information restore the technological process, skill display, and working scene of the design and production of "intangible cultural heritage" resources. The interactive way of live reproduction is easy to guide visitors to the theme of "intangible cultural heritage" and inspire interest.

2.3 360° Virtual Browsing

As a display mode closest to real objects, 360° virtual browsing is widely used in virtual museum exhibits and e-commerce platform goods display, achieving the best communication and exchange between users and exhibits [2]. 360° virtual display can clearly see the form of each angle and the details of each part of the exhibits, so as to achieve a sense of real life and immersive experience. Users can freely observe the details of the exhibits by 360° rotation, realize the virtual reproduction, knowledge visualization and interactive operation of "intangible cultural heritage" [3], and enhance the initiative of users in learning.

2.4 Combination of Narrative and Interactive Experience Increases the User's Operational Viscosity

The significance of user viscosity for APP products is not only the improvement of user activity, but also the improvement of brand value and the impact of payment conversion rate [4].

For example, in the pottery-making APP, users can practice the main technological processes in the process of making "ceramic art" through interactive games. The produced "ceramic artworks" can also be auctioned, and the money from the auction can buy more materials and other production materials. The improvement of production factors can make the production of "ceramic artworks" more beautiful, and then auction more money to form a growing incentive user experience mechanism. Only incentive factors can fully mobilize the enthusiasm of users [5], so as to improve the user stickiness of products, as shown in Fig. 2.

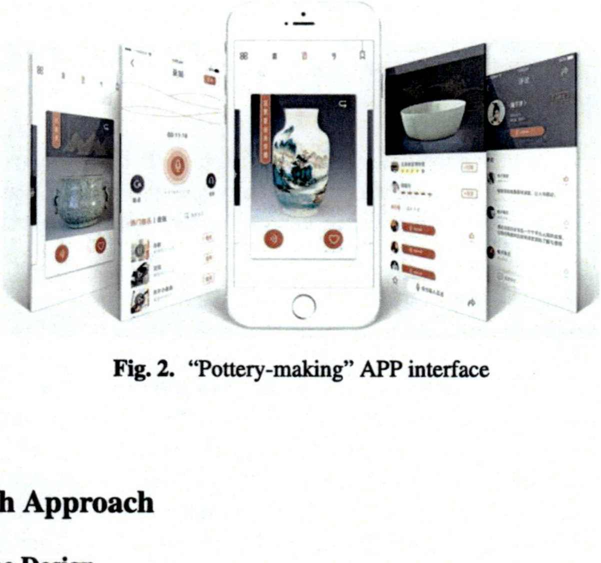

Fig. 2. "Pottery-making" APP interface

3 Research Approach

3.1 Prototype Design

When launching the application, first enter the welcome page to guide users into the world of art intangible cultural heritage. The four function buttons at the bottom of the home page are linked to "culture", "mall", "community", "creative space". In the era of mobile information, users want to obtain information more quickly. The "waterfall flow" layout is adopted to allow users to scan from top to bottom and quickly filter out the content of interest to conduct in-depth browsing, as shown in Fig. 3.

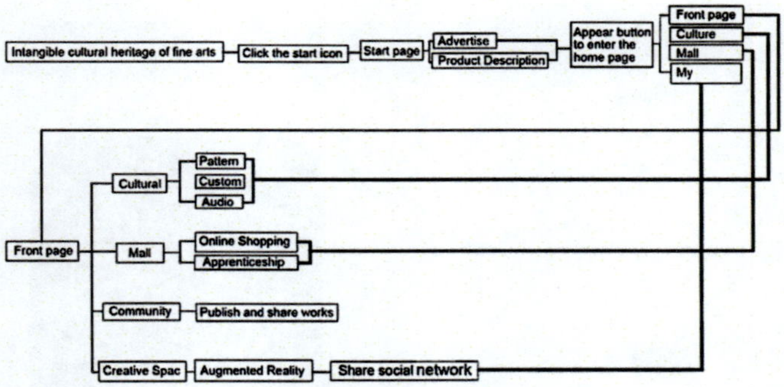

Fig. 3. Flowchart of the APP

Topology diagrams clearly show the pages and page relationships involved in the application and can clarify content, structure and relationships, allowing the prototyping of the application to take shape. Next, the production of a low-fidelity prototype of the application is entered at the architectural level. The low-fidelity prototype of the application is shown in Fig. 4.

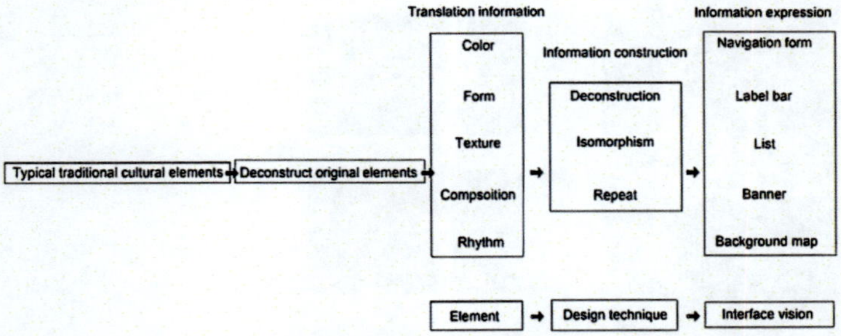

Fig. 4. Topology of the page

3.2 User Interface Visual Design

The first is the design of the application welcome page Fig. 5, where each time the software is opened, there is a representative thematic launch page for the artistic and cultural product intangible cultural heritage. The user completes the interactive use by logging in and registering. The second is the design of the app page. For the colour design of the application, the main colors chosen are dark red and dark blue. The icon elements in the application use artistic patterns of intangible cultural heritage. The decorative figurative elements were drawn in terms of both aesthetics and meaning. It is full of traditional culture and allows the user to receive more visual information in the shortest possible time, as shown in Fig. 6.

a b

Fig. 5. Design of the welcome page

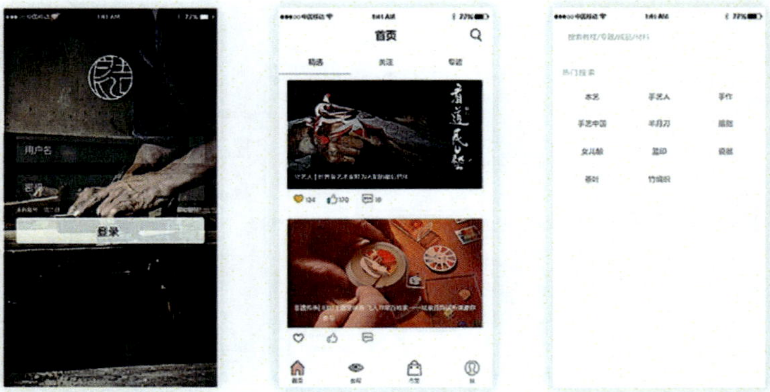

Fig. 6. Design and implementation of the homepage of the application

4 Conclusion

This paper designs an app based on the intangible cultural heritage of art and culture, and integrates product system theory, software development theory and art design theory into the design practice of the app. In the design of the app interface for the heritage and innovation of "intangible cultural heritage", the product functions, user experience and visual effects of the interface should be fully integrated, and the connotations of national culture should be interactively disseminated in the process of user experience through internet thinking and marketization. Only through the combination of science and culture, art and technology, and the use of digital technology, information technology and artistic means, can the heritage and innovation of national culture be protected, developed, inherited and innovated to the greatest extent.

Acknowledgements. This research is Supported by Key Lab or Intelligent and Green Flexographic Printing (KLIGFP-01).

References

1. Ye, D.: Study on regional cultural characteristics of tourism product design. Packag. Eng. **32**(16), 134–137 (2011)
2. Meng, F.: The application of virtual reality technology to modern exhibition art. J. Tianjin Acad. Fine Arts **2**, 76 (2012)
3. Huang, Y., Tan, G.: Research on China intangible cultural heritage protection and development of digital. J. Huazhong Normal Univ. (Humanit. Soc. Sci.) **3**, 49–55 (2012)
4. Zhang, X.: Guide to Approaches of Creative Thinking. Central Compilation & Translation Press, Beijing (2008)

Design and Application of an APP for Intangible Cultural Heritage Based on Cultural Translation

Li Ma[1](✉) and Wenliang Su[2](✉)

[1] Department of Printing and Packaging Engineering, Shanghai Publishing and Printing College, Shanghai, China
865798692@qq.com
[2] Guangxi Vocational and Technical College, Guangxi, China
1534142615@qq.com

Abstract. Based on the concept of cultural translation, it is proposed to extract typical cultural elements from the material, behavioral and spiritual layers of intangible cultural heritage. Design practical APP to promote the dissemination of intangible cultural heritage.

Keywords: Intangible cultural heritage · Design elements · Cultural translation · User experience · Application program design

1 Traditional Culture Communication in the Context of Mobile Internet

1.1 Application of Traditional Cultural Elements in APPs

The emergence of the mobile Internet era provides an opportunity for traditional culture to be "activated". Through the mobile internet platform, traditional culture has emerged in a form of communication that is more relevant to the life of the public. Its expression is also enriched, gradually evolving into new media interaction design [1]. On the one hand, the application allows users to understand and learn the essence of traditional culture; on the other hand, the combination of application and traditional cultural IP allows everyone to participate in the innovation and inheritance of culture, thus achieving a win-win situation for both cultural and commercial value.

1.2 Problems Existing in Traditional Cultural APPs

The direct migration of traditional cultural contents and elements into the design of the interface will make users feel raw and fragmented, and will not reflect the true traditional aesthetics. Another problem is that some product designs imitate the design structure and interaction forms of APPs in other fields, ignoring their inherent traditional cultural features and spiritual connotations, and ignoring the public's understanding of cultural elements and cognitive differences in interface operation, thus reducing the public's tolerance and recognition of the promotion of traditional culture.

M. Xu et al. (Eds.): CACPP 2022, LNEE 991, pp. 112–116, 2023.
https://doi.org/10.1007/978-981-19-9024-3_16

2 Concept of Cultural Translation

Cultural translation as defined in this paper is the process of extracting information, designing expressions, reorganizing and transmitting traditional culture based on the concept of translation [2]. The translation of traditional culture is the process of transforming traditional cultural elements from the original system to the new system. This process uses traditional cultural symbols as a link, the mobile internet as a carrier and the interface as a form of translation. The core of traditional culture translation is the translation of three levels: the material level, behavioural level and spiritual level. The derivation of translation levels is shown in Fig. 1. A typical example of a translation application is Mercedes Benz. In order to enter the Chinese market, it skillfully transliterates its English trademark "Mercedes Benz" into "Mercedes Benz" according to the characteristics of Chinese culture and the rules of the use of Chinese characters, thus successfully completing the cultural output of its brand values in China.

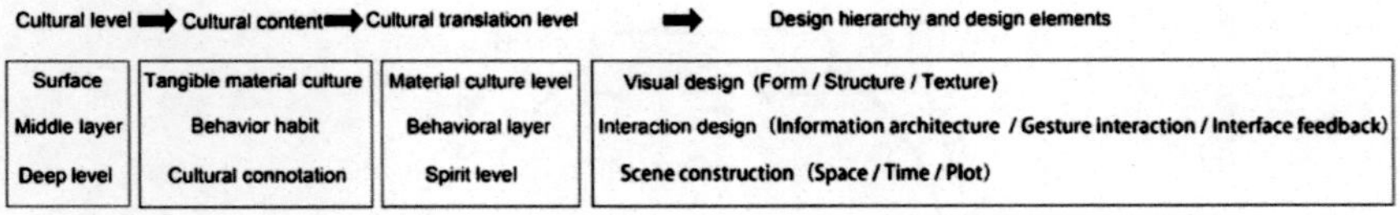

Fig. 1. Traditional cultural translation level deduction

3 Levels of Traditional Cultural Transcription and Design Applications

3.1 Physical Translation of Traditional Culture

Surface culture can also be called material culture, which is the use of materials by human beings. The public can understand the aesthetic value of the material layer of traditional culture through visual languages, such as couplets and red lanterns in the theme of New Year festivals. When translating the material layer of traditional culture, first of all, determine the traditional cultural theme and design orientation of the product to be developed [3]. Then, typical elements with unique symbolic meanings are selected from the complex traditional cultural system. Finally, determine the cultural elements that are more in line with the theme and perform visual translation. The screening process of traditional cultural elements is shown in Fig. 2.

3.2 Behavioral Translation of Traditional Culture

Behavioral habits are the actions that people naturally implement in their subconsciousness [4]. Behavioral cultural translation interprets people's natural behaviors as the design of interactive gestures, animation effects, and operational feedback in the application interface through simulation and guidance, and uses people's habitual behaviors as the elements to guide the user's cognitive interface [5]. According to the user's familiar

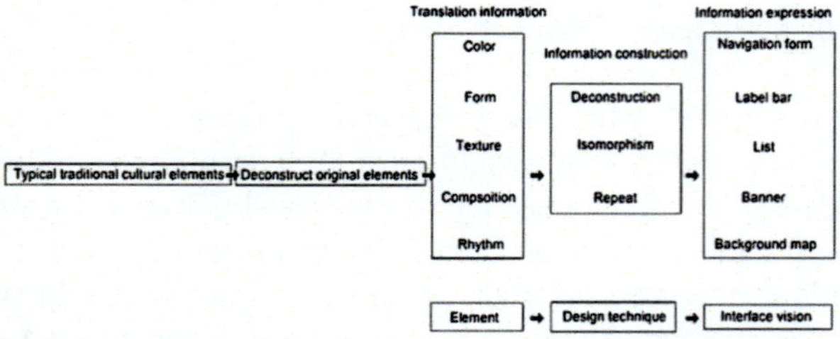

Fig. 2. Physical layer translation of traditional culture

natural behavior habits, the interface interaction design is carried out to bring users a natural and smooth operation experience, so as to achieve the goal of improving product consistency, ease of use, and ease of learning. Figure 3 shows the translation of traditional culture at the behavioral level.

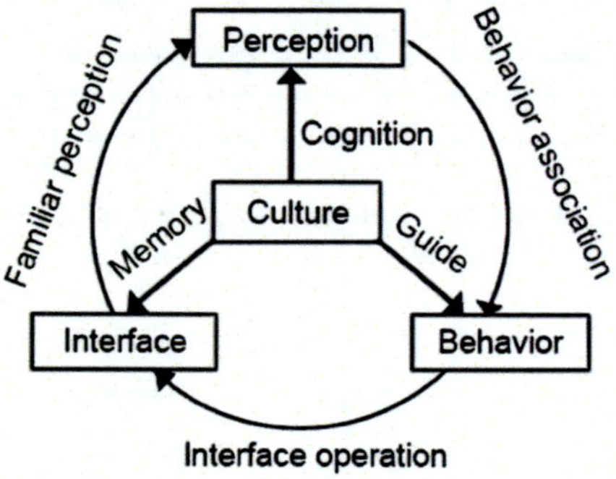

Fig. 3. Behavioral layer translation of traditional culture

3.3 Spiritual Translation of Traditional Culture

In mobile app design, the spiritual layer of translation is the integration of traditional cultural connotations into the scene imagery of the app. Firstly, the most representative and subjective traditional cultural spirits are selected from the wide variety of traditional cultures and the information is extracted for transcreation. Then, through storytelling, narrative, emotional and environmental shaping methods, the scene imagery is conceived to translate the spirit of traditional culture into a sense of space, time, interactive effects, and sensory experiences such as light and sound in the scene design, thus enhancing the sense of integration when the public experiences it, which is shown in Fig. 4. Dunhuang Culture is characterized by the history and culture of Dunhuang and is deduced through the symbolic features and stories in the interface. The icons show the beauty of Dunhuang culture in the form of lines and surfaces with the appearance features of sunset, hump, and grottoes, as shown in Fig. 5.

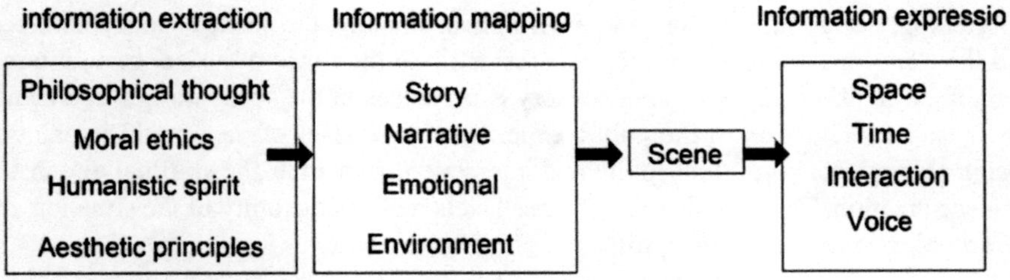

Fig. 4. Conceptual layer translation of traditional culture

Fig. 5. Dunhuang culture APP interface design

4 Conclusion

With the development of the mobile Internet, users' requirements for product user experience, cultural connotation, and artistic aesthetics are becoming more and more complex. More and more traditional cultural elements will be transformed into application design, and the translation forms will become richer. The simple visual translation of cultural elements can no longer be completed, which fully meets the needs of users. The material, behavioral and spiritual aspects of traditional culture can be transformed into application interface visual design, interaction design, and scene images, and then traditional culture can be transformed from shallow to deep, from outside to inside. In mobile Internet products, the user experience, artistic value, and commercial value of products are maximized. In the process of transformation, choose the most representative and subjective traditional cultural spirit, and extract translation information. Then, through storytelling, narration, emotion, and environment shaping, the scene is conceptualized, and the traditional cultural spirit is transformed into the sense of space, time, interaction effect, light, sound, and other sensory experiences in the scene design to enhance the sense of integration in the public experience. The level of vision and interaction design also affects users' perception and interaction, as well as the spiritual dimension in scene creation. The combination of three levels realizes the unity of the situation and the combination of form and spirit.

Acknowledgements. This research is Supported by Key Lab or Intelligent and Green Flexographic Printing(KLIGFP-01)1.Research on the Training Path of High quality Technical and Skilled Talents in Higher Vocational Architecture Major in Ethnic Areas Based on the Integration of Industry, Learning and Innovation (No.GXGZJG2021B118).Research on the Cultural Inheritance and Innovation of Hechi Rural Residential Buildings Based on Traditional Garden Architectural Aesthetics (No. 2022KY1426).Research and Practice on the Integrated Cultivation of "Production, Teaching, Research and Innovation" of Higher Vocational Architectural Design Talents under the Background of Rural Revitalization (No.GXGZJG2022A033).

References

1. Zhao, F.: Application of Chinese traditional cultural symbols in interface design of mobile terminal (2013)
2. Zhang, H.: The application of cultural translation in the context of Chinese contemporary art. Art Educ. (10), 28–29 (2013)
3. Wang, J.: The research about traditional culture APPS. Jiangnan University, Wuxi (2016)
4. Li, Z.R.: Based on the cultural translation of map design. J. Graph. **2018**(4), 629–634
5. Ma, J.: The value of Chinese traditional cultural semiotics. Cult. Vientiane (11), 227–228 (2011)

Printing Engineering Technology

Fabrication and Performance of Spherical $Ni(OH)_2$ Electrode Based on Screen Printing

Huirong Ye, Qiang Zhang, Qian Tu, Xianran Li, Xinyu Sun, Ting Guo, Xuejun Tian, and Liangzhe Chen(✉)

School of Electronic Information Engineering, Jingchu University of Technology, Jingmen, China
chen_lz1991@126.com

Abstract. As a kind of energy storage material with great application potential, the actual specific capacitance of $Ni(OH)_2$ is much lower than its theoretical specific capacitance, which is mainly due to poor morphology. Morphology regulation is an effective way to tackle this issue. Herein, spherical $Ni(OH)_2$ was prepared by a modified coprecipitation method, and the effect of urea content on surface morphology was investigated. Subsequently, $Ni(OH)_2$ inks were printed on foamed nickel via a stencil to form the electrodes, and their electrochemical performances were investigated. The results show that urea content has a great influence on the morphology of $Ni(OH)_2$ nanoballs, and the optimal morphology is achieved when the molar of urea and Ni^{2+} is 10:1. Additional, the screen-printed electrode has a higher specific capacitance (821 F/g) than that of other electrodes, and acceptable capacitance retention of 61.0% is performed after 800 continuous charge/discharge cycles. In conclusion, this paper has proposed a new route for preparing the $Ni(OH)_2$ nanoball with a controllable shape, and the combination of screen printing can realize the electrode production on a large scale, which is conducive to promoting the application and development of energy storage materials and printed devices.

Keywords: Spherical · $Ni(OH)_2$ · Screen printing · Morphology regulation

1 Introduction

In the last few years, there is an urgent need for the development of renewable energy storage to meet the growing electronics. As an excellent energy storage system, supercapacitors (SCs) are more attractive due to their fast charge/discharge, long cycle life, and environmental friendliness [1–3]. However, the major challenge of low energy density still needs to figure out. More recently, $Ni(OH)_2$ has been a promising cathode material for supercapacitors owing to its larger specific capacitance, abundant source, and large power density [4, 5], but the actual specific capacitance is much lower than the theoretical specific capacitance, which is mainly due to the poor surface structure. The morphology regulation is deemed as an effective way to tackle this issue, and the spherical morphology is competitive in contrast with others in early reports [6]. Hence, it is significant to design and develop a nanostructured $Ni(OH)_2$ nanoball with a facile route.

M. Xu et al. (Eds.): CACPP 2022, LNEE 991, pp. 119–125, 2023.
https://doi.org/10.1007/978-981-19-9024-3_17

In this paper, a modified coprecipitation method is put forward to prepare the spherical $Ni(OH)_2$, and the molar ratio of urea and Ni^{2+} was investigated to realize the morphological regulation. Then, the screen printing technology was employed to fabricate the $Ni(OH)_2$ electrodes, which will perform outstanding electrochemical properties. To sum up, the $Ni(OH)_2$ with controllable morphology is a potential candidate, that can promote the application and development of energy storage materials and printed devices.

2 Experiment

2.1 Chemicals and Reagents

Potassium hydroxide (KOH), hydrochloric acid (HCl), nickel chloride ($NiCl_2{\cdot}6H_2O$), polytetrafluoroethylene (PTFE), urea, acetone, and ethanol were purchased from Shanghai Macklin Biochemical Co., Ltd. All chemical reagents were of analytical purity and used without further processing. The activated carbon (AC) was bought from Nanjing/Jiangsu XFNANO Materials Tech Co., Ltd. Nickel foam (The thickness is 1 mm) was received from Cyber Electro Chemical Materials. Acetylene black was obtained from Yilongsheng Energy Technology Co., Ltd.

2.2 Preparation of the $Ni(OH)_2$ Nanoball

In short, the nanostructured $Ni(OH)_2$ nanoball was prepared via a modified coprecipitation method. Firstly, urea and $NiCl_2{\cdot}6H_2O$ were dissolved in deionized water and transferred to a round-bottomed flask with a condensing return pipe, followed by heating at 105°C for 48 h. Then, the resultant was treated with filtration and the residue was washed with H_2O and ethanol alternately. The $Ni(OH)_2$ nanoball was obtained after drying in a vacuum for 24 h. According to the molar ratio of urea and $NiCl_2{\cdot}6H_2O$ (7.5:1, 10:1, and 12.5:1), the resulting $Ni(OH)_2$ was indexed as $Ni(OH)_2$-7.5, $Ni(OH)_2$-10 and $Ni(OH)_2$-12.5, respectively.

2.3 Fabrication of the Screen-Printed $Ni(OH)_2$ Electrodes

The $Ni(OH)_2$ electrodes were fabricated utilizing the screen printing method, which can be described as three steps. In step one, 10 wt.% of PTFE/ethanol was added dropwise to the mixture of 80 wt.% of $Ni(OH)_2$ compound or AC and 10 wt.% of acetylene black to form a uniform ink. In step two, the obtained ink was forced through a stencil (80 meshes) and printed on the nickel foam with a square pattern (1×1 cm). In step three, the electrode was treated by drying in a vacuum and pressing under 8 MPa.

2.4 Characterization

The scanning electron microscope (SEM, Zeiss SIGMA 500, Germany) and energy-dispersive X-ray spectroscopy (EDS, Bruker Quantax, Germany) were used to characterize the morphology and the surface elemental content, respectively. The X-ray diffractometer (XRD, RIGAKU Miniflex600, Japan) with Cu Kα radiation ($\lambda = 1.5406$ Å) was employed to identify the crystal structure, and the Fourier transformed infrared spectrometer (FT-IR, ThermoFisher Nicolet 5700, Waltham, USA) was utilized to analyze the surface functional group information. The specific surface area was studied by a Brunauer-Emmett-Teller method (BET, Micromeritics ASAP2000, Boynton Beach, FL, USA) at 77K. All electrochemical measurements (cyclic voltammetry (CV), galvanostatic charge-discharge cycling (GCD), and cycling stability) were performed at an electrochemical workstation (CorrTest CS350H, Wuhan, China).

3 Result and Discussion

3.1 Characterization of the Spherical $Ni(OH)_2$

Figure 1 shows the SEM images of $Ni(OH)_2$-7.5, $Ni(OH)_2$-10 and $Ni(OH)_2$-12.5. The nanostructured balls are clear in all $Ni(OH)_2$ samples and the differences in surface morphology can be found. Obviously, the surface of $Ni(OH)_2$-7.5 nanoball (Fig. 1a and d) is relatively dense with sparse pores, and this morphology goes against the ion migration in the electrolyte. With the increase of urea, more and more tiny nanosheets emerge on the surface, resulting the abundant channels (Fig. 1b and c). Compared with the $Ni(OH)_2$-10 (Fig. 1e), the nanoball of $Ni(OH)_2$-12.5 are more irregular due to the excess of urea (Fig. 1f). EDS image of $Ni(OH)_2$-10 is displayed in Fig. 1g, in which a large number of Ni, C and O elements as well as a small number of N element are existed on the surface, confirming the successful preparation of $Ni(OH)_2$. Figure 1h indicates the crystal structure of the as-prepared $Ni(OH)_2$-10. The diffraction peaks at $2\theta = 12.4$, 24.9, 33.4 and 59.4° can be ascribed to the lattice planes of (003), (006), (101) and (110), respectively, which is in line with the rhombohedral phase of α-$Ni(OH)_2$ (JCPDS Card no. 38-0715 [7, 8].

The FT-IR spectra of $Ni(OH)_2$-10 are created in Fig. 2 (a). Apparently, two strong bands at 3448 and 1632 cm^{-1} are ascribed to the O-H vibration from the -OH groups or H_2O molecules, while the weak peak at 635 cm^{-1} is assigned to the Ni-O bending vibration. The result of FT-IR spectra confirms the successful preparation of $Ni(OH)_2$ [9]. As nanostructured materials, the larger the area surface, the higher the specific capacitance. The N_2 adsorption/desorption isotherms and the corresponding pore size distribution of $Ni(OH)_2$-10 can be observed in Fig. 2(b) and (c). Surprisingly, a large BET-specific surface area of 119 m^2/g is achieved in $Ni(OH)_2$-10, and the average pore size is ~ 3.4 nm, demonstrating the mesoporous characteristics for $Ni(OH)_2$-10.

3.2 Electrochemical Performances

In order to investigate the electrochemical performances, the as-prepared electrodes were performed by CV, GCD, and cycling stability in a three-electrode system. Figure 3(a) shows the CV curve at a scan rate of 10 mV/s with a pair of redox peaks, demonstrating the typical Faradic reaction from $Ni(OH)_2$. In contrast with $Ni(OH)_2$-7.5 and $Ni(OH)_2$-12.5 electrodes, $Ni(OH)_2$-10 has the largest integral area of the CV curve, implying the great specific capacitance. The GCD curves in Fig. 3(b) support this point of view, in which the $Ni(OH)_2$-10 electrode has a longer discharge time than that of others.

Fig. 1. SEM images of (a) and (d) $Ni(OH)_2$-7.5, (b) and (e) $Ni(OH)_2$-10, (c) and (f) $Ni(OH)_2$-12.5; (g) EDS image of $Ni(OH)_2$-10, inset shows the elemental content; (h) XRD pattern of $Ni(OH)_2$-10

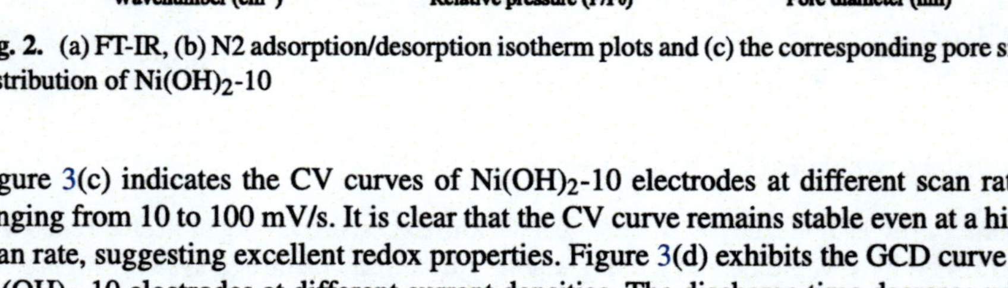

Fig. 2. (a) FT-IR, (b) N2 adsorption/desorption isotherm plots and (c) the corresponding pore size distribution of $Ni(OH)_2$-10

Figure 3(c) indicates the CV curves of $Ni(OH)_2$-10 electrodes at different scan rates ranging from 10 to 100 mV/s. It is clear that the CV curve remains stable even at a high scan rate, suggesting excellent redox properties. Figure 3(d) exhibits the GCD curve of $Ni(OH)_2$-10 electrodes at different current densities. The discharge time decrease with the increase of current density, and the voltage platforms in all curves are because of the presence of the Faradic reaction. The specific capacitance (C) of electrodes can be calculated according to the following formula [10]:

$$C = \frac{I \times \Delta t}{m \times \Delta V} \tag{1}$$

where m is the mass of the active materials (g), ΔV is the potential window (V), I and Δt are the discharge current (I) and time (s), respectively.

The calculation results are illustrated in Fig. 3(e) and listed in Table 1. It is expected that the specific capacitance of $Ni(OH)_2$-10 at 1 A/g (876 F/g) is much larger than that of $Ni(OH)_2$-7.5 (709 F/g) and $Ni(OH)_2$-12.5 (667 F/g), as well as the capacitance retention (48.4% of $Ni(OH)_2$-10 versus 22.0% of $Ni(OH)_2$-7.5 and 30.7% of $Ni(OH)_2$-12.5), demonstrating the optimal morphology of $Ni(OH)_2$-10. Furthermore, the cycling stability of $Ni(OH)_2$-10 is explored and shown in Fig. 3(f). After 800 continuous charge/discharge cycles, the $Ni(OH)_2$-10 electrode remains a specific capacitance of 387.4 F/g with a retention rate of 61.0%, manifesting the ball-like morphology not only can enhance the specific capacitance of $Ni(OH)_2$ but also improve its stability.

Fig. 3. (a) CV curve at a scan rate of 10 mV/s and (b) GCD curve at a current density of 1A/g for $Ni(OH)_2$-7.5, $Ni(OH)_2$-10, and Ni(OH)2-12.5 electrodes, (c) CV curve of $Ni(OH)_2$-10 electrode at different scan rates, (d) GCD curve of $Ni(OH)_2$-10 electrode at different current densities, (e) rate performance of $Ni(OH)_2$-10 electrode at different current densities and (f) cycling stability of $Ni(OH)_2$-10 electrode at 10 A/g after 800 cycles

4 Conclusions

In summary, three-dimensional $Ni(OH)_2$ with a spherical morphology is synthesized through a facile and modified coprecipitation method. The surface morphology of $Ni(OH)_2$ nanoball can be regulated by adjusting the addition of urea content and the optimal texture is achieved in $Ni(OH)_2$-10, which has a large BET-specific surface area

Table 1. Corresponding specific capacitance of $Ni(OH)_2$-7.5, $Ni(OH)_2$-10, and $Ni(OH)_2$-12.5 electrodes at various current densities

Samples	Current density (A/g)					
	1	2	3	5	10	20
$Ni(OH)_2$-7.5	709	701	693	625	474	156
$Ni(OH)_2$-10	876	859	798	730	635	424
$Ni(OH)_2$-12.5	667	624	605	557	452	205

of 119 m^2/g with an average pore size of ~ 3.4 nm. Additional, screen printing technology is employed to fabricate the electrodes. The screen-printed $Ni(OH)_2$-10 electrode has a higher specific capacitance (821F/g) than that of $Ni(OH)_2$-7.5 and $Ni(OH)_2$-12.5 and holds acceptable capacitance retention of 61.0% after 800 continuous charge/discharge cycles. This enhancement in electrochemical performance comes from the abundant pore structure resulting from the well-designed spherical morphology. It is concluded that this paper provides a new perspective for the promotion of $Ni(OH)_2$-based nanomaterial in electrochemical performance, which can be applied to other nanostructured materials and facilitate the development of energy storage materials and printed devices.

Acknowledgements. This work was funded by the Jingmen Science and Technology Project (2022YFZD045 and 2021YFYB119), the Research Project of Jingchu University of Technology (YB202205 and YY202206), Scientific Research Team of Jingchu University of Technology (TD202101) and College Students' Innovation and Entrepreneurship Training Program (KC2022022).

References

1. Zhu, Q., Zhao, D., Cheng, M., et al.: A new view of supercapacitors: integrated supercapacitors. Adv. Energy Mater. **9**, 1901081 (2019)
2. Choudhary, N., Li, C., Moore, J., et al.: Asymmetric supercapacitor electrodes and devices. Adv. Mater. **29**, 1605336 (2017)
3. Tu, Q., Li, X., Xiong, Z., et al.: Screen-printed advanced all-solid-state symmetric supercapacitor using activated carbon on flexible nickel foam. J. Energy Storage **53**, 105211 (2022)
4. Natarajan, S., Ulaganathan, M., Aravindan, V.: Building next-generation supercapacitors with battery type $Ni(OH)_2$. J. Mater. Chem. A **9**, 15542–15585 (2021)
5. Han, C., Si, H., Sang, S., et al.: Carbon dots doped with $Ni(OH)_2$ as thin-film electrodes for supercapacitors. ACS Appl. Nano Mater. **3**, 12106–12114 (2020)
6. Xu, L., Zhao, Y., Lian, J., et al.: Morphology controlled preparation of $ZnCo_2O_4$ nanostructures for asymmetric supercapacitor with ultrahigh energy density. Energy **123**, 296–304 (2017)
7. Zhang, J., Liu, S., Pan, G., et al.: A 3D hierarchical porous α-$Ni(OH)_2$/graphite nanosheet composite as an electrode material for supercapacitors. J. Mater. Chem. A **2**, 1524–1529 (2014)

8. Chai, H., Peng, X., Liu, T., et al.: High-performance supercapacitors based on conductive graphene combined with $Ni(OH)_2$ nanoflakes. RSC Adv. **7**, 36617–36622 (2017)
9. Li, B., Cao, H.: ZnO@graphene composite with enhanced performance for the removal of dye from water. J. Mater. Chem. **21**, 3346–3349 (2011)
10. Qiu, H., An, S., Sun, X., et al.: Excellent performance MWCNTs-GONRs/$Ni(OH)_2$ electrode for outstanding supercapacitors. Ceram. Int. **45**, 18422–18429 (2019)

Research Progress in Carbon Nanotube Thin Film Transistors by Printing Technologies

Suyun Wang[1,2], Nianjie Zhang[1], Shengzhen Liu[1], Lijuan Liang[1](✉), Zhaohui Yu[1], Lianfang Li[1], Beiqing Huang[1], Xianfu Wei[1], and Jianwen Zhao[2](✉)

[1] School of Printing and Packaging Engineering, Beijing Institute of Graphic Communication, Beijing, China
lianglijuan@bigc.edu.cn

[2] Suzhou Institute of Nano-Tech and Nano-Bionics (SINANO), Chinese Academy of Sciences, Suzhou, China
jwzhao2011@sinano.ac.cn

Abstract. In recent years, electronic products have continued to develop rapidly toward the trend of thinness, personalization, flexibility, high performance, high density, and multi-functional integration, and the development and research of new semiconductor materials and field-effect transistors (FETs) have become a hot topic of research today. With its excellent electrical, mechanical, and chemical properties, carbon nanotubes can be combined with printed electronics to become one of the most ideal channel materials for next-generation chips and flexible electronics. This paper outlines the separation methods of semiconductor carbon nanotubes and the research progress of printed carbon nanotube thin-film transistors in the fields of Internet of Things, artificial intelligence, and wearable electronics, and summarizes the challenges and opportunities for printed carbon nanotube thin-film transistor devices.

Keywords: Carbon nanotubes · Thin-film transistors · Printing electronics technologies · Separation methods

1 Introduction

Silicon based complementary metal oxide semiconductor (CMOS) transistors have reached their physical limits, and further channel reduction in silicon-based transistor devices can lead to problems such as short-channel effects, gate oxide tunneling and high power consumption. Semiconductor companies have been focusing on ways to continuously reduce transistor size to improve chip performance. On the other hand, scientists have been trying to find semiconductor materials and processing techniques that can replace silicon to develop next-generation transistor integrated circuits [1–5]. Also with the widespread development of the Internet of Things (IoT), there is an increasing demand for flexible electronics with low power consumption, high computing power, and good value for money. The development and research of high-performance semiconductor materials and transistor devices has become a hot topic of research today.

M. Xu et al. (Eds.): CACPP 2022, LNEE 991, pp. 126–139, 2023.
https://doi.org/10.1007/978-981-19-9024-3_18

Since the discovery of carbon nanotubes in 1991, the unique advantages of carbon nanotubes have attracted the attention and research of scientists. Semiconductor carbon nanotubes are an excellent one-dimensional semiconductor material with high electron and hole carrier mobility, thermal conductivity, excellent mechanical flexibility, and superior physical and chemical stability, and are considered to be one of the most ideal semiconductor materials for building next-generation chips and flexible electronics [6]. Compared to organic semiconductors (low mobility) and metal oxides (which require high temperature 300–500 °C processing to maintain stability and thus are restricted to most flexible substrates), semiconducting carbon nanotubes exhibit higher high mobility as well as lower processing temperatures, providing a viable strategy for the development of next-generation flexible digital integrated circuits [7].

In this paper, we will briefly introduce the development of printed carbon nanotube thin film transistors in the last decade or so from the aspects of semiconductor carbon nanotube separation, applications of printed carbon nanotube thin film transistor devices in the fields of Internet of Things, artificial intelligence and wearable electronics, as well as the opportunities and challenges faced.

2 Semiconductor-Type Carbon Nanotube Separation Methods

Commercially available carbon nanotubes are a mixture of semiconductor carbon nanotubes and metallic carbon nanotubes, and commercially available carbon nanotubes are not suitable for direct construction of thin-film transistor devices. In order to construct carbon nanotube thin-film transistor devices, a key problem needs to be solved first, which is the removal of metallic carbon nanotubes from commercial carbon nanotubes [8]. The presence of metallic carbon nanotubes in the transistor channel tends to cause the transistor device to fail to turn off, so the commercial carbon nanotubes need to be separated and purified to enable the construction of high-performance transistor electronics [9, 10]. In order to make commercial carbon nanotube powders available for application in electronic devices, two steps, coarse purification and selective separation and purification of the synthesized carbon nanotube powders, are usually required [11]. Coarse purification refers to the removal of other amorphous carbon materials (catalyst carriers, catalysts, amorphous carbon, etc.) from the powder. Selective separation and purification refers to the selective removal of metallic carbon nanotubes from carbon nanotubes by physical and chemical methods to obtain high purity semiconductor carbon nanotubes, as shown in Fig. 1, the SEM image of the high-density carbon nanotube film made after separation and purification [12]. These methods including selective chemical vapor deposition [13], density gradient ultracentrifugation (DGU) [14], gel chromatography separation [15], conjugated polymer wrapping [16], electrophoresis [17], two-phase extraction [18], and free radical reaction method, etc. The following is a brief description of the techniques for the preparation and separation of semiconductor carbon nanotubes.

2.1 Chemical Vapor Deposition Methods

The basic principle of chemical vapor deposition (CVD) is shown in Fig. 2, where carbon nanotubes are generated by the decomposition of carbon source gas flow over the surface

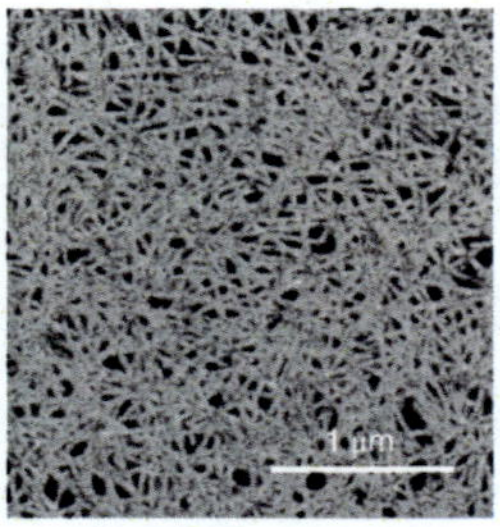

Fig. 1. SEM image of the high-density carbon nanotube film

of catalysts (e.g. Fe, Co, etc.) at high temperatures. Chemical vapor deposition method has become the most widely studied method for synthesis of carbon nanotubes due to its simple equipment and easy batch production. Wu et al. [19] synthesized (6, 5) and (7, 5) semiconductor carbon nanotubes preferentially at 700 and 800 °C using Co-MgO as catalyst, because the strong metal-support interaction between Co nanoparticles and MgO support facilitates the growth of semiconductor carbon nanotubes. In addition, Prof. Jin Zhang's group and Prof. Yan Li's group at Peking University designed and synthesized a variety of high-performance alloy catalysts to achieve selective synthesis of specific one-handed carbon nanotubes [20]. These works have laid a solid foundation for the selective synthesis of high-quality semiconductor carbon nanotubes.

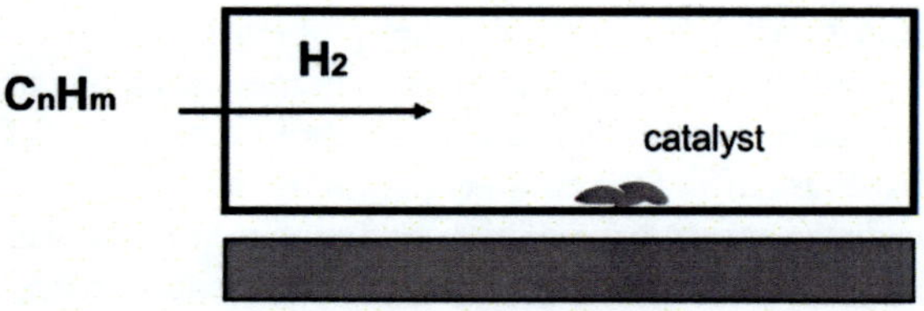

Fig. 2. Chemical vapor deposition schematic

2.2 Density Gradient Ultracentrifugation Methods

Density gradient ultracentrifugation (DGU) is a technique widely used in biology and medicine. Its principle [14] is to stratify substances of different densities by high-speed centrifugation. The advantages of this method are good separation, wide applicability and excellent dispersion characteristics. The first proposal of using DGU to separate semiconductor-type carbon nanotubes was made by Mark C. Hersam's group at Northwestern University [21]. They used DGU technique to obtain carbon nanotubes with different tube diameters from single-stranded DNA dispersed carbon nanotube solutions (as shown in Fig. 3(a)). Divided into three groups from top to bottom, the first group is the isolated semiconducting carbon nanotubes with different chirality, showing clear bands of different colors. The second group is gray, colorless, and consists of carbon nanotubes with larger density and wide distribution (no bands). The third group is the unseparated carbon nanotubes and insoluble material that precipitated at the bottom of the centrifuge tube. Batch preparation of semiconductor carbon nanotubes, even single-chiral semiconductor carbon nanotubes, can be achieved after several DGU separations and purifications. For example, Nanointegris has been selling high-quality semiconductor carbon nanotube materials purified by this method since 2009. Although DGU can obtain high purity semiconductor carbon nanotubes, the high cost required for this method (expensive ultra-high speed centrifuge, time-consuming, high price of density gradient agent) affects its industrial application.

2.3 Gel Chromatography Separation Methods

Gel chromatography separation is achieved by the difference in size and shape of the sample molecules. The method uses gel as a filler, and the larger molecules can only enter the part of the condensed pores with larger pore size, while the smaller molecules can enter more gel particles. The larger molecules travel a shorter distance within the gel column, so the larger molecules are retained in the column for a shorter time and are separated first, followed by the smaller molecules [22–24]. Structural separation of carbon nanotubes using gel chromatography is also one of the current simple and efficient methods with low cost and the capability of large-scale industrialization. As shown in Fig. 3(b), Yohei Yomogida et al. [25] used mixed surfactant gel chromatography to achieve high throughput separation of high-purity, single chiral single-walled carbon nanotubes. A mixed surfactant consisting of sodium dodecyl sulfate (SDS), sodium cholate (SC), and sodium deoxycholate (DOC) successfully separated seven single chiral single-walled carbon nanotubes. Compared with other reported methods, this separation method has the advantages of simplicity, low cost, high purity, high yield and high throughput, and this industrial-scale single-chiral separation can be applied to other single-chiral single-walled carbon nanotubes.

2.4 Conjugated Polymer Wrapping Methods

The conjugated semiconductor polymer wrapping method [26, 27] (e.g., Fig. 3c) is based on the selective wrapping and adsorption of single-walled carbon nanotubes of different structures by the conjugated polymer molecules themselves, which ultimately results in highly pure semiconducting carbon nanotubes. The conjugated polymer itself has good semiconductor properties and does not substantially degrade the performance of carbon nanotube thin-film transistor electronics without removing the conjugated polymer after separation [28]. As shown in Fig. 3(d), Liu et al. [29] proposed a hybrid extractor strategy to selectively disperse semiconductor carbon nanotubes in high yield and purity. The hybrid extractor system consists of two conjugated molecules with different binding abilities to carbon nanotubes, which are related to their molecular rigidity. Several hybrid systems such as P3DDT, PCO (PCz), and 4HP were used for detailed evaluation of semiconductor carbon nanotube sorting. The dispersion yields of P3DDT and PCz for semiconductor carbon nanotubes can be greatly improved up to 400% with little effect on the purity of carbon nanotubes.

Fig. 3. (a) Nanotubes observed in three regions after 10.5 h of centrifugation (b) High throughput, single chiral separation, optical absorption spectra of isolated (9,4) (red) and (10,3) SWCNTs (blue), (c) Illustration of polymer sorting (d) Chemical structure diagram of PCO and P3DDT, and schematic diagram of s-SWNTs sorting by hybrid extraction system

2.5 Electrophoresis Methods

Electrophoresis (EPD) is mainly used to separate and purify biomolecules, and since carbon nanotubes and biomolecules have similar dimensions on the scale, scientists have applied this technique to separate semiconductor carbon nanotubes. Within a current field, both semiconductor carbon nanotubes and metallic carbon nanotubes can induce the production of electric dipoles, with one end accumulating a positive charge and the other a negative charge. Through the high frequency alternating electric field, the frequency of change of metal carbon nanotubes is much greater than that of semiconductor carbon nanotubes, so that the polarized metal carbon nanotubes are adsorbed on the electrodes, while the semiconductor carbon nanotubes are left in the solution thus achieving separation. However, the experimental process of electrophoresis is more complicated and the space for adsorption of carbon nanotubes is limited, thus the yield is low and it is difficult to achieve industrial-scale production. As shown in Fig. 4, Lakshmanan et al. [30] deposited semiconductor carbon nanotubes on platinum electrodes by EPD in phosphate solution and with a system that allows monitoring when the semiconductor carbon nanotubes are deposited and form electrical contacts with the metal, this technique provides ideas and methods for integrating vertical single-walled carbon nanotubes in the EPD process.

2.6 Two-Phase Extraction Methods

Two-phase extraction is a novel method for separating single-chiral carbon nanotubes proposed by Zheng Ming's group at the Institute of Standards and Technology in recent years. The two phases consist of polyethylene glycol (PEG) and dextran solution (DX),

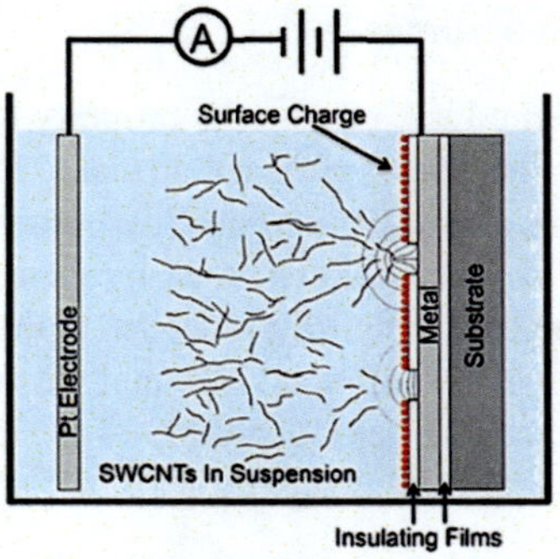

Fig. 4. Schematic diagram of EPD of carbon nanotubes deposited on metal through a nanoscale window

which are mixed and layered according to their different hydrophobicity. Then the carbon nanotubes with different chirality will be covered by surfactants and thus the hydrophobicity will be changed, the more hydrophobic carbon nanotubes will move up to the PEG solution and the hydrophilic carbon nanotubes will move down to the DX solution, thus achieving the structural separation of carbon nanotubes. The two-phase extraction method requires repeating the experimental steps, which is more tedious and relatively less reproducible. As shown in Fig. 5(a)(b), Li et al. [31] utilized a two-stage aqueous two-phase extraction (ATPE) technique in which the parent (initially located in the bottom phase) was divided into successive top-phase fractions (T1-Tn) by continuously adding hydrochloric acid in the first stage. Between each acid addition, the top phase (Tn) is extracted and replaced by a new simulated top phase. In the second stage, the top phase fraction (Tn) was added to the fresh bottom phase to re-establish the two-phase system, and then the successive addition of hydrochloric acid was repeated to obtain a new top phase fraction (TnTm) to isolate five different semiconductor carbon nanotube species with a diameter of 1.41 nm, and it was suggested that the use of alkanes for inside-filling would result in uniformly dispersed semiconductor carbon nanotubes and the method improved the yield.

Fig. 5. (a) Step-by-step diagram of a two-phase extraction method consisting of two stages and (b) spectral absorption diagram of the SWCNTs obtained by isolation

2.7 Free Radical Reaction Methods

The free radical reaction method is based on the property that the chemical activity of metal single-walled carbon nanotubes is greater than that of semiconductor single-walled carbon nanotubes. A free radical initiator is added to a homogeneously dispersed carbon nanotube solution, and free radicals are generated by means of ultrasound or heating, which selectively react against the metal single-walled carbon nanotubes, thereby eliminating them and then achieving separation. This method requires attention to control the reaction conditions, and thus the reproducibility of the experiment is relatively low.

3 Applications

With the continuous development of semiconductor carbon nanotube separation technology and printing technology, the performance of printed carbon nanotube thin-film transistor devices (especially the switching ratio and mobility of the devices, etc.) has been substantially improved. Researchers have constructed printed carbon nanotube thin-film transistor devices with various structures and different polarities on various substrates by different printing techniques and explored their applications in emerging fields (e.g., Internet of Things, artificial intelligence, aerospace, etc.). This chapter will briefly introduce the applications of carbon nanotube thin-film transistors on these fields through the successful cases that have been achieved by some research groups.

3.1 Ultra-low-Power, Low-Voltage Transistor Devices and Circuits for the Internet of Things

There is a growing demand for ultra-low operating voltage ($\leq$0.5V) and ultra-low power ($<$1nW/gate) thin-film transistor electronics in many emerging applications in the Internet of Things (IoT). Carbon nanotube thin-film transistor devices are not only capable of meeting the requirements of low-voltage driving, but also have ultra-low power consumption (e.g., by allowing them to operate in the deep subthreshold interval). For example, in 2020, Luis et al. [32] used a solution method to self-assemble monolayer organophosphorus molecules to provide the dielectric properties of ultrathin alumina, selectively deposited semiconductor carbon nanotube films by inkjet printing technique, and then encapsulated to be able to obtain bipolar transistor devices with excellent symmetry, and experimentally and theoretically investigated the effect of subthreshold slope and flat-band voltage on the operating voltage and power, etc. of the devices. The digital circuits constructed with bipolar printed carbon nanotube transistors with low subthreshold swing have similar complementary characteristics as in the inverters and NAND gates prepared in Fig. 6(a) and (b). They exhibit high noise tolerance, high voltage gain (80), ultra-low power consumption, and high operating frequency (100 kHz) characteristics at ultra-low operating voltage ($\leq$0.5V). In 2022, Jianwen Zhao's group [33] investigated low-power fully printed flexible carbon nanotube thin-film transistor electronics and CMOS circuits based on PHPS-derived dielectric layers. The prepared devices exhibited excellent electrical properties, such as a switching ratio of ~ 10^6, a small subthreshold swing (70–110 mV/dec), a high mobility (6.3 $cm^2V^{-1}s^{-1}$), and good stability at a low operating voltage of ± 1 V for the devices. In addition, by adjusting the semiconductor carbon nanotube ink concentration, the threshold voltage of the transistor can be adjusted to turn it into an enhanced device; then the fully printed flexible carbon nanotube CMOS circuit was successfully prepared by transducing the P-type transistor to N-type transistor by electron doping technique (using ethanolamine as electron dopant) (Fig. 6c). Its static power consumption can reach 0.05 pW/μm, voltage gain up to 45 (at $V_{DD} = 0.7V$) and large noise tolerance (about 27).

Fig. 6. (a) Optical images of the inverter and NAND gate circuits and results of the static power consumption of this operation compared to those reported at this stage [32] (b) Optical pictures of CMOS-like inverter gates and VTCs of inverters packaged in PMMA/CYTOP, gain values derived from VTCs in operation at different VDD values, VTCs of inverters not packaged in TFTs at different VDD values [34] (c) Fully printed flexible sc-SWCNTs TFTs and CMOS inverters circuit diagrams, schematics, static power consumption, voltage gain and noise margins [33]

3.2 Carbon Based Synaptic Transistor Devices for Artificial Intelligence

In 2019, Zhe-Nan Bao's group at Stanford University reported an all-inkjet printed stretchable carbon nanotube transistor array [35] that stretches to 20% strain in-plane (as shown in Fig. 7a). The stretchable field-effect transistor array was printed using intrinsically stretchable electronic materials, i.e., ionic PEDOT: PSS, ionic PVDF-HFP, and a network of semiconductor carbon nanotubes. The transistors are stable under atmospheric conditions with mobility up to 0.3 $cm^2V^{-1}s^{-1}$ and switching ratios up to 10^4. The unit channel current of 0.2 $mAcm^{-1}$ at operating voltages as low as 1 V is due to the ionic nature of their printed gate dielectrics. Transistors with a double-layer capacitive dielectric can simulate the synaptic behavior of neurons, and the slow-moving ionic properties of the gate dielectric accurately simulate short-time range plasticity. Excitatory and inhibitory postsynaptic currents can be generated under pulsed electrical stimulation at very small input voltages (40 mV). Due to the low operating voltage, stretchability, portability, stability in high air/water environments and synaptic properties, the device is expected to be directly applied in areas such as brain-computer interfaces.2021,

Chiu's group [36] developed a high-performance and multimodal $CsBi_3I_{10}$/SWCNT opto-synaptic transistor (Fig. 7b). The operating voltage range of $\pm$ 2V has an on-state current and off-state current of 5×10^6 and 6×10^{-12}A, respectively. The device successfully imitated not only the typical synaptic function but also the 'Pavlovian dog' function as well as the optical storage (only 3V erase voltage and 0.2s duration) and logic operations. In conclusion, multimimetic printed carbon nanotube transistor devices show great potential in the fields of neurobionics, storage components, and logic operations.

Fig. 7. (a) Schematic diagram of an array of intrinsically stretchable transistors and a cross-section of the device, with a set of optical images of a large area of transistors being bent in the hand [35] (b) Application of multimodal printed carbon nanotube transistors in neuronal biomimicry, storage elements, and logic operations [36]

3.3 Stretchable Transistors Carbon Nanotubes and Circuits for Wearable Electronics

With the increasing demand for wearable and implantable electronics, flexible stretchable electronics have recently attracted the interest of many scientists.In 2020, Min Zhang's group [37] designed a stretchable carbon nanotube thin-film transistor with high drive, high mobility, small size, and low power (Fig. 8a). A field-effect mobility of up to 221 $cm^2V^{-1}s^{-1}$ and a current density of 810μA mm^{-1} were obtained at a low driving voltage of 1 V. Moreover, the stretchable carbon nanotube thin-film transistors could withstand 50% repetitive stretching for 2000 times, indicating their promising application in stretchable circuits and wearable electronic devices. Nishio et al. [38] also designed a stretchable carbon nanotube transistor that could meet the requirements of low-voltage operation and good mechanical stretchability for wearable devices (shown in Fig. 8b). They added a local strain suppression layer on top of the active layer of an all-carbon nanotube stretchable thin-film transistor. The carbon nanotube thin-film transistors prepared by this method showed no degradation of the carbon nanotube transistor source current even when 35% external stretching strain was applied. Stretchable all-carbon nanotube integrated circuits such as inverters and ring oscillators have low supply voltages (0.7 V) and are also capable of biaxial stretching, and this work demonstrates the potential of strain-sensitive stretchable devices at low voltages.

Fig. 8. (a) [37] and (b) [38] is high-performance stretchable electronics

3.4 Irradiation-Resistant Electronic Devices and Circuits for Space Applications

Integrated circuits for some special applications require not only superior electrical properties but also certain special classes of functions, such as those used in space and nuclear industries, which require strong radiation resistance. Carbon nanotubes have super strong carbon-carbon double bonds and ultra-thin tubular structure, which makes them highly resistant to irradiation and are the most promising semiconductor materials for building irradiation-resistant electronic devices. Theoretically, carbon-based electronics have superb resistance to irradiation, but the developed carbon-based electronic devices and circuits are far from the expected resistance to irradiation, which is caused by the damage of the gate oxide insulator (e.g., isolation layer or substrate oxide) of the device under irradiation and the accumulation of a large amount of charge at the surface interface of the device as well as on the substrate [39]. In order to improve the radiation resistance of carbon-based electronic devices and circuits, various radiation hardening techniques have been developed to protect transistors from damage. For example, in 2020, Jianwen Zhao's group and Zhiyong Zhang-Lianmao Peng's group collaborated to investigate a radiation reinforcement technique to improve the irradiation resistance of carbon nanotube thin-film transistor devices and circuits [40], which uses semiconductor carbon nanotubes as the channel material, ionic gel as the gate, and polyimide as the substrate. The total irradiation dose that this FET can withstand is up to 15 Mard at a dose rate of 66.7 rad s^{-1}, which is significantly higher than the tolerance of silicon-based transistors (1 Mard). A similarly high tolerance is achieved by using this device to fabricate CMOS-like inverters (as shown in Fig. 9). In addition, radiation-damaged thin-film transistor devices can be annealed at a moderate temperature of 100 °C for 10 min to recover their performance. This high radiation tolerance and repairability of printed carbon nanotube transistor devices and circuit technology provides a viable strategy for developing transistors and integrated circuits that are immune to radiation damage.

4 Summary and Prospects

With the continuous development of semiconductor carbon nanotube separation technology, high-purity semiconductor carbon nanotubes (99.99%) as well as single-chiral

Fig. 9. TID-dependent property evolution of ion gel CNT FETs and CMOS-like inverters fabricated on polyimide substrates. (a) Schematic representation of radiation-induced interfacial defects on a silicon substrate, (b) Schematic diagram of CNT FETs damaged by Co-60 γ-ray irradiation, (c) Transmission characteristics of CNT FETs at Vds = 0.25V for different TIDs, (d) Leakage current versus TIDs, (e) Vds = 0.25V and Vgs = 1V, with 4 Mrad irradiation, Six device leakage statistics, (f) Optical images of CNT FETs prepared on PI substrates, (g) VDD = 3V, CMOS-like inverter VTC curves at different TIDs, (h) Excess voltage of CMOS-like inverters as a function of TID for ionic gel CNT FETs at Vmin and VDD = 3V

carbon nanotubes can be prepared on a large scale, which, together with the continuous transformation and upgrading of printed electronics technology and equipment, has led to a substantial improvement in the performance of printed carbon nanotube thin-film transistor devices and circuits, and has shown great potential for emerging applications in the Internet of Things, artificial intelligence, wearable electronics and irradiation-resistant They have shown great potential for applications in emerging fields such as the Internet of Things, artificial intelligence, wearable electronics, and irradiation resistance. Despite these unique properties and promising prospects, there are still many challenges in semiconductor carbon nanotube materials to fully exploit the advantages of carbon nanotubes, such as how to reduce traps on the surface of semiconductor carbon nanotubes and how to reduce or remove residual impurities introduced during solution-based binning, thin film preparation and device construction. In terms of electronic device applications, how to prepare large-area, density-tunable semiconductor carbon nanotube films, how to prepare low power consumption, low operating voltage and how to effectively control subthreshold voltage, etc. are the key issues that limit whether printed carbon nanotube thin-film transistor devices can be truly applied to these emerging fields. We believe that with the joint efforts of researchers, printed thin film transistor devices and circuits will shine in these emerging fields.

Acknowledgements. This research is supported by Preparation of high-resolution DNA dielectric film arrays and their application in fully printed flexible memory (04190120001/007), and Beijing Natural Science Foundation (No.2202018), General Project of Science and Technology Plan of Beijing Municipal Education Commission (No.KM202010015004 and No.KM202110015008), National Natural Science Foundation of China (N0.21604005), Basic Research General Project of Beijing Printing Institute (No.Eb202001), Research and development of intelligent packaging solutions for cultural relics (Ed202001), Construction and application transformation of cross-media cloud platform for anti-counterfeit traceability of printing and packaging (27170121005).

References

1. Zhao, C.Y., Zhong, D.L., Han, J., Liu, L.J., Zhang, Z.Y., Peng, L.M.: Exploring the performance limit of carbon nanotube network film field-effect transistors for digital integrated circuit applications. Adv. Funct. Mater. **29**(16) (2019)
2. Yasunaga, Y.: Why and how ITRS worked to recover the breakdown of 'scaling law' in 2000s-structural frame analysis of Si-CMOS semiconductor technologies. IEEE Trans. Eng. Manag. **68**(4), 1179–1194 (2021)
3. Jang, S., Kim, B., Geier, M.L., Hersam, M.C., Dodabalapur, A.: Short channel field-effect-transistors with inkjet-printed semiconducting carbon nanotubes. Small **11**(41), 5505–5509 (2015)
4. Zhao, C.Y., et al.: Strengthened complementary metal-oxide-semiconductor logic for small-band-gap semiconductor-based high-performance and low-power application. ACS Nano **14**(11), 15267–15275 (2020)
5. Qu, M., Li, H., Liu, R., Zhang, S.-L., Qiu, Z.-J.: Interaction of bipolaron with the H2O/O-2 redox couple causes current hysteresis in organic thin-film transistors. Nat. Commun. **5** (2014)
6. Goh, G.L., Saengchairat, N., Agarwala, S., Yeong, W.Y., Tran, T.: Sessile droplets containing carbon nanotubes: a study of evaporation dynamics and CNT alignment for printed electronics. Nanoscale **11**(22), 10603–10614 (2019)
7. Honda, W., Arie, T., Akita, S., Takei, K.: Bendable CMOS digital and analog circuits monolithically integrated with a temperature sensor. Adv. Mater. Technol. **1**(5) (2016)
8. Zou, J., Zhang, Q.: Advances and frontiers in single-walled carbon nanotube electronics. Adv. Sci. **8**(23) (2021)
9. Vidu, R., Rahman, M., Mahmoudi, M., Enachescu, M., Poteca, T.D., Opris, I.: Nanostructures: a platform for brain repair and augmentation. Front. Syst. Neurosci. **8**, 91 (2014)
10. Yao, J., Qiu, S., Jin, H.H., Li, Q.W.: Research on length sorting of semiconducting carbon nanotubes and the performance of thin film transistors. Semicond. Optoelectron. **42**(03), 348–352 (2021)
11. Wei, X.J., Li, S.L., Wang, W.K., Zhang, X., Zhou, W.Y., Xie, S.S., Liu, H.P.: Recent advances in structure separation of single-wall carbon nanotubes and their application in optics, electronics, and optoelectronics. Adv. Sci. **9**(14) (2022)
12. Tang, J., et al.: Flexible CMOS integrated circuits based on carbon nanotubes with sub-10 ns stage delays. Nat. Electron. **1**(3), 191–196 (2018)
13. Novikov, I.V., Khabushev, E.M., Krasnikov, D.V., Bubis, A.V., Goldt, A.E., Shandakov, S.D., Nasibulin, A.G.: Residence time effect on single-walled carbon nanotube synthesis in an aerosol CVD reactor. Chem. Eng. J. **420**(1) (2021)
14. Wang, J.Y., Lei, T.: Enrichment of high-purity large-diameter semiconducting single-walled carbon nanotubes. Nanoscale **14**(4), 1096–1106 (2022)

15. Rezaei, P., Pfefferle, L.D., Frey, D.D.: Metal-semiconductor sorting of large-diameter single-wall carbon nanotubes by pH-dependent binding to a hydrophobic-interaction adsorbent. Carbon **175**, 112–123 (2021)
16. Xu, L., et al.: Enantiomeric separation of semiconducting single-walled carbon nanotubes by acid cleavable chiral polyfluorene. ACS Nano **15**(3), 4699–4709 (2021)
17. Jia, M.L., Yi, C.H., Han, Y.K., Wang, L., Li, X., Xu, G.L., He, K., Li, N.C., Hou, Y.X., Wang, Z.G., Zhu, Y.H., Zhang, Y.A., Hu, M.Z., Sun, R., Tong, P.F., Yang, J.W., Hu, Y.G., Wang, Z.X., Li, W.M., Li, W.J., Wei, L., Yang, C.L., Chen, M.: Hierarchical network enabled flexible textile pressure sensor with ultrabroad response range and high-temperature resistance. Adv. Sci. **9**(14) (2022)
18. Li, Y., Lam, K.H., Hou, X.H.: CNT-modified two-phase manganese hexacyanoferrate as a superior cathode for sodium-ion batteries. Inorg. Chem. Front. **8**(7), 1819–1830 (2021)
19. Wu, Q., Qiu, L., Zhang, L., Liu, H., Ma, R., Xie, P., Liu, R., Hou, P., Ding, F., Liu, C., He, M.: Temperature-dependent selective nucleation of single-walled carbon nanotubes from stabilized catalyst nanoparticles. Chem. Eng. J. **431**(4) (2022)
20. Zhang, S., Kang, L., Wang, X., Tong, L., Yang, L., Wang, Z., Qi, K., Deng, S., Li, Q., Bai, X., Ding, F., Zhang, J.: Arrays of horizontal carbon nanotubes of controlled chirality grown using designed catalysts. Nature **543**(7644), 234–238 (2017)
21. Arnold, M.S., Green, A.A., Hulvat, J.F., Stupp, S.I., Hersam, M.C.: Sorting carbon nanotubes by electronic structure using density differentiation. Nat. Nanotechnol. **1**(1), 60–65 (2006)
22. Yang, D.H., Hu, J.W., Liu, H.P., Li, S.L., Su, W., Li, Q., Zhou, N.G., Wang, Y.C., Zhou, W.Y., Xie, S.S., Kataura, H.: Structure sorting of large-diameter carbon nanotubes by NaOH tuning the interactions between nanotubes and gel. Adv. Funct. Mater. **27**(40) (2017)
23. Flavel, B.S., Moore, K.E., Pfohl, M., Kappes, M.M., Hennrich, F.: Separation of single-walled carbon nanotubes with a gel permeation chromatography system. ACS Nano **8**(2), 1817–1826 (2014)
24. Liu, H.P., Nishide, D., Tanaka, T., Kataura, H.: Large-scale single-chirality separation of single-wall carbon nanotubes by simple gel chromatography. Nat. Commun. **2** (2011)
25. Yomogida, Y., Tanaka, T., Zhang, M., Yudasaka, M., Wei, X., Kataura, H.: Industrial-scale separation of high-purity single-chirality single-wall carbon nanotubes for biological imaging. Nat. Commun. **7** (2016)
26. Gao, W., Xu, W.Y., Ye, J., Liu, T.T., Wang, J.K., Tan, H.W., Lin, Y., Tange, M., Sun, D.F., Wu, L.Z., Okazaki, T., Yang, Y.J., Zhang, Z.Y., Zhao, J.W., Cui, Z., Ma, C.Q.: Selective dispersion of large-diameter semiconducting carbon nanotubes by functionalized conjugated dendritic oligothiophenes for use in printed thin film transistors. Adv. Funct. Mater. **27**(44) (2017)
27. Wang, J., Lei, T.: Separation of semiconducting carbon nanotubes using conjugated polymer wrapping. Polymers **12**(7), 1548 (2020)
28. Yao, J., et al.: Rapid annealing and cooling induced surface cleaning of semiconducting carbon nanotubes for high-performance thin-film transistors. Carbon **184**, 764–771 (2021)
29. Liu, D., Li, P., Yu, X., Gu, J., Han, J., Zhang, S., Li, H., Jin, H., Qiu, S., Li, Q., Zhang, J.: A mixed-extractor strategy for efficient sorting of semiconducting single-walled carbon nanotubes. Adv. Mater. **29**(8) (2017)
30. Lakshmanan, S., Kanwal, A., Liu, S., Patlolla, A., Iqbal, Z., Mitra, S., Thomas, G.A., Fagan, J.A., Farrow, R.C.: Improved electrophoretic deposition of vertical single wall carbon nanotubes with nanoscopic electrostatic lenses. Micromachines **11**(3) (2020)
31. Li, H., et al.: Separation of specific single-enantiomer single-wall carbon nanotubes in the large-diameter regime. ACS Nano **14**(1), 948–963 (2020)
32. Portilla, L., et al.: Ambipolar deep-subthreshold printed-carbon-nanotube transistors for ultralow-voltage and ultralow-power electronics. ACS Nano **14**(10), 14036–14046 (2020)

33. Li, X., Wang, X., Deng, J., Li, M., Shao, S., Zhao, J.: Printed carbon nanotube thin film transistors based on perhydropolysilazane-derived dielectrics for low power flexible electronics. Carbon **191**, 267–276 (2022)
34. Portilla, L., Zhao, J., Zhao, J., Occhipinti, L.G., Pecunia, V.: Ambipolar carbon nanotube transistors with hybrid nanodielectric for low-voltage CMOS-like electronics. Nano Futures **5**(2) (2021)
35. Molina-Lopez, F., et al.: Inkjet-printed stretchable and low voltage synaptic transistor array. Nat. Commun. **10**(1), 2676 (2019)
36. Li, M., Xiong, Z., Shao, S., Shao, L., Han, S.T., Wang, H., Zhao, J.: Multimodal optoelectronic neuromorphic electronics based on lead-free perovskite-mixed carbon nanotubes. Carbon **176**, 592–601 (2021)
37. Huang, W., Jiao, H., Huang, Q., Zhang, J., Zhang, M.: Ultra-high drivability, high-mobility, low-voltage and high-integration intrinsically stretchable transistors. Nanoscale **12**(46), 23546–23555 (2020)
38. Nishio, Y., Hirotani, J., Kishimoto, S., Kataura, H., Ohno, Y.: Low-voltage operable and strain-insensitive stretchable all-carbon nanotube integrated circuits with local strain suppression layer. Adv. Electron. Mater. **7**(1) (2020)
39. Fleetwood, D.M.: Evolution of total ionizing dose effects in MOS devices with Moore's law scaling. IEEE Trans. Nucl. Sci. **65**(8), 1465–1481 (2018)
40. Zhu, M., et al.: Radiation-hardened and repairable integrated circuits based on carbon nanotube transistors with ion gel gates. Nat. Electron. **3**(10), 622–629 (2020)

Simulation of Ink Droplet Spreading Based on XFlow

Li'e Ma(✉), Hongli Xu, Bingbing Hu, Jie Liu, and Qiang Wang

Faculty of Printing, Packaging and Digital Media Engineering, Xi'an University of Technology, Shaanxi, China
malie@xaut.edu.cn

Abstract. Droplet spreading on solid surface is commonly seen in coating, 3D printing, inkjet printing, automobile surface spray decoration and pesticide spraying on crops, etc. In practical applications, droplet spreading, retraction, fragmentation and secondary spreading will occur after the droplet colliding with the solid surface, which are usually affected by the droplet viscosity, velocity, surface tension and other factors. Therefore, the quantitative study of droplet spreading process on the solid surface is of great significance to many industrial fields. Based on the new generation of CFD software XFlow, the mathematical model of ink droplet is established by taking the droplet spreading factor as the research target. Through simulation analysis of the droplet spreading process on solid surface and the flow field, the influence law of viscosity, velocity and surface tension on ink droplet spreading is given. Compared with droplet viscosity and surface tension, droplet impact velocity has greater influence on spreading factor.

Keywords: Ink droplet · Spreading factor · Flow field analysis

1 Introduction

The process of droplet hitting substrate and spreading is widespread in nature such as inkjet printing, 3D printing, spraying, microfluidic technology and so on [1–3]. According to different industry needs, there are different requirements for the spread characteristics. In the coating process represented by inkjet printing, slit coating and drip coating [4–6], it is necessary to ensure that droplets hit the substrate surface with high precision and complete the stable spread of droplets. Wijshoff et al. [7] studied the droplet formation in inkjet printing and droplet dynamics in the process of droplet spreading. Sun et al. [8] studied the spreading characteristics of droplet impact on moving and stationary planes, as well as the phenomenon of droplet deposition on aluminum cooling surface. Huang et al. [9] used lattice Boltzmann simulation method to study and analyzed the influence of the microstructure on the surface of hydrophobic substrate on the wetting of the wall surface. Mehran et al. [10] conducted experiments on droplets of different volumes and the size of the texture surface, and studied the impact dynamics of droplets on the texture surface at different impact speeds. It is of great significance to accurately control the quality of printing products by analyzing the influence of droplet's physical

M. Xu et al. (Eds.): CACPP 2022, LNEE 991, pp. 140–145, 2023.
https://doi.org/10.1007/978-981-19-9024-3_19

characteristics on the spreading result of the substrate. In this paper, the spread factor is introduced to characterize the spread degree of droplet. Simulation is carried out by fluid simulation software XFlow, and the influence law of droplet viscosity, velocity and surface tension on the spread factor and the time for droplet spreading to reach stability are obtained.

2 Governing Equation

The governing equations involved in this paper are mass conservation and momentum conservation equation. Since momentum is a vector, the momentum conservation equation can be decomposed into three coordinate axes for three-dimensional problems to obtain the change law of velocity field and pressure field. The mathematical expression of the mass conservation equation [11] is:

$$\frac{\partial \overline{p}}{\partial t} + \nabla \cdot (\overline{\rho}\vec{v}) = 0 \tag{1}$$

where $\overline{\rho}$, $\vec{v}$ respectively represent the average density and motion velocity of each phase.

The mathematical expression of momentum conservation equation [11] is:

$$\frac{\partial \overline{\rho}\vec{v}}{\partial t} + \nabla \cdot (\overline{\rho}\vec{v}\vec{v}) = \overline{\rho}\vec{g} - \nabla p + \nabla\left(-\frac{2}{3}\mu\nabla \cdot \overline{v}\right) + \nabla \cdot (2\mu S) \tag{2}$$

where $\vec{g}$, p, μ and S respectively represent gravitational acceleration, pressure, dynamic viscosity and deformation rate tensor.

3 Fluid Simulation and Analysis

3.1 Flow Field Analysis

Droplet Spreading Process. The moment when the droplet touches the dry solid wall is defined as the initial point 0μs, and the 500μs is defined as the end point. The point 0–500μs is divided into six equal parts for analysis. As can be seen from Fig. 1, the droplet spreads and diffuses on the wall surface, presenting an axisymmetric distribution, and stops moving when it reaches the maximum spreading diameter. As time goes on, the droplet spreading speed is slower and slower because the impacting kinetic energy decreases under the dissipation of viscous force and surface tension. When inertia force, viscous force and surface tension reach equilibrium, the droplet reaches the maximum spread diameter and remains unchanged.

Fluid Velocity Field Analysis. As can be seen from Fig. 2, the velocity field presents an axisymmetric distribution, and as time goes by, the intensity of the velocity field gradually weakens, and the droplet neck has a large injection velocity. The kinetic energy of impact is constantly consumed by viscous force and surface tension, resulting in a continuous decrease of kinetic energy of impact, that is, the droplet velocity decreases. It is considered that the larger neck jet velocity is one of the reasons for jet formation.

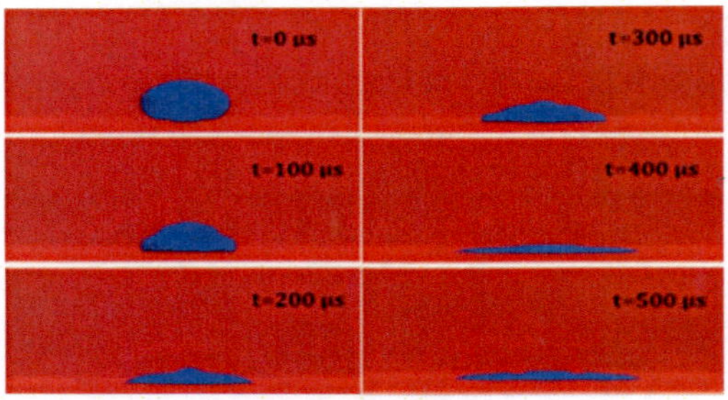

(a) (b)

Fig. 1. Spreading process of ink droplets on dry wall surface

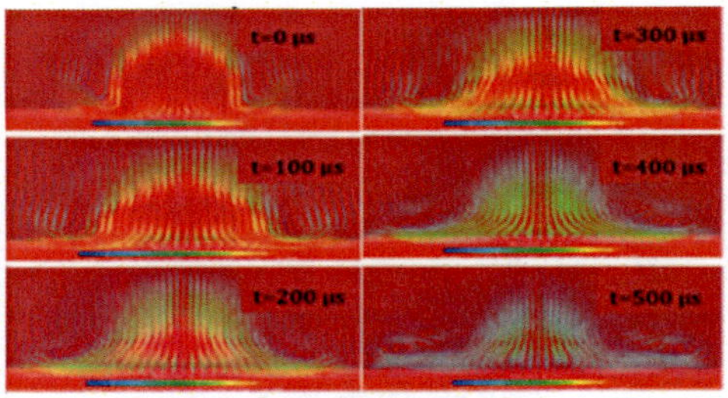

(a) (b)

Fig. 2. Velocity field of ink droplet spreading on dry wall surface

Fluid Pressure Field Analysis. As can be seen from Fig. 3, the pressure field formed by droplet impacting the dry wall presents an axisymmetric distribution, and the strength of the pressure field gradually decreases as time goes by, which is caused by the viscous force between the droplet and the wall and the dissipation of the surface tension of the droplet. At the initial stage of impact, the maximum pressure appears in the impact center, and then gradually shifts to the neck of the droplet. Because of this high pressure difference the droplet breaks through the action of surface tension and viscous force, produces jet flow, and further generates secondary droplet.

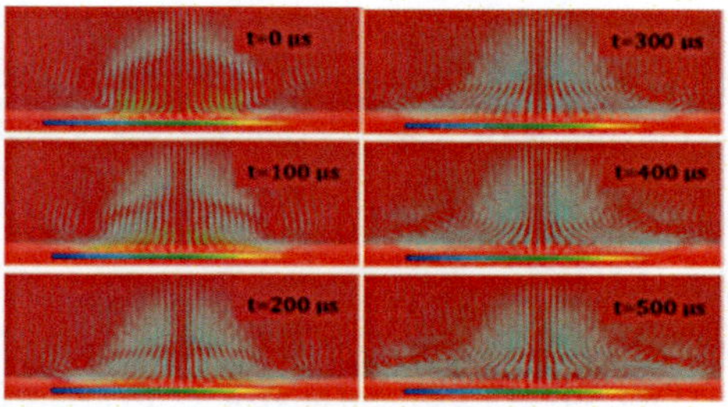

(a) (b)

Fig. 3. Pressure field of ink droplet spreading on dry wall surface

3.2 Influence of Different Parameters on Spreading Result

In order to study the influence of different parameters on spreading factor, the following specific parameter is changed with other parameters unchanged.

Effect of Droplet Viscosity. The spreading factors varing with time under different viscosity (0.002, 0.003, 0.010 Pa•s) are shown in Fig. 4.

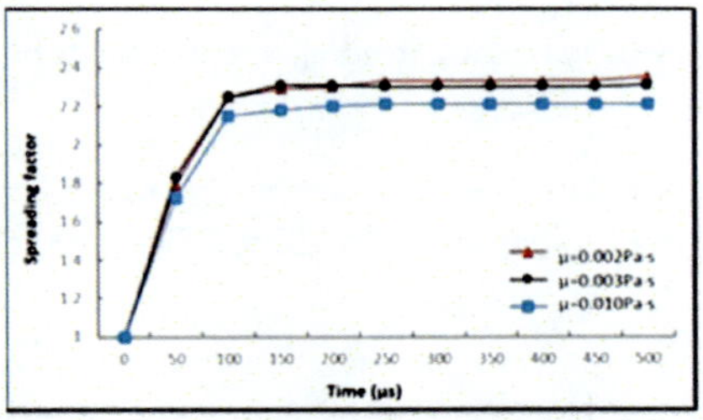

Fig. 4. Effect of viscosity on the spreading of ink droplets on the drying wall

As can be seen from Fig. 4, the spreading factor (Ratio of droplet spread diameter to droplet diameter) increases gradually over time, but the increasing speed decreases gradually. With the increase of the coefficient of motion viscosity, the spreading factor decreases gradually, and the time to reach the maximum spreading factor increases correspondingly. At the initial stage of impact, the impact kinetic energy plays a dominant role, leading to a rapid increase in the spread factor. However, the impact kinetic energy decreases continuously under the dissipation of viscous force, surface tension and gravity. At the later stage of impact, the viscous force plays a dominant role, leading to a gradual decrease in the spread rate. The larger the viscosity coefficient of the droplet is, the stronger the viscous dissipation effect between the droplet and the surface is. When the impact velocity of the droplet is unchanged, that is, the impact kinetic energy is unchanged, the viscous dissipation increases, and the energy used for spreading of the droplet decreases, leading to the gradual decrease of spreading factor and the corresponding increase of the time to reach the maximum spreading factor.

Effect of Droplet Velocity. The spreading factors varing with time under different impact velocity U (1, 2, 3m/s) are shown in Fig. 5.

As can be seen from Fig. 5, within the research scope of this section, the spreading factor increases gradually with the increase of droplet impact velocity. When the droplet impact velocity increases, the impact kinetic energy increases. Although the energy dissipation also increases at this time, the influence of the increase of kinetic energy plays a dominant role, making the droplet has more energy for spreading.

Therefore, the spreading factor increases, and the maximum spreading factor also increases accordingly. In addition, the spreading factor increases gradually with the passage of time, but the increasing speed gradually decreases, because the viscous force plays a dominant role in the late impact period, and the spreading gradually tends to be stable with the increase of viscous dissipation.

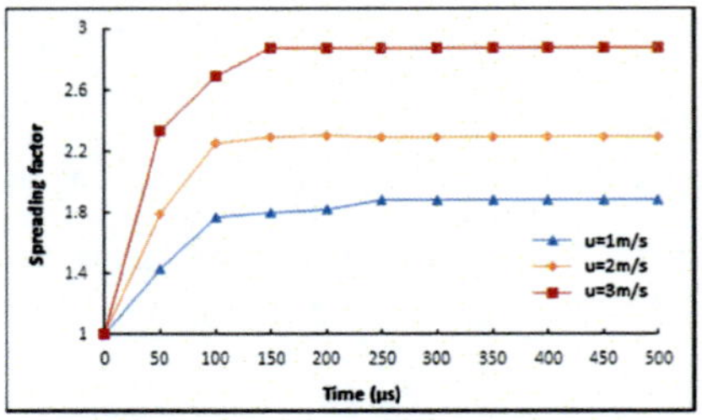

Fig. 5. Effect of velocity on the spreading of ink droplets on the drying wall

Effect of Droplet Surface Tension. The spreading factors varing with time under different surface tension coefficient σ (0.01, 0.025, 0.04N/m) are shown in Fig. 6.

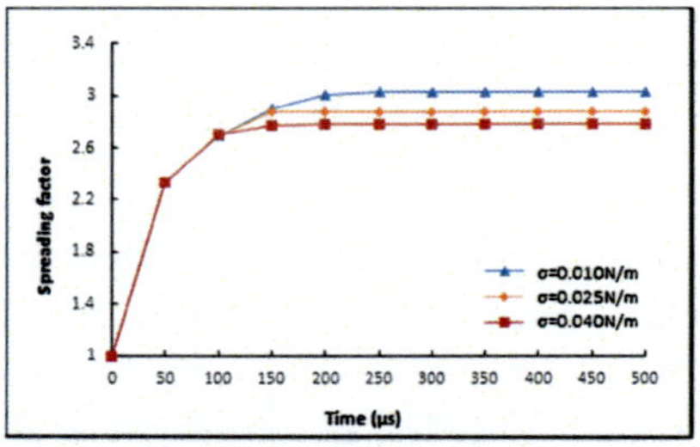

Fig. 6. Effect of surface tension on the spreading of ink droplets on the drying wall

As can be seen from Fig. 6, within the scope of study in this section, the spread factor gradually increases with the progress of time. At the initial stage of impact, the spread factor increases at a faster rate, while at the later stage it is slower. The reasons have been explained in Sect. 3.2. The spreading factor decreases with the increase of surface tension coefficient. When the surface tension coefficient is increased, the more surface tension the droplet needs to overcome to spread forward, and when the impact kinetic energy is unchanged, the greater the surface tension, the more severe the dissipation, resulting in the decrease of the energy used for droplet spread, the spread factor and the maximum spread diameter.

4 Conclusions

In this paper, XFlow fluid simulation software was used to conduct numerical research on the movement process of droplet impacting on the dry wall, and the influence law of droplet viscosity, speed, surface tension and other parameters on spreading factor are studied, and the conclusions are as follows.

The velocity field, pressure field and vorticity field are symmetrically distributed, and the field strength decreases gradually with the passing of time due to the dissipation of viscous force and surface tension. At the initial stage of impact, the maximum pressure appears in the impact center, and then transfers to the droplet neck. The pressure difference formed in the neck is the main reason for jet formation. Two symmetrical vortices are formed at the spreading edge of the droplet, which is conducive to the formation of secondary droplet.

In this study, with the increase of droplet viscosity and surface tension, the spread factor gradually decreases, and the time to reach the maximum spread factor increases accordingly. With the increase of droplet impact velocity, the spread factor increases gradually, and the time to reach the maximum spread factor increases correspondingly. Compared with droplet viscosity and surface tension, droplet impact velocity has greater influence on spreading factor.

Acknowledgements. This research is supported by the Technology Innovation Leading Program of Shaanxi Province (No.2020QFY03–04).

References

1. Van Dam, D.B., Le Clerc, C.: Experimental study of the impact of an ink-jet printed droplet on a solid substrate. Phys. Fluids **16**(9), 3403–3414 (2004)
2. Horacek, B., Kiger, K.T., Kim, J.: Single nozzle spray cooling heat transfer mechanisms. Int. J. Heat Mass Transf. **48**(8), 1425–1438 (2005)
3. Liang, X., Zhang, J.Y.Z., Li, M., Wang, K., Luo, G.S.: Dynamic interfacial tension and adsorption kinetics of nonionic surfactants during microfluidic droplet formation process. Chem. Eng. J. **445**, 136658 (2022)
4. Seung, H.K., San, K., Jong, W.L., Dong, K.S., Ko, H.S.: Study on fall velocity of continuously ejected micro inkjet droplet. J. Mech. Sci. Technol. **34**(8), 1–5 (2020)
5. Akbarzadeh, V., Hrymak, A.N.: Coupled fluid-particle modeling of a slot die coating system. AIChE J. **62**(6), 1933–1939 (2016)
6. Ryu, J., Lee, K., Yun, J., Yu, H., Lee, J., Jang, J.: Paintable carbon-based perovskite solar cells with engineered perovskite/carbon interface using carbon nanotubes dripping method. Small **13**(38), 1701225 (2017)
7. Wijshoff, H.: Drop dynamics in the inkjet printing process. Curr. Opin. Colloid Interface Sci. **36**, 20–27 (2017)
8. Sun, Z.C., Xu, J., Wu, T.Y., Li, D., Qian, C.L., Zhao, X.B.: Freezing deposition characteristics of liquid droplets impinging on cold aluminum surface. J. Eng. Thermophys. **39**(08), 1780–1785 (2018)
9. Huang, Q.G., Pan, G.: Numerical simulation of wettability of hydrophobic surface based on lattice Boltzmann method. Funct. Mater. **46**(10), 10023–10028 (2015)
10. Mehran, A., Reza, M.: Droplet impact on superhydrophobic surfaces fully decorated with cylindrical macrotextures. J. Colloid Interface Sci. **509**, 422–431 (2018)
11. Zou, G.W., H,Z., G,X.: National Defense Industry Press, 128–134 (2013)

Structure Design of Braille Puzzles Based on 3D Printing Technology

Chunmei Li(✉)

Department of Graphic Communication, Shanghai Publishing and Printing College, Shanghai, China
1410625618@qq.com

Abstract. There are so many researchers who studied how to help blind people to recognize the world using Braille produced by 3D printing technology. The Braille puzzles based on 3D printing technology are convenient for the beginners to learn Braille writing and can be used repeatedly. But spelling with 3D-printed blind spots isn't as easy as it is for the rest of the population because of the limited vision of blind people. So it is necessary to design a suit of convenient Braille puzzle board. Based on the principle of 3D printed Braille that it should be used easily by the blind people, the structures of raised dots and flat dots which are easy to be assembled have been designed in this paper. A storage board is also designed to accommodate these dots and at the same time these dots are easily in and out of the board. Three kinds of single-word board structures are listed and the path of spelling the braille dots is planned, which is convenient for the blind people to complete the fast spelling. The optimal structure and path were selected and printed by photosensitive resin. The experimental results show that the designed structure can improve the speed and accuracy of Braille spelling when using these Braille puzzles.

Keyword: 3D printing · Braille puzzles · Structure design

1 Introduction

Braille is a very important tool for the visually impaired people to acquire knowledges from the outside world or to express opinions to others. Many researchers are working hard to help the blind people to communicate with other people. In recent year 3D printing technology is well developed and it can satisfy the requirement of Braille dots perfectly. 3D printing technology has brought great convenience to the blind people. For example, Braille maps [1], medical tablets [2], Braille-encoded Intraoral Films [3], keyboards, books, labels, etc. can be printed by 3D printer. Most researches have focused on how to print [4, 5] or recognize [6] or display [7, 8] Braille. But the most important thing for visually impaired people is to learn how to read and write the Braille.

There are many defects of the traditional handwritten braille: the rules of reading and writing are not consistent; the order of reading and writing is not consistent; reading and writing cannot be done at the same time. In [9] a new design to study Braille by 3D

M. Xu et al. (Eds.): CACPP 2022, LNEE 991, pp. 146–152, 2023.
https://doi.org/10.1007/978-981-19-9024-3_20

printing puzzle which can be used repeatedly is approached. These 3D printing puzzles are very convenient for teachers to teach Braille for the blind people. But it is difficult to the blind people to put the Braille dots on the correct place.

So it is necessary to design a suit of Braille puzzle board which is easy for the blind people to put the dots on the exact location. In this paper the structures of raised dots and flat dots have been designed. A storage board is also designed to accommodate these dots and at the same time these dots are easily in and out of the board. Three kinds of the structure of a cell box are listed and the path of spelling the braille dots is planned, which is convenient for the blind people to complete the fast spelling. The optimal structure and path were selected and printed by photosensitive resin. The experimental results show that the designed structure can improve the speed and accuracy of Braille spelling by using these Braille puzzles.

2 Methods

2.1 Structures of Raised Dots and Flat Dots

A braille cell has six raised or flat dots in a systematic arrangement with two columns of three lines as shown in Fig. 1. The standard parameters of Chinese Braille [10] are shown in Table 1. Each Chinese character can be composed by three cells. Each cell can be spelled with six parts which have raised dot or not.

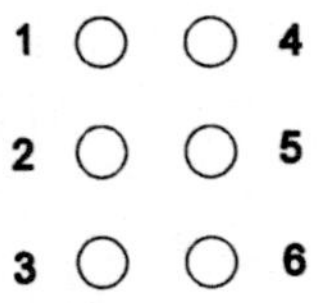

Fig. 1. Six-dot cell

Table 1. Standard parameters of Chinese Braille

Parameters	Dimensions (mm)
Dot height	0.2 ~ 0.5
Diameter of dot	1 ~ 1.6
Dot spacing	2.2 ~ 2.8
Cell spacing	3.5 ~ 4.0
Row spacing	5 ~ 6

Because the raised or flat dots are highly repeatable a new 3D printing Braille puzzle is approached in [9]. These 3D printing puzzles are very convenient for teachers to teach Braille for the blind people because teachers can see the dots and boxes and arrange them in rules. But it is difficult to the blind people to pick up the Braille dots and then put them into the cell box on the correct place. In this paper the structures of the raised dots and flat dots are improved as shown in Figs. 2 and 3. The cylinder is usable to make the dot passing through smoothly, and the cube under a Braille dot are used to orient to the correct position in the cell box which will be shown in Sect. 2.2.

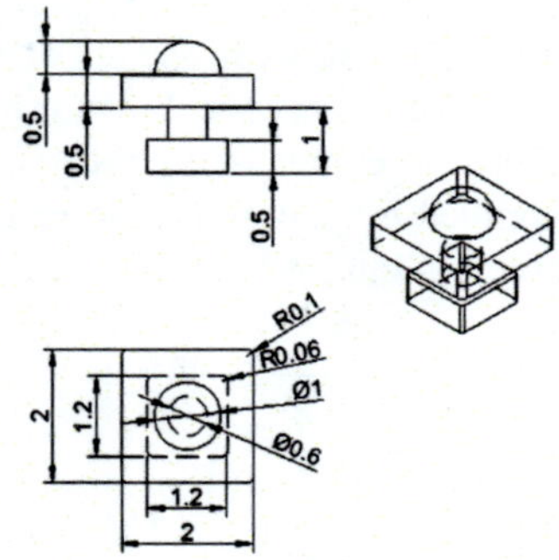

Fig. 2. Structure of a raised dot

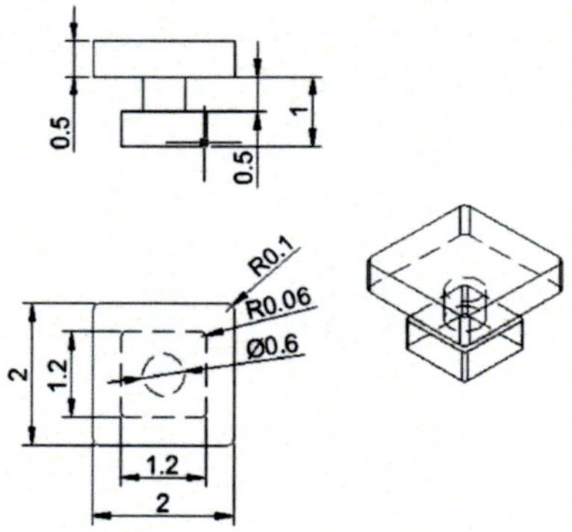

Fig. 3. Structure of a flat part

2.2 Structure of a Cell Box

Each cell box can be divided into six parts as shown in Fig. 4(a). So that the blind people can put raised dots or flat dots in a cell box. As the Fig. 4(b), (c), (d) shown there are three kinds of routes designed in this paper.

The red lines divide the box into six parts. The blue arrow is the direction of the road that a dot gets into the box. There have three entrances in route(d) and only one entrance in route(b) and route(c). The six dots are put into the box from the entrance one by one and the sequence list as in Table 2. One step is defined as half the width of a dot. Then the steps of each dot should to walk to the correct position can be calculated. So the total steps of six dots of each route are shown in Table 2. Route(d) has the minimum number of steps but it has three entrances. The more entrances of a route have the more difficult alignment to blind people. So Route(d) is abandoned. The total steps of route(b) are less than the total steps of route(c). But the sequence of route(c) is exactly alike the order of the writing of blind people. So the route(c) is chosen as the route of the final cell box.

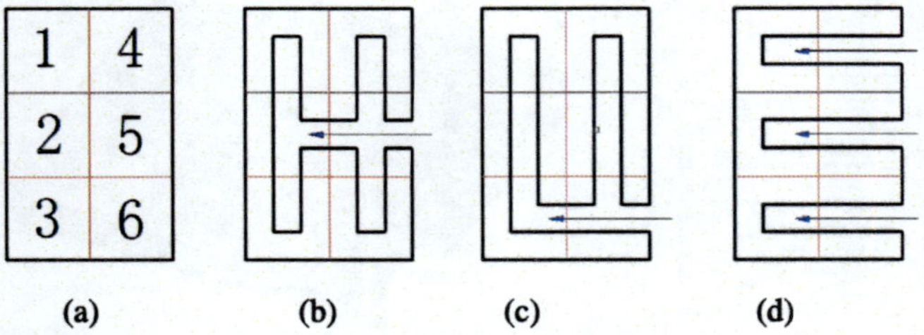

Fig. 4. Three routes of the spelling methods

Table 2. Comparison of three Routes

	Route(b)	Route(c)	Route(d)
Sequence list	1-3-2-4-6-5	1-2-3-4-5-6	1-4-2-5-3-6
Total steps	20	24	12
Entrance number	1	1	3

According the size of a dot in Figs. 2 and 3 the structure of a cell box of route(c) is designed as shown in Fig. 5. There have small gaps between the cell box and dots so that the dots can be moved smoothly in the box.

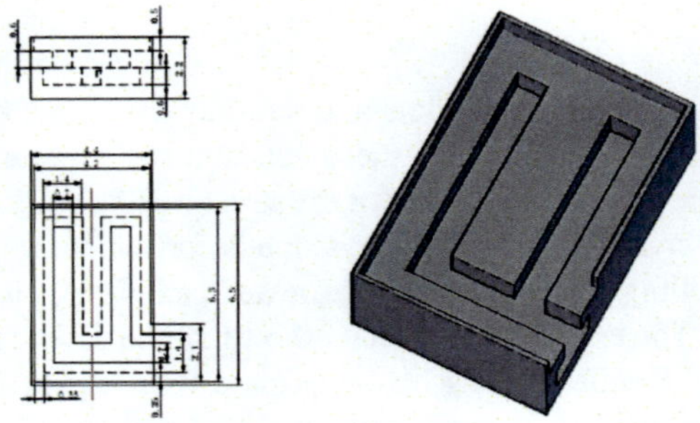

Fig. 5. Structure of a cell box

2.3 Structure of a Storage Board

The blind people can't see the dots. It is convenient to the blind people if the dots are regularly arranged in a special board. So the structure of a storage board is designed as shown in Fig. 6. It has 9 columns and 9 lines. It can storage 72 dots. The board has an outlet for docking with a cell box. And the structure of the board is similar to the structure of the cell box.

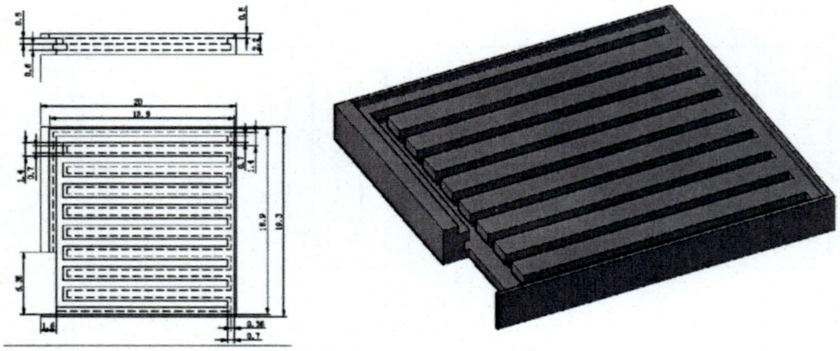

Fig. 6. Structure of a storage board

3 Experimental Results

The raised dots, flat dots, the cell box and the storage board were printed by photosensitive resin which is the suitable material that has been proved in [9]. The Layer thickness is 0.014mm. The cost is 83 YUAN with 1 storage board, 2 cell boxes, 15 raised dots and 15 flat dots. Since the size of the traditional Braille is too small to be picked up so the four times the size of the dots are printed to test. The time spend from Fig. 7(a)–(b) is 30.2s. The time spend from Fig. 7(c)–(d) is 11.5s. The experimental results show that the improved structures of Braille puzzles can increase the speed of Braille spelling.

4 Conclusions

In this paper the structures of raised dots and flat dots have been designed. A storage board is also designed to accommodate these dots and at the same time these dots are easily in and out of the board. Three kinds of the structure of a cell box are listed and the route of spelling the braille dots is planned, which is convenient for the blind people to complete the fast spelling. The optimal structure and path were selected and printed by photosensitive resin. The experimental results show that the new designed structure can improve the speed of Braille spelling by using these Braille puzzles. In the future the new Braille puzzles may be tried to popularize in the special school for the blind people.

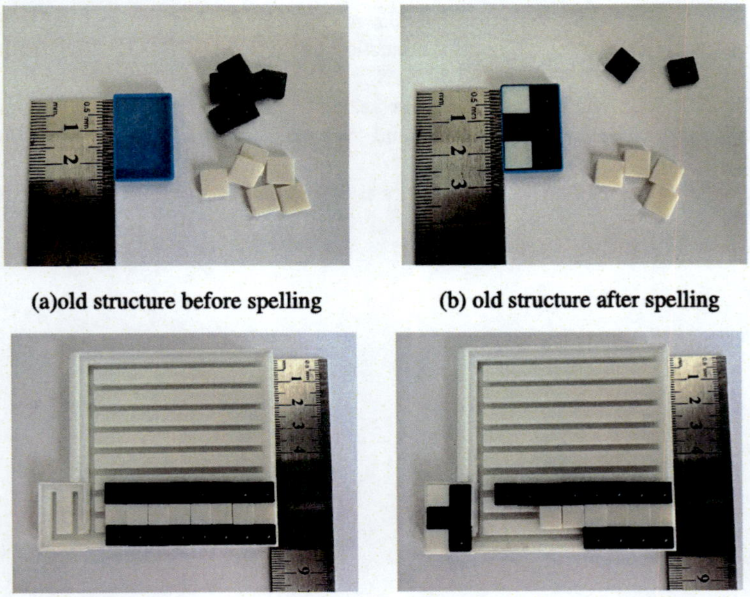

(a)old structure before spelling (b) old structure after spelling

(c) new structure before spelling (d) new structure after spelling

Fig. 7. Experimental results

Acknowledgements. This study is supported by Key Lab of Intelligent and Green Flexographic Printing (KLIGFP-03).

References

1. Holloway, L., Marriott, K., Butler, M. et al.: Accessible maps for the blind: comparing 3D printed models with tactile graphics. In: Proceedings of the 2018 CHI Conference on Human Factors in Computing Systems, vol. 198 (2018)
2. Awad, A., et al.: 3D printed tablets (printlets) with braille and moon patterns for visually impaired patients. Pharmaceutics **12**, 172 (2020). https://doi.org/10.3390/pharmaceutics12020172
3. Eleftheriadis, G.K., Fatouros, D.G.: Haptic evaluation of 3D-printed braille-encoded intraoral films. Eur. J. Pharm. Sci. **157**, 105605 (2021)
4. Loconsole, C. et al.: An experimental study on fused-deposition-modeling technology as an alternative method for low-cost braille printing. In: Advances in Intelligent Systems and Computing, vol. 500 (2016). https://doi.org/10.1007/978-3-319-41962-6_18
5. Jo, W. et al.: 3D printed tactile pattern formation on paper with thermal reflow method. R. Soc. Chem. **2014**(4), 31764–31770 (2014)
6. Samonte, M.J., Laurente, E.D. Magno, K.M., Perez, C.: Braille3D: using haptic and voice feedback for braille recognition and 3D printing for the blind. IOP Conf. Ser.: Mater. Sci. Eng. (2019). https://doi.org/10.1088/1757-899x/482/1/012027
7. Yang, W. et al.: A survey on tactile displays for visually impaired people. IEEE Trans. Haptics (2021). https://doi.org/10.1109/TOH.2021.3085915

8. Aryan, R.E., Doshi, S.: Refreshable braille module using CAM actuated mechanism. In: International Conference on Design, Automation, and Control (ICDAC 2020) (2020). https://doi.org/10.1088/1757-899X/1123/1/012028
9. Li, C., Zheng, L., Xiao, Y.: Application research of 3D printing technology in braille. In: Interdisciplinary Research for Printing and Packaging, pp. 207–213. Springer, Singapore (2022)
10. Chinese Braille: GB/T 15720-2008. National Standards of the People's Republic of China (2008)

Study on Morphology of Quantum Dots Films Prepared by Inkjet Printing

Linhong Huang, Wan Zhang, Xianfu Wei, Yongjian Wu, and Beiqing Huang(✉)

School of Printing and Packaging Engineering, Beijing Institute of Graphic Communication, Beijing, China
571478757@qq.com

Abstract. In this research, the quantum dot films, which were improved by solvent selecting and ratio adjusting, controlling the droplet spacing and inkjet voltage for the coffee ring suppression, were fabricated using inkjet printing technology. As a result, the coffee ring was eliminated by mixing three organic solvents and adjusting the ratios between three solvents with different boiling points. Therefore, the coffee ring phenomenon was reduced when the volume ratio of cyclohexylbenzene/butyl benzoate/n-octane in the QD ink was 7: 3: 5. The uniformity and completeness of the film surface were developed by controlling the droplet spacing. When the droplet spacing was 60 μm, the film surface was more uniform and complete. To improve the film surface, the control of the inkjet voltage regulated the droplet injection and droplet shape. When the inkjet voltage was 15 V, not only was the inkjet effect more stable, but the ejected dropleshape was complete. The research of quantum dot films provides technical support for the preparation of QLED and QLED displays.

Keywords: Quantum dot film · Inkjet printing · Solvent ratio · Droplet spacing · Inkjet voltage

1 Introduction

Quantum dot (QD) is widely concerned and developed in the display industry, owing to outstanding luminous efficiency, stable optical property, excellent color purity and convenient preparation process [1]. Inkjet printing technology provides more possibility for the preparation of Quantum dot light-emitting diode (QLED). Industrialization of large area QLED displays would be realized by inkjet printing technology in the future [2, 3].

However, as the light-emitting layer of QLED, the preparation process and the uniformity of QD films are still facing great challenges. Firstly, QD ink should be successfully sprayed from a nozzle to form a stable droplet, which then falls on the printed substrate to form QD films with a uniform surface. The injection of QD ink was evaluated by the Reynolds number Re, the Weber number We, and the inverse Ohnesorge number Z [4]. It is necessary to consider not only the solubility of solvent in QD materials, but rheological parameters such as viscosity and surface tension for the preparation of QD

M. Xu et al. (Eds.): CACPP 2022, LNEE 991, pp. 153–162, 2023.
https://doi.org/10.1007/978-981-19-9024-3_21

ink [5]. Secondly, the printing parameters such as droplet size, droplet spacing, printing voltage, inkjet height, temperature and printing environment will affect the apparent quality of QD films, thus affecting the work performance of QLED. Therefore, it is important to select the appropriate parameters for inkjet printing to achieve uniform QD films and suppress the coffee ring effect [6, 7]. Furthermore, the morphology of QD films would affect the hole transport layer (HTL) [8] and the efficiency of QLED. In conclusion, the preparation process and the morphology of QD films have been explored more intensively to expand the application of wide area QLED displays.

2 Experimental Section

2.1 Material and Instrument

The red (CdSe/CdS/ZnS) QDs, octane, n-butyl benzoate, cyclohexyl benzene(CHB), Fuljifilm DMP2800 inkjet printer, Confocal laser scanning microscope(CLSM), AFM((Bruker Multimode 8), surface tension meter(KRUSS k100), rheogoniometer(AR2000ex), magnetic mixer(HJ-6).

2.2 QD Ink Preparation

The QDs with the diameter of 2–10 nm were purchased from Beida Jubang Science & Technology, which has great luminous efficiency of above 85%. The QDs were dispersed in CHB: n-butylbenzoate: octane at a concentration of 8 mg/ml to form the ink.

The QD ink was printed on the ITO substrate by inkjet printer, followed by annealing at 100 °C for 10 min to form QD films.

The surface morphology of QD films was characterized using CLSM and AFM.

3 Result and Discussion

3.1 Effect of Solvent Ratio

QD ink was prepared by mixing two kinds of high boiling point solvents and low boiling point solvent. The high boiling point solvents, n-butylbenzoate and CHB, were chosen to adjust the surface tension and viscosity of QD ink for the stability of inkjet printing. The low boiling point solvent, octane, was chosen as a excellent solvent to dissolve the QDs. The physical properties of solvents are tabulated in Table 1.

The optimal proportion of QD ink was produced by adjusting the ratio of three solvents. Table 2(a) presents the experimental parameters of varying the ratio of high boiling point solvents and fixing the ratio of low boiling point solvent. It can be seen from Fig. 1 (a) that when the ratio of CHB and n-butyl benzoate was changed, the surface tension of QD ink decreased with the increasing ratio of n-butyl benzoate. To determine the uniformity of the QD films, the equation:

$$H = H1 / H2 (\%) \quad (1)$$

Table 1. Physical properties of cyclohexyl benzene, n-butylbenzoate and octane

Solvent	Boiling point/°C	Density/(g*cm^{-3})	Surface tension/(mN*m^{-1})	Viscosity/cp
Cyclohexyl benzene(CHB)	240	0.95	34.5	2.70
n-Butyl benzoate	250	1.01	30.2	4.25
Octane	26	0.70	20.9	0.52

It is used to describe the films morphology [9]. We assume that H represents the uniformity of the film (%), and H1 is the height from CLSM to film edge, H2 is the height from CLSM to the middle of the film. The closer the H value is to 1, the better the uniformity of the film is. Table 2 (a) shows that when the ratio of cyclohexylbenzene: n-butyl benzoate: n-octane is 7: 3: 5, the film uniformity is 96.03%, indicating that QD film of this ratio is uniform. Table 2 (b) shows the parameters for high boiling point solvents of 7: 3, changing the ratio of low boiling point solvents. With the addition of octane, the surface tension of QD ink and the H value decreased continuously, indicating that the uniformity of the QD films is affected by octane and the surface of the films becomes more uneven, as shown in Figs. 1 (b) and 2 (b).

Table 2. Surface tension and uniformity H of solvents with different ratios

(a)					
CHB:n-Butyl benzoate:Octane	9:1:5	8:2:5	7:3:5	6:4:5	5:5:5
Surface tension/(mN*m^{-1})	27.05	26.99	26.89	26.71	26.09
H/%	89.88	89.55	96.03	95.88	92.42
(b)					
CHB:n-Butyl benzoate:Octane	7:3:5	7:3:6	7:3:7	7:3:5	
Surface tension/(mN*m^{-1})	26.89	26.33	26.09	25.63	
H/%	96.03	96.84	93.27	83.22	

From Fig. 3(a), it can be seen that the ratio of n-butyl benzoate increased while the ratio of octane remained unchanged, the surface of the quantum dot film became uneven, indicating a decline in the integrity of the quantum dots. With the addition of n-butyl benzoate, the aggregation of QDs on the film surface is also noticeable. Therefore, QD films print more effectively in the ratios of 9: 1: 5, 8: 2: 5, and 7: 3: 5. When the ratio is 7: 3: 5, the edge of the QD film is the smallest, indicating that the coffee ring effect is the weakest.

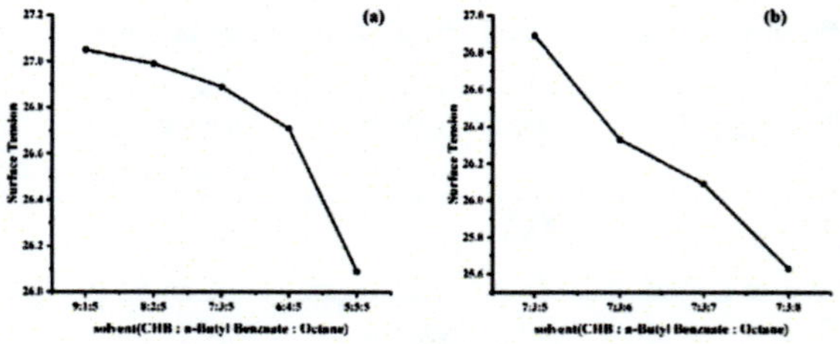

Fig. 1. Surface tension of QD ink with different solvent ratios

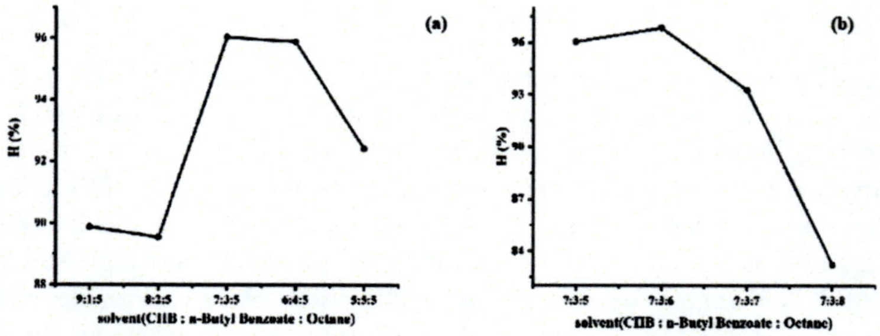

Fig. 2. H value of QD ink with different solvent ratios

Figure 3(b) shows that as the ratio of octane increases, the edges of the QD films thicken and the coffee ring effect is evident. QD films with ratios of 7: 3: 5 and 7: 3: 6 look better with more prominent film edges and a weaker coffee ring effect.

Fig. 3. Morphology of QD films at different ratios

The surface morphology of these three proportions of QD films was observed by AFM. Figure 4 shows that when the solvent ratio was 7: 3: 5, the roughness of the quantum dot film was 3.10 nm. However, when the ratio of CHB and n-butyl benzoate was changed to 9: 1, it showed a faster rate of evaporation, which prevented QDs from reaching the edge when moving from center to edge. This can be attributed to the fact that the boiling point of CHB is lower than n-butyl benzoate when the proportion of CHB increases. This increases the roughness, ra = 5.38 nm. As the volume of octane increases, the evaporation rate of octane becomes faster because of its low boiling point and surface tension. At the same time, with the increase of difference in surface tension, the evaporation rate of the droplet edge is increased, leading to QDs remaining on the edge and the edge width becomes thicker. Hence, when the ratio was 7: 3: 6, the roughness increased, ra = 4.30 nm.

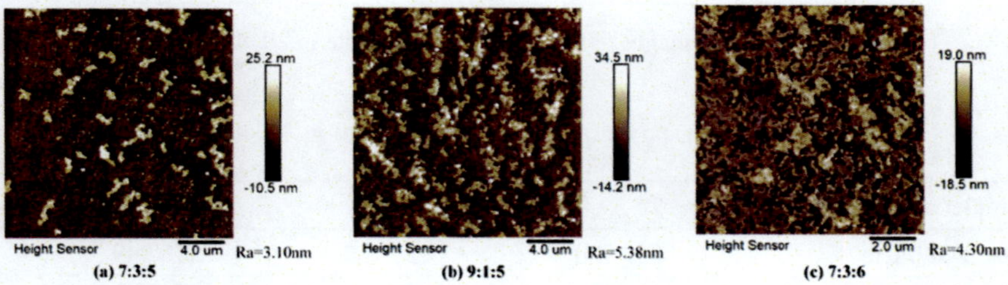

Fig. 4. AFM morphology of quantum dot films with ratios of 7: 3: 5, 9: 1: 5, 7: 3: 6

3.2 Effect of Droplet Spacing

The droplet spacing is the distance between the centers of the drops, as shown in Fig. 5.

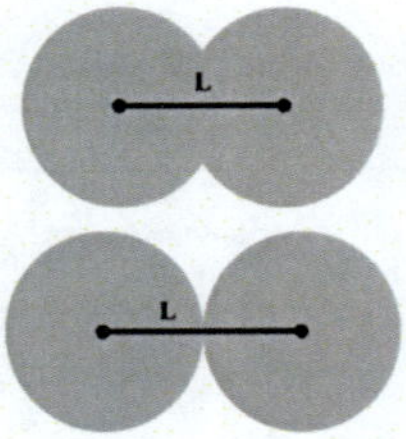

Fig. 5. Drop spacing image

Figure 6 shows the microscope image of the QD line films with the ratio of 7: 3: 5 observed at the spacing of 20 μm, 40 μm and 60 μm. As shown in Fig. 6 and Table 3, the effect of the coffee ring on the edge of the line film was evident when the spacing was 20 μm, the line width was 219.700 μm, and the H was 95.39%. When the spacing was 40 μm, the edge of the line film became thinner with a faint coffee ring, the line width was 133.182 μm, and the H was 99.11%. When the spacing increased to 60 μm, the coffee ring was almost invisible near the edge of the line film, indicating that the coffee ring phenomenon was effectively suppressed, at this time the line width was 118.949 μm, and the H was 96.05%.

The experiment with line films at 20, 40, and 60 μm shows that as the droplet spacing increases, the line films have a weaker coffee ring and a narrower line width. Figure 7 shows that when the spacing was between 20 and 60 μm, the line width decreased as the

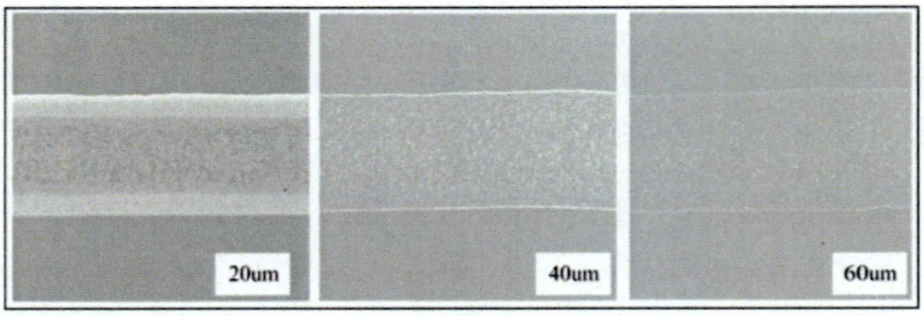

Fig. 6. Microscope images of the films of the QD line in 20, 40 and 60 μm

Table 3. Line width and uniformity of QD line films in 20, 40 and 60 μm

Droplet spacing/μm	20	40	60
Line width/μm	219.700	133.182	118.949
H/%	95.39	99.11	96.05

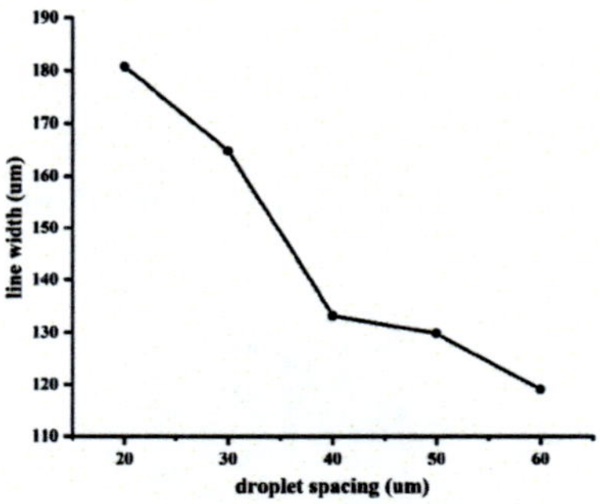

Fig. 7. Line width images with different droplet spacing

droplet spacing increased. Consequently, it is necessary not only to adjust the droplet spacing to improve the uniformity of the QD films, but also to obtain a thin line film to ensure uniform light emission performance of the device in the manufacture of QLED.

Figure 8 shows that when the spacing was adjusted to 80 and 100 μm, the line film began to lose its integrity. When the spacing was 80 μm, the line film appeared droplet-shaped, with both sides of the edge wavy. When the spacing was 100 μm, the entire droplet shapes and wide spacings can be clearly seen. Here the line film was deformed on both sides of the edge, which was no longer complete. Therefore, when adjusting the spacing to achieve the best QD line films, in addition to the completeness of the line shape, the droplet spacing adjustment range for the uniformity of the QD films and the line width setting must also be taken into account.

3.3 Effect of Inkjet Voltage

The voltage of inkjet printing affects the droplet injection [10] and the shape of the QD line film. The printing voltage was set to 15, 18, 21 and 24 V. Injection of droplets of

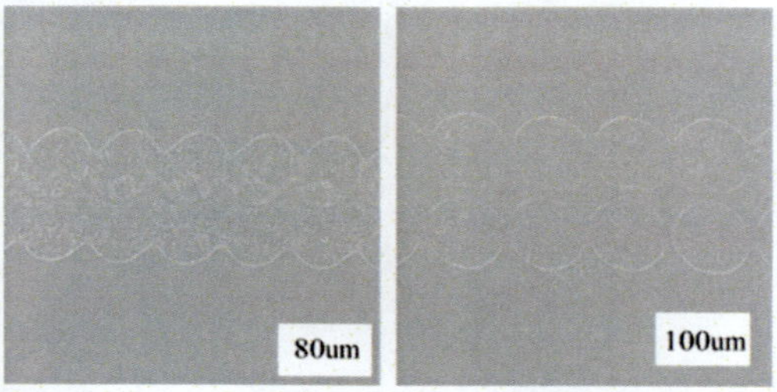

Fig. 8. Images of line films with spacing of 80 μm and 100 μm

various voltages was observed by the camera system of the Fuljifilm DMP2800 inkjet printer.

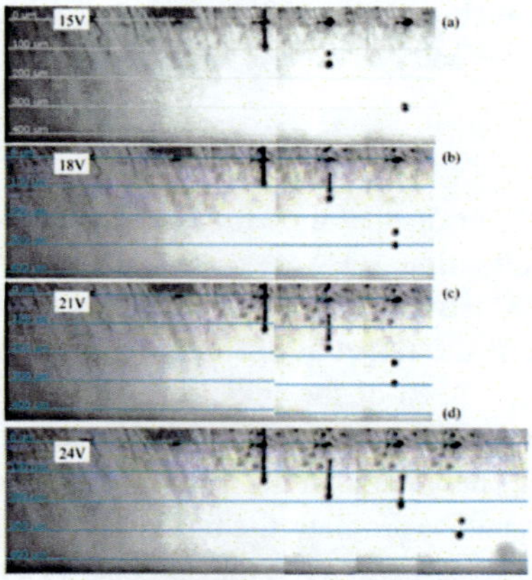

Fig. 9. Injection of droplets with different inkjet voltages

Table 4. Injection time and velocity of the different inkjet voltages

Voltage /(V)	15V	18V	21V	24V
Time/(s)	1.75	1.70	1.57	1.54
v/(μm/s)	114.29	117.65	127.39	129.87

Figure 9(a) shows the droplet injection at 15 V. QD ink was ejected from the position of 0 μm, with satellite droplets appearing at the 100 μm position. At.

300 μm, the satellite droplet merged with the main droplet into a single drop. When the voltage was 18 V, there was a tail at 0–100 μm. After 100 μm, the tail formed a descending satellite drop along with the main droplet, as shown in Fig. 9 (b). In Fig. 9(c), the trailing became longer when the voltage was 21 V. The trailing appeared at 0–100 μm and retracted at 100–200 μm. Satellite droplet formed at 200 μm descending with the main droplet. Figure 9 (d) shows that at the voltage of 24 V, the dropping condition of

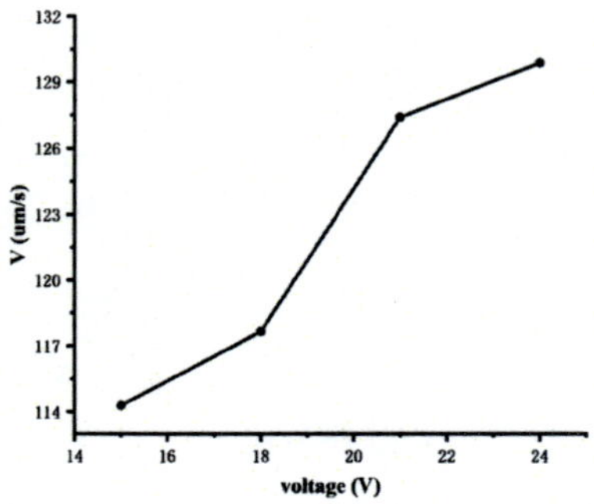

Fig. 10. Images of the relationship between inkjet voltage and inkjet speed

the droplet was similar to that at 21 V. However, the trailing tail merged with the main droplet during retraction, causing a drop of its trajectory falling without verticality. The formula for calculating the velocity of drop is as follows:

$$v = (s - s')/(t - t') \tag{2}$$

Let s = 400 μm, which means that the droplet drops to the position of 400 μm, at which point the corresponding time is t. Set s′ = 200 μm, which means that the droplet drops to the position of 200 μm, at which point the corresponding time is t′. Table 4 shows the time and speed of the inkjet at different voltages. Figure 10 illustrates that as the voltage increases the speed of the inkjet also increases. As the inkjet voltage increased, the electrostatic pressure at the nozzle was increased, followed by the higher injection speed of droplets. The volume of the sprayed droplets increased with the onset of the droplet trailing and satellite droplets, the trailing length increased with increasing voltage. When the voltage increased to 24 V, the descending state of the satellite droplet was not in the same direction as the main droplet, indicating that the injection was unstable for the higher voltage.

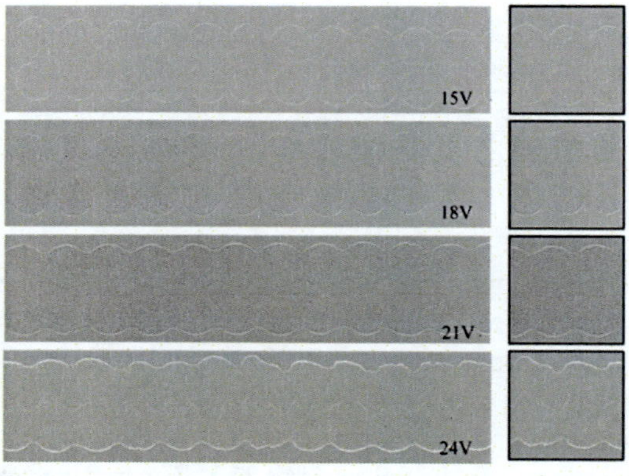

Fig. 11. Inkjet printed QD films images at different voltages

Figure 11 shows the image of the QD line film at different voltages when the printing spacing is 100 μm, in which the morphology between droplets can be observed well. As the voltage increased, the shape of the QD droplets began to deform due to the trailing effect. At the voltage of 24 V, the QD droplets began fusing with each other and barely showed the shape of QD droplets as well as the line shape distortion.

Therefore, in order to perform inkjet printing, it is important to consider the effect of the injection on the film morphology when adjusting the inkjet voltage. When the voltage is too high, the injection pressure the increases, making the injection speed too fast to produce trailing and satellite droplets, and the fusion between the droplets affects the integrity and uniformity of the QD film. When the voltage is too low, the injection pressure is too small, so that the QD ink does not have enough motivation to be ejected from the nozzle to complete the inkjet printing process.

4 Conclusion

When exploring the effects of solvent ratio, droplet spacing, and inkjet voltage on the preparation of QD films, the coffee ring phenomenon was weakened. The uniformity, integrity and stability of inkjet printing of the films have been optimized. Here the coffee ring effect has been suppressed at the solvent ratios of 7: 3: 5, 9: 1: 5 and 7: 3: 6. The morphology of the QD films at the ratio of 7: 3: 5 was better in the roughness test. The line width decreases with the increasing droplet spacing. The uniformity and integrity of the QD films were damaged when the droplet spacing was increased to 100 μm. The best morphology of the QD films was achieved in 60 μm. Good injection conditions and inkjet effect were accomplished at 15 and 18 V. There was the best inkjet condition without trailing and satellite droplets at 15 V. In summary, the coffee ring phenomenon could be attenuated to obtain complete and uniform QD films when the ratios of QD ink was 7: 3: 5, the droplet spacing was 60 μm and the inkjet voltage was 15 V. The improvements in the ink engineering, droplet spacing and inkjet voltage in this study have taken the uniform and inkjet stability of the QD films to the next level, laying the technical foundation for the preparation of QLED devices and confirming the application prospects of the QLED display prepared by inkjet printing.

Acknowledgements. This study is funded by BIGC Project (Nos. Ea202203, Ee202205).

References

1. Shirasaki, Y., Supran, G.J., Bawendi, M.G., et al.: Emergence of colloidal quantum-dot light emitting technologies. Nat. Photonics **7**(1), 13–23 (2013)
2. Jiang, C.B., Mu, L., Zou, J., et al.: Full-color quantum dots active matrix display fabricated by ink-jet printing. Chin. Sci. **60**(10), 1349–1355 (2017)
3. Chen, P.Y., Chen, C.C., Lin, Y.S. et al.: 41-3: 17.3 inch UHD Resolution AMOLED panel fabricated by ink jet printing process. In: SID Symposium Digest of Technical Papers, vol. 51, No. 1, pp. 595–598 (2020)
4. Coenen, M.J.J., Slaats, T.M.W.L., Eggenhuisen, T.M., et al.: Inkjet printing the three organic functional layers of two-colored organic light emitting diodes. Thin Solid Films **583**, 194–200 (2015)

5. Krainer, S., Smit, C., Hirn, U.: The effect of viscosity and surface tension on inkjet printed picoliter dots. RSC Adv. **9**(54), 31708–31719 (2019)
6. Weon, B.M., Je, J.H.: Fingering inside the coffee ring. Phys. Rev. E **87**(1), 013003 (2013)
7. Cao, B., et al.: Inkjet printing organic electroluminescent device film optimization research progress. Spectrosc. Spectral Anal. **40**(10), 3071–3076 (2020)
8. Xiong, X., et al.: Inkjet printing cadmium-based green quantum dot light emitting diode and its interface. J. Light Emitting **40**(10), 1274–1280 (2019)
9. Xiao-xiao, L.E.I., Yun, Y.E., Nan, L.I.N., En-guo, C.H.E.N., Tai-liang, G.U.O.: Morphology controlling of quantum dots thin films prepared by inkjet printing. Acta Photonica Sinica **48**(6), 616001 (2019)
10. Bo, S., Zheng-ning, T.: Parameter analysis of droplet ejection of piezoelectric ink-jet printing. Packag. Eng. **32**(19), 93–96 (2011)

Research on Image Quality Control Technology of Pad Printing

Yingjie Xu(✉), Wan Zhang, Xianfu Wei, and Beiqing Huang

School of Printing and Packaging Engineering, Beijing Institute of Graphic Communication, Beijing, China
xuyingjie@bigc.edu.cn

Abstract. Objective: In this paper, the factors that affect the printing quality of pad printing images are studied, and the pad printing products with excellent printing quality are obtained. Methods: Using electronic depth of field microscope to measure the dot cell depth of different pre exposure plates on the solid plate and dot area ratio, the effect of cell depth on the transfer process of pad printing ink was analyzed; Use the software Photoshop to draw the measurement and control strip, then which are used to make the printing plate and proofing, at last the spectrophotometer is used to test the dot area rate of the sample, analyze the difference between the output dot area rate and the tested dot area rate, establish the compensation curve of the image original in the software Photoshop. Results: With the extension of pre exposure time, the cell depth at the same dot area ratio on the printing plate decreased; the cell depth at different dot area ratio on the printing plate decreases gradually with the decrease of dot area ratio. The dot area ratio of 60–100% and less than 30% has an approximate linear relationship with the cell depth, and the cell depth at the dot area ratio of 30–60% has little change. For the unadjusted image originals, there is a large gap between the actual printed dot area rate and the output dot area rate, which is also the main reason for the printing quality problems. After the compensation curve adjustment of the image originals, the density value of the measurement and control strip obtained from proofing is basically consistent with the expected value. Conclusion: If the thickness of the transferred ink layer is too thick, it will be difficult to completely dry, which will lead to printing failure. Therefore, controlling the cell depth at the solid plate of the printed matter is an important means of printing quality control. The cell depth at the solid plate should be controlled at about 35 microns; the printing quality of the image originals adjusted by the compensation curve is greatly improved.

Keywords: Pad printing · Image quality · Compensation curve · Dot cell

1 Introduction

Pad printing expands the application range of traditional printing. Pad printing can replace some screen printing, labeling and heat transfer printing. To make a printing plate for pad printing, a printing pattern is etched on the surface of a steel plate or a photopolymer film. During printing, the blade of the ink scraper scrapes the ink into the

M. Xu et al. (Eds.): CACPP 2022, LNEE 991, pp. 163–169, 2023.
https://doi.org/10.1007/978-981-19-9024-3_22

dot cell, and the remaining ink is scraped back into the ink tank. During pad printing, the flexible rubber blocks transfers the ink from the printing plate to the substrate [1–4].

Pad printing can print on various uneven surfaces. Compared with screen printing, pad printing ink layer is very thin, which is a wet stamping method, so it is suitable for printing on substrates of various materials, shapes and sizes. The substrate generally has a common characteristic: the printing surface is uneven. The printing graphics are bound to be deformed in the process of pad printing, which will cause printing failures such as dot gain, resulting in the decline of printing image quality. Therefore, in order to compensate the deformation of the pad printing blocks, it is necessary to study how to control the image quality of pad printing [5, 6].

It is necessary to calculate the dot gain, through the test of the proofing sheet, the image deformation can be corrected by the graphics software to improve the printing effect [7, 8].

2 Experiments

2.1 Experiment Materials and Equipment

Pad printing plate (Jinan Yingtong pad printing Engineering, China), Pad printing ink (Sun Chemical, UK), Pad printing adhesive blocks (Jinan Yingtong pad printing Engineering, China), Laser Image-setter (6120, screen, Japan), Pad printing machine (2011, Kin Ngai Engineering China), Spectrophotometer (CI60, x-rite, USA).

2.2 Sample Preparation

Preparation of pad printing plate. Closely fit the color separation film with the pre-coated photosensitive layer of the pad printing plate, fix it with transparent tape, and extract air. Place the printed plate pasted with the film on the platform, and vacuum. Set the exposure time for exposure operation.

Preparation of pad printing samples. First, the image and text area of the printing plate are filled with ink, then scrape off the redundant ink with an ink scraper blade. The pad printing blocks is rolled evenly on the printing plate surface, and the ink adheres to the pad printing blocks. When the pad printing blocks is pressed to the surface of the substrate, the ink adheres to the surface of the substrate.

3 Result and Analysis

3.1 Influence of Pre Exposure Time on Printing Image Quality

Effect of pre exposure time on dot cell depth of printing plate. The software Photoshop is used to draw the measurement and control strip, and the size of each square of the measurement and control strip is 6 × 6mm, the dot area ratio on the measurement and control strip is 100%, 90%, 80%, 70%, 60%, 50%, 40%, 30%, 20%, 10% and 5% respectively. Set the number of screen ruling for the color chip as 200 lpi; The pre exposure time of the plate printer is set to 23s, 32S, 41s and 50s respectively, and the depth of each printing plate dot cell is measured with the electronic depth of field microscope. The measurement results are shown in the following Fig. 1.

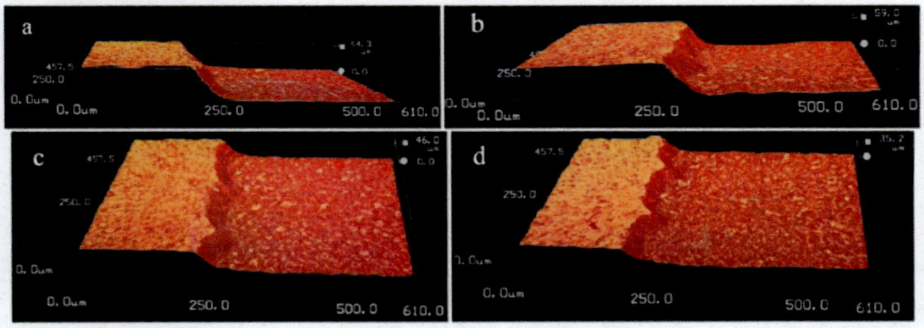

Fig. 1. Schematic diagram of dot cell depth at the solid plate of printing plate made with pre exposure time of 23S (a), 32S (b), 41S (c), 50S (d)

As the Fig. 2 shown, with the extension of pre exposure time, the solid plate dot cell depth of the printing plate gradually decreases in a linear relationship, and the slope is $k = (64.3 - 35.2)\ \mu\ M/(50 - 23)\ s \approx 1.1\ \mu\ m/s$.

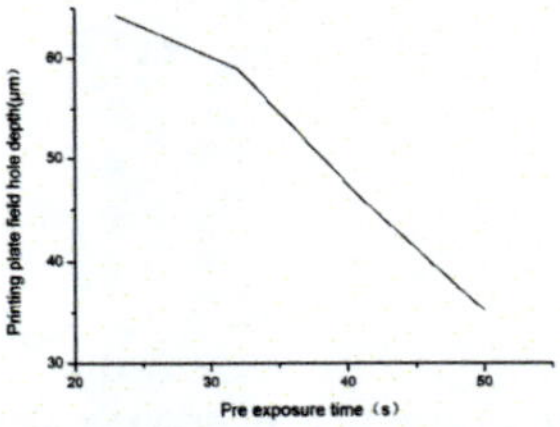

Fig. 2. Relationship between pre exposure time and printing plate solid plate dot cell depth

For this reason, as the dot area ratio decreases, the dot cell depth will also decrease. The Fig. 3 shows the dot cell depth at different dot area ratios of the printing plate with a pre exposure time of 50s.

As the Fig. 4 shown, when the dot area ratio is 60–100% or below 30%, the dot cell depth decreases significantly, showing a linear decline; between 30–60% of the dot area, the dot cell depth changes little.

Effect of dot cell depth on printing image quality. When the dot cell depth is more than 20 microns, the density value of print will not change too much with the change of dot cell depth. The main factor affecting the density value is the dot area ratio.

When the dot cell depth of the solid plate is too depth, due to the high thickness of the ink layer to be transferred, there will be stickiness and image blur caused by incomplete drying, which will affect the printing quality (Fig. 5). Therefore, controlling the dot cell depth of the solid plate density is an important method to control the printing quality. According to the relationship between the dot cell depth of the solid plate and

Fig. 3. Schematic diagram of dot cell at 90% (a), 80% (b), 70% c), 60% (d), 50% (e), 40% (f), 30% (g), 20% (h), 10% (i), 5% (j) dot area ratio of printing plate made with pre exposure time of 50s

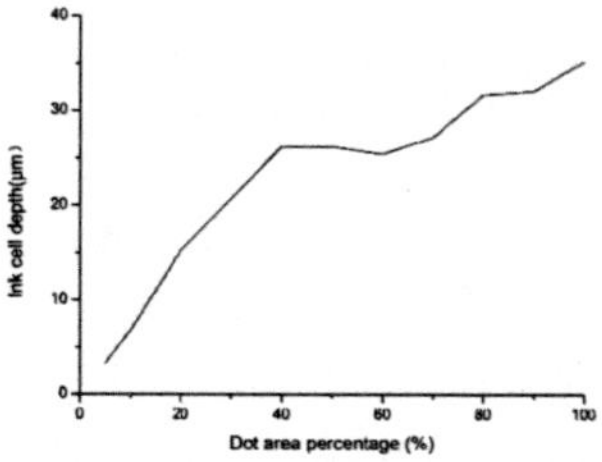

Fig. 4. Relationship between dot area ratio and dot cell depth.

the printing quality, it is concluded that the appropriate dot cell depth of the solid plate should be about 35 microns.

It can be seen from the Fig. 6 that when the ink accumulates (picture on the left), the printed matter will lose details and appear serious color deviation.

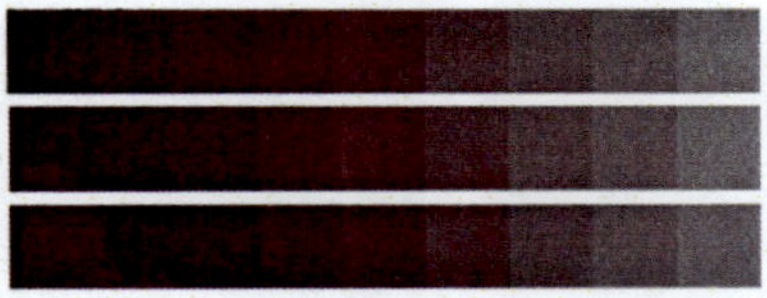

Fig. 5. Influence of dot cell depth on printing effect of measurement and control strip (dot cell depth 35μm (a), 45μm (b), 60μm (c))

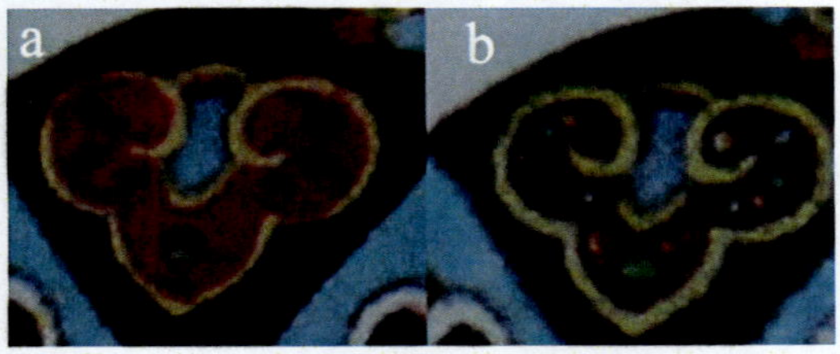

Fig. 6. Ink accumulation has an impact on printing quality (picture on the left)

3.2 Establishment of Color Density Compensation Curve for Image Originals

The software Photoshop is used to draw the measurement and control strip, and the size of each square of the measurement and control strip is 7 × 7mm, the dot area ratio on the measurement and control strip is 100%, 90%, 80%, 70%, 60%, 50%, 40%, 30%, 20%, 10% and 5% respectively, and the number of screens ruling for the color chip is sets as 200lpi; The pre exposure time of the plate printer is sets to 50s. Make a printing plate and use the printing plate for proofing. Use a densimeter to test the dot area ratio of the sample, and compare it with the output dot area ratio (Table 1).

Table 1. Comparison between output value and measured value of dot area ratio

Output value (%)	100	90	80	70	60	50	40	30	20	10	5
Measured value (%)	100	95	85	69	55	45	30	25	11	3	0

It can be seen from the Fig. 7 that there is a gap between the actual printed dot area rate and the output dot area rate, which is also a main reason for the deformation of printed image.

Find the intersection with the abscissa, through make a vertical line from the intersection of the marking lines of points 90, 80, 70, 60, 50, 40, 30, 20, 10 and 5 on the ordinate and the measuring curve to the abscissa.

Open the printing original with Photoshop, and select image - adjustment - curve, as shown in the Fig. 8.

Take the ordinate value and intersection value in Table 2 as the input value and output value in the options shown in Fig. 8 left to obtain the curve shown in the Fig. 8 right, which is the compensation curve of the printed original. The image printed with this compensation curve is closed to the original (Table 3).

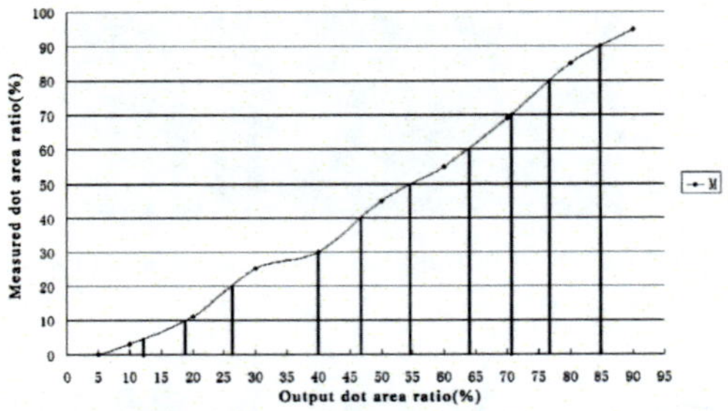

Fig. 7. Measured value of output dot area ratio

Table 2. Abscissa value of intersection point of longitudinal marking and measuring curve

Ordinate value (%)	100	90	80	70	60	50	40	30	20	10	5
Intersection value (%)	100	85	76	70	64	55	47	40	26	19	13

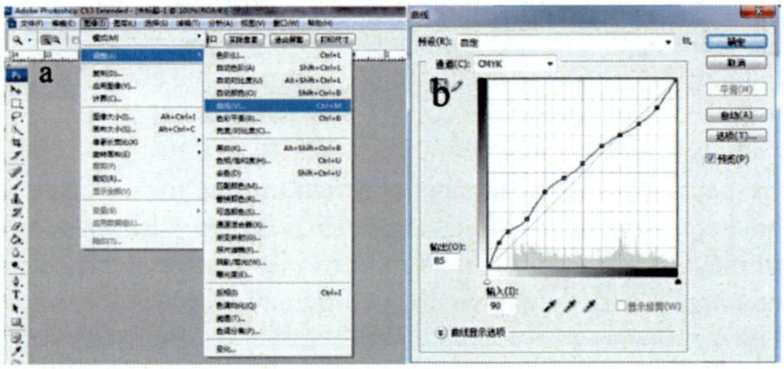

Fig. 8. Select image - adjust – curve (a); set the compensation curve of the original image with the Photoshop adjustment curve function (b)

Table 3. Corrected output value

Expected value (%)	100	90	80	70	60	50	40	30	20	10	5
Measured value (%)	100	89	83	68	57	52	44	36	22	10	5

The most important factor affecting the quality of printing image is the deviation of dot output. By adjusting the compensation curve of the original, the image gradation level is richer, and the color reproduction result of printing image is improved (Fig. 9).

4 Conclusions

With the extension of pre exposure time, the cell depth at the same dot area ratio on the printing plate decreased; the cell depth at different dot area ratio on the printing plate

Fig. 9. (a) (b) Uncorrected print (left) and corrected print (right)

decreases gradually with the decrease of dot area ratio. The dot area ratio is 60–100% or below 30%, the dot cell depth decreases significantly, showing a linear decline, between 30–60% of the dot area, the dot cell depth changes little. If the thickness of the transferred ink layer is too thick, it will be difficult to completely dry, which will lead to printing failure. Therefore, controlling the cell depth at the solid plate of the printed matter is an important means of printing quality control. The cell depth at the solid plate should be controlled at about 35 microns. For the unadjusted image originals, there is a large gap between the actual printed dot area rate and the output dot area rate, which is also the main reason for the printing quality problems. After the compensation curve adjustment of the image originals, the density value of the measurement and control strip obtained from proofing is basically consistent with the expected value. The compensation curve of the original, the image gradation level is richer, and the color reproduction result of printing image is improved, the printing quality of the image originals adjusted by the compensation curve is greatly improved.

Acknowledgements. This study is funded by BIGC Project.

References

1. Norberg, O., Andersson, M., Kruse, B.: The influence of paper properties on color reproduction and color management. In: NIP and Digital Fabrication Conference (2003)
2. Li, S., Manabe, Y., Chihara, K.: Accurately estimating reflectance parameters for color and gloss reproduction. Comput. Vis. Image Underst. **187**(2), 308–316 (2009)
3. Wu, Y.J.: The effect of substrate properties on print attributes for gravure printing from proof to press. Dissertations and Theses Grad works (2008)
4. Hermanski, M., Kohn, K.-U.: An adaptive spectral compensation algorithm for avoiding flexural vibration of printing cylinders. J. Sound Vib. **187**(2), 185–193 (1995)
5. Jing, J., Hatakeyama, Y., Takama, Y., Hirota, K.: Printing pressure state inspection system based on fuzzy inference. JACIII **12** (2008)
6. Wu, S., Wang, Y., Chai, C., Bai, J.: Research on the methods of the measurement of printing pressure in offset press. In: International Conference on Measuring Technology and Mechatronics Automation, vol. 1, pp. 342–344 (2009)
7. Mizutani, F., Takeuchi, S., Yoshida, T.: The relation between dot gain and rubber blanket in offset printing. J. Printing Sci. Technol. **27**(3), 299–303 (1990)
8. Prince, R.J.: Spraying cylinders for press damage. Am. Printer **126**(4), 32 (2009)

Research Progress on Thin Film Transistors Fabricated with Printing Technologies

Nianjie Zhang, Suyun Wang, Shengzhen Liu, Lijuan Liang(✉), Zhaohui Yu, Lianfang Li, Beiqing Huang, and Xianfu Wei

School of Printing and Packaging Engineering, Beijing Institute of Graphic Communication, Beijing, China

350186950@qq.com

Abstract. In recent years, with the rise and rapid advancement of printed electronics technology in recent years, the fabrication of large-scale, inexpensive, high-precision transistor devices has gradually moved into the center of attention for transistor research. However, there are still some technical bottlenecks in the process of realising the wide application of printed technology in electronic devices. This paper introduces the composition and main structure of transistors, the characteristics and advantages and disadvantages of different printing processes, as well as the materials commonly used for each structure in the preparation of transistors, and reviews the relevant research progress in recent years, and finally looks forward to the development prospects of the printed method for the preparation of transistors.

Keywords: Organic thin film transistor · Printing process · Organic semiconductor · Dielectric material · Conductive electrode

1 Introduction

On 23 December 1947, a team of scientists at Bell Labs, including Shockley, unveiled one of the most important inventions of the 20th century: the first real transistor. Since then, mankind has entered the rapidly evolving age of electronics. The processes used for the preparation of conventional transistor devices are often very complex, to the detriment of the development of transistors towards higher performance, and difficult to apply on a large scale in various fields [1]. Therefore, low-cost and diverse device preparation methods have always been the goal pursued. In recent years, the printing method has gradually come into the limelight. Compared with traditional processes, the advantages of the printing method, such as low cost, easy preparation of large areas and flexible use of materials, have turned it into one of the most promising technologies [2] with great potential for the construction of flexible devices and the greening of processes.

M. Xu et al. (Eds.): CACPP 2022, LNEE 991, pp. 170–181, 2023.
https://doi.org/10.1007/978-981-19-9024-3_23

2 Overview of Transistors

2.1 Composition and Classification of Transistors

A transistor consists of Sourse Eletrodes (S), Drain Eletrodes (D), Gate Eletrodes (G), an Active Layer and a Dielectric Layer. The device is a three-terminal device in which the source/drain electrodes are in direct contact with the active semiconductor layer and the gate electrodes are separated from the semiconductor layer by an insulating dielectric layer.

Among them, the Filded Effect Transistors (FETs) are the more typical types of unipolar transistors. Thin Film Transistors (TFTs), as a branch of FETs, are used for signal conversion/amplification, logic switching, voltage regulation, rectification, supply of drive current, etc. and are now used in a wide range of applications.

2.2 Transistor Structures

Transistors have different structures depending on the position of the gate electrode and the source/drain electrode. According to the relationship between the gate electrode and the source/drain electrode in a transistor, when the gate electrode is above the source/drain electrode, the transistor can be referred to as a top gate structure, and when it is below, it is referred to as a bottom gate structure. Also, according to the position of the source/drain electrode in relation to the semiconductor layer, when the source/drain electrode is located above the semiconductor layer it is called top contact and when it is located below the semiconductor layer it is called bottom contact. Common transistor structures are shown in Fig. 1 and are mainly divided into four types: bottom-gate bottom contact, bottom-gate top contact, top-gate bottom contact and top-gate top contact.

Fig. 1. Schematic diagram of transistor structure

In general, different types of transistor devices have different properties and strengths and weaknesses. As the semiconductor layer is exposed to air for a long time, bottom-gate structure devices are more susceptible to damage due to interference from water vapour and various gases in the air, for which reason we need to deposit passivation layers on the semiconductor layer to prevent this. In the case of top-gate devices, the insulating layer is located on top of the semiconductor layer and is passivated, so that it is not damaged by airborne gases. However, due to the deposition of the insulating layer, the transistor device performance is affected to a certain extent accordingly.

3 Fabrication of Transistors by Printing

At present, the more widely used printing methods in the preparation of organic thin film transistors are mainly: inkjet printing, aerosol jet printing, screen printing, reverse offset printing, etc., [3]. The following analysis summarizes the characteristics and advantages and disadvantages of various printing methods, and reviews some research advances in the preparation of transistors by this method in recent years.

3.1 Inkjet Printing

Inkjet printing is a contact-free, pressure-free, plate-free printing, the specific process is: the ink pump with a certain pressure, from the nozzle jets out a continuous stream of ink, this stream of ink in the printing head filled with high voltage in the metal tube to obtain electrostatic, become an orderly arrangement of ink droplets flow. The ink is then shot onto the substrate by means of thermal or acoustic control, etc., to complete the printing.

The advantages of inkjet printing are its independence from any material, its versatility, and the high precision and quality of the printed devices. In 2018, Qing Zhang et al. proposed a self-aligning inkjet printing process for the fabrication of small-channel metal oxide thin-film transistors. The method selectively deposited ink in the hydrophilic channel region by an inkjet printer and accurately deposited the semiconductor layer above the gate electrode to obtain excellent performance with a switching ratio of 108 and a maximum mobility of 3.3 $cm2V^{-1}s^{-1}$ [4]. In 2019, Kim et al. used a piezoelectric and dot-drop inkjet printer to in situ deposit poly(lysine) (PLL) with a direct impregnation technique patterned deposition of semiconductor single-walled carbon nanotubes. And the relationship between the patterning size of PLL and the electrical properties of the device was investigated, with experimental results showing that the yield of the TFT array reached 92.2% and the mobility reached 3.87 ± 0.70 $cm2V^{-1}s^{-1}$ as the pattern size decreased [5]. In 2019, Molina-Lopez F et al. prepared compressible and low-voltage synaptic transistor arrays by inkjet printing of polymers and carbon nanotubes and patterning from solutions, which exhibited excellent transistor performance. The devices had an average mobility of 27 $cm2V^{-1}s^{-1}$ and a maximum mobility of up to 30 $cm2V^{-1}s^{-1}$ with a switching ratio greater than 104 (e.g. Fig. 2) [6]. Kwon et al. ink-jet printed different types of silver nanoparticle electrodes such as source/drain and gate to prepare dual-gate organic transistors and to achieve scalable 3D integration on plastic foil. The 3D stacking of three complementary transistors was then combined to create a programmable 3D logic array, i.e. "p-type to n-type" and "n-type to p-type" [7]. In 2020, Francesca et al. investigated charge transport in a field-effect transistor based on an inkjet-printed semiconductor-type single-walled carbon nanotube (s-SWCNT) network, where the sparse network provided effective charge percolation for band transport when partial nanotubes with a small band gap were effectively interconnected, with a maximum saturation mobility of up to 20.2 $cm2V^{-1}s^{-1}$ [8]. Despite the current widespread use of inkjet printing, many issues, such as ink selection and the simplification of complex processing, are still awaiting resolution and will be the focus of future work [1].

Fig. 2. Compressible and low-voltage synaptic transistor arrays fabricated by inkjet printing

3.2 Aerosol Jet Printing

Aerosol jet printing, also known as air jet printing, refers to the technique of depositing aerosols onto the surface of a substrate by spraying them onto the substrate after they have been gathered by a surrounding gas at the nozzle. Generally speaking, aerosol jet printing can be based on different ink viscosities, and aerosols can be produced by two methods, namely pneumatic atomisation and ultrasonic atomization [9]. Compared to other printing methods, this printing method has advantages such as high resolution and a wide range of ink viscosities used for printing, and has good prospects for development. 2019, Lu et al. successfully demonstrated the structure of 1D-2D thin film transistors (1D-2D TFTs) were prepared. After testing, the on/off ratio was 3.5×10^5 and the channel mobility was up to 10.7 cm^2/Vs, showing excellent performance [10]. In 2020, Lu S et al. overcame the challenges associated with print uniformity and demonstrated an aerosol jet printing process that produced devices with a hole mobility of $\mu_h = 12.5$ cm^2/Vs and relative standard deviations as small as 4%. The impact of a commonly overlooked dimension in CNT aerosol jet printing is revealed, and a practical strategy for large-scale, high uniformity implementation of CNT-TFT is provided (e.g. Fig. 3) [11]. However, aerosol spraying technology also has some pressing issues that need to be addressed. For example, the printing process requires a large number of complex parameters to be adjusted in order to achieve the desired print result; the size of the spray droplets also plays a more important role. In addition to this, sheath gas flow, carrier gas flow and printing speed can all have a significant impact on print quality. Due to these limitations, aerosol injection is not yet available as a universal printing technology.

Fig. 3. Devices fabricated by aerosol jet printing process

3.3 Screen Printing

Screen printing, a non-contact printing technique, is a printing process in which the ink is pressed out with a certain force, the ink enters the mesh opening of the stencil, and the substrate is bonded to it after scraping off the excess ink with a squeegee. The advantages of this printing process are that it can be printed on top of a variety of substrates and requires less ink, while allowing large areas to be processed. In 2019, Zare Bidoky et al. reported a simple and high-resolution patterning process for the fabrication of electrolyte-gated transistors and circuits using photogel-linked ion gels and stencil screen printing, a technique that can be used to create circuits on narrow to 10m semiconductor channels to form 0.5-1m thick ion-gel films, with the resulting n-type ZnO EGTs showing high electron mobility (>2cm^2/Vs) and on/off current ratios (>10^5). A new idea for screen printing and photogel coupling strategies [12]. In 2019, Peter Andersson et al. based fully printed large-scale integrated circuits with organic electrochemical transistors on a reduced transistor size by a screen printing process, thereby minimising the number of terminals required to drive a monolithic integrated fully printed electrochromic display. Such logic gates typically have high and low output voltages above 1.25V and below 0.15V, respectively, and low and high input voltages of 0.1V and 1.3V, respectively (e.g. Fig. 4) [13]. In 2020, Marzieh Zabihipour et al. demonstrated the potential of screen printing methods for the mass production of organic electrochemical transistors by fabricating fully screen printed OECTs on flexible polyethylene terephthalate substrates with an area of 1 mm^2, with more than 100 OECTs forming a circuit that requires only 20% of the total area of the printed A4 size paper The total production pass rate is 99.7% [14]. In 2021, Torricelli, F et al. prepared electrolyte gated transistors for performance-enhancing bioelectronics by screen printing techniques. The fabrication of gate electrodes by masking of printed deposited metals does not require photoresists and the functionalisation of the electrodes can be achieved by immersing the electrodes in solution or by delivering the solution to the electrodes using PDMS wells or microfluidic cells. This process allows local bioelectric signal conduction and amplification, prompting an improved signal-to-noise ratio [15]. Although screen printing can be printed on a variety of substrates and the type of ink used is flexible, the process still has limitations due to the thicker films prepared and the low resolution.

Fig. 4. Organic electrochemical transistor fabricated by screen printing a) The structure diagram of the fully printed transistor b) The microscope image of the fully printed transistor c) The transfer characteristic curve of the fully printed transistor d) The switching characteristics of the fully printed transistor

3.4 Reverse Offset Printing

Reverse offset printing is a new printing technique that has emerged in recent years. Its printing process is divided into three main steps: coating, pattern and transfer. Its ability to provide good thickness control and edge definition gives this technique an advantage that other printing techniques do not have, namely higher resolution [16].

Therefore, in the future, this technology is expected to be widely used in the field of fine integrated circuits. In 2018, Y. Kusaka et al. demonstrated a fully printed with MoOx source/drain electrodes and ZrOx plasma mask layer based on metal complex inks, which were coated on polydimethylsiloxane (PDMS) surfaces using a reverse offset printing (ROP) process after ionophore treatment IGZO-TFT with an average mobility of 0.17 cm^2/Vs [17]. In 2019, Jaakko Leppäniemi et al. tested In_2O_3 inks using a lab-scale roll-to-roll compatible (R2R) reverse offset printing (ROP) device and a flexible electroplated nickel overprint machine. The final thin-film transistors (TFTs) were produced with average mobilities of 3.1 cm^2/Vs and 3.5 cm^2/Vs and switching ratios of up to 108 and 107, respectively (e.g. Fig. 5) [18].

In 2021, Sneck Asko et al. obtained Al-InzOs thin-film transistors based on reverse offset printing (ROP) with sacrificial polymer resist and vacuum deposition plus ultrasonic peeling by printing polymethyl methacrylate (PMMA), polyvinylpyrrolidone (PVP) and poly-4-vinyl phenol (PVPh) on the surface of flexible substrates, respectively, and Transparent metal mesh conductors and oxide TFTs were prepared for source/drain applications with a metal mesh conductor resistance of 35 Ω, a percentage of transparent area of 85% and a mobility of 1 cm^2/Vs [19].

Fig. 5. Thin film transistor prepared by reverse offset printing

4 Prepare Materials for Transistors

In general, the main materials required for printed transistors can generally be divided into electrode materials, semiconductor materials and dielectric materials, with different types of materials performing different functions in the device. The differences in materials, meanwhile, directly determine the level of device performance and have an important impact on transistor devices.

4.1 Electrode Materials

Electrode materials play the role of a medium in transistor devices and can control the direction of carrier movement in semiconductors [20]. Currently, the most commonly used printing process for the preparation of printed electrodes is inkjet printing. As the conductive inks required for inkjet printing need to have good stability as well as good rheological properties, this places requirements on the choice of electrode materials.

Currently, commonly used electrode materials include poly-3,4-ethylenedioxythiophene (PEDOT), graphene, and conductive metals such as Au and Ag. PEDOT has high value due to its simple structure and high conductivity. In 2019, Brian Schmatz et al. doped polystyrene sulfonic acid (PSS) in PEDOT to prepare a uniform and stable suspension and used PEDOT:PSS as an electrode to prepare a transistor device with a completely green solvent for solution treatment (e.g. Fig. 6a), which greatly reduced the environmental pollution [21]. Graphene, as an inorganic non-metallic material, also has high electrical conductivity and chemical stability, and by combining it with other solvents, good conductive inks can be prepared [22]. In 2018, Aditi R. Naik et al. demonstrated direct printing of high-resolution graphene patterns, using a printing process to apply graphene to the source/drain electrodes of transistors to obtain devices with high mobility (e.g. Fig. 6b) [23]. In recent years, as the price of metal Au has moved up, resulting in the rising cost of transistors prepared using Au electrodes, a nano-scale silver nanowires (AgNWs) composed of Ag has received attention for this reason. In addition to the good conductivity of Ag, AgNWs also have excellent light transmission and are seen as the most likely material to replace traditional ITO electrodes. In 2020, TAO WAN et al. designed a rapid forming technique by inkjet printing as a way to rapidly prepare complex AgNWs patterns (e.g. Fig. 6c), which have features such as high resolution and flexible size adjustment, and have a wide scope for the preparation of devices [24].

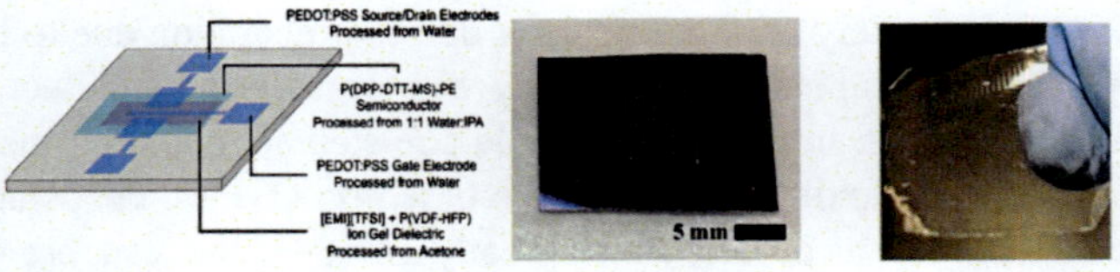

a. Transistor block diagram for full solution processing b. Large-area images of graphene electrodes c. Physical map of AgNW device

Fig. 6. Devices prepared from electrode materials

4.2 Semiconductor Materials

As the semiconductor material plays a decisive role in the magnitude of carrier concentration and carrier transport channels, the selection of this material is critical for the performance of transistor devices. Currently, commonly used semiconductor materials include 2,7-dioctyl[1]benzothieno[3,2-b]benzothiophene (C8-BTBT), 6,13-bis(triisopropylsilylene)-pentacene (TIPS-pentacene), etc. C8-BTBT, as a high performance organic semiconductor with high hole transport properties and strong self-assembly capability. In 2021, Xiaochen Fang et al. investigated a method combining inkjet printing technology and melt processing technology as a way to achieve simple preparation of liquid crystal (LC) films for high performance integrated circuits The method resulted in high quality LC films of C8-BTBT (e.g. Fig. 7a) [25]. Currently one of the most frequently used small molecule semiconductor materials in transistor preparation, it has good solubility.In 2019, Bingyao Shao et al. prepared a gas sensor based on TIPS-pentacene OTFTs by spin-coating and further optimised the properties of TIPS-pentacene films by selecting a suitable organic solvent (e.g. Fig. 7b) [26]. This provides a novel idea for the preparation of high-quality semiconductor-based sensors.

Fig. 7. Devices fabricated from semiconductor materials

4.3 Dielectric Materials

What is determined by the dielectric material is the state of charge aggregation between the semiconductor layer and the dielectric layer, which also plays a crucial role in the performance of the device. In recent years, the mainstream dielectric materials mainly include poly(methyl methacrylate) (PMMA), poly(methylsesquioxane) (PMSQ), SiO_2, Al_2O_3, etc. PMMA is widely used in transistor device preparation due to its own good physical properties and its better compatibility with organic semiconductor materials. In 2019, Shihui Hou et al. introduced a simple one-step spin-coating method for the preparation of nitrogen dioxide (NO_2) sensors based on OTFTs, choosing PMMA as the dielectric material for the device, which greatly improved the response of the device to NO_2 concentration (e.g. Fig. 8a) [27]. SiO_2 has now been the preferred inorganic dielectric material because of its good dielectric strength, coupled with its mature and well-established preparation process. In 2021, Qingqing Sun et al. proposed a strategy for preparing multilayer electronic patterns and devices by using low-temperature catalytic solution-treated SiO_2 (LCSS) films as a dielectric for layer-by-layer printing (e.g. Fig. 8b) [28]. This medium has excellent printing performance and, at the same time, good stability. This strategy is an important reference for the preparation of low-cost, efficient and stable fabrication of 3D electronic devices.

5 Summary and Outlook

With the advancement of transistor device research and the rapid development of printed electronics, the advantages of printing methods have made high-performance, low-cost

Fig. 8. Devices fabricated from dielectric materials

transistor devices possible. Several printing processes are currently used to prepare transistor devices, such as inkjet printing, aerosol jet printing, screen printing and reverse offset printing, all of which have their own characteristics as well as unique advantages and disadvantages. In general, the high operability and high accuracy of inkjet printing has received more attention from various fields than the limitations of the complex parameters of the aerosol jet method and the impact of the low resolution of the screen printing method. However, at present, there are still many problems for inkjet printing technology to prepare large-scale transistor devices, such as how to further simplify the complex processing and how to choose the right ink. In the future, a further trend in the preparation of high performance transistor devices by printing method is to achieve large scale preparation of electronic devices with low cost and high precision by printing technology. It is believed that in the near future, the printing method will show a broad prospect in various fields.

Acknowledgements. This research is supported by Beijing Natural Science Foundation (No.2202018), General Project of Beijing Municipal Education Commission Science and Technology Program (No. KM202010015004), Research and development of intelligent packaging for cultural relics(Ed202001), Construction and application transformation of cross media cloud platform for printing and packaging anti-counterfeiting and traceability(27170121005), National Natural Science Foundation of China (No.21604005), the general project of fundamental research of BIGC(No.Eb202001), and the general project of science and technology of Beijing Municipal Education Commission (No.KM202110015008).

References

1. Chung, S., Cho, K., Lee, T.: Recent progress in inkjet-printed thin-film transistors. Adv. Sci. **6**(6), 1801445 (2019). https://doi.org/10.1002/advs.201801445
2. Park, J.W., Kang, B.H., Kim, H.J.: A review of low-temperature solution-processed metal oxide thin-film transistors for flexible electronics. Adv. Funct. Mater. **30**(20), 1904632 (2020). https://doi.org/10.1002/adfm.201904632
3. Fukuda, K., Someya, T.: Recent progress in the development of printed thin-film transistors and circuits with high-resolution printing technology. Adv. Mater. **29**(25), 1602736 (2017). https://doi.org/10.1002/adma.201602736
4. Zhang, Q., Shao, S., Chen, Z.: High-resolution inkjet-printed oxide thin-film transistors with a self-aligned fine channel bank structure. ACS Appl. Mater. Interfaces **10**(18), 15847–15854 (2018). https://doi.org/10.1021/acsami.8b02390
5. Kim, H., Seo, J., Seong, N.: P-29: Solution-processed single-walled carbon nanotube thin film transistors in-situ patterned by inkjet-printing of surface treatment material. SID Symp. Dig. Tech. Pap. **50**(1), 1321–1324 (2019). https://doi.org/10.1002/sdtp.13178
6. Molina-Lopez, F., Gao, T.Z., Kraft, U.: Inkjet-printed stretchable and low voltage synaptic transistor array. Nat. Commun. **10**(1), 1–10 (2019). https://doi.org/10.1038/s41467-019-10569-3
7. Kwon, J., Takeda, Y., Shiwaku, R.: Three-dimensional monolithic integration in flexible printed organic transistors. Nat. Commun. **10**(1), 1 (2019). https://doi.org/10.1038/s41467-018-07904-5
8. Scuratti, F., Salazar-Rios, J.M., Luzio, A.: Charge transport in high-mobility field-effect transistors based on inkjet printed random networks of polymer wrapped single-walled carbon nanotubes. Adv. Funct. Mater. **31**(5), 2006895 (2021). https://doi.org/10.1002/adfm.202006895
9. Abdolmaleki, H., Kidmose, P., Agarwala, S.: Droplet-based techniques for printing of functional inks for flexible physical sensors. Adv. Mater. **33**(20), 2006792 (2021). https://doi.org/10.1002/adma.202006792
10. Lu, S., Cardenas, J.A., Worsley, R.: Flexible, print-in-place 1D–2D thin-film transistors using aerosol jet printing. ACS Nano **13**(10), 11263–11272 (2019). https://doi.org/10.1021/acsnano.9b04337
11. Lu, S., Zheng, J., Cardenas, J.A.: Uniform and stable aerosol jet printing of carbon nanotube thin-film transistors by ink temperature control. ACS Appl. Mater. Interfaces **12**(38), 43083–43089 (2020). https://doi.org/10.1021/acsami.0c12046
12. Zare, B.F., Tang, B., Ma, R.: Sub-3 V ZnO electrolyte-gated transistors and circuits with screen-printed and photo-crosslinked ion gel gate dielectrics: new routes to improved performance. Adv. Funct. Mater. **30**(20), 1902028 (2020). https://doi.org/10.1002/adfm.201902028
13. Andersson, E.P., Lassnig, R., Strandberg, J.: All-printed large-scale integrated circuits based on organic electrochemical transistors. Nat. Commun. **10**(1), 1–9 (2019). https://doi.org/10.1038/s41467-019-13079-4
14. Zabihipour, M., Lassnig, R., Strandberg, J.: High yield manufacturing of fully screen-printed organic electrochemical transistors. NPJ Flexible Electron. **4**(1), 1–8 (2020). https://doi.org/10.1038/s41528-020-0078-9
15. Torricelli, F., Adrahtas, D.Z., Bao, Z.: Electrolyte-gated transistors for enhanced performance bioelectronics. Nat. Rev. Methods Primers. **1**(1), 1–24 (2021). https://doi.org/10.1038/s43586-021-00065-8
16. Matsui, H., Takeda, Y., Tokito, S.: Flexible and printed organic transistors: from materials to integrated circuits. Org. Electron. **75**, 105432 (2019). https://doi.org/10.1016/j.orgel.2019.105432

17. Kusaka, Y., Shirakawa, N., Ogura, S.: Reverse offset printing of semidried metal acetylacetonate layers and its application to a solution-processed IGZO TFT fabrication. ACS Appl. Mater. Interfaces **10**(29), 24339–24343 (2018). https://doi.org/10.1021/acsami.8b07465
18. Leppäniemi, J., Sneck, A., Kusaka, Y.: Reverse-offset printing of metal-nitrate-based metal oxide semiconductor ink for flexible TFTs. Adv. Electron. Mater. **5**(8), 1900272 (2019). https://doi.org/10.1002/aelm.201900272
19. Sneck, A., Ailas, H., Gao, F.: Reverse-offset printing of polymer resist ink for micrometer-level patterning of metal and metal-oxide layers. ACS Appl. Mater. Interfaces **13**(35), 41782–41790 (2021). https://doi.org/10.1021/acsami.1c08126
20. Wang, L., Zhang, Q., Chang, L.: Electrochemically driven giant resistive switching in perovskite nickelates heterostructures. Adv. Electron. Mater. **3**(10), 1700321 (2017). https://doi.org/10.1002/aelm.201700321
21. Schmatz, B., Lang, A.W., Reynolds, J.R.: Fully printed organic electrochemical transistors from green solvents. Adv. Funct. Mater. **29**(44), 1905266 (2019). https://doi.org/10.1002/adfm.201905266
22. Chandran, A., Joshi, T., Sharma, I.: Monolayer graphene electrodes as alignment layer for ferroelectric liquid crystal devices. J. Mol. Liq. **279**, 294–298 (2019). https://doi.org/10.1016/j.molliq.2019.01.140
23. Naik, A.R., Kim, J.J., Usluer, O.: Direct printing of graphene electrodes for high-performance organic inverters. ACS Appl. Mater. Interfaces **10**(18), 15988–15995 (2018). https://doi.org/10.1021/acsami.8b01302
24. Wan, T., Guan, P., Guan, X.: Facile patterning of silver nanowires with controlled polarities via inkjet-assisted manipulation of interface adhesion. ACS Appl. Mater. Interfaces **12**(30), 34086–34094 (2020). https://doi.org/10.1021/acsami.0c07950
25. Fang, X., Shi, J., Zhang, X.: patterning liquid crystalline organic semiconductors via inkjet printing for high-performance transistor arrays and circuits. Adv. Funct. Mater. **31**(21), 2100237 (2021). https://doi.org/10.1002/adfm.202100237
26. Shao, B., Liu, Y., Zhuang, X.: Crystallinity and grain boundary control of TIPS-pentacene in organic thin-film transistors for the ultra-high sensitive detection of NO_2. J. Mater. Chem. C **7**(33), 10196–10202 (2019). https://doi.org/10.1039/c9tc01219b
27. Hou, S., Yu, J., Zhuang, X.: Phase separation of P3HT/PMMA blend film for forming semiconducting and dielectric layers in organic thin-film transistors for high-sensitivity NO_2 detection. ACS Appl. Mater. Interfaces **11**(47), 44521–44527 (2019). https://doi.org/10.1021/acsami.9b15651
28. Sun, Q., Gao, T., Li, X.: Layer-by-layer printing strategy for high-performance flexible electronic devices with low-temperature catalyzed solution-processed SiO_2. Small Methods **5**(8), 2100263 (2021). https://doi.org/10.1002/smtd.202100263

Research on Influence of Vibration on Rubber Surface Friction

Jiandong Lu(✉), Gaimei Zhang, Xiaoli Song, and Lizheng Zhang

School of Printing and Packaging Engineering, Beijing Institute of Graphic Communication, Beijing, China
lujiandong@bigc.edu.cn

Abstract. Friction and vibration are common in mechanical systems. In terms of friction, there is also friction on the surface of rubber components, in addition to the friction between metal components. The friction on the surface of rubber components sometimes affects the production quality. Because the vibration produced by mechanical system will affect the friction between components, it is necessary to study the change of rubber surface friction under the vibration. In this paper, an experimental study about the effect of vibration on rubber surface friction is carried out. An experimental platform is set up independently to measure the friction changes under different vibration parameters. The experimental results show that vibration can reduce the friction of rubber surface, and the greater the amplitude, the smaller the friction. The vibration frequency is also related to friction. These results can provide the support for explaining the change of surface friction of rubber components during mechanical movement.

Keywords: Friction · Rubber surface · Vibration · Normal vibration · Tangential vibration

1 Introduction

Friction is common in the mechanical system. In addition to the friction between metal components, there is also friction on the surface of rubber components in the mechanical system, such as blanket cylinders, rubber rollers for printing and rubber plates for relief printing in the printing mechanical system. The friction between printing components and the friction between the printing components and the substrates can affect the printing quality [1].

Vibration is also widespread in mechanical systems. Existing research results have shown that vibration can reduce friction interface. The friction on the non lubricated vibration interface will decrease compared with the friction on the interface without vibration [2–6]. According to different vibration directions, scholars have studied the effects of normal vibration perpendicular to the interface and tangential vibration parallel to the interface on friction.

M. Xu et al. (Eds.): CACPP 2022, LNEE 991, pp. 182–187, 2023.
https://doi.org/10.1007/978-981-19-9024-3_24

In respect of normal vibration, Chowdhury introduces normal vibration on the interface of various materials. The experimental results show that the friction decreases more and more obviously with the increase of normal amplitude [7]. Yoo proves that the friction reduction effect of normal vibration is greater than that of tangential vibration through experiments [8]. Shi finds that the friction between the FFM tip and the substrate decreases to zero in the microscopic experiment, after the normal amplitude increases [9].

In terms of tangential vibration, Popov uses ultrasonic waves to generate high-frequency vibration. And the experimental results show that the ultrasonic vibration can reduce the friction [10]. Gutowski also measure that the tangential vibration can reduce the friction between metal contact surfaces in the experiment [11].

However, more researches are focused on the change of friction between metal and metal materials, while less research is focused on the influence of vibration on the interface friction of rubber materials. Therefore, it is necessary to study the influence of vibration on rubber interface friction. The research results can clarify the effect of vibration on friction of the rubber components during the operation of mechanical system. The results also can provide relevant support for analyzing the causes of the problems caused by friction of the rubber components.

In this paper, a experimental device is designed independently. The influence of vibration on the friction of rubber surface is studied experimentally, based on this experimental device.

2 Experimental Device and Materials

A piece of rubber is pasted on the bottom of a metal sliding block. The rubber is mainly made of rubber insulating materials. The sliding block is placed on the vibrator. The slider moves on the vibrator under the pull of the horizontal tension machine. The pulling speed of the tension machine is constant, so the average sliding speed of the slider is constant. Because the slider moves at a uniform rate on the vibrator, the force value collected by the sensor of the horizontal tension machine is equal to the average friction force between the rubber and the vibrator.

The signal generator inputs a sinusoidal vibration signal to the vibrator. Under the excitation of this signal, the vibrator generates normal vibration and tangential vibration. The schematic diagram of the experimental device is shown in Fig. 1. Two laser vibrometers are used to measure the amplitude of the vibrator: laser vibrometer 1 measures the normal vibration amplitude of the vibrator, laser vibrometer 2 measures the tangential vibration amplitude of the vibrator. The physical diagram is shown in Fig. 2. The details of the experimental materials and device are shown in the Table 1.

During the experiment, the friction force with no vibration between the interfaces is measured at first. Then, the normal vibration signal is input, and the vibrator begins to produce vibration. As the voltage value of the input normal vibration signal increases under fixed frequency conditions, the amplitude of the vibrator increases. At this time, the change of friction between interfaces is measured, and the influence of vibration on interface friction is observed. The conditions for all the experiment is about 40% in humidity, temperature at 26 °C or so.

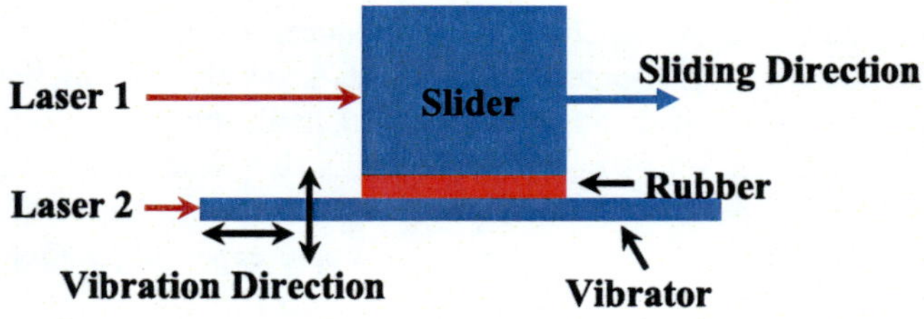

Fig. 1. Schematic diagram of the experimental device

Fig. 2. Physical diagram of the experimental device

Table 1. Details of the experimental materials and device

Materials and device	Details
Slider (stainless steel)	Quality: 0.2kg
Rubber mat (rubber)	Diameter: 30mm, Height: 3mm
Vibrator (stainless steel)	Length: 10mm, Width: 10mm, Height: 3mm

3 Results and Analysis

3.1 Relationship Between Friction and Time

Under the condition of 3000Hz, the normal amplitude A_n of the vibrator is changed, and the influence of vibration on the friction between the interfaces is measured under different normal amplitude. The average horizontal sliding speed of the slider is 2mm/s. The friction between the rubber mat and the vibrator is represented by F_x. The experimental results are shown in Fig. 3.

According to Fig. 3, when $A_n = 0.056\mu m$ and $A_t = 0.018\mu m$, the effect of friction reduction is not obvious. However, as the normal amplitude increases, the friction between the interfaces decreases more and more obviously. The above data suggests that vibration does reduce friction between interfaces.

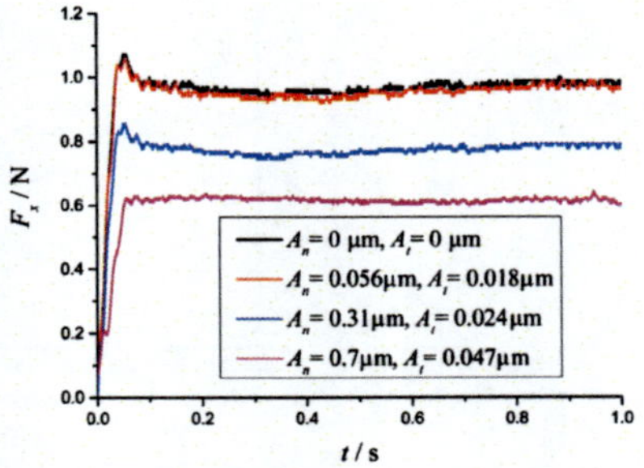

Fig. 3. Relation between friction and time

In order to achieve the research goal, the variation of friction between interfaces under more vibration parameters was measured. In addition, in order to better analyze the effect of vibration on friction, the dimensionless ratio F_L/F_0 is introduced. F_L is the average value of sliding friction under vibration. F_0 is the average value of sliding friction without vibration. The smaller the value F_L/F_0, the more obvious the friction reduction effect.

3.2 Relationship Between Friction and Amplitude

Firstly, the vibrator frequency f is set to be 750Hz, and the average horizontal sliding speed of the slider is 2mm/s. Under the condition of a fixed frequency of 750Hz, the normal amplitude of the vibrator is changed. F_L/F_0 is shown in Fig. 4. Figure 4(a) shows the relation between F_L/F_0 and the normal amplitude. Figure 4(b) shows the relation between F_L/F_0 and the tangential amplitude.

Secondly, the friction between the rubber and the vibrator is measured at 1000Hz. The average horizontal sliding speed of the slider is also 2mm/s. The experimental results are shown in Fig. 5. Figure 5(a) shows the relation between F_L/F_0 and the normal amplitude. Figure 5(b) shows the relation between F_L/F_0 and the tangential amplitude.

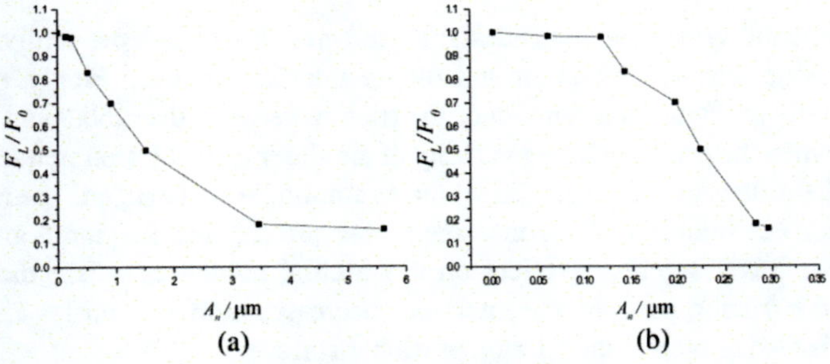

Fig. 4. Relation between friction and amplitude at 750Hz

According to Figs. 4 and 5, the influence of the vibration on the friction is analyzed. With the increase of the excitation signal voltage, the normal and tangential amplitudes

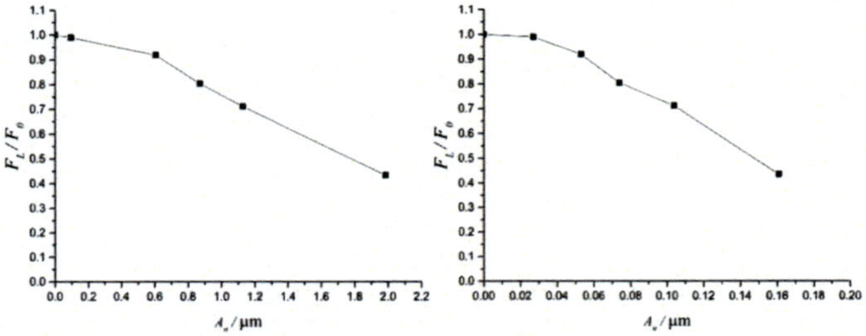

(a) (b)

Fig. 5. Relation between friction and amplitude at 1000Hz

of the vibrator increase. It can be seen from Figs. 4 and 5 that the four curves are the variation of interface friction with amplitude when the vibrator vibrates at a fixed frequency of 750Hz and 1000Hz respectively. All the four curves show that, when the vibration frequency remains constant, the friction between the interfaces becomes smaller and smaller with the increase of the normal and tangential amplitude of the vibrator. The above data show that the amplitude is one of the parameters that affect the friction of the rubber interface.

The influence of vibration frequency on friction can also be seen from Figs. 4 and 5. The test point $A_n = 2.0\mu m$ in Figs. 4(a) and 5(a) is taken as an example. F_L/F_O at 750Hz is close to that at 1000Hz at $A_n = 2.0\mu m$. This shows that the influence of normal vibration frequency on friction is not obvious in the range of 750–100Hz, when the normal amplitude is the same. In addition, the test point $A_n = 0.15\mu m$ in Figs. 4(b) and 5(b) is also taken as an example to determine the influence of tangential vibration frequency on friction. F_L/F_O at 1000Hz is much lower than that at 750Hz when $A_n = 2.0\mu m$. This shows that the influence of tangential vibration frequency on friction is obvious in the range of 750–100Hz, when the tangential amplitude is the same. Therefore, the vibration frequency can also affect the friction.

4 Conclusions

The experimental system, which is designed and constructed independently, can reflect the relationship between vibration and rubber interface friction. Firstly, it is proved experimentally that vibration can reduce the friction between the rubber and the mental. Secondly, when the average sliding velocity of the slider and vibration frequency remain unchanged, the normal and tangential vibration amplitude are changed. The results show that vibration can reduce the friction of the rubber surface, and the vibration parameters of amplitude has a significant effect on the friction reduction. When the amplitude increases, the friction between the interfaces decreases more obviously. The vibration frequency is also related to the friction of rubber surface.

Acknowledgement. This work was supported by the Key Research Project of Beijing Institute of Graphic and Communication (No. Ea201002), by the Scientific Research Team Project of Beijing Institute of Graphic and Communication (No. Eb202104) and by School Platform Project of Beijing Institute of Graphic and Communication (No. Eb202201).

References

1. Guang, L.: Influence of friction coefficient in printing process. Shanghai Packag. **2006**(06), 45 (2006)
2. Vanossi, A., Manini, N., Urbakh, M., et al.: Modeling friction: from nanoscale to mesoscale. Rev. Mod. Phys. **85**(2), 529–552 (2013)
3. Hesjedal, T., Behme, G.: The origin of ultrasound-induced friction reduction in microscopic mechanical contacts. IEEE Trans. Ultrason. Ferroelectr. Freq. Control **49**(3), 356–364 (2002)
4. Holl, H.J., Meindlhumer, M., Simader, V., et al.: Experimental investigation of friction reduction: by superimposed vibrations. Mater. Today: Proc. **5**(13), 26615–26621 (2018)
5. Qu, J.W.Y., Zhou, N.: Experimental study of air squeeze effect on high-frequency friction contact. Tribol. Int. **43**(11), 2190–2195 (2010)
6. Gnecco, E., Socoliuc, A., Maier, S. et al.: Dynamic superlubricity on insulating and conductive surfaces in ultra-high vacuum and ambient environment. Nanotechnology **20**(2), 025501 (2009)
7. Chowdhury, M.A., Helali, M.: The effect of amplitude of vibration on the coefficient of friction for different materials. Tribol. Int. **41**(4), 307–314 (2008)
8. Yoo, S.S., Kim, D.E.: Effects of vibration frequency and amplitude on friction reduction and wear characteristics of silicon. Tribol. Int. **94**, 198–206 (2016)
9. Shi, S., Guo, D., Luo, J.: Micro/atomic-scale vibration induced superlubricity. Friction **9**(5), 1163–1174 (2021)
10. Popov, V.L., Starcevic, J., Filippov, A.E.: Influence of ultrasonic in-plane oscillations on static and sliding friction and intrinsic length scale of dry friction processes. Tribol. Lett. **39**(1), 25–30 (2010)
11. Gutowski, P., Leus, M.: The effect of longitudinal tangential vibrations on friction and driving forces in sliding motion. Tribol. Int. **55**, 108–118 (2012)

Hydrodynamic Analysis of Coating Stability in Slot-Die Coating Processes

Li'e Ma(✉), Qiang Wang, Shanhui Liu, Hongli Xu, and Zhengyang Guo

Faculty of Printing, Packing and Digital Media Engineering, Xi'an University of Technology, Shaanxi, China
malie@xaut.edu.cn

Abstract. Slot-Die coating, has been widely applied in flexible electronics, functional films, micro-nano manufacturing and circuit production because of its fast production speed, low cost, and large-area coating available. In this paper, the fluid dynamic finite element analytical methodology was used to analyze the initial flow field of the slot-die coating produced by the anode slurry of lithium-ion battery on the copper foil substrate. Based on the theoretical analysis of the Navier-Stokes equations on the dynamics of the initial flow field at the coating head, a theoretical model of the flow rate, the flow relationship, and various pressure differences between the upstream and downstream of the coating beads was established. The flow mechanism of the coating layer was also revealed. Considering the inertia effect, the coating stability was predicted by applying the viscous capillary model. This study can contribute to the theoretical foundation for further improvement of the stability of slit coating.

Keywords: Slot-die coating · Two-phase flow · Hydrodynamic

1 Introduction

Slot-die coating (SDC) is an advanced predictive coating technology in which all the fluid fed into the extrusion die head forms a coating on the substrate [1]. In the actual process, the uniformity, stability, edge, and surface effect of coating solution are affected by the rheological properties of coating solution, which directly determines the quality of coating [2]. Schmit et al. [3] studied the stability of the coating edge in the process of lithium-ion battery cathode paste extrusion coating and discovered the phenomenon that intermittent coating and continuous coating process led to the issue of thick edge. Ruschak [4] analyzed the initial flow field at the coating head and proposed a viscous capillary model. On this basis, Higgins [5] considered the influence of Couette flow and Poiseuille flow and predicted the minimum coating thickness achieved under different parameters. Lee et al. [6] used the VC model to study the SDC process of Newtonian and non-Newtonian fluids through frequency response analysis and analyzed the effects of process parameters on fluid behavior, such as lip structure, material properties, operating conditions, etc. In this paper, a mathematical model is established to simulate and verify the initial process of coating graphite anode slurry of Lithium-ion battery, analyze the

M. Xu et al. (Eds.): CACPP 2022, LNEE 991, pp. 188–196, 2023.
https://doi.org/10.1007/978-981-19-9024-3_25

stability of the slurry coating process, observe the fluid flow characteristics at the initial coating stage, and study the influence of different process parameters on coating stability. This research aims to provide theoretical support for coating process optimization.

2 Establishment of Mathematical Model and Theoretical Analysis

The slurry is transported to the reservoir through the feeding channel and then flows equably from the slit. When the slurry reaches the substrate, due to the relative speed between the coating head and the substrate, the slurry on the substrate will be stabilized after the temporal accumulation and will form the meniscus, which then transform into the liquid bridge because of the relative motion between the coating of the slit and the substrate. The liquid held between the upper lip of the slit and the substrate is called the coating bead whose flow field is shown in Fig. 1.

Fig. 1. Schematic of slot coating

After the coating is stable, the shape of the coating bead is constant and the flow process can be regarded as the steady flow of the viscous incompressible fluid. The flow field of the coating bead is analyzed by applying the differential motion equation of the viscous incompressible fluid, namely, the $N - S$ equation, the flow process is three-dimensionally steady.

The state of the fluid in this paper is laminar flow. Because the coating width is much larger than the gap height, the direction Z is mainly considered. So we can get,

$$u = \frac{y^2}{2\mu} \cdot \frac{\partial p}{\partial x} + C_1 \cdot z + C_2 \tag{1}$$

Approximately assuming that $\frac{\partial p}{\partial x} = \frac{d_p}{d_x} = \frac{\Delta p}{l}$, among them:

$$\begin{cases} \text{upstream: } \frac{\Delta p}{l} = \frac{p_2 - p_3}{l_1} \\ \text{downstream: } \frac{\Delta p}{l} = \frac{p_4 - p_5}{l_2} \end{cases} \tag{2}$$

In formula (1), C_1 and C_2 are integral constants, which can be solved by taking into specific boundary conditions. The fluid-solid interface is a non-slip boundary condition.

Therefore, when $z = 0$, $u = 0$; when $z = h_m$, $u = U_c$. Taking them into formula (1) we obtain:

$$C_2 = 0, C_1 = \frac{U_c}{h_m} \cdot \left(1 - \frac{\Delta p}{2\mu l}\right) \tag{3}$$

$$u = \frac{U_c}{h_m} \cdot z - \frac{z \cdot (h_m - z)}{2\mu} \cdot \frac{\Delta p}{l} \tag{4}$$

Flow changes during the coating process can be represented as Q.

$$Q = \int_0^H u \cdot b \cdot \mathrm{d}z \tag{5}$$

Taking Eq. (4) into Eq. (5) we obtain:

$$Q = -\frac{bH^3}{12\mu} \cdot \frac{\Delta p}{l} + \frac{bH}{2} \cdot U_c \tag{6}$$

After the coating is stable, the upstream shape of the coating bead is also stable. The flow rate of the upstream coating bead is 0. The amount of fluid entering at the inlet is equal to the flow increment of the coating downstream, that is:

$$Q_{\text{up}} = 0, \quad Q_{\text{down}} = U_c \cdot b \cdot h_m \tag{7}$$

Substituting Eqs. (2) and (7) into Eq. (6), we simplify the model to a 2D model. Two pressure differences can be obtained as follows:

$$\begin{cases} p_2 - p_3 = -\frac{6l_1\mu U_c}{H^2} \\ p_4 - p_5 = -\frac{12l_2\mu U_c}{H^3}\left(h_m - \frac{H}{2}\right) \end{cases} \tag{8}$$

The pressure difference upstream the coating bead is determined by the Young-Laplace equation. The shape upstream of the coating bead is approximately an arc. The static contact angle between the upstream of the coating bead and the substrate is θ. The static contact angle between the upstream of the coating bead die and the substrate is ϕ. Therefore, the pressure drop here is:

$$p_1 - p_2 = -\frac{\sigma}{H}(\cos\theta + \cos\phi) \tag{9}$$

For the pressure difference at the meniscus downstream of the coating bead, it can be derived from the Landau–Levich film equation [7]:

$$p_5 - p_6 = -1.34\left(\frac{\mu U_c}{\sigma}\right)^{2/3}\frac{\sigma}{h} \tag{10}$$

For the stabilized coating, since the coating is thin and exposed to air, the pressure can be considered to be evenly uniform. In previous studies, in the viscous capillary model we assumed that the pressure difference at the exit of the slit was 0. That is $p_4 - p_3 = 0$. Therefore, the length l_1 can be deduced.

$$l_1 = \frac{H^2}{6\mu U_c}\left\{p_6 - p_1 - 1.34\left(\frac{\mu U_c}{\sigma}\right)^{2/3}\frac{\sigma}{h_m} - \frac{12 l_2 \mu U_c}{H^3}\left(\frac{H}{2} - h_m\right) - \frac{\sigma}{H}(\cos\theta + \cos\phi)\right\} \quad (11)$$

When the upstream length is close to 0, it is easy for the air to enter the coating beads intermittently, which leading to uneven coating or even coating defects [8]. In most theoretical studies, $l_1 = 0$ is considered as the critical condition and minimum coating condition for coating stability, where the value of l_1 is related to the pressure difference around the coating bead. To ensure the accuracy of the model, the pressure difference at the exit of the gap, i.e., $p_4 - p_3$ should also be considered.

The flow process between p_3 and p_4 in the slot exit is approximately Couette flow. Its pressure changes basic equation is $\frac{dp}{dx} = \mu \frac{d^2 u}{dz^2}$. After integrating this equation, the formula is as follows,

$$u = \frac{1}{2\mu}\frac{dp}{dx}z^2 + C_1 z + C_2 \quad (12)$$

Its boundary conditions are: $z = 0, u = U_c$; $z = H, u = 0$. Taking them into Eq. (12), we get,

$$u = \frac{1}{2\mu}\frac{dp}{dx}\left(z^2 - H\right) - \frac{U_c \cdot z}{H} + U_c \quad (13)$$

Then the flow rate of the section is as follows,

$$Q = \int_0^H u dz = \frac{U_c H}{2} - \frac{H^3}{12\mu}\frac{dp}{dx} = U_c \cdot h_m \quad (14)$$

Approximately $\frac{dp}{dx} = \frac{\Delta p}{W}$, $\Delta p = p_4 - p_3$, Then,

$$p_3 - p_4 = \frac{12\mu U_c W}{H^3}\left(\frac{H}{2} - h_m\right) \quad (15)$$

After considering this pressure difference, the length l_1' upstream the coating bead can be deduced as:

$$l_1' = l_1 + \frac{2W}{H}\left(\frac{H}{2} - h_m\right) \quad (16)$$

3 Numerical Simulation

3.1 Simulation Parameter Setting

In the coating flow field, the Reynolds number is 0.0435, so the laminar flow model is selected, assuming that the viscosity of the slurry does not change. The material parameters of the anode slurry, the geometric parameters of the die and the process parameters are shown in Table 1, in which the slurry inlet velocity is respectively selected as 0.020, 0.035, and 0.050 m/s, and the coating speed is selected as 0.10, 0.15 and 0.20 m/s to study the influence of different process parameters on the coating stability [10].

3.2 Analysis of Simulation Results

In this paper, the finite element analysis is performed using Fluent19.0, and the simulation parameters are set according to the coating process parameters in Table 1. When the coating speed is set as 0.15 m/s, Fig. 2 a, b and c are the fluid flow states from the start of coating to the steady state when the slurry inlet velocity is 0.020 m/s, 0.035 m/s and 0.050 m/s, respectively. And it can be observed from the figure that when the inlet velocity is 0.020 m/s, the coating bead is the smallest. So there is insufficient material supply or even a fracture at the downstream; when the inlet velocity is 0.035 m/s, the entire coating process is relatively stable, and the coating can reach a stable state; When the inlet velocity is 0.050 m/s, slurry accumulation occurs at the upstream and the coating is too thick at the downstream, hence it is difficult for the coating process to reach a stable state.

Table 1. Main parameters

Parameter	Viscosity	Density	Surface tension	Slit size	Coating distance	Inlet velocity	Coating speed
Unit	Pa·s	Kg/m^3	N·m	mm	mm	m/s	m/s
Symbol	ρ	μ	σ	w	H	V	U_c
Numerical value	1	1450	0.0417	0.55	0.20	0.020 0.035 0.050	0.10 0.15 0.20

When the inlet velocity is set as 0.035 m/s, Fig. 2 d and e show the fluid flow states from the start of coating to the steady state when the coating velocity is 0.10 m/s and 0.20 m/s, respectively. It can be observed from the figure that when the coating speed is 0.10 m/s, the slurry accumulates at the upstream and the coating is too thick at the downstream; When the coating speed is 0.20 m/s, the coating bead is the smallest and it is difficult to achieve a stable coating state, and there is insufficient material supply or even a fracture in the downstream.

(a) V=0.020m/s , U_c=0.15 m/s

(b) V=0.035 m/s , U_c=0.15 m/s

(c) V=0.050 m/s , U_c=0.15 m/s

(d) V=0.035 m/s , U_c=0.10 m/s

(e) V=0.035 m/s , U_c=0.20 m/s

Fig. 2. Variation of slurry flow with time under different process parameters

It can be observed from Fig. 3 that if the coating speed is too high or if the inlet speed is too low, the length of the upstream of the coating bead will be less than 0, which will easily lead to gas entrainment into the coating, resulting in bubbles. At the same time, the coating layer will be unevenly distributed and the coating will be too thin. When the coating speed is too low or the inlet speed is too high, slurry accumulation occurs upstream the coating bead, resulting in large change in the pressure difference across the coating bead, and the coating layer is too thick and uneven, which cannot meet the process requirements. When the inlet velocity is 0.035 m/s and the coating velocity is 0.15 m/s, the coating can quickly reach a stable state. When the coating tends to be stable, the length of the upstream of the coating bead reaches the minimum and is greater than 0, and the slurry volume at the upstream also reaches the minimum.

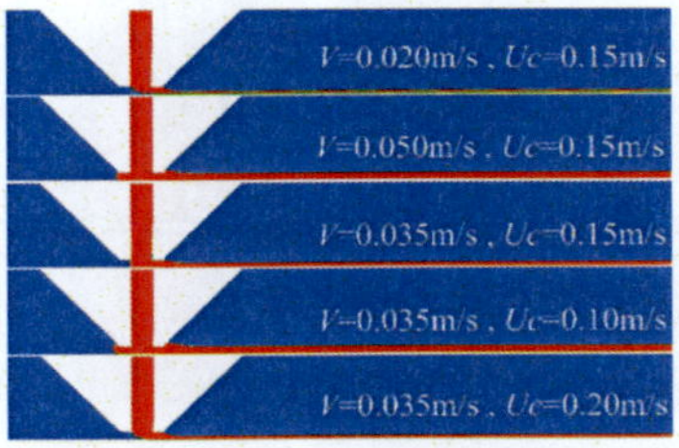

Fig. 3. Fluid distribution in stable coating state with different process parameters

The change of the volume fraction upstream of the coating bead can be monitored, as shown in Fig. 4 a. After about 0.3s, the volume fraction of the slurry hardly changes, that is, the coating process reaches a stable state. At the same time, the change of static pressure at the inlet is calculated, as shown in Fig. 4 b, and the change trend is consistent with the change trend of the volume fraction of the slurry.

After comparing several groups of simulation data with the calculated data, it is found that when the static pressure at the inlet no longer changes, the volume upstream the coating beads reaches the minimum. That is, the length l_1 reaches the minimum. Combining the two curves together as shown in Fig. 4 c, we can observe that the two have the same change trend with time.

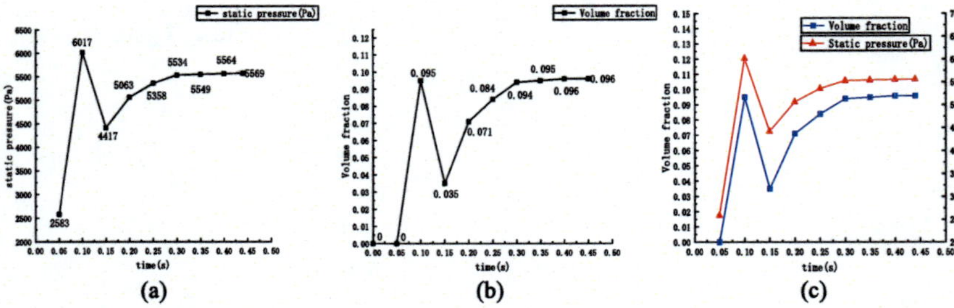

Fig. 4. Static pressure at the inlet and volume fraction upstream the coating beads with time

The inlet static pressure changes with time in the case of Fig. 2 a, c, d, e, are calculated, and the calculation results are shown in Fig. 5 a, b, c, d respectively.

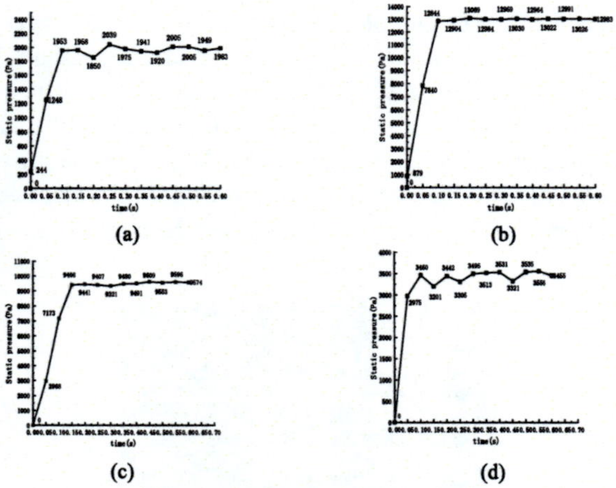

Fig. 5. Curve of inlet static pressure changes with time under different process parameters

It can be seen from the figure that when the inlet speed is too low or the coating speed is too high, the static pressure at the inlet changes little but it is difficult to achieve stability; When the inlet speed is too high or the coating speed is too low, the static pressure at the inlet changes rapidly. Though stability is achieved rapidly, the value is large, resulting in excessive pressure on the substrate and easy accumulation of slurry. Therefore, a relatively stable coating process can be obtained only by selecting appropriate process parameters.

4 Conclusion

The *N-S* equation is applied and simplified in terms of the actual boundary conditions in this paper to theoretically analyze the upstream section and deduce the flow velocity and flow distribution formulas in the upstream and downstream sections. Using the viscous capillary model, the stability of the coating is predicted considering the influence of inertial effect. By the aforementioned numerical and simulation analysis, the conclusions are as follows.

1) When the inlet speed is 0.035 m/s, the coating layer can quickly reach a stable state. It is also found that when the coating speed is 0.15 m/s, the coating effect is the best, and the coating layer can quickly reach a steady state.
2) The variation law of the length at l_1 position is the same as that of the static pressure at the entrance. Therefore, the pressure change at the entrance can be observed and measured. When the static pressure change tends to be stable, the coating layer reaches a stable state.

Acknowledgements. This research is supported by The National Key Research and Development Program of China (2019YFB1707200), The Technology Innovation Leading Program of Shaanxi Province (No.2020QFY03-04 and No.2020QFY03-08) and The Key Research and Development Program of Shaanxi Province (No.2020ZDLGY14-06).

References

1. Lee, J., Kim, S., Lee, C.: Large area electrolyte coating through surface and interface engineering in roll-to-roll slot-die coating process. J. Ind. Eng. Chem. **76**, 443–449 (2019)
2. Bajaj, M., Prakash, R., Pasquali, M.: A computational study of the effect of viscoelasticity on slot coating flow of dilute polymer solutions. J. Non-Newtonian Fluid Mech. **149**(1), 104–123 (2008)
3. Schmitt, M., Scharfer, P., Schabel, W.: Slot die coating of lithium-ion battery electrodes: investigations on edge effect issues for stripe and pattern coatings. J. Coat. Technol. Res. **11**(1), 57–63 (2013). https://doi.org/10.1007/s11998-013-9498-y
4. Ruschak, K.J.: Limiting flow in a pre-metered coating device. Chem. Eng. Sci. **31**(11), 1057–1060 (1976)
5. Higgins, B.G., Scriven, L.: Capillary pressure and viscous pressure drop set bounds on coating bead operability. Chem. Eng. Sci. **35**(3), 673–682 (1980)

6. Lee, S.H., Koh, H.J., Ryu, B.K., Kim, S.J., Jung, H.W., Hyun, J.C.: Operability coating windows and frequency response in slot coating flows from a viscocapillary model. Chem. Eng. Sci. **66**(21), 4953–4959 (2011)
7. Landau, L., Levich, B.: Dragging of a liquid by a moving plate. Dyn. Curved Fronts **17**(2), 141–153 (1988)
8. Jang, I., Song, S.: A model for prediction of minimum coating thickness in high speed slot coating. Int. J. Heat Fluid Flow **40**, 180–185 (2013)
9. Hirt, C.W., Nichols, B.D.: Volume of fluid (VOF) method for the dynamics of free boundaries. J. Comput. Phys. **39**, 201–225 (1981)
10. Wu, X.K.: Simulation study on initial flow field of lithium-ion battery slurry slit coating. Chin. J. Power Sources **42**(04), 500–503 (2018)

Verification and Adjustment Method of ICC Applied in Xerographic Digital Printing

Wenjun Guan(✉)

Printing Training Center, Shanghai Publishing and Printing College, Shanghai, China
461014112@qq.com

Abstract. Purposes: The application of ICC in xerographic digital printing is very common, but the generated ICC may not necessarily match the color target in practical application. If the ICC curve is adjusted to match the color target, the operator needs to have a lot of experience. And this method is very inefficient. Accurately verifying the color quality of the ICC and finding ways to adjust it is important to obtain the best ICC. This improves production efficiency and ensures product color accuracy and consistency. Method: Generate an ICC conforming to the Fogra51 standard, use the Color Tool to complete the verification, and find the adjustment method. Result: Even ICCs generated after device color calibration is done may not meet Fogra51 requirements. This shows that the default state of the device does not meet the ICC production requirements of Fogra51. Conclusion: The ICC generated according to different color targets can be verified to judge whether further adjustment is needed. By adjusting the toner density, the device state can be directly changed to generate an ICC that fully complies with the color target.

Keywords: Xerographic digital printing · ICC profile · Toner concentration

1 Introduction

As the field of xerographic digital printing has become wider and wider, the requirements for printing color quality have become higher and higher. Compared with the multi-link color management method of traditional printing, the digital printing color management method will be simpler, more efficient and more durable [1]. In the entire digital printing process, the most critical link to ensure the accuracy of colors is color management [2]. Color management is to ensure the color quality of printed products through three links: equipment calibration, ICC profile (hereinafter referred to as "ICC") and color conversion.

2 Research Purpose and Current Research Situation

2.1 Research Purpose

ICC is the core of color management, but in actual production, the application of ICC is often ineffective. When the ICC application does not work well, the common method

M. Xu et al. (Eds.): CACPP 2022, LNEE 991, pp. 197–206, 2023.
https://doi.org/10.1007/978-981-19-9024-3_26

is that the operator adjusts the ICC curve by experience to meet the requirements. And sometimes it is repeated many times. This method relies too much on work experience, which is not conducive to the development trend of high-speed mass production of advanced digital printing, and is not conducive to improving the production efficiency of enterprises. Only through data quantification to realize process operation, thereby reducing the dependence on manual experience, can digital printing have high color quality while producing at high speed. The actual operation process of color management must have a standardized process and improve the professional awareness of operators to ensure the stable operation of the standardized system [3].

This article will verify the generated ICC by comparing the color sample standards to judge its effectiveness in practical applications. Analyze the reasons that affect the poor application of ICC, and find the adjustment method.

2.2 Current Research Situation

In the past five years, the color management of digital printing in China is mainly based on the principle of color management and the operation method of generating ICC, and there is very little research and analysis on the application effect of ICC. Some studies involving color pre-inspection mentioned that pre-inspection of color management effects can improve production efficiency and reduce costs from the source [4], but the research is limited to the color verification of screen proofing [5]. To sum up, the existing domestic research fails to conduct a more in-depth exploration according to the development trend and characteristics of digital printing.

3 Reasons and Solutions for Poor ICC Application

3.1 Reasons

When digital printing is produced, an objective and authoritative international color certification standard is usually chosen as the color target. International color certification standards have their corresponding ICC files that can be used as output profiles. Therefore, device ICC generation as a source profile (or input profile) is particularly important.

Printing press is unstable. He premise of generating the device ICC is that the digital printing press should be in a stable state. It can be seen from Tables 1 and 2 that if the printing press is unstable, even if the color calibration is performed, the four-color field density value will still fluctuate greatly (the maximum difference in Table 1 is 0.098). The field density values are relatively stable (the maximum difference in Table 2 is 0.019).

Inaccurate parameter settings. When the printing press is in steady state, use the specific software to generate the device ICC. Due to insufficient understanding of the color target, the omission or wrong setting of important parameters in the software (for example, the ISO standard corresponding to the color target, paper requirements, test methods, etc.) will lead to calculation errors when ICC is generated.

Table 1. Calibration when the press is unstable

Field density	First	Second	Third
C	1.704	1.682	1.607
M	1.494	1.485	1.406
Y	0.979	0.984	0.96
K	1.812	1.782	1.88

Table 2. Calibration when the press is stable

Field density	First	Second	Third
C	**1.671**	1.671	1.673
M	1.435	1.454	1.456
Y	0.988	0.989	0.988
K	1.837	1.825	1.828

Unable to grasp the characteristics of the device ICC. In practical applications, if you do not understand the characteristics of the ICC of the device, it will lead to color problems that cannot be judged, so that you cannot find the correct adjustment method. Repeated corrections that rely on human experience reduce color quality and productivity.

3.2 Solutions

Achieving a stable machine state. Digital printing machines have higher requirements on the temperature and humidity of the production environment. Temperature can affect the fusing and the color performance of the printing press. Humidity affects paper properties, as well as the effects of the fusing and the color performance. Temperature and humidity have an effect on the fusing and color performance. Temperature of 25 °C ± 2 °C and humidity of 50% ± 5% indoors are better [6].

Understanding color targets. Understanding color target properties is critical for device ICC generation. When generating device ICC, you need to select the correct paper and test chart for output, and select the correct measurement method to collect color information. When applying device ICC, you need to select the profile corresponding to the color target for analog output. At present, the common international color certification standards are mainly ISO 12647 in Europe and ISO 15339 in the United States. Each color certification includes a series of specifications for the realization of ISO standards (such as: Fogra51, GRACoL 2013, etc.). Choosing the correct test chart, paper type, measurement method, profile, etc. when producing a device ICC can also help the device find the best device state.

Verify device ICC color quality. In order to avoid using unqualified device ICC or using device ICC in inappropriate device state, it is necessary to perform color quality verification on the generated device ICC to check whether the application effect of the device ICC can match the target color, and find the appropriate device state when match the color target. The device ICC color quality verification process is shown in Fig. 1.

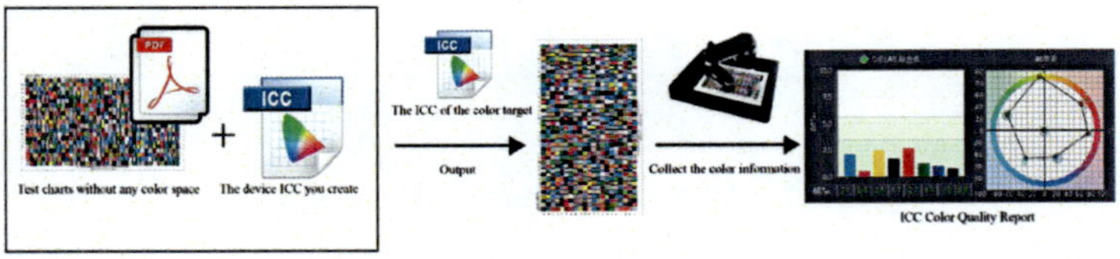

Fig. 1. Method of verifing the device ICC

4 ICC Verification and Adjustment

4.1 Explanation

The ΔE_{00} is the best formula that reflects the color difference at present, which makes up for the difference in the sensitivity of the human eye to different colors [7]. The ΔE_{00} formula is applied to this experiment because it can more accurately reflect human visual perception through data.

It should be noted that the ICC generated in this experiment is the device ICC.

4.2 Steps to Verify

Step 1: Preparing for generating the ICC. After the temperature and humidity of the environment where the machine is located are stable, the digital printing press is warmed up to ensure its stable state. Then, device color calibration is done via DFE.

Step 2: Generating the ICC. The detailed data of Fogra51 are shown in Table 3. Select an IT8.7/4 chart suitable for the ISO 12647-2 standard and output it, turning off the color management and color conversion functions. The paper used in the test is 157 g/m^2 coated paper (the ΔE00 value of the paper white of the paper and the Fogra51 paper is 0.8). Through the X-Rite i1iSis2 scanner to collect the color information and the information is entered into the Color Toolbox software. The parameter settings are shown in Figs. 2 and 3.

Step 3: ICC color quality verification. In the current device state, import the generated ICC and PSOcoated_v3.icc into the printing process. When the former is used as the source profile and the latter is used as the output profile, the IT8.7/4 chart is output and the color information is collected again. The color data obtained at this time is compared with the digital printing standard data of Fogra51 and the color quality report is generated (Fig. 4).

Table 3. Fogra51 data

Properties	Standards
Printing method	Offset printing
ISO standard	ISO 12647-2
Paper type	Glossy coated paper
Paper white (lab)	95, 1.5, – 6
Measurement mode	M1
ICC profile	PSOcoated_v3.icc
Digital printing comparison standards	Digital PSD2018

Fig. 2. Choose the correct chart

It can be seen from Figs. 5 and 6 that the color quality difference between not using the ICC and using the ICC is obvious. The color of the product output without this ICC has a larger deviation from the target, while the color of the product using this ICC is significantly closer to the target.

4.3 Adjusting the ICC

Figure 6 shows that although a good color effect has been obtained, the color quality has not yet reached the requirements of Fogra51, and the color quality of cyan is slightly worse (ΔE_{00} value is more than 3).

When this ICC was generated, the device state was stable and the parameters were set correctly, but the best color quality was still not obtained. This indicates that the current device state does not match the state required by the color target. When the software cannot continue to adjust, it is necessary to adjust the hardware.

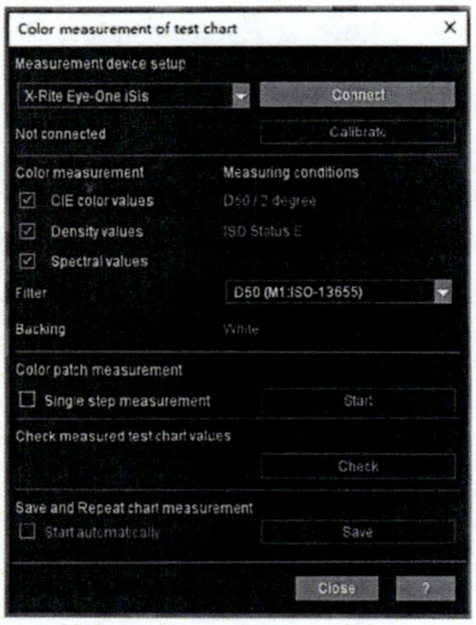

Fig. 3. Choose the correct instrument and filter

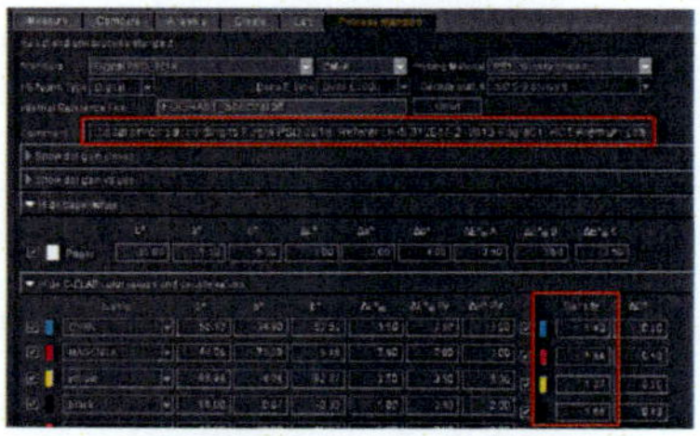

Fig. 4. Digital PSD2018 properties

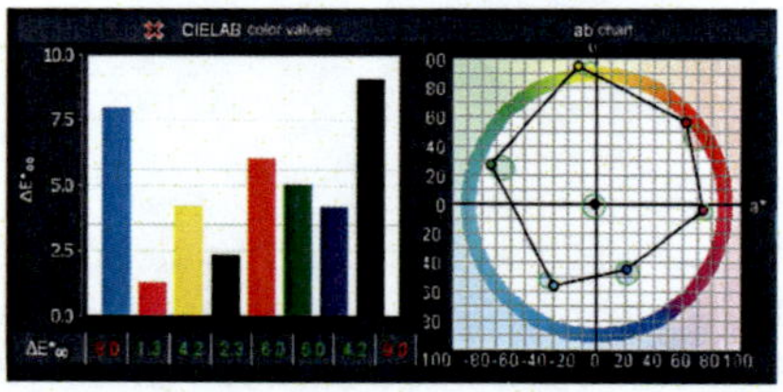

Fig. 5. Color quality without the device ICC

Relationship between toner concentration and field density. The essence of device color calibration is the change of color density value. It can be seen from Fig. 7 that the field density value of cyan, magenta, yellow and black increase with the increase of toner concentration. After the color calibration is completed, the four-color field density value must be recorded, and then through the adjustment of the toner concentration, the digital printing press can reach the state that match the target.

Relationship between toner concentration and lab value. The density of the toner will affect the thickness of the toner transferred to the paper, and the change of the toner thickness will directly affect the toner's absorption and reflection of color light.

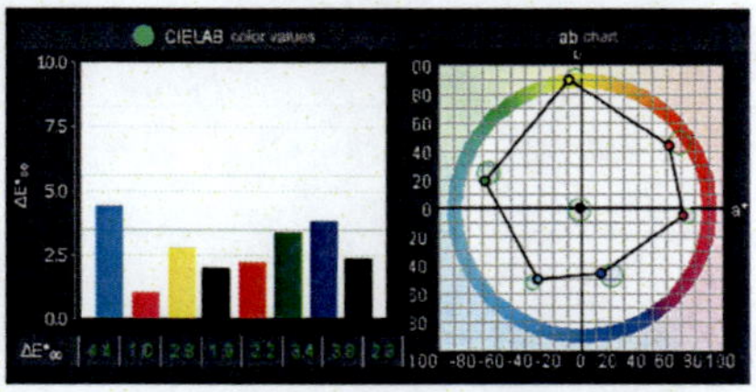

Fig. 6. Color quality with the device ICC

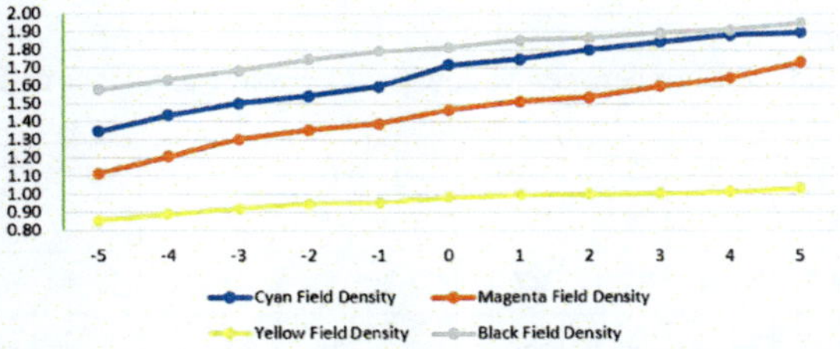

Fig. 7. Field density corresponding to the different toner concentrations in four colors

Thicker toner absorbs more light. As can be seen from Tables 4, 5, 6 and 7, when the toner concentration increases, the luminance values (L values) of cyan, magenta, yellow and black all decrease. Therefore, you can decide to increase or decrease the toner concentration of a certain color according to the L value. For example, by increasing the toner concentration, the L value can be lowered.

Table 4. Relationship between cyan field lab value and toner concentration

Toner concentration	− 5	− 4	− 3	− 2	− 1	0	1	2	3	4	5
L	55.1	55.4	54.9	54.6	54.4	53.8	53.4	53.1	52.6	52.8	53.5
a	− 33.2	− 33.0	− 32.9	− 31.0	− 32.7	− 29.6	− 29.7	− 28.9	− 28.9	− 28.8	− 27.8
b	− 50.0	− 50.2	− 49.9	− 49.5	− 49.7	− 49.6	− 49.0	− 49.0	− 48.4	− 46.0	− 47.0

Adjusting the ICC. It is known from Fig. 6 that ΔE_{00} value of cyan is 4.4, and cyan contains green and blue, and the ΔE_{00} value of these two colors also exceeds 3. Therefore, in order to achieve a better device ICC, it is necessary to adjust the cyan toner concentration.

The cyan L value of specified in the Digital PSD2018 standard is 56.12 (known from Fig. 4), while the current cyan L value is 51.9 (known from Fig. 8). Therefore, it is necessary to lower the current cyan toner concentration to increase the L value so that the cyan L value can be closer to 56.12. It is known from Fig. 7 that the density of cyan

Table 5. Relationship between magenta field lab value and toner concentration

Toner concentration	− 5	− 4	− 3	− 2	− 1	0	1	2	3	4	5
L	52.3	51.4	50.4	49.4	49.3	48.3	48.0	47.9	47.6	46.8	47.2
a	65.3	67.9	67.1	71.1	72.5	74.6	73.9	74.1	73.9	73.5	71.9
b	−7.7	−7.7	−7.9	−6.3	−5.7	−4.8	−4.8	−5.0	−4.2	−4.1	−3.9

Table 6. Relationship between yellow field lab value and toner concentration

Toner concentration	− 5	− 4	− 3	− 2	− 1	0	1	2	3	4	5
L	87.7	87.9	87.1	87.2	87.3	86.9	87.5	87.3	87.3	86.9	87.5
a	−7.9	−8.3	−7.1	−7.4	−7.8	−7.9	−7.6	−7.4	−7.5	−7.0	−7.9
b	82.5	85.2	87.1	89.8	89.2	86.4	85.2	90.5	90.1	89.4	90.9

Table 7. Relationship between black field lab value and toner concentration

Toner concentration	− 5	− 4	− 3	− 2	− 1	0	1	2	3	4	5
L	18.4	16.7	15.1	14.1	13.5	13.3	14.0	13.6	13.2	13.0	12.8
a	−0.2	−0.3	−0.2	−0.2	0.0	0.0	−0.2	−0.2	−0.2	−0.2	−0.4
b	−0.3	−0.1	−0.4	−0.2	−0.7	−0.7	−0.6	−0.8	−1.0	−0.8	−0.7

toner has an obvious effect on its density, so the test can be performed first by reducing the density of cyan toner by 1 level.

4.4 Results of the Experiment

It can be seen from Fig. 9 that when the cyan toner density decreases, the field density value of cyan decreases significantly. Although the field density of magenta, yellow, and black varies slightly, the ΔE_{00} value of all colors is less than 3. At this point, the generated ICC has fully complied with the the color target. Table 8 shows the four-color field density values under the current equipment state: C: 1.598; M: 1.45; Y: 0.989; K: 1.814. This set of data will be used as a reference for the device state when the ICC is applied in the future.

Fig. 8. Detailed data on the color quality of the ICC

Fig. 9. Optimized ICC color quality

Table 8. Field density values of colors before and after adjusting the toner concentration

Color	Before	After
C	1.673	1.598
M	1.456	1.45
Y	0.988	0.989
K	1.828	1.814

5 Conclusion

The test results prove that the adjustment method can obtain the best quality and best ICC by means of data quantification, without relying on manual experience.

Verify the generated ICC, and judge whether the toner density needs to be adjusted according to the ΔE_{00} value and the L value in the color quality report. If the L value is to be decreased, increase the toner density; otherwise, decrease the toner density. Thereby obtaining the best device state and obtaining the best ICC. Be sure to record the four-color solid density value at this time. When the ICC is applied, the device state, that is, the color density, needs to be adjusted according to the recorded four-color density values. To increase the color density value, increase the toner density, otherwise, decrease the toner density. It should be noted that the value between the adjustment of the toner density and the color solid density value will be affected by the environment, equipment hardware, etc. In the face of a new device and a new color target, a new ICC must be generated and verified and adjusted.

References

1. Chen, Y.: Talking about the color control process of digital printing. Print. Field **2017**(05), 18–21 (2017)
2. Li, G.: Application of color management in digital printing design. Art Eval. **2017**(05), 173–176 (2017)
3. Zhao, C.: Application of color management in digital proofing and printing production. Screen Print. **2021**(09), 57–60 (2021)
4. Zhang, H., Tian, Z.: Exploration of printing color preflight. Print. Field **2018**(12), 30–31 (2018)
5. Gao, X.: Quality inspection and control technology of digital printing. Print Today **2019**(04), 61–64 (2018)
6. Liu, H., Mo, R.: A brief introduction to the importance of color management in digital printing presses. Digit. Print. **2018**(03), 33–35 (2018)
7. Kaifeng, W., Li, H.: Application of CIE2000 color difference formula in paint color difference evaluation. Shanghai Coat. **48**(06), 27–30 (2010)

Research on Collaborative Application of Parametric Design and 3D Printing Based on Complex Shape Packaging Container

Shengyuan Zhao(✉)

Art and Engineering Institute, Tianjin Vocational Institute, Tianjin 300410, China
13906920@qq.com

Abstract. The collaborative application characteristics of parametric design and 3D printing in complex shape packaging containers are studied. By comparing and summarizing the advantages and disadvantages and causes of the three key factors of 'scheme formulation (modeling), material selection and process, data acquisition (data modeling) and model generation' in the collaborative application of parametric design and 3D printing in the prototype development and customization process of packaging containers. The characteristics and optimization inspiration of the collaborative application of parametric design and 3D printing to packaging containers are analyzed with cases. This paper discusses how to solve the problem of container modeling and diversified customization of fine structure in parametric design of packaging container, and Lists some suggestions for future research topics.

Keywords: Complex shape · Packaging container · Parametric design · 3D printing technology · 3D printing materials

1 Introduction

Under the guidance of 'human-machine collaboration industry 5.0', 'China intelligent manufacturing 2035' and 'The Belt and Road Initiative', '3D printing technology' has not only been deeply developed and applied in high-tech key areas such as aerospace, construction, medical treatment, new energy, cultural relics protection, etc., but also in fashion industries such as jewellery, clothing and footwear. Compared with the traditional industrial mold opening method, it is limited by printing materials, printing speed and accuracy, supporting software and hardware equipment, process, cost and other issues, and cannot be mass-produced, which restrict the development of the 3D industry. However, with the development of digitalization, the packaging manufacturing of daily FMCG ranks also has a good application prospect under the coordinated development of parametric design and 3D printing technology.

M. Xu et al. (Eds.): CACPP 2022, LNEE 991, pp. 207–216, 2023.
https://doi.org/10.1007/978-981-19-9024-3_27

2 Collaborative Application of Parametric Design and 3D Printing, Analysis of Key Factors, Advantages, Disadvantages and Causes of Packaging Container (Prototype Development and Customization) Process

Packaging containers are mainly classified according to the use of commodities, such as cosmetics, washing supplies, medicines, food, and daily necessities. Classify bottles, cups, cans, pipes, basins, buckets, pots, etc. according to the modeling features. The main process of packaging container manufacturing is: scheme formulation (including capacity calculation and cost budget), material selection and process, data acquisition (data modeling) and model generation, data model modification and model post-processing, determination of scheme production, etc. The collaborative application of parametric design and 3D printing to packaging modeling model mainly depends on three key factors: scheme formulation (modeling), material selection and process, data acquisition (data modeling) and model generation. The following is mainly analyzed from these three aspects.

2.1 Parametric Design in Collaboration with 3D Printing for Packaging Containers (Prototype Development and Customization) Analysis of Strengths and Weaknesses in the Formulation Phase

Parametric design is parametric design, that is, the design of all aspects and elements of the variable, that is, the design is controlled by the parameters: when changing the value of the parameters, the design results will automatically change [1].

The advantage is that the same parameter model can generate multiple records at the same time after inputting different variables, which greatly improves work efficiency. Specific visual effects can be randomly generated by adjusting parameter variables. The previous design idea was subverted. The original idea was to first conceive the shape and structure of the packaging container, then draw a sketch, and then model and shape. The parametric design can first build the logical relationship of parameters, and then modify the parameters and logical adjustment to determine the shape and then 3D printing. The disadvantage is that due to randomness, data replication is more difficult to achieve, and the complexity of design and structure is relatively high, which makes it difficult to open the mold and cannot be put into production.

2.2 Parametric Design and 3D Printing, Applied to Packaging Containers (Prototype Development and Customization) Material Selection and Process Stage Advantages and Disadvantages Analysis

Different from the traditional manufacturing process, additive manufacturing technology (i.e., 3D printing) is based on a digital model, and is manufactured by slicing three-dimensional parts and stacking materials layer by layer [2]. The 3D printing equipment and technology are mainly introduced by Stratasys, a professional 3D printing equipment manufacturer in the United States, supplemented by domestic research and development. The mainstream print technologies suitable for package container design mainly include

FDM and SLA, and others include DLP, SLS, 3DP, etc. FDM (Fused Deposition Modeling) FDM (Fused Deposition Modeling) Fused deposition manufacturing molding, plastic filaments after melting layer by layer printing molding, are very suitable for hollow vertical packaging containers without supporting materials; SLA (Stereolithography), the use of liquid photosensitive resin under UV irradiation can be quickly cured and formed, more durable than FDM delicate; DLP (Digital Light Procession) digital light projection technology, with digital light source in the form of surface light in liquid photosensitive resin, is faster than SLA, delicate, suitable for small size carving parts; SLS (Selected Laser Sintering) laser selective sintering molding, infrared laser through the supersonic powder feeding layer by layer adhesion, raw materials must be powder (plastic powder, ceramic powder, metal powder, etc.), especially suitable for ceramic, clay, smooth or coarse texture packaging containers such as non-closed modeling; 3DP color powder stacking technology, raw materials are gypsum powder plus resin, which can print color.

The advantage lies in the continuous improvement and development of 3D printing materials and equipment. There are many types of materials and low cost, mainly including plastics, ceramics, paper, gypsum, metal and glass. The following three points are summarized.

Plastic material. Packaging container prototype development, the largest use of 3D printing material is 'plastic'. Suitable categories by physical properties are: engineering plastics, photosensitive resin, thermosetting plastics, bioplastics, polymer gel, etc. [3]. The plastic used in 3D printing technology is melted into liquid or powder, and has fluidity. In the later stage of molding, it will pass, polymerize, solidify, solidify and so on. Disadvantages include cracking, yellowing, etc.

In recent years, 4D printing has set off a boom. 4D printing is actually a shape memory polymer. PCL-biodegradable polyester with shape memory function can be applied to products requiring extrusion and rebound function. It can be customized and biodegradable, so it is more environmentally friendly. Biodegradable materials, such as polylactic acid (PLA), are suitable for small-batch production of disposable biomedical products, meal boxes, and portable cosmetic containers.

Ceramic and plaster materials. The biggest advantage of industrial selective laser sintering technology (SLS) for finer investment casting lies in its wide selection of materials, such as nylon, wax, resin-coated sand (coated sand) ABS, poly carbonates, and even metal powders and ceramic powders. Especially it is suitable for wine containers and washing supplies such as porcelain, metal or glass material simulation. The disadvantage is that the technology that 3D printing ceramic materials need to break through is that the shrinkage rate is large, generally around 30%. This is not easy to control, and it is still very difficult to ensure that the printing accuracy is controlled to 0.1 mm.

Composite paper. The advantage lies in the organic integration of 3D printing with the traditional printing industry, and the realization of small batch production of paper or composite paper. Mcor is the world's first company to have a desktop paper 3D printer as a model manufacturing company. It has developed selective deposition layering (SDL) technology, which can print relatively large-scale packaging product prototypes, print multiple components separately and then glue them. It is particularly suitable for food packaging container development and small-batch production. Domestic has developed 3D printing flexible technology, the disadvantage is still 3D printing accuracy.

2.3 Collaborative Parametric Design and 3D Printing for Data Acquisition (Data Modeling) and Model Generation of Packaging Containers (Prototype Development and Customization)

Its advantages are first reflected in the rapid and accurate design of complex shapes. Traditional product packaging container design and development, using gypsum or sludge as raw materials for model design, has the disadvantages of time-consuming and laborious, imprecise details, difficult modification and lack of texture; the traditional porcelain container needs clay to mold, dry, mold, billet, firing and other complex processes; metal containers using CNC technology requires polishing, painting and other post-processing, has high cost; the auxiliary three-dimensional software design which is convenient for client demonstration realizes the simulation of precise material, modeling and space and can be observed from multiple angles. Convenient parametric data modification can realize both virtual simulation and fast physical simulation. 3D printing technology brings a more intuitive touch to the simulation design of packaging container modeling.

After mold processing, can be directly affixed color printing packaging, simulation terminal finished products. In 2012, 'Shanghai Jahwa' R&D Center, the largest daily chemical giant in China, purchased an Objet Eden350V 3D printer from Stratasys, a famous 3D printing manufacturer in the United States, and successfully opened the parametric design and development of cosmetic packaging containers. It is commendable that the printer can support 18 different materials, and transparent material printing, greatly meet the special needs of cosmetics packaging containers, appearance, feel, capacity (such as liquid level line) and other aspects of rapid verification. It optimizes the initial draft screening and internal evaluation of the program, and can print the mold within 24 h, shortening the development cycle of 70–90% packaging containers.

The following are the main processes of 3D printing in the product prototype design and development stage. First of all, according to the design draft, the parametric modeling method is used to simulate the three-dimensional modeling of the container by using three-dimensional software such as CAD, rhinoceros and Maya, and the OBJ and STL format files are output. The data is input into the Object studio processing software to detect whether the data model has a 'damaged surface' and cannot be printed, and the ratio or specific size of the output model to the object is set, and then input into the Cura slice printing software, automatically or manually add support, perform cross-section layered data scanning and print preview; finally, select suitable materials to quickly print the physical simulation, trim the printed parts in the later stage, compare the printed physical objects, data models and the original design draft, and revise them repeatedly. In addition, UG software "shell extraction" processing command not only greatly saves printing materials, but also is very suitable for hollow container modeling structure; For large-size objects, it can also be printed automatically in blocks to realize shape disassembly.

3 Analyze the Characteristics and Optimization of Parametric Design and 3D Printing Applied to Complex Packaging Modeling

Complex form packaging is evolved from the basic geometry. Such as cosmetics packaging design, to meet the portability and vertical placement, the shape used is mostly small and simple columnar, cubic geometry, other abstract sphere evolution. Too simplified design has long been unable to meet the aesthetic needs of the public for novelty and difference. The collaborative application of parametric design and 3D printing can meet the public's pursuit of complex beauty. The following analyzes the characteristics and optimization inspiration of the coordinated development of the two.

3.1 Parametric Design and 3D Printing Collaborative Application Packaging Container, Can Present a Complex Fine Art, Personalized Features

The modeling structure is too simple, looks rigid and lacks humanization, loses the exquisite and surprise feeling. People prefer to be moderately complex, functional, and fun. Inducted in the famous American design psychologist Donald Norman 'design psychology–coexistence with complexity' theory [4]. A honey packaging by Serbian designer Tamara Mihajlovic, inspired by the honeycomb structure, adopts an irregular section shape as noble and luxurious as diamond. Spine Vodka, a vodka bottle designed by German designer Johannes Schulz, contains sophisticated human skeleton components. Both cases can be summarized as parametric design style. Patrick Schumacher explained that parametricism, like modernist design style, is not a design tool, but a design style, a new art style based on computer language with strong recognition [5]. Mark Gage believes that parametric style is a digital, soft, continuous surface style.

Texture generation is combined with the user's emotional experience [6]. Regarding packaging containers with a complex appearance, according to 183 questionnaires, young people and children between 6 and 28 years of age prefer parametric design and consider it high-tech or artistic; a small number of people do not like, thinking cold cumbersome, meaningless. This article lists six typical textures in which the interviewee selects preferences in a pre-defined parameterized texture. Among the top three textures, 'flow sense' is the most, accounting for 28.37%; followed by 'rule, repeat', accounting for 19.31%; finally, 'semi-stereoscopic, hard edge', accounting for 5.1%. It is found that people like complex shapes, soft flowing rhythm and strong touch. The preset parametric texture style sample is shown in Fig. 1.

Parametric design of Rhino software plug-in Grasshopperis based on programming logic modeling visualization, through Grasshopper curve dry winding, lattice gradient interference, random flow curve, gradient Tyson polygon, wave gradient skin, bidirectional flow gradient skin, Cosn curve texture, etc. Adjust variables, change density, gradient, rhythm state, associate color and cell size, and imitate weaving and origami folding polygons.

Fig. 1. Preset parameterized texture style sample

3.2 Although 3D Printing Technology Has Certain Limitations, It Can Bring Some Optimization Inspiration to Packaging Modeling When Cooperating with Parametric Design

For example, the surface roughness of the 3D printing model is mainly adjusted by adjusting the scanning speed and extrusion diameter. There are often problems of insufficient scanning speed and printing accuracy. To achieve a better smooth curing effect, you need to reduce the scanning speed. However, if the FDM technology of ABS and PLA (polylactic acid: biodegradable material) is used to increase the speed, special hairy texture effects can be achieved, such as animal hair and rock surface.

Nylon materials are prone to warpage and fracture. Polymaker combined 3D printing technology to improve PLA materials and improve their mechanical strength, and developed PolyMax ductile consumables to solve the problem. If you want to achieve a larger mesh elastic texture structure (as Fig. 2 shows), liquid polymer materials can be used. This defect can be used in the parametric design of packaging containers. Irregular curvatures or warped angles are used in modeling, such as folded laminated skin (as Fig. 3 shows), petals, and eaves.

4 Conclusions

With regard to the collaborative application of parametric design and 3D printing, how to better realize the diversification of fine morphology and structure of packaging modeling and the future research and development trend are expounded from the following four aspects.

First, 3D printing materials and technologies suitable for complex packaging containers are constantly refurbished. From paper to ceramic and transparent silica gel printing, the quality and details of various polymer materials, metal materials, ceramic materials and composite materials have been greatly improved and improved.4D memory polymer degradable printing material, which is a sub-micron grid with basic layer composition structure (see Fig. 4), has the characteristics that the temperature changes with time, changing its geometric shape and optical characteristics, such as the shape

Fig. 2. Hollow modeling of the Grasshopper plug-in

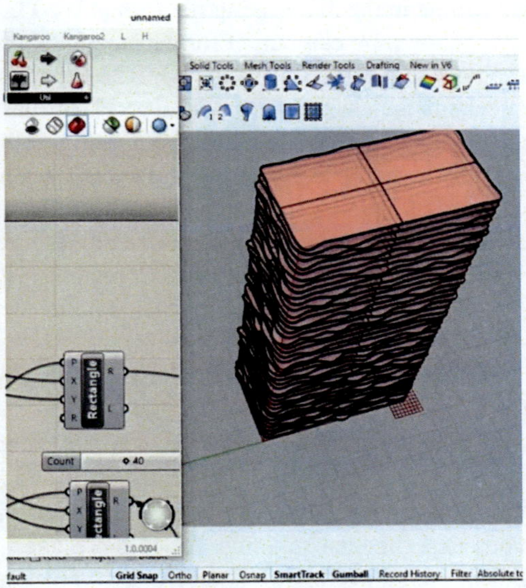

Fig. 3. Folded skin texture of architecture building software

memory polymer sub-micron 4D printing developed by the research team of Singapore University of Science and Technology Design (SUTD) [7]. In the future, it can be applied to complex micro-elastic structures such as spacers, folding, weaving, and hollowing in packaging, but the cost price is relatively high.

Second, Due to the powerful processing function of 3D printing equipment, the model accuracy of parametric design is improved. 3D printing can imitate or even surpass the fine difficulty that traditional handicrafts can achieve, so that the subtle morphological structure of packaging modeling can be diversified. Most parametric modeling 3D model softwares use triangular mesh, whose structure is very flexible, with combination of

Fig. 4. Submicron grid structure

various forms, diamond, square, polygon, and such information is easy to be processed by computer. Although the surface of the curved layer is fine, the triangle is always a plane, not a surface [8].

The most common format in the 3D printing file format is STL. Most of the parameterized 3D modeling data software can output formats are STL, obj and vrml formats, and the more popular conversion format software 3D trans Vidia, TransMagic, etc. The AMF format not only retains the characteristics of the STL format of the curved mesh structure, but also has the advanced nature of the parametric design software and 3D printer equipment, which is perfectly explained by the new functions. For example, different colors, different types of materials, creation of mesh structures, and other detailed and complex internal structures (just one of the unique advantages of additive manufacturing) can all be processed in AMF format files. Compared with the plane triangle used in STL format, the AMF format of surface triangle is more accurate and concise in recording surface data.

Three, Parametric design and 3D printing collaboration, contribute to the complex shape modeling fine structure diversification.

With the emergence of new parametric software, complex structures, such as integrated structure, bionic structure, lightweight lattice structure and hollow lattice structure, which can realize parametric design, can be put into packaging production, and it still needs continuous experimental research. Since the biggest advantage of additive manufacturing lies in the realization of a variety of complex structures, structural design innovation and optimization is the core of additive manufacturing deepening applications [9]. This is well reflected by parametric design, hollow structure, buffer structure, folding structure, packaging internal barrier structure, bionic structure, fold texture, etc. In the structural design research for additive manufacturing, such as hollow lattice structure as a lightweight, multi-functional structure, has become a key research field of additive manufacturing [10]. There are natural hollow structures in nature, such as bones, sponge honeycombs and spider webs[11]. Hollow lattice structure can achieve engineering strength, toughness, durability, static and dynamic performance and cost balance. Complex mathematical topology can be obtained by designing specific parameters. According to the different filling methods of lattice elements, lattice structure can be divided into regular lattice, conformal lattice and random lattice [12]. At present, foreign engineering softwares such as PTC CREO, Siemens NX, Autodesk Netfabb,

Materialise Magics, and domestic CAD, etc. can serve the functions of 3D printers and are most suitable for parametric design of hollow structures. Magics 'main function are model repair, building support, slicing, etc., while 3-matic's main function are to generate complex surface textures and lightweight construction. Visual parametric design software is the most flexible and widely used Grasshopper plug-in for Rhino software.

Fourth, we Improve the quality optimization problems such as chromaticity, whiteness, roughness, transparency, roughness and mechanical principle after 3D printing.

3D printing color reproduction problem: Compared to the traditional manual coloring, dip dyeing, electroplating paint, nanometer mirror spraying, we can use full-color printing, solid color printing (FDM), multi-color printing, local gradient color printing, which make us more efficient and convenient at work. Stratasys Ltd. Currently offers 500,000 colors. To prevent color interpenetration, when two colors or two materials neighbour, we create a print object–Wipe tower. The light-cured SLA is mainly white and translucent.

Model product surface roughness can be used sandpaper grinding, ketone polishing, soaking PLA polishing solution, the appearance of sandblasting treatment. In particular, metal 3D printing can be 'double laser technology' noise reduction.

In parameterized design, we must be based on the principle of mechanical requirements of the shape, whether the technology can meet the printing quality. The mechanical principles of different fiber characteristics have different anisotropy. For example, the greater the bending and inclined angle or the higher vertical distance often produce problems, and the greater the vertical diameter, the greater the failure strain.

In terms of the discussion above, the two synergistically are applied to the future development trend of packaging containers. With the rapid development of 4D memory polymer elastic materials, the combination of 3D printing with virtual reality and augmented reality, the optimization of 3D printing quality, and the continuous reduction of 3D printing material costs, 3D printing has a better application prospect. The application fields of hollowing out, simulated paper folding, bamboo or wool weaving, scalability, lightweight, and stiffness optimization in packaging containers [13] can be further studied. In addition, many optimization problems after 3D printing model forming, such as paper-based full-color 3D printing, powder-based full-color 3D printing, plastic-based full-color 3D printing color reproduction quality evaluation research, are also one of the future research hotspots [14].

5 Concluding Remarks

3D printing technology is an important project of industrial 5.0, 2035 intelligent manufacturing in China. At present, 3D printing is still mainly used in prototype research and development, followed by personalized customization and small batch production. The development of complex product structure by parametric design collaborative 3D printing technology has become a key research field of additive manufacturing. It is a production mode that improves production efficiency and effectively controls costs. Although due to the defects of 3D printing technology and its materials, there are some limitations in the coordination with parametric design, but compared with traditional modeling, parametric design is good at parametric variation, which helps the virtual simulation of product packaging; compared with traditional mold opening, 3D printing technology is better at solving complex and personalized structural components, helping product forming and customization. They complement each other and develop in coordination, which greatly enriches and satisfies the product packaging shape and structure, and enhances the aesthetic appreciation of popular art and the quality of life.

Acknowledgements. Phased Research Results of Tianjin 2010 Philosophical and Social Science Research Plan Funding Project 'Practical Research on School-enterprise Cooperation to Improve the Core Competence of Tianjin Characteristic Cultural Tourism Products' (Approval No. TJYY10-2-449).

References

1. Bian, J.: Parametric Investigation of the Product Modeling Design. China Academy of Art, Hangzhou (2012)
2. Paolini, A., Kollmannsberger, S., Rank, E.: Additive manufacturing in construction: a review on processes, applications, and digital planning methods. Addit. Manuf. **30**, 100894 (2019)
3. Wang, Y.: Rapid Prototyping and Vacuum Casting Technology and Application, pp. 137–139. Xi'an Jiaotong University Press, Xi'an (2014)
4. Norman, D.: Design Psychology-Coexistence with Complexity, pp. 157–158. CITIC Press, Beijing (2015)
5. Zhu, G.: Dialectical exploration of "parameterism" and "parametric design." Archit. Cult. **8**, 184–185 (2019)
6. Wu, C., Gao, T., Meng, Y.: Parametric product design model based on positive experience. Packag. Eng. **42**(06), 142–150 (2021)
7. Wang, L., Leng, J., Du, S.: Research and application progress of 4D printed shape memory polymers and their composites. J. Harbin Inst. Technol. **52**(06), 227–244 (2020)
8. Hudi, L., Kuman, M.: 3D Printing: From Imagination to Reality. Translated by the Expert Group of Sadie Research Institute, pp. 118–119. CITIC Press, Beijing (2013)
9. Lu, B.: Additive manufacturing technology-present and future. China Mech. Eng. **31**(1), 19–23 (2020)
10. Helou, M., Kara, S.: Design, analysis and manufacturing of lattice structures: an overview. Int. J. Comput. Integr. Manuf. **31**(3), 243–261 (2018)
11. Zhang, Q., Yang, X., Li, P., et al.: Bioinspired engineering of honey comb structure—using nature to inspire human innovation. Prog. Mater Sci. **74**, 332–400 (2015)
12. Tao, W., Leu, M.C.: Design of lattice structure for additive manufacturing. In: 2016 International Symposium on Flexible Automation (ISFA), pp. 325–332. IEEE, Cleveland, OH, USA (2016)
13. Hu, J.: Design and Optimization of Complex Topology for 3D Printing. Dalian University of Technology, Dalian (2021)
14. Wang, X.C., Chen, C., Yuan, J.P., et al.: Color reproduction accuracy promotion of 3D-printed surfaces based on microscopic image analysis. Int. J. Pattern Recognit. Artif. Intell. **34**(1), 2054004 (2020)

Conductive Electrode Quality Research Based on Screen Printing Technology

Yingmei Zhou(✉) and Junwei Qiao

Printing and Packaging Department, Shanghai Publishing and Printing College, Shanghai, China
47131532@qq.com

Abstract. With the development of printed electronics on application, printing technology made more and more effect on conductive electrode. The printed conductive electrode has obvious advantages that it could apply to EL, heating film element, printed microchips, flexible sensors and RFID. Due to the conductive electrode functional application, printing technology including screen printing, Inkjet printing and flexography printing usually is chosen for its green, low cost and convenient. Specially screen printing technology has widely used. Commonly the screen printing quality is related with ink viscosity, line width, printing pressure, printing speed, roughness and so on. In order to improve conductive electrode quality, the study focuses on some parameters like squeegee Shore hardness, type of mesh, mesh count and ink viscosity, etc. This study compared line width and roughness under same screen mesh count. Through high mesh count, the thinner line was printed and good conductive property was got. According to the experimental measurement of surface roughness and width, which showed ink viscosity more important than other option and the conductive electrode of nano-silver showed good bending and higher conductive ability.

Keywords: Printed electronics · RFIID · Screen printing · Shore hardness · Roughness

1 Introduction

Printed electronics is the mixture of printing and electric technology. Compared with traditional etching technique, printed electronics has the advantage with low cost, flexible, green and so on, which is concerned as the essential way to produce electronic components. The screen printing technology is an essential option. Even though many researches on how to control the printing technology to make sure the output quality, there are still some unthoughtful issues for those study.

Liu [1] built one physical model to compare the quality influence factors and gave order with screen parameters, squeegee pressure, off-screen gap, ink viscosity, squeegee angle and squeegee speed. In fact, these parameters should be calculated in certain condition such as the machine type, screen mesh. Li [2] showed us about the fixed blade position to reduce the impact factor in movement process to control the image quality. Tian [3] set one experimental to look for the rules based on the orthogonal test

M. Xu et al. (Eds.): CACPP 2022, LNEE 991, pp. 217–222, 2023.
https://doi.org/10.1007/978-981-19-9024-3_28

analysis. Qiao [4] improved effectively the print quality by neural network. Among these papers, lots of parameters compared or ranked. However, the printing machine types are different, there are various requirements in ink viscosity, mesh material, mesh count, squeegee hardness, etc.

2 Experimental and Methods

2.1 Screen Printing Quality Key Parameters

According to the basic screen printing technology theory, the components should work together to complete the process. Screen printing machine includes so many components that we should get the main keys.

The mesh count and the ink viscosity are related with images output requirement. In order to make sure high quality of the different products, many experiments will be compared by screen printing process.

The ink viscosity decided the conductive electrode resistance because of different fluid movement power. Its unit is Pa s, which will fluctuate with the substrate humid. In order to overcome the bad quality, the soft or hard squeegee should be choosing for the different ink viscosity in Fig. 1.

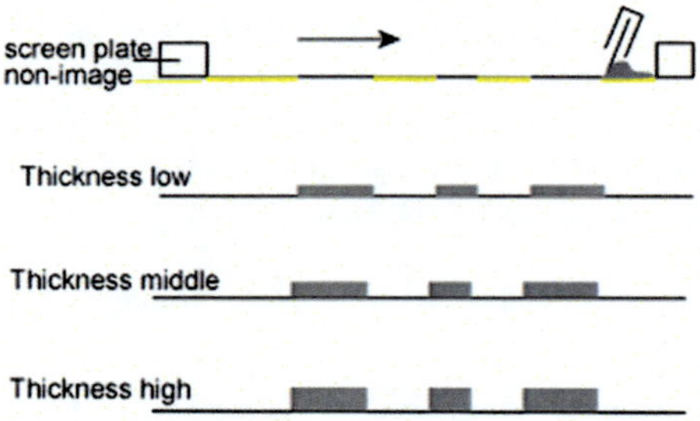

Fig. 1. Plate surface roughness measurement principle

The mesh count decides the resolution. It means the space between the rows and the columns. People waved the mesh with various patterns. There are many options such as mesh count, the thicker, the bigger gap which influenced the ink throughput.

The screen roughness describes the screen surface uniform. Usually, it is a reference for check the printing quality as Rz, which measured by the surface curve tool as the function in Fig. 2.

$$\mathrm{Rz} = \frac{\sum_{i=1}^{5} y_{pi}}{5} + \frac{\sum_{i=1}^{5} y_{vi}}{5} \tag{1}$$

In the function, y_{pi} means the number i peak value of rough sketches. Y_{vi} means the number i valley value of rough sketches. Rz value comes from mesh count and screen size, which the rougher the screen is, the high value Rz is. If the gap between screen plate

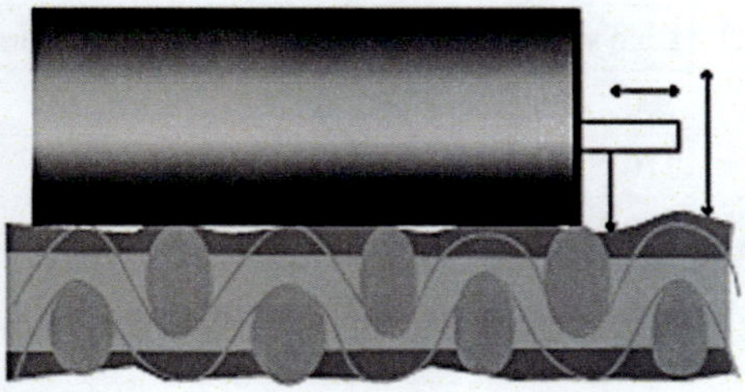

Fig. 2. Plate surface roughness measurement principle

and substrate is too big, the ink will fluid into the substrate to build images with zigzag line or dirty. The ink surface roughness will influence conductive electrode resistance.

This study focuses on the parameter to measure conductive electrode width and density as well. The main aim is to find the data relationship and use the density tool to look for the quality rules among them.

The density is a complex concept. It reflects the light throughout. The function means D in (2), the I_o in (2) means the tool light throughout, and I_r in (2) means the substrate surface light throughout. The reflect rate

$$R = \frac{I_R}{I_O} \quad D = \lg\frac{1}{R} \tag{2}$$

The research focuses on screen machine printing process [5] with parameters such as squeegees, mesh count, roughness, density and ink viscosity to do more application on the personalize products. It is very suitable for the personal printing quality guarantee and data reference.

2.2 Experimental Design and Measurement

The experimental set at least four kinds of mesh count with different width, ink thickness and squeegee hardness in Table 1.

According to designed experimental, these measurements for the surface roughness and density result will be put into Excel. The printing quality of metal mesh is best, but it is very expensive and needs custom-made. The mesh count can be made with 500n/inch. The nylon cost lower and has higher rate of elongation, the polyester mesh has high stretch and good print resistance. The polyester material is also fit for high resolution printing. Usually, the squeegee material is judged by the Shore hardness. The squeegee hardness should be matched with the ink type. The thicker ink needs the softer squeegee. Or changing squeegee angle could get the perfect quality. In order to get the extract result, the experimental data was measured by density tool, line surface roughness, line width and ink viscosity.

Ink viscosity depends on the ink type and squeegee material. Seven types of viscosity are prepared. In some way, the soft is suitable for higher viscosity, and the hard for lower

Table 1. Seven screen parameters setting

Mesh count (number/in.)	Line width (um)	Ink thickness type	Squeegee (shore hardness)
250	200	L1	60
300	150	L2	70
420	100	L3	85
420	50	L4	85
500	20	L5	90
500	10	L6	90
500	10	L7	90

Table 2. Seven kinds of line viscosity and plate roughness

No	Ink viscosity (mPa s)	Plate roughness
L1	2506	14.7
L2	2238	15
L3	1895	11.6
L4	1523	10.9
L5	1065	9.65
L6	890	10
L7	659	7.83

viscosity or adjust the squeegee angle for better result. Ink viscosity and screen plate roughness in Table 2 influenced the density and electrode line width and resistance after bending.

3 Result and Analysis

3.1 Parameters Comparison

The thicker ink we prepared, the more roughness line we got. L5 ink is special and its roughness is so low. So compared the density help the study continued with the special number high to 1.79 in Fig. 3. This data 1.79 means printing ink here should be very uniform.

According to Fig. 4, the ideal line width is above the measured line under the amount with the ink viscosity 1065 mPa s. L5 is at the turning point, which tells us the same situation with density and roughness comparison. Thinking of the conductive electrode application field, the test continued to measure these lines resistance respectively. The conductive electrode resistance below thickness 1065 was lower than above the thickness 1065. Due to the line conductive property, the line width was so thin that it could broke

Fig. 3. Roughness and density comparison

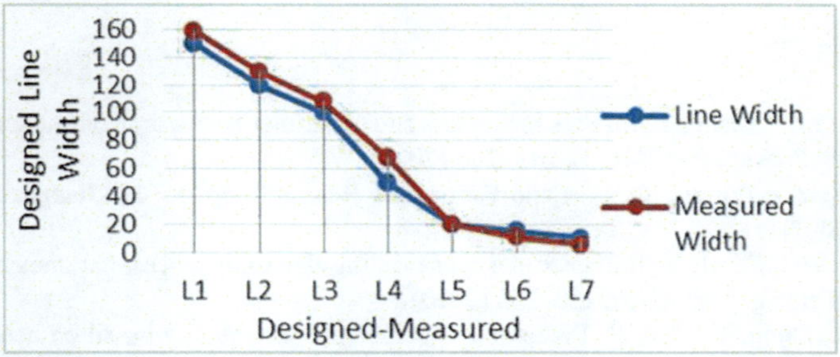

Fig. 4. Comparison of designed and output

Fig. 5. Electrode appearance of seven lines

and could not be the used for conductive electrode. This condition was very suitable for the conductive resistance in Fig. 5.

4 Conclusion

This paper compared the ink viscosity, line width, conductive ability, squeegee hardness, plate roughness and density. We found the plate roughness fluctuated with the ink thickness, and the conductive electrode density changed along with the ink viscosity too. The roughness and density keep changing in harmony, and L5 with ink viscosity 1065 mPa s had low roughness and high density value.

According to this study, we found the ink viscosity, screen printing plate roughness and optical density made more effect on electrode conductive ability with nano silver ink.

Acknowledgements. The paper is supported by Key Laboratory of National Press and Publication Administration: Green Platemaking and Standardization for Flexographic Printing (Project: ZBKT201706).

References

1. Liu, S., Li, Y., Zhang, Y.: Influencing factors and experimental research of fine screen printing electrons. Packag. Eng. **41**(13), 243–250 (2020)
2. Li, F.: Screen Printing Process and Parameters Research, pp. 15–20. Jiangnan University, Wuxi (2007)
3. Tian, Y., Li, Y., Yuan, Y.: Influence of screen printing electronic process parameters on printing quality. Packag. Eng. **41**(5), 250–258 (2020)
4. Qiao, H., Dong, Y., Gao, F.: Prediction of screen printing quality based on neural network. Electron. Process Technol. **42**(5), 281–284 (2021)
5. Tang, Z., Li, F., Weng, X., An, J.: Study on resolution characteristics of screen printing. Packag. Eng. **27**(2), 90–93 (2006)
6. Li, J.: Research on Fine Screen Printing Technology, pp. 16–18. Jiangnan University, Wuxi (2009)
7. Liu, F.: Research on Key Technologies of Screen Printing Electronics Based on TRIZ Flow Analysis, pp. 38–47. Beijing Institute of Graphic Communication, Beijing (2019)
8. Ti, Y., Li, Y., et al.: Influence of screen printing electronic process parameters on printing quality. Packag. Eng. **41**(5), 250–259 (2020)
9. Tang, J.: Transparent conductive graphene films. Progr. Chem. **24**(4), 501–511 (2012)
10. Lee, S.W., Yabuuchi, N., Gallant, B.M., et al.: High-power lithium batteries from functionalized carbon-nanotube electrodes. Nat. Nanotechol. **5**(7), 531–537 (2010)
11. Wang, T., Farajollahi, M., Choi, Y.S., et al.: Electroactive polymers for sensing. Interface Focus **6**(4), 1–19 (2016)
12. Hui, Y., Chen, T., Wang, H., et al.: One-pot rapid synthesis of high aspect ratio silver nanowires for transparent conductive electrodes. Mater. Res. Bull. **102**, 79–85 (2018)
13. Guo, C.F., Ren, Z.: Flexible transparent conductive based on metal nanowire networks. Mater. Today **18**(3), 143–154 (2015)
14. Han, K., Xie, M., Zhang, L., et al.: Fully solution processed semi-transparent perovskite solar cells with spray-coated silver nanowire/ZnO composite top electrode. Sol. Energy Mater. Sol. Cells **185**, 399–405 (2018)
15. Lan, W., Chen, Y., Yang, Z., et al.: Ultraflexible transparent film heater made of Ag nanowire/PVA composite for rapid-response thermotherapy pads. ACS Appl. Mater. Interfaces **9**(7), 6644–6651 (2017)

Analysis of Drying Characteristics of Suspension Oven Substrate Based on CFD

Qiumin Wu(✉), Teng Liu, and Xinkang Jiao

Faculty of Printing, Packing Engineering and Digital Media Technology, Xi'an University of Technology, Xi'an, China
wuqiumin@xaut.edu.cn

Abstract. In order to study the effect of different slit sizes in the suspension oven on the drying characteristics of the substrate surface, a model of the suspension oven was established according to the actual geometric size, and the flow field was simulated by Fluent software, and the turbulent flow was used the $SST\ k - \omega$ model do numerically calculate. The research results show that with the increase of the slit width of the air nozzle, the pressure on the surface of the substrate and the wind velocity increase, the maximum temperature of the substrate surface will decrease, and the increased pressure will also lead to the increase of the bending deformation of the suspended substrate.

Keywords: Suspension oven · Numerical simulation · Hot air drying · Drying characteristics

1 Introduction

The suspension drying technology was proposed by Roy Downhawn, the founder of ASI (Advance Systems Inc) in the United States in 1970. The prosperity of the printing industry in the United States in the 1980s significantly promoted the development of suspension drying oven technology [1]. Domestic and foreign scholars began to apply numerical simulation methods to study the suspension oven. Yi et al. analyzed the effect of substrate movement speed and temperature on the drying performance of substrates inside the suspension oven by constructing CFD models [2]. Chang et al. studied the phenomenon of substrate fluctuation between the upper and lower air nozzles due to tension [3]. Moretti determined the tension effect of the substrate by establishing partial differential control equations [4]. The homogeneous flow characteristics as well as the blocking characteristics of the air nozzle were studied by Li et al. [5–7]. Cheng Qianju et al. improved the poor drying quality of the pole piece due to the uneven distribution of jet pressure during the drying process by designing the discharge pressure relief hole on the nozzle [8].

In this paper, the analytical model from the surface of the substrate to the air outlet part is established and analyzed by numerical calculation. The drying performance of the substrate surface under different parameters is analyzed by varying the size of the air outlet, and the change law of the pressure and temperature characteristics of the substrate surface is finally determined with the change in the size and dimension of the air outlet, and the optimal nozzle design can be determined according to this law.

M. Xu et al. (Eds.): CACPP 2022, LNEE 991, pp. 223–229, 2023.
https://doi.org/10.1007/978-981-19-9024-3_29

2 Model Simulation

2.1 Model Analysis

According to the reference data provided by an enterprise, a three-dimensional model of the suspension oven is established. When the oven is drying, the substrate is transferred in a sinusoidal forward motion between the upper and lower air nozzles. In order to reduce the computational effort, the model is simplified here, as shown in Fig. 1.

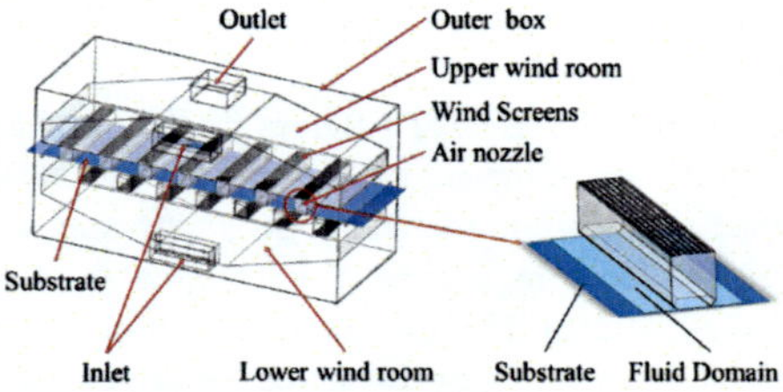

Fig. 1. Schematic diagram of suspension oven and simplified model

The simulation calculation of the simplified model is carried out using Fluent software. The size of the hot air inlet is designed to be 2.5 mm × 200 mm, the overall length of the model is 400 mm, the width is 200 mm, and the height of the air nozzle from the substrate is 5 mm. Ignore the internal temperature conduction of the substrate, and set its surface as a thermostatic wall. The inlet wind velocity is set to 8 m/s, and the hot air is blown in the inner 45° direction, the hot air temperature is 90 °C, and the substrate moving speed is 20 m/min. The rest of the conditions are set by default.

2.2 Mathematical Model of Flow Field Simulation

In this paper, the main focus is on the drying effect of the hot air flow on the surface of the substrate, so $SST\ k - \omega$ model is chosen as the calculation model [9], which is more suitable for low Reynolds number models, and the transmission of wall shear stress is considered in the definition of turbulent viscosity. In addition, the use of this computational model requires at least one mesh node near the wall, which means $y^+ = 1$. Therefore, it is necessary to set the boundary layer height according to the Reynolds number and the value during the meshing, which makes the model more accurate in solving the wall properties. The equations of turbulent kinetic energy k and specific dissipation rate ω are as follows:

$$\frac{\partial}{\partial t}(\rho k) + \frac{\partial}{\partial x_i}(\rho k u_i) = \frac{\partial}{\partial x_j}\left(\Gamma_k \frac{\partial k}{\partial x_j}\right) + G_k - Y_k \tag{1}$$

$$\frac{\partial}{\partial t}(\rho \omega) + \frac{\partial}{\partial x_i}(\rho \omega u_i) = \frac{\partial}{\partial x_i}\left(\Gamma_\omega \frac{\partial \omega}{\partial x_j}\right) + G_\omega - Y_\omega + D_\omega \tag{2}$$

In the above equation, G_k is the generic term of turbulent kinetic energy k due to the mean velocity gradient, G_ω is the generic term of specific dissipation rate ω, Γ_k and

Γ_ω is the effective dispersion rate of turbulent kinetic energy k and specific dissipation rate ω, respectively. Y_k and Y_ω is the dissipation term of turbulent kinetic energy k and specific dissipation rate ω, respectively. D_ω is the cross-diffusion term.

2.3 Flow Field Simulation Analysis

Figure 2 shows the fluid domain boundary conditions and meshing. The air nozzle slit inlet is set as the velocity inlet, and the substrate surface is a thermostatic wall surface. Set the air outlet as the pressure outlet. The rest of the walls are adiabatic walls. The fluid domain is divided by hexahedra, and the mesh near the walls and inlets is relatively tight.

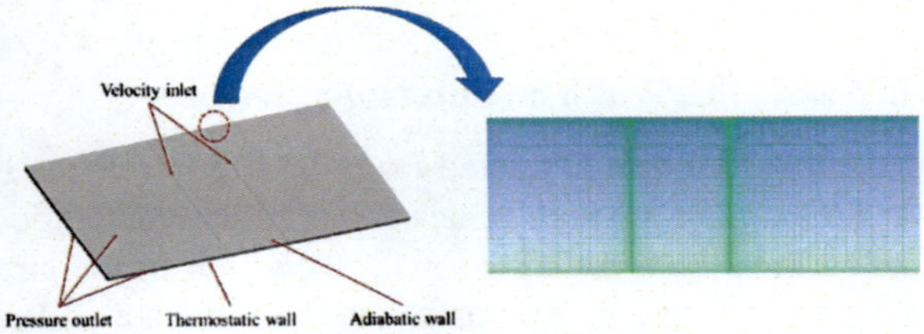

Fig. 2. Fluid domain boundary conditions and meshing

Due to the complexity of calculating the flow field, a pseudo-steady-state method is used to increase the convergence, and the calculation results are shown in Fig. 3. Figure 3(a) shows the trace diagram of the flow field, and it can be seen from the figure that the hot air flow between the two air nozzles is exceptionally complex. Figure 3(b), (c), and (d) show the temperature, pressure, and wind velocity on the surface of the substrate, respectively. As can be seen from the graphs, the values are larger in the vicinity of the air nozzle slit and smaller in the rest of the area. In the area between the two slits, there is a certain degree of oscillation, which affects the drying quality.

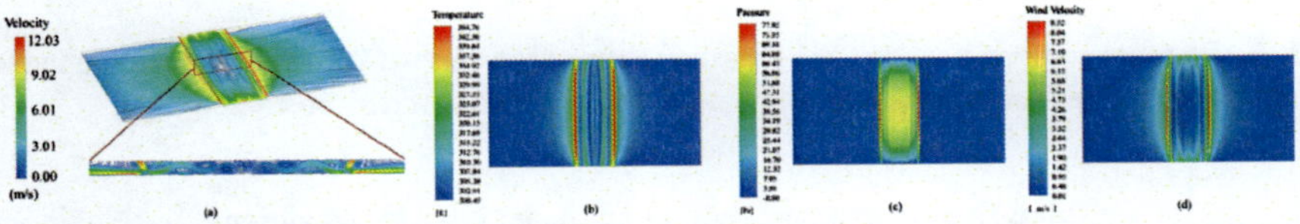

Fig. 3. Flow field simulation analysis chart. (a) The trace diagram of the flow field. (b) Substrate surface temperature cloud map. (c) Pressure cloud map of substrate surface. (d) Cloud map of hot wind velocity on the surface of the substrate

3 Study on the Effect of Air Nozzle Slit Width Size on the Drying Characteristics of Substrates

As shown in Fig. 4, data extraction is performed at L1, L2, L3, and L4 in order to study the variation of pressure, temperature and wind velocity in different regions of the model.

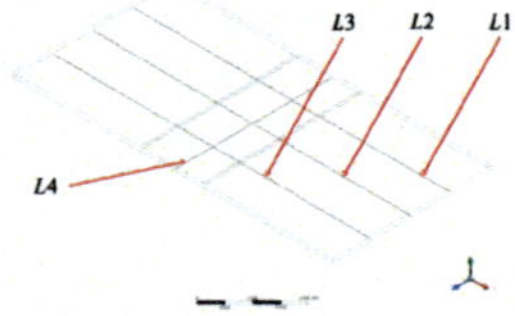

Fig. 4. Schematic diagram of the data collection location

Assuming the air inlet width size of W. This experiment designed a total of four specifications of air nozzle slit width size, respectively, W = 2.5 mm, W = 3.5 mm, W = 4.5 mm, W = 5.5 mm. Selected different slit sizes for simulation calculations, and the analysis results are as follows.

3.1 Effect of Air Nozzle Slit Width on Substrate Pressure

A comparison of the pressure data extracted from the four reference lines L1–L4 are shown in Fig. 5, and maximum pressure values under different conditions were showed in Table 1. As can be seen from the figure, with the width of the air nozzle slit size gradually increasing, the pressure on the surface of the substrate will increase, and the maximum pressure difference between the two slits part of the substrate surface will gradually decrease.

From the data in Fig. 5(d), it can be seen that as the slit size increases, the pressure on the surface of the substrate increases, and the difference between the pressure on the middle part and the edge part also increases, which means that the bending deformation of the substrate increases.

Fig. 5. Pressure comparison chart. (a) Pressure comparison chart at L1. (b) Pressure comparison chart at L2. (c) Pressure comparison chart at L3. (d) Pressure comparison chart at L4

Table 1. Maximum pressure values under different conditions

W/mm	2.5	3.5	4.5	5.5
Maximum pressure/Pa	77.38	101.12	130.35	163.05

3.2 Effect of Air Nozzle Slit Width on Substrate Temperature

Figure 6(a) shows the temperature change curve of the surface of the substrate at L1, L2, and L3. It can be seen from the curve in the figure that the temperature near the slit of the air nozzle is the highest and gradually decreases to both sides, and the existence of the complex vortex airflow between the two slits leads to complex temperature change and oscillation in this part.

Figure 6(b) shows the temperature change curve at L2 with different slit width sizes. The peak temperature of the substrate surface decreases gradually with the gradual increase of the slit width. Table 2 shows maximum temperature values under different conditions.

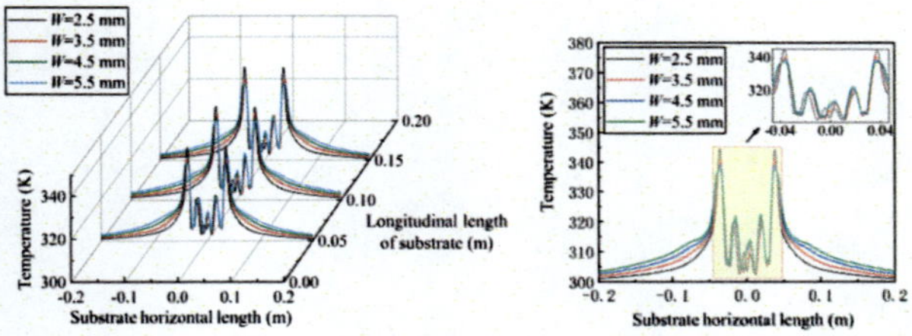

(a) (b)

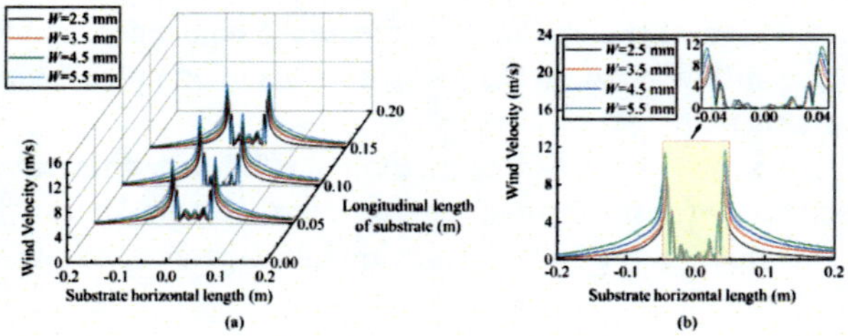
W=2.5 mm
W=3.5 mm
W=4.5 mm
W=5.5 mm
Wind Velocity (m/s)
Substrate horizontal length (m)
Longitudinal length of substrate (m)
(a)
(b)

5. Li, X.J., Gao, D.R., Yang, Z.B., et al.: Study of uniform characteristics of wind velocity field of dryer for lithium battery pole piece. J. Mach. Des. **28**(08), 77–81 (2011)
6. Li, X.J.: Research on the Flow Field Characters of Lithium Battery Pole Piece Drying Cabinets. Yanshan University (2012)
7. Li, X.J., Gao, D.R., Wang, H.S.: Study on experiment and simulation of the flow characteristics of air nozzles used in lithium battery film drying device. J. Mech. Eng. **51**(24), 105–111 (2015)
8. Cheng, Q.J., He, S.Q., Hu, H., et al.: Jet pressure distribution optimization in air nozzle of lithium battery coating oven. Packag. Eng. **40**(05), 180–186 (2019)

Study on Quality Evaluation and Optimization Scheme of White Ink in Flexography

Yan Liu(✉) and Chunyan Bai

Printing and Packaging Engineering Department, Shanghai Publishing and Printing College, Shanghai, China
sppcly@163.com

Abstract. This paper mainly studies the quality evaluation of white ink in flexography and the optimization scheme of printing effect. White ink not only affects the covering power of the package, but also affects the realization of color ink. This paper expounds the influence of white ink on printing quality from two aspects of opacity and leveling, and discusses the feasibility of optimizing white ink by improving the anilox engraving technology and controlling the process parameters of ink transfer process.

Keywords: White ink · Opacity · Flexography

1 Introduction

Flexography has unique advantages in environmental protection and low cost [1]. With people's attention to environmental protection, the continuous improvement of flexography technology and the growth of new market demand, especially because of the efforts of people engaged in flexography, the grade of flexographic printing products has continued to improve, the application scope of flexography is rapidly increasing, flexible packaging market is also gradually expanding.

In flexible packaging, the inner printing is mostly used, and white ink is often used as the primer for inner printing. The printing quality of white ink not only affects the covering power of packaging, but also affects the realization of color ink [2]. How to use less white ink to print flexible packaging products with better hiding power is of great significance to the flexible packaging industry. This paper expounds the influence of white ink printing on printing quality from its printing performance, discusses the improvement scheme of white ink printing, in order to provide some reference for white ink printing in flexible packaging industry.

2 Quality Evaluation of White Ink

The main component of white ink is titanium dioxide. Titanium dioxide is a white opaque pigment with no visible light absorption, high refractive index, and can be made into appropriately sized particles during the production process for maximum opacity

M. Xu et al. (Eds.): CACPP 2022, LNEE 991, pp. 230–235, 2023.
https://doi.org/10.1007/978-981-19-9024-3_30

and whiteness. The main component of titanium dioxide is titanium dioxide, which can absorb ultraviolet rays and has relatively high durability [3]. Just as the color of an oil painting relies on a well-covered primed canvas to maintain vibrancy, the color of a flexible packaging depends on an opaque white ink base to improve color accuracy and hiding power. In the flexographic printing process, opaque white ink is the basis for graphic imaging on a variety of transparent substrates, and it is an important background for brands to express their carefully crafted images. To get high-quality flexible products, white ink must have sufficient opacity and high-quality leveling properties.

2.1 Opacity of White Ink

The opacity of white ink is the ability of the white ink layer to cover. The covering power of white ink refers to the ability of ink film to cover the background color of the painted surface after uniform printing [4]. Covering power is an important performance of white ink. The covering power of ink essentially depends on the ratio of the refractive index of pigment and connecting material. When the ratio is 1, the pigment is transparent. A ratio greater than 1 indicates that the pigment is opaque, which means it has masking power [5]. In practice, a spectrophotometer can be used to measure the opacity of printed matter. Calibrate the instrument before measuring using standard whiteboard and standard blackboard. A standard whiteboard equals 100% opacity and a standard blackboard equals 0% opacity [6]. When measuring, the white printed matter is covered on the black solid for measurement, and the data read is the opacity of the printed matter. Poor opacity of white ink will darken the color of the print and will not mask the contents of the package. As shown in Fig. 1, the image on the right has better color expression due to the higher opacity of white ink.

Fig. 1. Different opacity of whit ink

2.2 Leveling of White Ink

White ink leveling is the ink on the substrate surface flow evenly enough luster without pinhole performance [7]. The opacity of white ink obtained by physical measurement of the amount of light transmitted by the ink layer is only a part of the opacity of white ink. The tightness or uniformity of the white ink layer determines the quality of the opacity.

The uniformity of ink is expressed as the density, smoothness and compactness of the ink layer [8]. If the ink layer formed during printing is not uniform, the ink layer will appear mottles, pinholes and orange peel appearance. In this case, even if the printing thickness of the white ink layer is increased, the print may still have a relatively low transparency.

Pinhole formation is mainly caused by bubbles in the white ink liquid. In the flexographic printing process, when the roller supplement ink, air will be introduced into the ink cavity through the mesh hole on the surface of the roller, and air will be retained in the mesh hole [6]. Therefore, a small amount of air is carried to the substrate along with the ink. When the ink dries, resulting in the formation of uneven ink surface layer, the performance of the pinhole, therefore, the uniformity of the ink layer is damaged, resulting in a reduction in opacity.

There are pinholes on the left and no pinholes on the right in Fig. 2. It can be seen from the comparison that the reflection of the ink layer without pinholes is uniform and very smooth, and the color looks relatively clean.

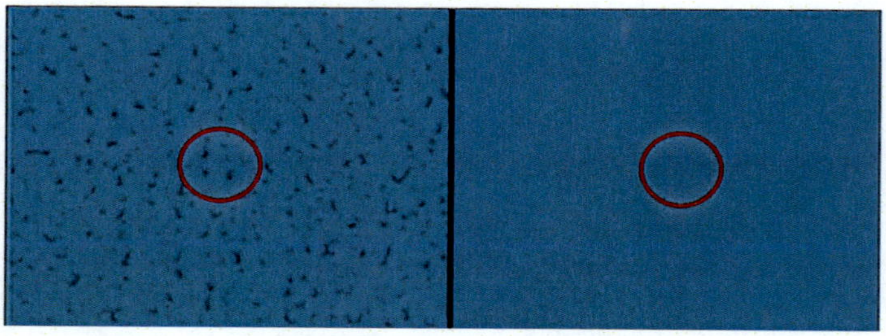

Fig. 2. Comparison of pinholes effect on the plate

3 Optimization of White Ink

White ink is the single most used ink in flexographic production. It is estimated in flexible packaging, a minimum 50% of a printer's ink spend is purely on white ink. Now with the increasing price of titanium oxide, it is more important than ever to get to best effect with the premise of cost control.

In order to improve the covering power of white ink, some printing enterprises solve the problem by printing white ink twice. Printing two times of white ink means 50% more white ink consumption compared to one-color white ink printing. Using too much white ink has many impacts on printing, such as slower printing speed, higher drying temperature, heavier packaging, excessive ink consumption, higher carbon emissions and so on. So conventional remedies by using more white ink, printing two layers or increasing the amount of white pigment are not the best solution.

Through the previous analysis, we have learned that opacity is an important indicator to characterize the covering power of white ink printing, but opacity measures the average calculated value of a region. How to make the effect of white ink printing in this region without "mottles" is the key to control white ink consumption. From the flexographic printing process, to improve the quality of white ink printing, it is necessary to improve and control the ink from the roller to the plate, and then from the plate to the substrate transfer process, as well as on the substrate surface (may be film surface, may also be ink surface) leveling.

3.1 Appropriate Anilox Engraving Specification

The function of the anilox roller is to transfer a certain amount of ink to the surface of the printing plate, as if each cell is a measuring cup. In flexographic printing, the standard anilox type for transferring large volumes of ink is the traditional regular hexagonal cell. In this ink transfer process, air will enter the ink in the form of bubbles. In the case of closed cells, the bubbles may not be completely replaced by the ink after entering the cells, nor will they be removed during the scraping process. Clear, so easy to produce obvious pinholes.

To improve the engraving process of the anilox roller is to improve the traditional closed cell into an open elongated cell structure. Open cell channel designs are generally 30–40% shallower than closed cell depths, with 66% less screen wall area, which means a larger area of ink will touch the plate when it touches the anilox roll, these inks transfer easily. In addition, the open cell structure allows ink to flow within its channels, avoiding the compressive turbulence of the squeegee. Due to the positive ink pressure within the cells and the escape path provided by the open channels for air, little air is trapped in the ink, so ink homogeneity remains unchanged. At present, some companies have used this open-cell anilox roll for production testing [6, 9]. According to the test practice, it is proved that under the same conditions of printing plate, tape, ink, viscosity, substrate, printing speed and printing machine, by improving the engraving process of the anilox roll, the use of open cells allows for higher opacity.

3.2 Appropriate Parameters of Plate Making and Printing Process

In the transfer process of the ink from anilox roller to printing plate, and then from printing plate to substrate, printing parameters such as anilox roller, plate tape, printing plate and printing speed are involved [10]. In order to judge the improvement of white ink printing quality by changing these parameters, three combination schemes were designed. In these three schemes, the printing speed and ink are kept constant, and the printing comparison is carried out by changing the number of screen lines of the anilox roller, the hardness of the printing plate and the plate tape, and the screen treatment of the printing plate surface. The printing parameters of the three processes are shown in Table 1. Process 1 is to evaluate the performance of the original standard printing process combination. Process 2 and 3 are the improvements.

In order to evaluate the effectiveness of the package in each process, the mottle and opacity data of white ink printing were used for analysis and comparison. The mottle factor was measured with a flexographic plate analyzer, and the opacity was measured with a spectrophotometer over white and black tile and confirmed with an opacimeter.

Table 1. Printing parameters of three processes

	Anilox (Lpcm)	Plate	Tape	Plate
Process 1	250	Med plate	Hard	No surface treatment
Process 2	360	Soft plate	Hard	Surface treatment
Process 3	250	Soft plate	Hard	Surface treatment

In process 1, we used a 250 Lpcm anilox roller with a medium-durometer plate, hard tape and no surface treatment. With this combination, the measured mottle factor was 20.09.

Process 2 used a 360 Lpcm anilox roller with a soft durometer plate, an affixing tape with the same hardness and applied a surface treatment. The result of these plate characteristics change is a dramatically reduction in mottle and pin holing. And a measured mottle factor is 0.7.

Process 3 used a lower volume anilox roll 250 Lpcm, a soft durometer plate and hard tape with a surface treatment. The measured mottle factor was 0.8, while the opacity was greatly improved at 56.3%. The results show that the process can improve both speckle factor and opacity. It can be seen that the effect of Process 3 is the most ideal.

This experiment shows that the number of screen lines of the anilox roll is not as high as possible, and choosing the appropriate number of screen lines can help reduce pinholes. In addition, when the number of screen lines of the anilox roller is determined, a printing plate with low hardness is used, and the surface of the printing plate is screened to obtain smoother and higher opacity white ink printing. In general, in white ink printing, the cell structure of the anilox roller, the number of screen lines, the performance of the printing plate and the plate-laying tape will all have an impact on the quality of white ink printing. Reasonable setting of these printing parameters can improve white ink printing.

4 Conclusion

In flexible packaging, white ink printing is a necessary and critical part. Achieving smooth, pinhole-free white ink coverage on substrates requires improved and controlled transfer of ink from anilox roll to plate, plate to substrate, and flow over the substrate surface. Flat. Specifically, in order to get high quality white ink printing, can be achieved by using the open net hole structure of anilox roll to decrease the amount of bubbles and improve the effective transfer of ink. Anilox volume should match the plate and scraper and the plate surface should be treatment. Enhance the mobility of ink, reduce pinhole phenomenon and improve ink in substrate or ink layer leveling.

References

1. Wang, Y.: Satellite Flexography. Cultural Development Press (2021)
2. Wang, Y.: The importance of white ink printing. CI Flexo Tech. **12**, 1 (2021)
3. Wang, L.: How much do you know about white ink. Screen Print. **12**, 37–38 (2014)

4. Zhang, Y., Zhou, Y.: Study on the effect of white ink on bar code quality on laser card paper. Chin. Packag. **5**, 46–49 (2017)
5. Gong, Y., Wei, X., Guo, C., Huang, B.: Studies on relations between dispersivity and covering property of white water-based ink. J. Beijing Inst. Graph. Commun. **12**, 10–12 (2014)
6. Fang, J.: White ink consumption test of anilox roller for different engraving processes. CI Flexo Tech. **6**, 35–41 (2021)
7. Zhang, Y.: Analysis on influencing factors of offset printing ink transfer. Guangdong Print. **2**, 46–49 (2014)
8. Chen, Y.: Discussion on the covering power of gravure white ink. Print. Technol. **29**, 51–52 (2004)
9. White Ink: The most critical color to control. https://www.flexography.org/industry-news/white-ink-the-most-critical-color-to-control/. Last accessed 18 June 2022
10. Whiter Whites: Achieving smooth, pinhole-free ink coverage on transparent film. https://www.flexography.org/industry-news/whiter-whites-achieving-smooth-pinhole-free-ink-coverage-on-transparent-film/. Last accessed 20 June 2022

Packaging Engineering Technology

Experimental Study on Leak Detection of Beverage Using Infrared Thermal Camera

Chuan Zhang, Shengwei Yang, and Enyin Fang(✉)

Department of Printing and Packaging Engineering, Shanghai Publishing and Printing College, Shanghai, China
fensarying@163.com

Abstract. Infrared thermal cameras are often used to detect gas and liquid leaks. The use of infrared thermal cameras can detect and warn of leaks in time. With the increase in demand for food delivery, beverages often leak during delivery due to bumps in delivery and damaged packaging. One way to solve this problem is to use infrared thermal cameras to detect leaks in beverages. In this study, an infrared thermal camera was used to monitor the leakage of hot and cold beverages in an insulated food delivery box. It was found that even the viewing angle of the infrared thermal camera was blocked by the top of the paper cup, the hot water leakage could be found within 120 s after the leakage began. Cold water leakage could be found within 60 s. However, the water leakage could not be detected by the data of the thermocouple. Infrared thermal cameras have broad application prospects in packaging liquid leakage monitoring.

Keywords: Infrared camera · Beverage · Leakage

1 Introduction

During the age of fast food consumption and the COVID-19 pandemic, the services transporting hot food to the consumers' doorsteps are increasing [1, 2]. Beverages in insulated food delivery boxes can easily leak due to bumps or damaged packaging during delivery, which seriously affects the experience of consumers. Infrared thermal cameras can obtain infrared thermal images of an area and detect gas or liquid leaks in time. Furthermore, the early warnings are possible. A large number of studies have focused on the use of infrared thermal cameras to monitor gas [3–6] and liquid [7–9] leaks. The infrared thermal cameras can also be used in insulated food delivery boxes to monitor the temperature in the box, and the leakage of beverages can be detected in time. In this study, an infrared thermal camera is used to observe the leakage of hot beverages and cold beverages in an insulated food delivery box, and the data of the thermocouple in the box are compared.

2 Experimental Method

As shown in Fig. 1, an infrared thermal camera is fixed on the top of a takeaway incubator box to record the thermal images inside the box. The size of the incubator is 44 × 32 ×

M. Xu et al. (Eds.): CACPP 2022, LNEE 991, pp. 239–243, 2023.
https://doi.org/10.1007/978-981-19-9024-3_31

30 cm. This type of box is often used for the delivery of hot or cold beverages such as coffee, tea. A thermocouple is placed inside the front panel of the box to record the inner temperature variation of the box. The thermocouple is 100 mm from the bottom of the box. The container of beverages used in this experiment is a paper cup with a volume of 400 ml. When the experiment starts, the bottom of the cup is pierced to create a 0.5 mm hole to simulate the paper cup being punctured and broken during transportation. For hot drinks, 80 °C hot water is used, and for cold drinks, a mixture of ice and water is used, the volume percentage of ice is less than 30%. After the experiment starts, the computer records the temperature variation measured by the thermocouple every two seconds. The thermal camera also records a thermal image every two seconds.

Fig. 1. Experimental setup

3 Results and Discussions

3.1 Results of Hot Beverage

A thermal image is picked every 30 s for analysis. From the 120 s thermal image in Fig. 2, it can be clearly observed that a prominent high temperature area appears above the paper cup. The temperature of the high temperature area is about 50 °C, which is significantly higher than the temperature of other areas in the box. As the experiment progresses, the high temperature area gradually expands. When the experiment went to the 360th second, a low temperature line appears at the edge of the high temperature area, which was due to the evaporation of hot water. The evaporation takes away the heat, so the low temperature line gradually obvious. The evaporation takes away the heat, so that the low temperature line is gradually obvious after 360 s. As fig the picture shows. As the experiment continues, the edge temperature of the leaking water continues to drop, and the temperature was clearly distinguished from the dry area.

The box temperature of two cases with and without holes on the bottom of the cup is recorded by thermocouple. As shown in Fig. 3, it can be observed from the two temperature variations that the data of the two cases are close, and it is impossible to

Fig. 2. Results of hot water

find the leakage of the paper cup through the temperature variation of the thermocouple. Therefore, some incubators with thermocouple sold and used on the market cannot issue an alarm for liquid leakage.

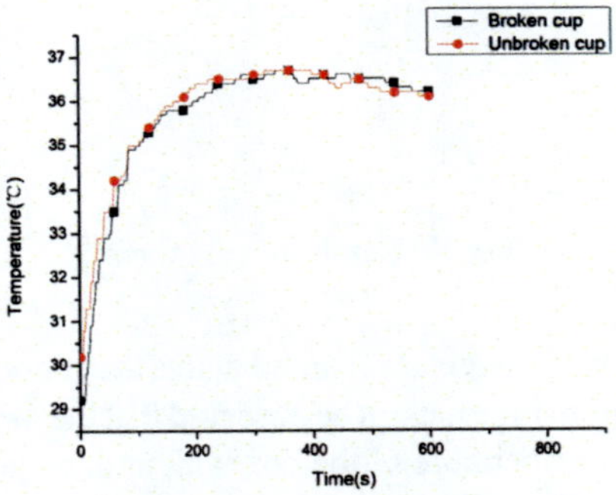

Fig. 3. Thermocouple results of hot water

3.2 Results of Cold Beverage

A thermal image is picked every 30 s for analysis. From Fig. 4, it can be seen that a prominent low temperature area appears above the paper cup in the 60 s thermal image, which is caused by cold water flowing out of the hole. The temperature of the cold water area is 10 °C which is significantly lower than that of other areas in the box. As the experiment progresses, the low temperature area gradually expands.

Fig. 4. Results of cold water

The inner temperature of the box is recorded with a thermocouple, and the paper cups were damaged and not. From Fig. 5, it can be observed that the data of the two temperature variations are close. The temperature of the box with undamaged cup placed in the box is about 1 °C higher than that of the damaged cup. As the experiment progresses, the temperature of the box in the two cases gradually approached. In the usage scenario, a temperature difference of 1 °C is not enough to judge the leakage of cold water in the package contents.

4 Conclusions

A thermal camera is placed in the takeaway incubator box to observe the temperature distribution to obtain a thermal image. In this way, the leakage of the paper cup in the box

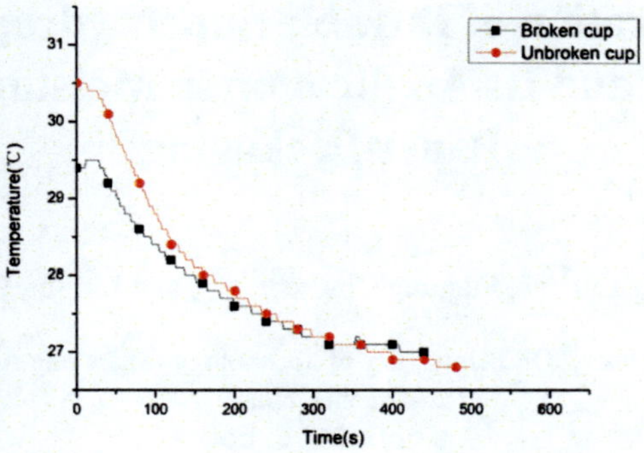
Broken cup
Unbroken cup
31
30
29
28
27
Temperature(℃)
0
100
200
300
400
500
600
Time(s)

Preparation of Durable Superhydrophobic Coating and Its Application in Moisture-Proof Paper Packaging

Chuang Liu[1,2,3], Fuqiang Chu[1,2,3,4](✉), and Liming Qin[1,2,3]

[1] Faculty of Light Industry, Qilu University of Technology (Shandong Academy of Sciences), Jinan, China
fqchu@126.com

[2] State Key Laboratory of Biobased Material and Green Papermaking, Qilu University of Technology (Shandong Academy of Sciences), Jinan, China

[3] Key Laboratory of Green Printing and Packaging Materials and Technology in Universities of Shandong, Faculty of Light Industry, Qilu University of Technology (Shandong Academy of Sciences), Jinan, China

[4] Kiev College, Qilu University of Technology (Shandong Academy of Sciences), Jinan 250353, China

Abstract. The purpose of this study is to improve the water and moisture resistance of the paper packaging, thereby improving the mechanical properties of the paper packaging, to open a larger market for its use. The self-adhesive was sprayed on the filter paper substrate as an adhesive, the silica nanoparticles were used to construct a rough structure, and PDMS (Polydimethylsiloxane) was used to reduce the surface energy. The prepared superhydrophobic filter paper has good hydrophobicity, the static contact angle of 4 μL droplets exceeds 152°, and the rolling angle is less than 5°. The superhydrophobic filter paper has good moisture resistance and still has good tensile strength in a high humidity environment. The super-hydrophobic filter paper is prepared by spraying method, the method is simple, and it can be mass-produced. Because of its good moisture-proof performance, the quality of the packaged objects can be protected for a long time, and the application scope of paper packaging is expanded.

Keywords: Durable · Superhydrophobic · Paper packaging · Moisture-proof

1 Introduction

Paper packaging is loved for its advantages of biodegradability, recyclability, renewability, flexibility, low cost, light weight, good mechanical strength, etc. Since the country promulgated the 'Plastic Restriction Order' [1, 2], it has brought new development opportunities to paper packaging [3]. Smithers Pira, a well-known institution, forecasts the demand of the paper packaging market industry in the next 5 years. Whether it is a developed area or an underdeveloped area, the demand for paper packaging will

M. Xu et al. (Eds.): CACPP 2022, LNEE 991, pp. 244–251, 2023.
https://doi.org/10.1007/978-981-19-9024-3_32

continue to grow in the next few years [4]. However, the defects of paper packaging cannot be ignored. On the one hand, the porous structure of the paper packaging material itself will lead to its poor barrier performance. On the other hand, it is rich in hydroxyl groups, which will make the paper packaging have strong hygroscopic ability [5]. This is an Achilles heel of paper packaging materials. Therefore, for the application of paper packaging, it is very necessary to enhance its moisture resistance.

The preparation of superhydrophobic surfaces has become a hot spot and has been focused in recent years. Its low surface tension and certain roughness can help the free rolling of water droplets. The superhydrophobic surface also has the characteristics of self-cleaning [6–9], anti-corrosion [8, 10], and anti-icing [11, 12], which is very suitable for the application in paper packaging. Spraying is one of the most common ways to prepare superhydrophobic surfaces. Usually, the micro-nano particles are dispersed in a solvent, and then the dispersion system is extruded from the container under pressure to form a mist dispersed on the surface of the substrate, and finally the stable adhesion of the coating is achieved by drying and curing [13].

In this paper, a layer of self-adhesive is first sprayed on the filter paper, and then the silica nanoparticles and PDMS (Polydimethylsiloxane) and its curing agent are mixed and dispersed in ethyl acetate. Finally, use a small spray gun to spray the paint and heat to cure. The preparation of superhydrophobic filter paper can be realized by spraying three layers. The prepared superhydrophobic filter paper can achieve a static contact angle of 4 μL water droplets greater than 152°; it has good friction resistance and moisture resistance, which can significantly improve the filter paper's resistance to water. The barrier properties of water vapor lay the foundation for the moisture-proof application of paper packaging.

2 Experimental

2.1 Experimental Materials and Instruments

Main materials: Super self-adhesive (Super 75 new formula), 3M China Co., Ltd.; Silica (15 nm), Taipeng Metal Materials Co., Ltd.; PDMS (Dow Corning Sylgard 184), Microfluidic Technology Co., Ltd.; Ethyl acetate (AR, $\geq$ 99.5%), Chengdu Kelong Chemical Co., Ltd.; Anhydrous ethanol ($\geq$ 99.7%), Sinopharm Chemical Reagent Co., Ltd.; Deionized water, used without further purification; Qualitative filter paper (Φ = 9 cm, medium speed), Hangzhou Special Paper Co., Ltd.

Main equipment: Electronic balance (JA-1003), Shanghai Hengping Instrument Co., Ltd.; Digital Magnetic Stirrer (DF-101S), Shanghai Zigui Instrument Co., Ltd.; CNC Ultrasonic Cleaner (KQ-600DE), Kunshan Ultrasonic Instrument Co., Ltd. Company; CNC six-inch plate heating plate (HP 380-PRO), Beijing Dalong Xingchuang Experimental Instrument Co., Ltd.; Electric Heating Constant Temperature Blast Drying Oven (DHG-9240), Gongyi Yuhua Instrument Co., Ltd.; Blue Brand LP- 130 Airbrush (0.5 mm), Yongkang Mutai Trading Co., Ltd.; Automatic Contact Angle Measuring Instrument (DSA100), Cluse Scientific Instruments (Shanghai) Co., Ltd.

2.2 Experiments

Preparation of spray coating. Weigh 5 g of silica nanoparticles with an electronic balance, add them in 100 ml of ethyl acetate solvent and ultrasonic for 10 min to make them uniformly dispersed. And then add PDMS (PDMS and its the ratio of curing agent is 10:1). Then the beaker is sealed with laboratory plastic wrap, stirred for 30 min with the help of a digital magnetic stirrer to make it evenly mixed, that is, the required uniform spray coating is formed.

Preparation of superhydrophobic filter paper. Cut the filter paper into a suitable square size (5 cm × 5 cm), first lightly spray a layer of No. 75 super self-adhesive on filter paper. After drying at 60 °C on a heater for 1 min, use an airbrush to spray the paint manually and evenly on the surface of the filter paper that has been sprayed with self-adhesive at 20 cm from the filter paper. One cycle is to spray one layer of self-adhesive and one layer of paint. The preparation of superhydrophobic filter paper can be realized by spraying three cycles. The coating amount is about 4 g/m^2. The preparation of superhydrophobic filter paper is shown in Fig. 1.

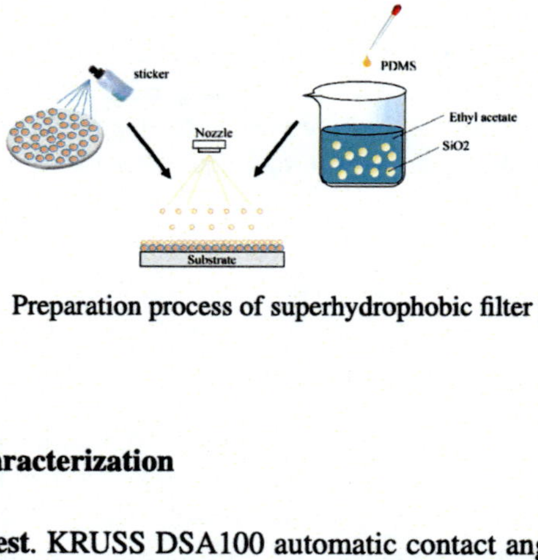

Fig. 1. Preparation process of superhydrophobic filter paper

2.3 Testing and Characterization

Surface wettability test. KRUSS DSA100 automatic contact angle measuring instrument was used to measure the static contact angle of water droplets on the surface of superhydrophobic filter paper. The test water droplet volume was 4 μL. Five parallel experiments were performed on each sample, and the average value was taken as the test result value.

SEM analysis. Cut the samples to be tested and adhere to the platform with conductive tape to spray gold, and use a cold field to transmit high-resolution scan electronic microscope to observe the surface of the ultra-cooked water filter paper.

Mechanical stability test. The mechanical stability of the superhydrophobic surface was tested by sandpapering experiments and characterized by the static contact angle.

Moisture test. Put the qualitative filter paper and superhydrophobic filter paper of the same size in a constant temperature and humidity airtight environment, change the relative humidity of the airtight environment to 50, 60, 70, 80, 90, 100%, and weigh after standing for 24 h. The relative moisture content of qualitative filter paper and superhydrophobic filter paper was measured to characterize their moisture resistance.

3 Results and Discussion

3.1 Wetting Performance Analysis

Influence of SiO_2 mass fraction on wetting properties. As shown in Fig. 2(a), when the mass fraction of SiO_2 is 0.5%, the static contact angle of a 4 μl water droplet is only 133.7°, and when the mass fraction of SiO_2 increases to 1%, the contact angle increases to 144°. But with the further increase of the mass fraction of SiO_2 in the system, the static contact angle shows a decreasing trend, and the hydrophobicity decreases. This may be because when the mass fraction of SiO_2 is too large, the phenomenon of cross-linking between particles will occur, which cannot provide sufficient roughness, so the surface contact angle of the hydrophobic filter paper decreases.

Influence of single-layer spraying amount on wetting performance. As shown in Fig. 2(b), when the spraying amount of each layer is 1ml, the static contact angle of a 4 μL water droplet is only 126.4°, which is because the spray thickness of a single layer is too thin to cover the self-adhesive layer. When it increased to 5 ml, the static contact angle increased to 145.9°, and continued to increase to 10 and 15 ml, and the static contact angle did not increase much. This is because after the self-adhesive layer is covered, if the single layer spraying amount is increased, the roughness of the surface of the filter paper will not change greatly.

Influence of the number of spray layers on the wetting performance. As shown in Fig. 2(c), when the number of sprayed layers is 3, the static contact angle of water droplets can achieve super-hydrophobicity. If the number of spray layers continues to increase, the increase of the static contact angle of water droplets is not obvious. Therefore, spraying three layers is the most ideal situation, and the static contact angle of a 4 μL water droplet can be as high as 152° or more.

3.2 SEM Analysis

The surface morphologies of the untreated filter paper and the modified superhydrophobic filter paper were observed by scanning electron microscopy, and the results are shown in Fig. 3.

As can be seen in Fig. 3a, the surface of the unmodified cotton fiber is smooth and neat, without particles attached. As the super-adhesive and silica nanoparticles accumulate more layers on the surface of the substrate, the particles adsorbed on the fiber surface the number of like substances is also increasing. Figure 3b shows the surface morphology of the superhydrophobic filter paper sprayed with three layers of self-adhesive and paint. The surface of the superhydrophobic filter paper has a dense and rough structure. At this time, the surface roughness of the filter paper is around the micron level. With the further magnification of the microscope, obvious changes in roughness are observed.

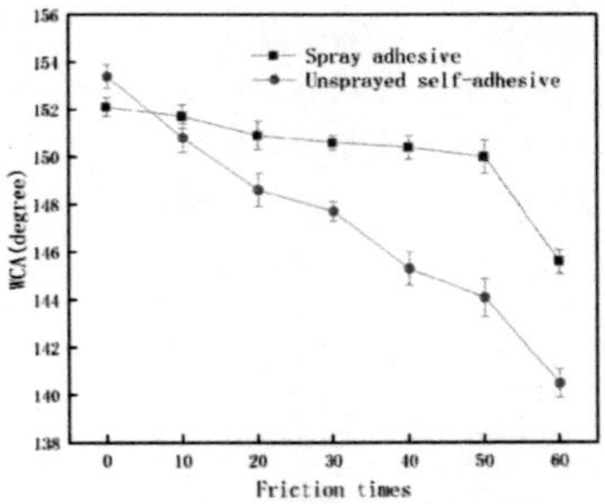

Fig. 4. Variation of surface contact angle of original qualitative filter paper and superhydrophobic filter paper with friction times

3.4 Analysis of Moisture-Proof Performance

The untreated qualitative filter paper and superhydrophobic filter paper were dried in an oven at 80 °C until their own mass no longer changed and recorded as dry weight. Then place it in a closed environment with constant temperature and humidity and change the relative humidity of the closed environment of 50, 60, 70, 80, 90, 100%. The untreated filter paper and the superhydrophobic filter paper after 24 h treatment under the corresponding humidity were weighed and recorded as the sample weight. The moisture content of the sample is calculated by formula 1, and formula 2 is the calculation method of relative moisture content.

$$MC\% = \frac{M1 - M}{M} \tag{1}$$

$$RMC\% = \frac{MC}{MC50\%RH} \tag{2}$$

Among them, MC is the moisture content of the sample, M1 is the weight of the sample, M is the dry weight of the sample, and RMC is the relative moisture content.

Fig. 5. Changes of relative moisture content of original qualitative filter paper and superhydrophobic filter paper under different relative humidity conditions

As shown in Fig. 5, when the relative humidity increased from 50 to 100%, the relative moisture content of the untreated qualitative filter paper increased to 573%. The relative moisture content of superhydrophobic filter paper increased to 151% only when the relative humidity increased to 100%. The superhydrophobic filter paper has excellent moisture resistance.

4 Conclusions

In conclusion, in this paper, super-hydrophobic filter paper was prepared by modifying the surface of qualitative filter paper by using common super self-adhesive and silica nanoparticles combined with a simple and convenient spraying method. The coating is suitable for a variety of paper substrates, such as filter paper, corrugated paper, white cardboard, A4 paper, etc. Good mechanical stability can withstand nearly 60 sandpaper polishing. The preparation method of the superhydrophobic surface is simple, low in cost and high in efficiency, and its excellent moisture-proof performance expands the application scope of paper packaging, laying a foundation for the industrial production of moisture-proof paper packaging.

Acknowledgements. This research is supported by Shandong Province Science and Technology Small and Medium-sized Enterprise Innovation Capability Improvement Project (NO. 2021TSGC1168).

References

1. Sangerlaub, S.: Extrusion coating of paper with poly (PHBV)—packaging related functional properties. Coatings (2019). https://doi.org/10.3390/coatings9070457
2. Nechita, P., Roman, M.: Review on polysaccharides used in coatings for food packaging papers. Coatings (2020). https://doi.org/10.3390/coatings10060566
3. Gao, H., Ding, C., Li, D.: Revision of talent training plan for light chemical engineering professionals based on green paper-based packaging materials. Packag. Eng. **41**(1), 137–139 (2020)
4. Kopacic, S., Walzl, A., Zankel, A.: Alginate and chitosan as a functional barrier for paper-based packaging materials. Coatings (2018). https://doi.org/10.3390/coatings8070235
5. Nakajima, A.: Transparent superhydrophobic thin films with self-cleaning properties. Langmuir, 7044–7047 (2000). https://doi.org/10.1021/la000155k
6. George, J.E., Rodrigues, V.R.M., Mathur, D., Chidangil, S., George, S.D.: Self-cleaning superhydrophobic surfaces with underwater superaerophobicity. Mater. Des., 8–18 (2016). https://doi.org/10.1016/j.matdes.2016.03.104
7. Furstner, R.: Wetting and self-cleaning properties of artificial superhydrophobic surfaces. Langmuir, 956–961 (2005). https://doi.org/10.1021/la0401011
8. Arunima, S.R., Deepa, M.J., Geethanjali, C.V., Saji, V.S., Shibli, S.M.A.: Tuning of hydrophobicity of WO_3-based hot-dip zinc coating with improved self-cleaning and anti-corrosion properties. Appl. Surf. Sci. (2020). https://doi.org/10.1016/j.apsusc.2020.146762
9. Zheng, S., Bellido-Aguilar, D.A., Huang, Y., Zeng, X., Zhang, Q., Chen, Z.: Mechanically robust hydrophobic bio-based epoxy coatings for anti-corrosion application. Surf. Coat. Technol., 43–50 (2019). https://doi.org/10.1016/j.surfcoat.2019.02.020

10. Chao, Q., Meng, L., Shuxian, C.: Anti-Icing characteristics of PTFE super hydrophobic coating on titanium alloy surface. J. Alloy. Compd. (2021). https://doi.org/10.1016/j.jallcom.2020.157907
11. Zhang, D., et al.: Electrospun fibrous membranes with dual-scaled porous structure: super hydrophobicity, super lipophilicity, excellent water adhesion, and anti-icing for highly efficient oil adsorption/separation. ACS Appl. Mater. Interfaces, 5073–5083(2019). https://doi.org/10.1021/acsami.8b19523
12. Kim, S., et al.: Multi-objective Bayesian optimization of super hydrophobic coatings on asphalt concrete surfaces. J. Comput. Des. Eng. (2019). https://doi.org/10.1016/j.jcde.2018.11.005,693-704
13. Zhang, J., Wang, D., Jiang, L., Xia, J., Bo, M., Yao, Z.: Mussel-inspired catechol-based chemistry for direct construction of super-hydrophilic and waterproof coatings on intrinsic hydrophobic surfaces. J. Appl. Polym. Sci. (2019). https://doi.org/10.1002/app.48013

Study on the Adsorption Performance of Graphene for One-Component Solvent

Lehao Lin, Baiqing Sun, Xiaoli Song, Peiyuan Zhu, Gaimei Zhang(✉), Jingjing Hu, and Zhihao Ren

Beijing Institute of Graphic Communication, Beijing, China
Zhang_gaimei@163.com

Abstract. Volatile organic compounds (VOCs) emitted in packaging and printing industry mainly are produced from solvent-based ink, adhesives and organic solvent for dilution during printing, drying, compounding and equipment cleaning process. The treatment technology of high concentration solvents includes adsorption recovery technology for single component VOCs and regenerative thermal oxidizer for multi-component solvents, in which the adsorption process is used. Therefore, the performance of adsorption materials is paid more attention. In this paper, graphene was used as the adsorbent, and its structure had a very good adsorption effect, attributing to its very large specific surface area and adjustable pore structure, macromolecular organic pollutants could easily react with the functional groups on the surface of graphene to form compounds with stable chemical performance, so that the organic pollutants could be removed. To explore the adsorption performance of graphene, the adsorption mechanism and the influencing factors of adsorption performance, the models of graphene and ethanol molecule were established based on material studio, and the adsorption simulation was carried out using multiple modules. The adsorption energy and system energy were obtained and the effectiveness of adsorption materials on adsorbate was analyzed. The results showed that adsorption simulation was an effective method to lay a theoretical foundation for the design and application of adsorption materials.

Keywords: Adsorption method · Graphene · VOCs · Adsorption energy

1 Introduction

The treatment technologies of volatile organic compounds (VOCs) emitted in packaging and printing industry includes recycling technology and destruction technology. It can be purified and recycled by physical methods such as adsorption, condensation and separation, or it is converted to CO_2 and H_2O through thermal catalytic oxidation, biodegradation, photo catalytic oxidation and plasma catalysis [1]. Adsorption method has been widely used in treatment of VOCs and adsorption material is paid attention more and more. Graphene has much excellent properties such as excellent electrical properties, good light transmittance, high mechanical strength and large specific surface

M. Xu et al. (Eds.): CACPP 2022, LNEE 991, pp. 252–260, 2023.
https://doi.org/10.1007/978-981-19-9024-3_33

area [2, 3]. The graphene with big adsorption capacity can play a pivotal role in the treatment of water pollution, it's recycling rate is high [4–6].

Simulation technology based on material studio (MS) is widely used in the fields of structural characterization of material, quantitative calculation of reaction energy and electronic structural properties [7, 8]. In this paper, the adsorption properties of graphene for toluene and ethanol are analyzed. It is valuable to the application of graphene into VOCs treatment.

2 Graphene Adsorption Simulation Based on Material Studio

2.1 Adsorbent and Modeling

The software used in this article was Material Studio 2020.The 3D Graphene model was imported and the calculations were performed. Figure 1 showed the graphene model obtained after supercell. The model of the ethanol molecule was built by adding oxygen, vinyl and hydrogen atoms, as shown in Fig. 2. The model of toluene was built by adding one carbon atom and multiple hydrogen atoms, as shown in Fig. 3.

Fig. 1. Graphene model

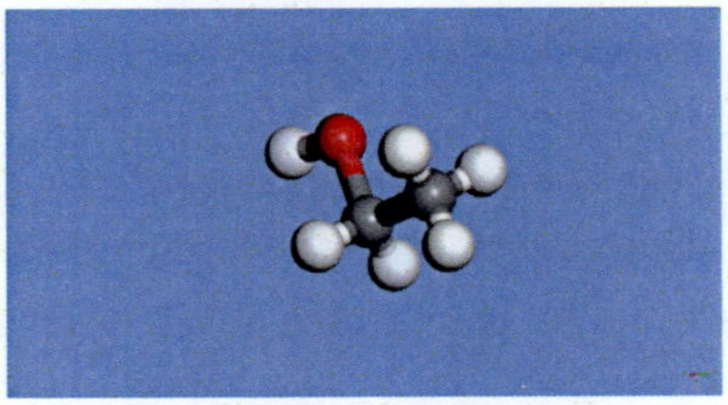

Fig. 2. Ethanol model

2.2 Adsorption Simulation

The graphene and ethanol models were combined and the ethanol was set to the designated position. The constraints of motion along X and Y direction were set for ethanol, so that the ethanol could only move along the Z direction.

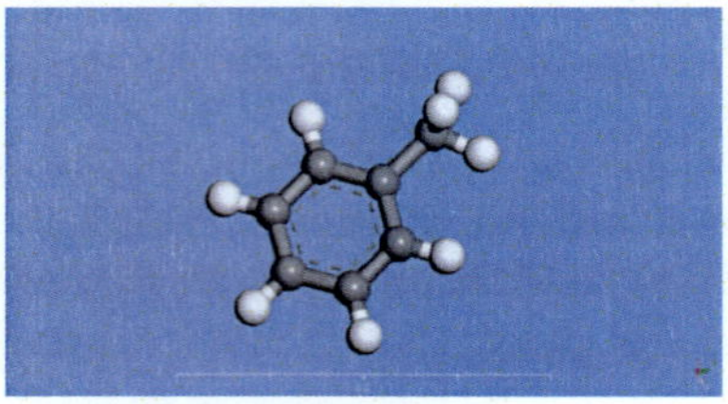

Fig. 3. Toluene model

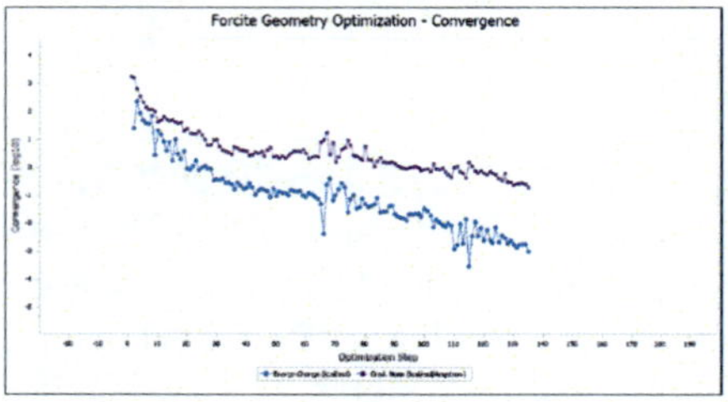

Fig. 4. Convergence change graph

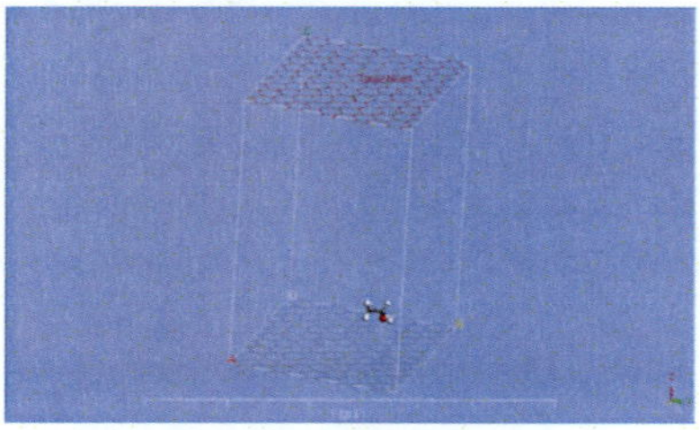

Fig. 5. Optimal adsorption position

The force field was set as compass through the Forcite and the convergence graph after running was shown in Fig. 4. Then the most stable adsorption point was found, the quantity of ethanol molecule was set as one, the force field was set as compass II. After running, the most stable adsorption point was shown in Fig. 5.

In order to explore the changes of the system caused by the change of quantity, the number of ethanol molecules was adjusted as ten, and it was calculated. It could be seen that the molecular position of ethanol on the surface of graphene had changed, as shown in Figs. 6 and 7, while the molecular position far away from the surface of graphene had hardly changed. It could be concluded that the best adsorption position was close to the surface of graphene. The adsorption of graphene to toluene was calculated, as shown in Fig. 8.

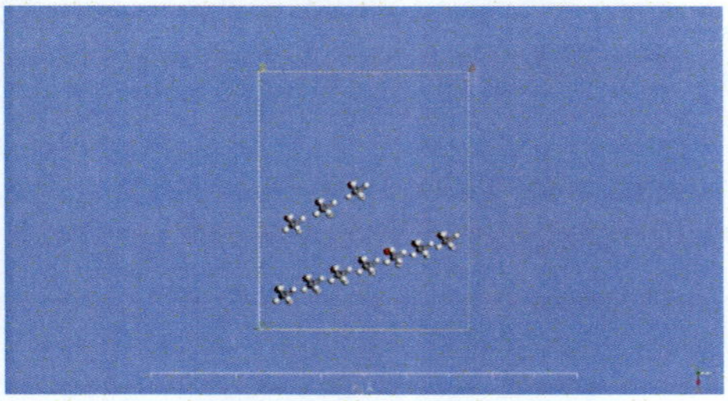

Fig. 6. Before calculation

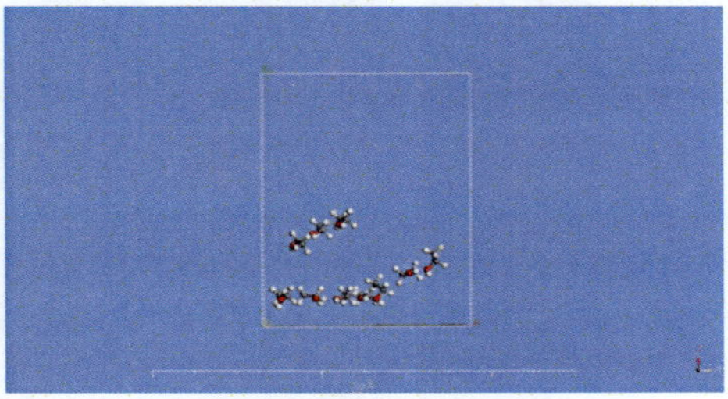

Fig. 7. After calculation

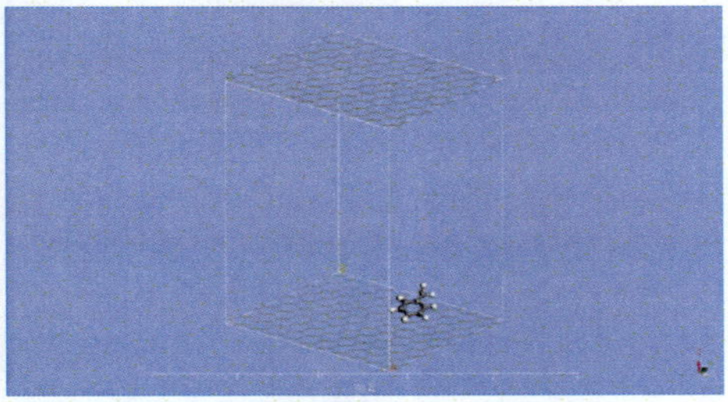

Fig. 8. Adsorption of graphene on toluene

3 Result and Analysis

3.1 Analysis of Graphene Adsorption Properties

The system energy of graphene to single ethanol molecule reduced to 10,290.03 kcal/mol from 10,920.75 kcal/mol after the adsorption reaching stability. The geometric optimization diagram was shown in Fig. 9. It could be seen that the energy of the system decreased gradually with the progress of the adsorption process, shown in Fig. 9. When the geometric optimization reached 80 steps, it tended to be stable. Figure 10 shows the most likely adsorption sites in adsorption. It could be seen that most of the sites were relatively close to the surface of graphene, and these sites were easier to be adsorbed.

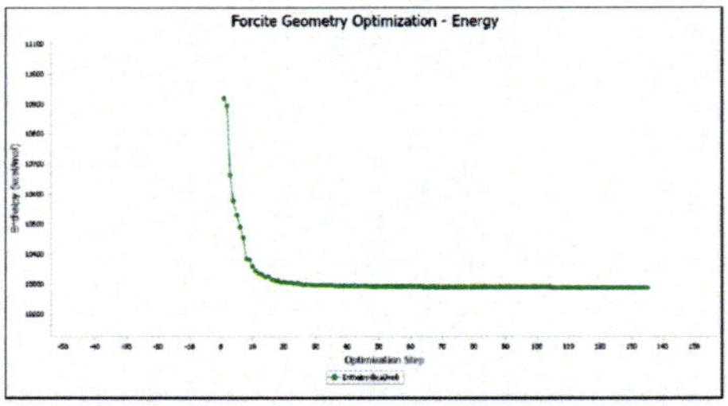

Fig. 9. Geometric optimization diagram of ethanol

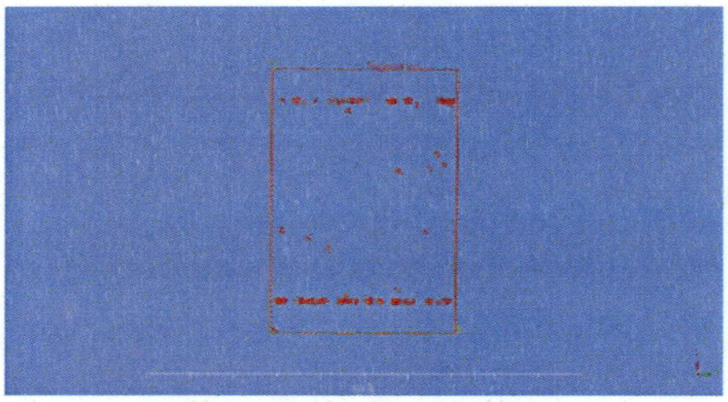

Fig. 10. Possible adsorption sites

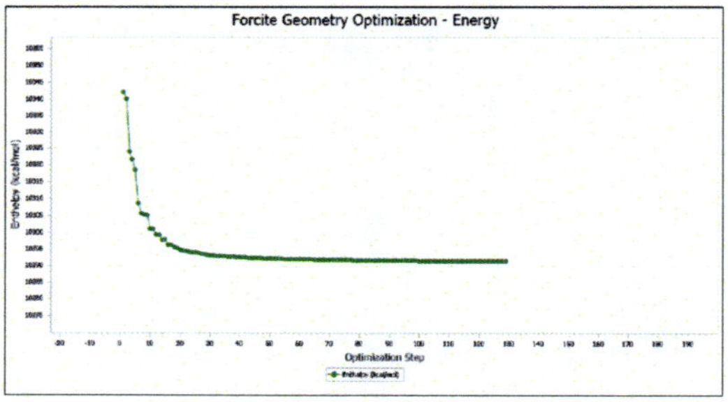

Fig. 11. Geometric optimization diagram of toluene

The energy of graphene and monotoluene molecular system reduced to 10,291.39 kcal/mol from 10,342.09 kcal/mol, shown in the Fig. 11. It could be seen that the total energy of the system with different adsorbates were different.

The adsorption energy of each position for one ethanol molecule was listed in Table 1. It could be seen that the adsorption energy was negative, which indicated that it was beneficial to the adsorption. The adsorption energy was the difference around the stabilized state.

The adsorption energy of graphene to the adsorbates with 10 ethanol molecules is showed in Table 2. The total energy of the system was much larger than that of single ethanol molecule. The value in the second row of the table was the optimal adsorption site, and the adsorption energy at this position was the smallest. The total systems energy contained single and 10 ethanol molecules are showed in Figs. 12 and 13 respectively.

Table 1. Adsorption energy at various positions of graphene

Structure	Graphene-1	Graphene-2	Graphene-3	Graphene-4	Graphene-5	Graphene-6	Graphene-7
Adsorption energy (kcal/mol)	− 81.11	− 80.68	− 80.45	− 80.21	− 79.99	− 79.09	− 75.78

It could be seen that the change of the total energy was periodic. The average energy first decreased gradually, and then tended to be constant, which indicated that the system reached a stable state.

Table 2. List of energy for ten ethanol molecules

Structures	Total energy (kcal/mol)	Adsorption energy (kcal/mol)	Rigid adsorption energy (kcal/mol)	Deformation energy (kcal/mol)
Ethanol	66.41			
Graphene-1	− 178.96	− 843.043	− 88.05	− 754.99
Graphene-2	− 176.59	− 840.68	− 86.29	− 754.38
Graphene-3	− 176.34	− 840.42	− 85.32	− 755.10
Graphene-4	− 173.60	− 837.69	− 83.06	− 754.63
Graphene-5	− 172.87	− 836.95	− 82.34	− 754.61
Graphene-6	− 171.03	− 835.11	− 82.22	− 752.89
Graphene-7	− 164.29	− 828.38	− 72.96	− 755.42
Graphene-8	− 163.81	− 827.89	− 72.68	− 755.22
Graphene-9	− 163.50	− 827.59	− 72.21	− 755.38
Graphene-10	− 161.77	− 825.86	− 70.27	− 755.58

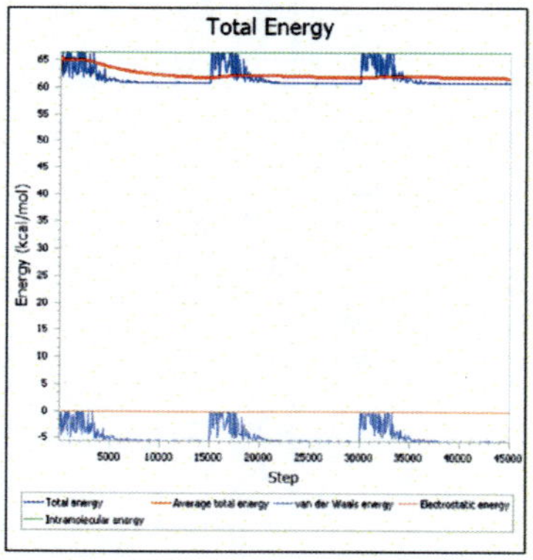

Fig. 12. System energy for one molecule

The energy distribution curves are shown in Fig. 14. Figure 14(a) and (b) showed the systems with ten and one ethanol molecule respectively. It could be seen that the most likely energy in these two cases were 81 kcal/mol. The vertical axis represented the probability of energy appearing.

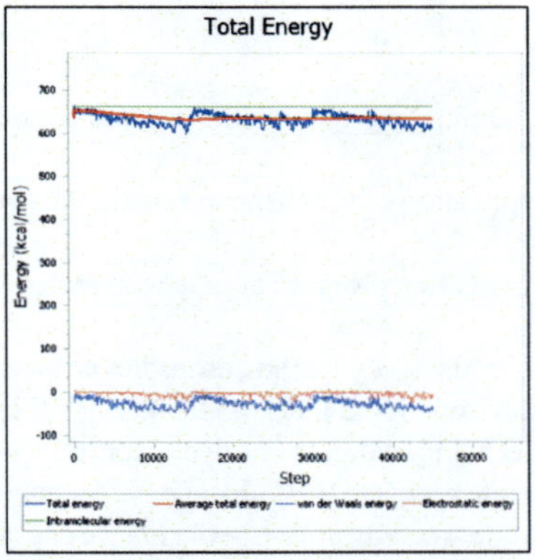

Fig. 13. System energy for ten molecules

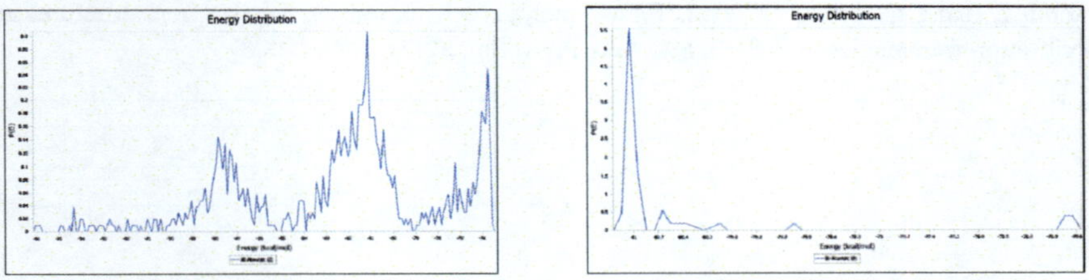

(a) Ten molecules

(b) One molecule

Fig. 14. Energy distribution probabilities

4 Conclusions

The adsorption simulation of graphene to two kind of VOCs gas molecules based on MS were observed, the conclusions were followed:

(1) The total energy of the system combined by the adsorption material graphene and the adsorbate ethanol molecules, after the adsorption is stable, the total energy decreases.
(2) When the quantity of adsorbate increases, the total energy of the system increases, and the adsorption time is longer than that of a single adsorbate. The process is fluctuating and periodic, gradually from fluctuation to stability.
(3) The system energy of different adsorbates is also different, but after the adsorption is stable, the total energy is also reduced.

Acknowledgements. This work was supported by the Key Research Project of Beijing Institute of Graphic and Communication (Ed202104, Eb202201).

References

1. Zhang, X., Xue, Z., Li, H., et al.: Ambient volatile organic compounds: pollution in China. J. Environ. Sci. **55**(5), 69–75 (2017)
2. Dou, P.: A review of the application of graphene in functional coatings. Shandong Chem. Ind. **4**, 63–64 (2017)
3. Zou, Y.: Research on the application progress of graphene materials in water treatment. Chem. Manag. **01**, 83–85 (2015)
4. Wu, Y., Luo, H., Wang, H., et al.: Study on the adsorption properties of modified graphene for methylene blue in water. Environ. Sci. **34**(11), 4333–4339 (2013)
5. Zhu, Y., Chen, Y., Peng, Y., et al.: Preparation and dye adsorption properties of reduced graphene oxide/iron tetroxide composite aerogels. J. Wuhan Univ. Technol. **2**, 181–185 (2020)
6. Cao, X., Li, F.: Research progress on adsorption performance of graphene aerogel in wastewater. Mater. Rev. **7**, 7020–7025 (2020)
7. Pan, R.: The application of material studio 7.0 molecular simulation software in the teaching of structural chemistry crystal structure. Chem. Educ. (Chin. Engl.) **12**, 73–77 (2018)
8. Zhu, Y., Yang, C., Wang, M., et al.: First-principles calculations on the electrical structures and vibration frequencies of β-Si_3N_4. Acta Phys. Sin. **57**(2), 1048 (2008)

Study on Compressive Strength Calculation of Corrugated Boxes Based on Printing Impact

Xuejiao Xing[1], Lijiang Huo[1](✉), and Huizhong Zhang[2]

[1] School of Light Industry and Chemical Engineering, Dalian Polytechnic University, Dalian, Liaoning, China
lijianghuo@163.com

[2] Tat Seng Packaging Group, Suzhou, Jiangsu, China

Abstract. This paper explores the effect of printing on the compressive strength of corrugated cartons and the safety factor "K" in GB/T6543 standard. Taking 25 kg 0201 dye box as the research object, to investigate the effects of three factors: printing shape, printing position and number of colors on the compressive strength of cartons. The existing reference data of estimated safety factor "K" was supplemented, and the "K" factor was further derived according to the $K = 1/\prod ai$ formula. The results show that under the same area, vertical printing has less influence on the "K" factor than horizontal printing; printing position near the bottom of the carton has less influence on the "K" factor than the top. Based on the above experiments, we can estimate the safety factor "K" for corrugated cartons after printing, which can be used in actual production to protect products and save costs.

Keywords: Printing impact · Compressive strength of corrugated cartons · K coefficient

1 Introduction

Different printing methods have different effects on the strength of corrugated cartons. In order to reduce the impact on the compressive strength of cartons and improve the safety of packaging in transportation and storage, it is important to study the effect of printing on the compressive strength of corrugated cartons.

The printing process, printing pressure, printing area and number of colors will all have an impact on the compressive strength of corrugated boxes. Flexo printing is widely used and has a low impact on the compressive strength of corrugated boxes, so this experimental study focuses on flexo printing [1].

2 Calculation of the Compressive Strength of Corrugated Boxes and Its "K" Factor

The following factors affect the compressive strength of corrugated boxes: printing openings, load duration, temperature and humidity, shock and vibration during transport,

M. Xu et al. (Eds.): CACPP 2022, LNEE 991, pp. 261–265, 2023.
https://doi.org/10.1007/978-981-19-9024-3_34

handling, etc. The "K" factor, also known as the deterioration factor of the carton, is used to compensate for the loss of compressive strength of corrugated boxes caused by the processing process and the storage environment. According to GB/T 6543-2008 Appendix D, the formula for calculating the compressive strength of a corrugated box is:

$$P = KG\left(\frac{H}{h} - 1\right) \times 9.8$$

In the formula, P is the compressive strength, N; K is the deterioration factor; G is the Single package weight, kg; H is the stacking height, m; h is the box height, m; H/h to take the integer digit.

In the calculation of the compressive strength of corrugated boxes, all the other values in the above equation are clear, except for the "K" factor. What we can see now is the formula proposed by Peng Guoxun: $K = 1/\prod ai$ (ai is the safe multiplier) [2]. In practice, it was found that these data could not be used in the actual design and production of corrugated boxes. Therefore, this paper further investigates the effect of printing method on the "K" factor through experiments.

3 Experimental Research

3.1 Experimental Subjects

China is a large import and export country, of which the import and export of chemical dyes accounted for a large proportion [3]. This paper therefore conducts experiments on a 25 kg BC-type double corrugated carton. The outer dimensions of the carton are chosen to be 370 mm × 272 mm × 466 mm and it is placed on a standard 1200 mm × 1000 mm pallet at 3 × 4.

3.2 Test Method

The experiments were carried out in 12 groups at 23 °C/50% [4]. The relevant national standards (GB-10739, GB-T4857.4, GB-T1539, GB-T6546), TAPPI standards and ISTA standards are implemented during the experiments.

3.3 Shape, Position and Color Arrangement of Printed Graphics on Cardboard Boxes

This experiment classified the cartons into three categories based on common graphic shapes: full-page printing, horizontal strip printing and vertical strip printing [5], as shown in Fig. 1(a, b, c). Equal printing area for horizontal and vertical strip. The printing position was divided into three parts: the top, middle and bottom, with an equal printing area of 370 × 150 mm^2, see Fig. 1(d, e, f). This experiment investigates the effect of single, two and three color printing on the compressive strength of the corrugated box. The printing colors and their positions are shown in Fig. 1(g, h, i).

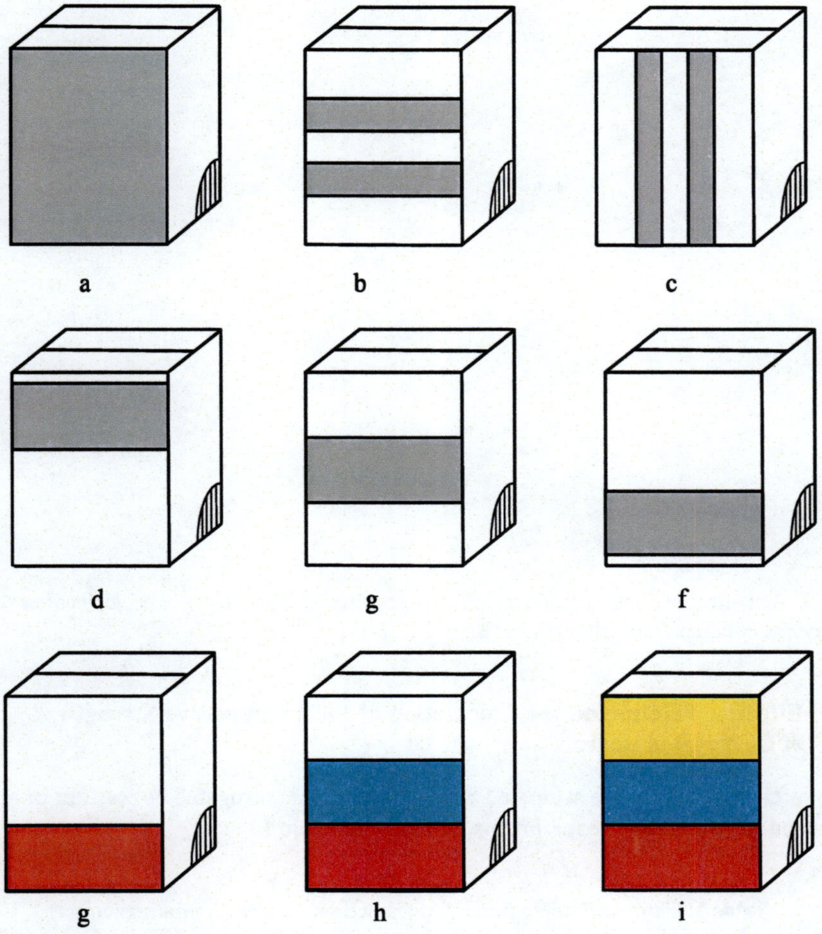

Fig. 1. Printing form (a) full-page, (b) horizontal strip, (c) vertical strip. Printing position (d) top, (e) middle, (f) bottom. Printing color quantity (g) single color, (h) two color, (i) three color

4 Results and Discussion

4.1 Effect of Printing Shape on the Compressive Strength of Corrugated Boxes

After averaging the 12 sets of experimental data, the effect of printing shape on the compressive strength of corrugated boxes is shown in Fig. 2(a). K-factor as a compensation factor, the amount of which is related to the decay rate of the compressive strength of the carton [6]. According to the equation $K = 1/\prod ai$, assuming that other conditions are ideal and only the effect of printing is considered, $K = 1/a1$ (a1 is the safe multiplier considering only the printing factor, a1 = 1-decay rate). For example, decay rate = 6%, $a1 = 1–6\% = 0.94$; $K = 1/0.94 = 1.06$. The effect of printing shape on compressive strength and k-factor is shown in Fig. 2(a); The effect of printing position is shown in Fig. 2 (b); The effect of the number of printing colors is shown in Fig. 2(c).

Fig. 2. The effect of printing shape (a), printing position (b) and number of printing colors (c) on decay rate of compressive strength and K

4.2 Effect of Printing on the Calculation of the Compressive Strength of Corrugated Boxes

The reference data for the estimated safety factor K for corrugated boxes after printing, obtained from the above experiments, is shown in Table 1.

Table 1. Estimated safety factor K reference data (based on printing impact)

Printing	Classification	Loss of compressive strength (%)	Multiplier-ai
Printed shapes	Horizontal-tape	8.0	0.920
	Vertical-tape	6.0	0.940
	Full-page	11.8	0.882
Printing position	Top	6.6	0.934
	Middle	1.2	0.988
	Bottom	4.3	0.957
Number of printing colors	Single-color	4.7	0.953
	Two-color	10.7	0.893
	Three-color	16.7	0.833

The compressive strength of a cardboard box is not affected by printing alone. Its variation is the result of various factors working together. According to the formula K $= 1/\prod$ai, The specific K-factor can be calculated by consulting the ai values of other influencing factors. For example, if a crossband printed carton with a mass of 25kg and a carton height of 466 mm is aligned and stacked for 30 days at a height of 3 m and 80% humidity (RH = 80%, ai = 0.68; stacked for 30 days, ai = 0.60; Aligned stacking, ai = 1) [2], the K-factor should be:

$$K = 1/\prod ai = 1/(0.68 \times 0.6 \times 0.92) = 2.66$$

The compressive strength of the corrugated box can then be designed according to GB/T 6543-2008 Appendix D as:

$$P = 2.94G\left(\frac{H}{h} - 1\right) \times 9.8 = 2.66 \times 25 \times \left(\frac{3000}{466} - 1\right) \times 9.8 = 3258.5\ (N)$$

However, if we calculate based on past reference data, taking a1 = 0.8, we get K = 3.06; P = 3748.5 N.

Design of corrugated boxes according to the estimated value P = 3258.5 N, after 30 days of stacking, there is no deformation or crushing of the cartons. It means that the choice of safety factor K = 2.66 has been able to meet the requirements of the strength of the transport packaging under this condition. If K = 3.06 is chosen for the design of the corrugated carton, the carton will be too strong, the manufacturing cost will increase and resources will be wasted, thus affecting the economic efficiency of the enterprise.

5 Conclusions

(1) Vertical tape printing has 2% less effect on the K-factor than horizontal tape printing. (2) The graphics should be printed in the middle of the box, and if they are to be printed on the edges, then they should be printed near the bottom rather than the top. (3) For the same print area, the rate of impact on K-factor is less when the printed panels are one whole piece than when they are two pieces apart. (4) Two-color printing decay rate of compressive strength is more than twice that of single-color printing. (5) The use of optimized data to estimate K values is more accurate and can be applied to actual production to both protect the product and save costs.

References

1. Liu, T., Tang, Z., Che, Y., et al.: Study on flexographic printing process of corrugated paper. Packag. Eng. **28**(1), 3 (2007)
2. Peng, G., Wu, Z., Wu, Y.: Corrugated Packaging Design. Printing Industry Press (2007)
3. Zhang, H.: Operational practices of using corrugated cartons for chemical dyestuff packaging. China Packag. **2**, 23–29 (2020)
4. Allaoui, S., Aboura, Z., Benzeggagh, M.L.: Effects of the environmental conditions on the mechanical behaviour of the corrugated cardboard. Compos. Sci. Technol. **69**(1), 104–110 (2009)
5. Liao, Y.: Effect of printing and opening on the compressive strength of corrugated boxes. Print. Technol. **24**, 3 (2008)
6. Chen, X.: Optimal design of some transport packaging technologies and safety factors. China Packag. Ind. **12**, 5 (2010)

Design of Structural Parameters and Its Effect on the Static Cushioning Performance of Paper Elliptic Porous Materials

Xiaoli Song(✉), Gaimei Zhang, Jiandong Lu, Yuqi Yao, and Jiacan Xu

School of Printing and Packaging Engineering, Beijing Institute of Graphic Communication, Beijing, China
songxiao109@163.com

Abstract. In order to improve the cushioning performance of cushioning materials and reduce the use of materials, the structural parameter is designed and the static cushioning performance of the porous materials made of paper with elliptic shape is analyzed in this paper. The cushioning coefficient and the stress-strain curve are investigated under the different structural parameters. The results show that the different structural parameters of the materials affect the static cushioning performance of such materials. The cushioning performance is not proportional to thickness, major-minor axis ratio or paper grammage. The yield value depends on the structural parameters. Thickness and paper grammage are proportional to the yield value, whereas the major-minor axis ratio is inversely proportional to the yield value, concluding that different structural parameters have different effects on yield value. Based on the test results, the paper elliptic cushioning material with a thickness of 40 mm, major-minor axis ratio of 3 and grammage of 110 g/m^2 has the best cushioning performance.

Keywords: Porous materials · Parameter design · Static cushioning performance · Cushioning packaging materials

1 Introduction

During transportation, product packaging is inevitably subjected to mechanical shocks and vibrations. Improper packaging easily causes product damage, resulting in huge economic loss for customers. Cushioning packaging materials absorb the energy generated by the impact, thereby effectively protecting the goods and avoiding damage [1–4]. Hence, cushioning packaging materials are widely used as packaging today. Existing cushioning packaging materials mainly include foam plastics and paper cushioning packaging materials. Foam plastics show good cushioning properties, but they do not degrade. Considering their harm to the environment, many countries have restricted the use of foam plastics. On the other hand, paper-based materials achieve a certain cushioning effect through structural changes. Moreover, they degrade easily and do not harm the environment [5–8]. Therefore, paper cushioning packaging materials have become a research focus. To improve the cushioning performance of such materials and to reduce

M. Xu et al. (Eds.): CACPP 2022, LNEE 991, pp. 266–272, 2023.
https://doi.org/10.1007/978-981-19-9024-3_35

cost, in this study, we discuss the pass design of paper-based elliptic porous materials and the effect of structural parameters on the static cushioning performance of paper elliptic porous materials [9, 10].

2 Processing of the Porous Materials Made of Paper with Elliptic Shape

2.1 Design of Pore Type of the Porous Materials

Figure 1 is a schematic diagram of the pass; a is the semi-major axis of the ellipse, and b is the semi-minor axis. Thus, the major–minor axis ratio is a/b. Here, this ratio is represented by r. The different r values represent how close the elliptical hole is to the circle. In general, the smaller the value of r is, the closer the hole shape is to the circle. When the value of r is 1, the hole is round.

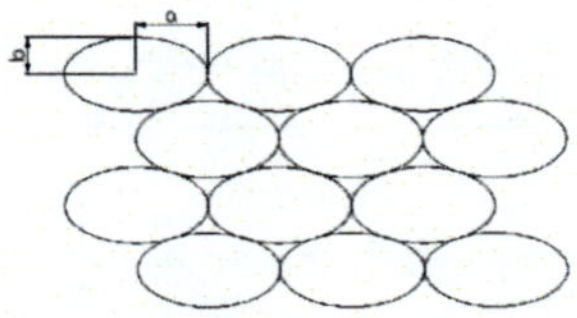

Fig. 1. Schematic diagram of the pass

2.2 Processing Technology of the Porous Materials Made of Paper with Diamond Shape

First, the paper is cut to a certain width. Then, the paper is soaked in water and placed in the mould, which is dried in a vacuum oven at 100 °C for approximately 10 min for corrugation. Then, multiple papers are bound together to obtain the paper elliptic porous material, as shown in Fig. 2.

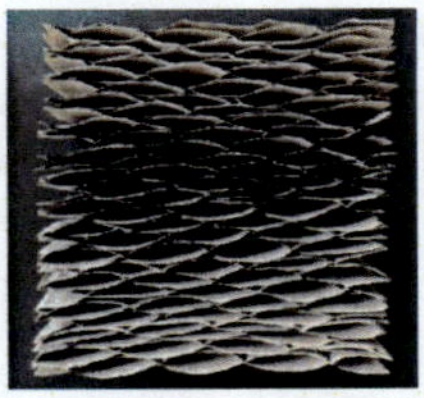

Fig. 2. Picture of a sample

3 Results and Analysis

3.1 Effect of Major–Minor Axis Ratio on the Static Cushioning Performance of Paper Elliptic Porous Materials

The grammage was maintained at 100 g/m^2 and the major-minor axis ratio was adjusted to be 2, 3 and 4. The cushioning materials with a height of 50 mm were prepared. And static compression tests were performed on the samples to obtain their stress-strain curves (Fig. 3).

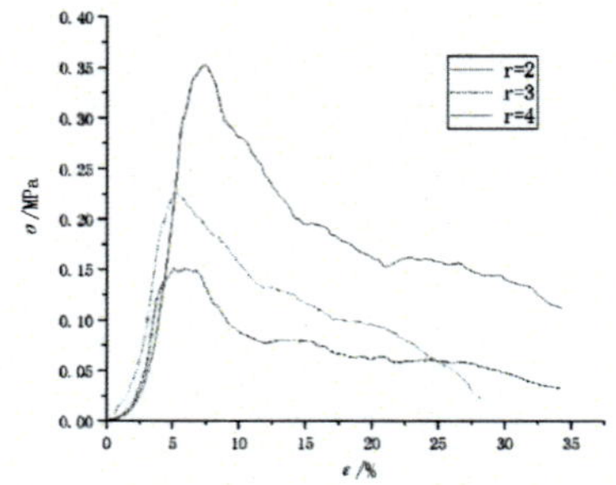

Fig. 3. Stress–strain curves of paper elliptic porous materials with different major–minor axis ratios

According to Fig. 3, the compression process of the paper elliptic porous materials can be divided into the linear elasticity stage, plastic stage and plateau stage. In the linear elasticity stage, the three curves of samples with different major-minor axis ratios show different slopes. Indicating that the two samples are less rigid, weaker and easier to deform in this stage, the curves of paper elliptic porous materials with the major–minor axis ratio of 2 and 4 coincide, showing a relatively small slope. In the plastic stage, the three samples showed different yield values: 0.353 at the linear elasticity stage, 0.227 at the plastic stage and 0.151 MPa at the plateau stage. According to Fig. 3, as the yield value decreases, the major-minor axis ratio increases gradually. When the ratio is 2, the material yield value is the smallest. Judging by the structure, the smaller the major–minor axis ratio, the closer the pass is to a circle. As in the plateau stage, the curves flatten, indicating that the materials are being crushed; the closer the material is to a circle, the greater the stiffness of the material and the more difficult it is to compress.

According to the data shown in Fig. 3, the cushioning coefficient–strain curves were drawn with ε as the x-axis and C as the y-axis, as shown in Fig. 4. According to Fig. 4, when the value of ε is less than or equal to 8%, the sample with the major-minor axis ratio of 4 shows the smallest cushioning coefficient, indicating that the material can absorb more energy under deformation, i.e., the best cushioning performance. When the value of ε is greater than or equal to 15% and the value of ε is less than or equal to 42%, the cushioning materials have the same result as above. When the value of ε is greater than and the value of ε is less than 15%, the sample with the major–minor axis ratio of 2 shows the smallest cushioning coefficient, i.e., under such circumstances, it absorbs the most energy and has the best cushioning properties. When the value of ε is greater than or equal to 42% and the value of ε is less than or equal to 70%, the cushioning materials have the same result as above. When the value of ε is greater than or equal to 70%, the sample with the major-minor axis ratio of 3 has the smallest cushioning coefficient. Figure 4 also indicates that the major-minor axis ratio is not proportional to the cushioning coefficient. The sample with the major-minor axis ratio of 3 has the smallest cushioning coefficient of 2.548.

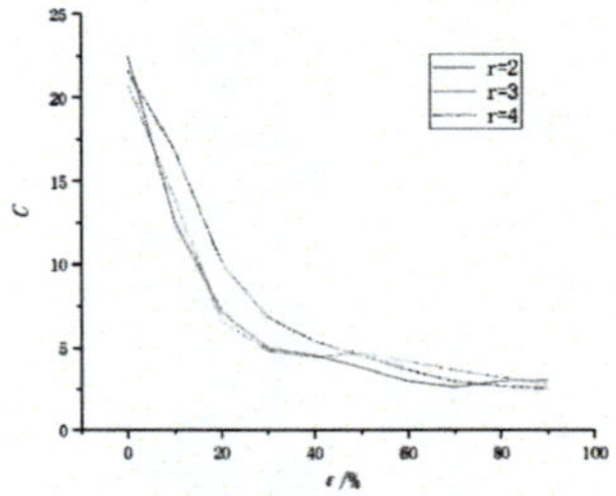

Fig. 4. Cushioning coefficient–strain curves of paper elliptic porous materials with different major-minor axis ratios

3.2 Effect of Paper Grammage on the Static Cushioning Performance of Paper Elliptic Porous Materials

The major-minor axis ratio was maintained at 3 and the original paper grammage was changed to 100, 110 and 170 g/m^2. The cushioning materials with a height of 50 mm were prepared. The samples were subjected to static compression tests to obtain their stress–strain curves (Fig. 5).

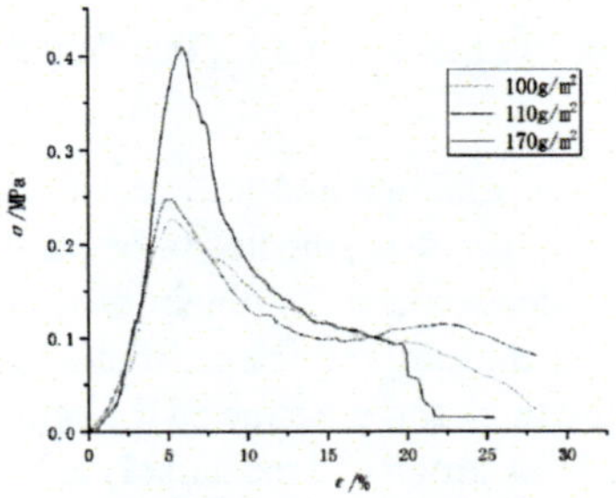

Fig.5. Stress–strain curves of paper elliptic porous materials with different grammage

According to Fig. 5, in the linear elasticity stage, the three curves coincide, indicating similar stiffness of the three samples in the linear elasticity stage. In the plastic stage, the three materials showed different yield values: 0.227, 0.248 and 0.410 MPa. With the increase in paper grammage, the yield value increased gradually. At the grammage of 170 g/m^2, the yield value of the material peaks, i.e., the material is the stiffest and most difficult to compress. When the value of ε is equal to 17%, the curve of the sample with the grammage of 170 g/m^2 reduces sharply and drops to 0 in a very short time. This phenomenon is mainly caused by the cracking of the material during compression, leading to a rapid decline in stress.

According to the data shown in Fig. 5, the cushioning coefficient–strain curves were drawn with ε as the x-axis and C as the y-axis, as shown in Fig. 6.

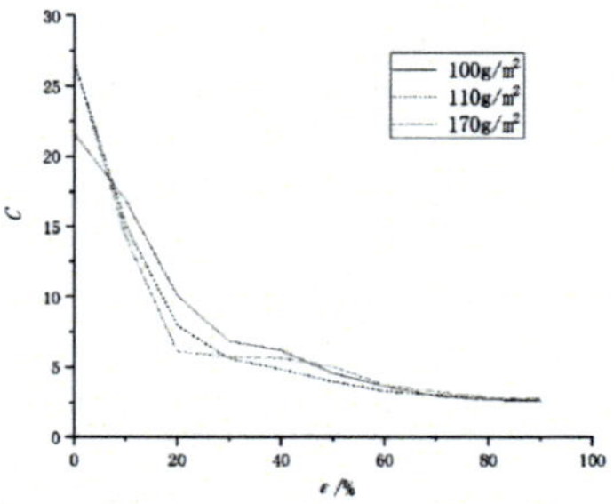

Fig. 6. Cushioning coefficient–strain curves of paper elliptic porous materials with different grammage

According to Fig. 6, when the value of ε is less than or equal to 7%, the sample with the grammage of 100 g/m^2 has the smallest cushioning coefficient. When the value of ε is greater than 7% and the value of ε is less than or equal to 25%, the sample with the grammage of 170 g/m^2 has the smallest cushioning coefficient. When the value of ε is greater than 25% and the value of ε is less than or equal to 64%, the sample with the grammage of 110 g/m^2 shows the smallest cushioning coefficient. According to Fig. 6, paper grammage is not proportional to the minimum cushioning coefficient. The sample with the grammage of 110 g/m^2 has the smallest cushioning coefficient of 2.616.

3.3 Effect of Thickness on the Static Cushioning Performance of Paper Elliptic Porous Materials

The major–minor axis ratio was maintained at 3 and the original grammage of 100 g/m^2 was unchanged. The thickness was adjusted to be 30, 40 and 50 mm. The samples were subjected to static compression tests to obtain their stress–strain curves (Fig. 7). According to Fig. 7, in the linear elasticity stage, thickness is not proportional to the slope. The curve of the 50 mm thick sample showed the smallest slope, i.e., in this stage, this sample is the least rigid, least strong and most likely to be deformed by compression. The three samples showed different yield values: 1.060, 0.848 and 0.227 MPa. As the thickness increases, the stress value at the yield point of the material decreases. When the thickness is 50 mm, the yield value is the smallest; that is, the material is more prone to deformation under external forces. The 50 mm thick sample showed the gentlest decline while the 30 mm thick sample showed the most rapid decline, where, after reaching the yield value, the curves continued to decrease, but with different decreasing trends at different thicknesses. In other words, the thinner the material, the easier it is to be crushed.

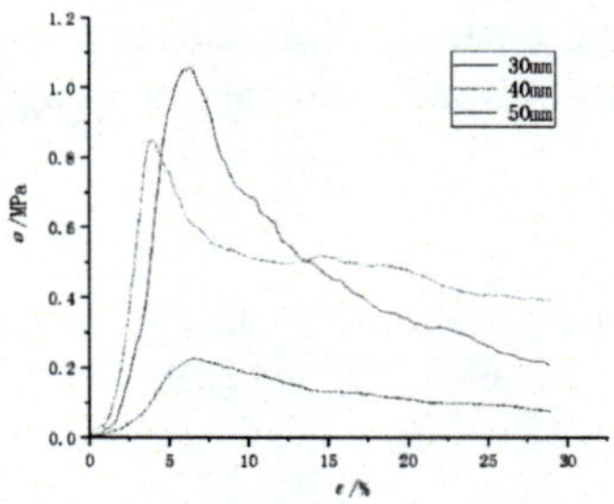

Fig. 7. Stress–strain curves of paper elliptic porous materials with different thickness

According to the data shown in Fig. 7, the cushioning coefficient–strain curves were drawn with ε as the x-axis and C as the y-axis, as shown in Fig. 8. According to Fig. 8, when the value of ε is less than or equal to 12%, the 50 mm thick sample has the smallest cushioning coefficient. When the value of ε is greater than 12% and the value of ε is less than or equal to 43%, the 30 mm thick sample shows the smallest cushioning coefficient. Moreover, according to Fig. 8, thickness is not proportional to the minimum cushioning coefficient. When the thickness is 40 mm, the paper elliptic porous materials show a minimum cushioning coefficient of 2.413.

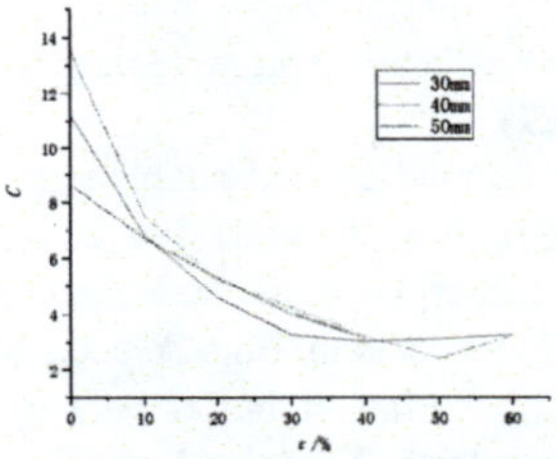

Fig. 8. Cushioning coefficient–strain curves of paper elliptic porous materials with different thickness

4 Conclusions

The following conclusions were drawn through data processing and research analysis:

- The static compression process of paper elliptic cushioning materials can be divided into the linear elasticity stage, plastic stage and plateau stage. The different structural parameters of the materials affect the static cushioning performance of such materials. The cushioning performance is not proportional to thickness, major-minor axis ratio or paper grammage.
- The yield value depends on the structural parameters. Thickness and paper grammage are proportional to the yield value, whereas the major-minor axis ratio is inversely proportional to the yield value, concluding that different structural parameters have different effects on yield value.

- Based on the test results, the paper elliptic cushioning material with a thickness of 40 mm, major-minor axis ratio of 3 and grammage of 110 g/m^2 has the best cushioning performance.

Acknowledgements. This work was supported by the Key Research Project of Beijing Institute of Graphic and Communication (No. Eb202201 and Ed202104).

References

1. Wang, Z., Peng, G.: Ecological design of military logistics packaging system. Eng. Pack. **35**(9), 140–146 (2014)
2. Kavermann, S.W., Bhattacharyya, D.: Experimental investigation of the static behaviour of a corrugated plywood sandwich core. Compos. Struct. **9**, 836–844 (2018)
3. Wu, T., Wu, J., Wang, H., Cai, J.: Research progress on technology economy and environmental impact assessment of buffer packaging materials. Eng. Pack. **42**(9),17–24 (2021)
4. Ding, Y., Qian, Y.: Simulation analysis of dynamic cushioning characteristics of integral packaging based on Ansys Workbench. Packag. Food Mach. **35**(11), 18–22 (2014)
5. Zhong, C., Yang, Y., Zhou, L.: Research status and prospect of e-commerce logistics packaging. J. Packag. **12**(05), 27–34 (2020)
6. Measures for the management of mail express packaging. Gazette State Counc. People's Republ. China **12**, 22–25 (2021)
7. Shen, Z., Chen, D., Luo, J.: Simulation and analysis of dropping impact acceleration of polyethylene foam buffer system. Eng. Pack. **37**(19), 128–131 (2016)
8. Castigioni, A., Castellani, L., Cuder, G., et al.: Relevant materials parameters in cushioning for EPS foams. Coll. Surf. A Physicochem. Eng. Aspects **534**(1), 71–77 (2017)
9. Kong, Z.: Research on the application of buffer packaging materials in the storage and transportation of agricultural products. Xinjiang Agric. Sci. Technol. **1**, 15–16 (2018)
10. Singh, S., Gaikwad, K., Lee, M., et al.: Thermally buffered corrugated packaging for preserving the freshness of mushrooms. Food Eng. **216**(1), 11–19 (2018)

Research on Temperature Monitoring Method of a New Type of Medical Carrying Case

Pengfei Cheng(✉), Shengwei Yang, and Chuan Zhang

Department of Printing and Packaging Engineering, Shanghai Publishing and Printing College, Shanghai, China
gustdashao@163.com

Abstract. Medical carrying case have always been an important carrier for blood, vaccine and organ transportation. With the spread of the new crown epidemic, the demand for medical carrying case has further increased. However, how to monitor the temperature of the contents of the medical carrying case has always been a difficult problem. More advanced medical carrying case will embed temperature sensors in the box to monitor the temperature. However, this method cannot truly reflect the temperature of the contents because it can only monitor the temperature of one point. Therefore, this study proposes to use multiple temperature sensors to realize flexible temperature measurement, which can realize contact temperature monitoring on the surface of the content, so as to reflect the temperature of the content more directly and avoid the denaturation of the content caused by high temperature.

Keywords: Medical carrying case · Temperature monitoring · Multiple temperature sensors · Contact monitoring

1 Introduction

The medical carrying case refers to the device used in the medical industry, which relies on the condensate and its own structure for refrigeration and heat preservation [1]. Some contents need to be transported within a specific range [2-4]. Currently, in order to meet this requirement, a temperature monitoring and recording module is generally equipped in the industry to judge the temperature status of the contents [5]. However, the single-point temperature monitoring function has certain limitations. It can only record the temperature within a certain range around the temperature sensor, but cannot record the temperature of the entire box, nor can it directly reflect the temperature of the contents, so it is possible that the contents reach the temperature of the box. It takes a long time and there is the possibility of degeneration [6]. In this study, a new temperature measurement method is proposed, which records the temperature through multi-point recording and flexible contact of the content, which can reflect the temperature of the content more intuitively and avoid the problem of content denaturation caused by high temperature.

M. Xu et al. (Eds.): CACPP 2022, LNEE 991, pp. 273–276, 2023.
https://doi.org/10.1007/978-981-19-9024-3_36

2 Experimental Method

In this study, a medical carrying case with a temperature sensor was used, the inner dimensions of the box were 53.5 × 33.5 × 32 cm, the outer dimensions were 59.5 × 40 × 40 cm, and the inner interlayer was made of PU polyurethane foam. The temperature sensor is in the middle of the front panel of the box, as shown in Fig. 1. Eight thermocouples are arranged in two rows on the top plate of the box which are powered by button batteries, and heavy objects are connected to the ends of the thermocouples so that the thermocouples can sag to the bottom of the box, as shown in Fig. 2. In this way, the medical carrying case is opened, and the contents are placed After entering the box, the drooping thermocouple can contact the contents with the falling of the box cover. Before the experiment starts, place enough cold storage boxes in the box, the laboratory temperature was measured between 34 and 36 °C during the experiment. A 500 ml water bag was used in this study to simulate a blood bag at a temperature of 36 °C.

After the start of the experiment, use a temperature recorder to record the temperature data of the eight thermocouples, and record the temperature data of the thermocouples on the medical carrying case each one minute until the temperature of the contents is stabilized to a certain temperature.

Fig. 1. Medical carrying case appearance

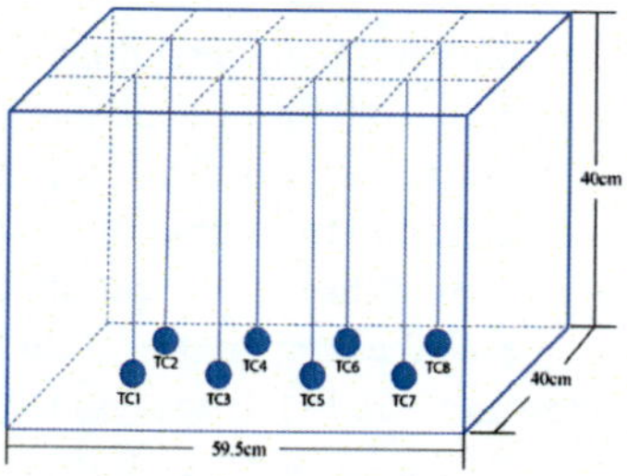

Fig. 2. Schematic diagram of adding thermocouple to medical carrying case

3 Results and Discussions

3.1 Temperature Comparison of Contacted and Uncontacted Thermocouples

Of the eight thermocouples, one behind the lid in contact with the contents is thermocouple TC4. Select another thermocouple (TC3) in contact with the contents to compare the temperature data with the temperature data of TC4, as shown in Fig. 3. As can be seen from the figure, at the beginning of the experiment, the temperature of the thermocouple in contact with the water bag was about 27 °C, while the temperature of the thermocouple not in contact with the water bag was about 2 °C. As the experiment progressed, the temperature of the water bag was gradually lowered, eventually dropping to 10 °C. During this process, using the temperature data obtained by the thermocouple in contact with the water bag, it is possible to monitor the temperature change of the water bag and observe whether the temperature of the water bag reaches a reasonable temperature within a reasonable time.

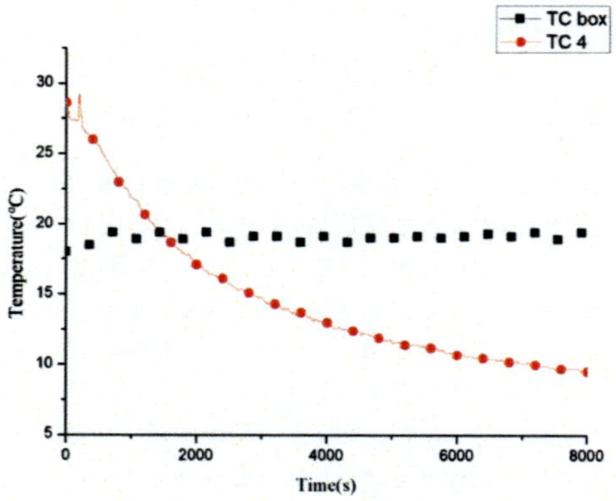
TC box
TC 4
Temperature(℃)
Time(s)
30
25
20
15
10
5
0
2000
4000
6000
8000

Mechanical Engineering Technology

Research on Accurate Positioning of Unwinding and Splicing Position Based on Curve Fitting

Zhijiang Yang[1](✉), Zeling Zhang[1], and Hui Wang[2]

[1] Xi'an Aerospace-Huayang Mechanical & Electrical Equipment Co., Ltd., Xi'an, China
zhijiangyang@163.com

[2] China Academy of Aerospace Liquid Propulsion Technology, Xi'an, China

Abstract. In view of the problem that the unwinding and splicing position cannot be accurately positioned due to the change of the radius of the material roll when the splicing axis of the unwinding unit of the flexographic printing machine is switched. In this paper, the rotation position data of the cantilever corresponding to the radius of the material roll is obtained by the simulation modeling of the CAD on the basis of keeping the angle of the main axis of the unwinding cantilever and the transitional roller unchanged. Secondly, the nonlinear functional relationship between the rotation angle and the radius is derived by geometric method, and the polynomial fitting algorithm of the rotation angle and radius of the cantilever is proposed by using the non-linear least square method. The optimal parameters of the nonlinear function are obtained by controlling the minimum sum of squared error, and finally tested and verified on the YRC1270-8 satellite flexographic printing machine. The results show that the proposed algorithm can solve the problem of determining the rotation angle of the discharge cantilever, and realize the precise positioning of the discharge and receiving position.

Keywords: Flexographic printing press · Non-linear least squares method · Polynomial fitting

1 Introduction

C.I Flexo Press "CINOVA" is the main press of Xi'an Aerospace-Hua Yang Mechanical & Electrical Equipment Co., Ltd, as the Fig. 1, which is the high end models and specialized for flexible packaging industry. The maximum mechanical speed 450 m/min, the application industry for plastic softy packaging, breathable film and non-woven etc.

When pre-driven adopted non-stop machine splicing method, in order to make sure cutter accurate cutter, which need to accurate positioning for new axis position.

At present, the table checked method is widely use at China domestic and abroad, the staff works at site after multiple measurement and tabulation, then manually input the new volume radius and rotation angle data to the touch screen, so that the unwind tower can rotate to the corresponding angle. In this process, unwind tower rotation angle size directly affects the success rate of feeding. In order to ensure the maximize the splicing success rate and printed products quality, it is necessary to control the precision of unwind splicing positioning.

M. Xu et al. (Eds.): CACPP 2022, LNEE 991, pp. 279–287, 2023.
https://doi.org/10.1007/978-981-19-9024-3_37

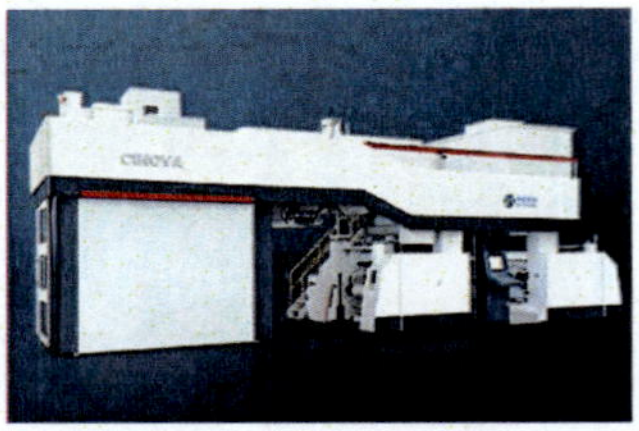

Fig. 1. CINOVA Flexo Press

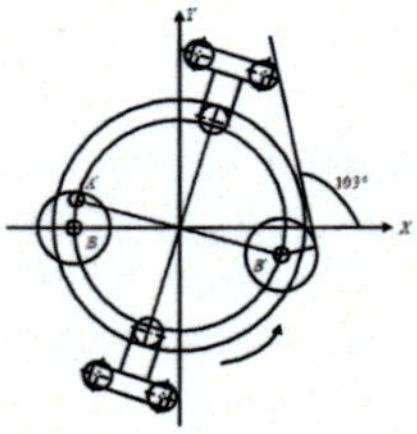

Fig. 2. Unwind tower mechanical structure drawing

New axis rotation position related with the angle between the web roll radius and pass roller, unwind tower main shaft and pass roller. The angle between unwind tower main axis and guide roller which decided by the mechanical structure, as Fig. 2 show in. Make sure the main axis level and guide roller the angle 103° not changed, new axis B rotate to B', the angle only corresponding with web radius.

Firstly, the rotation angle data corresponding to the web radius by CAD drawing, and the corresponding relationship is obtained by geometric derivation method. Secondly, polynomial is used as the curve fitting [1] basis function and the least square method is used to approximate the rotation angle. Finally, test and validation are carried out in the production site. Unwind tower rotation and running well, it will save more labor cost and measure hardware cost compare with before.

2 Unwind Tower Rotation Position Mould and Derivation of Geometric Relation

2.1 Unwind Tower Rotation Position Mould

CAXA electronic drawing was used to unwind splicing model, as show in Fig. 3. Which obtained actual web roll radius corresponding rotation angle accuracy data, which is the supplementary angle for below photo α.

The actual web roll radius and corresponding rotation angle precision data to collect 6 groups as Table 1.

2.2 Unwind Tower Derivation of Geometric Relation

From above data it cannot obtain the relation between radius and splicing angle, firstly derivation the geometric relation.

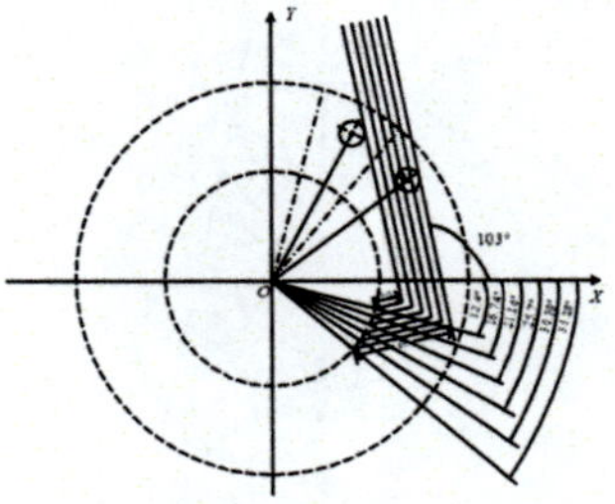

Fig. 3. CAXA electronic drawing unwind splicing mould

Table 1. Web roll radius and position angle CAXA simulation data

Web radius r/(mm)	125	150	175	200	225	250
Position angle/(°)	169.75	167.6	165.44	163.26	161.06	158.84
Web radius r/(mm)	275	300	325	350	375	400
Position angle/(°)	156.58	154.3	151.98	149.61	147.19	144.72

Figure 4 unwind tower geometrical relation and Fig. 5 unwind tower geometrical derivation flow chart. Use geometric relation and inverse function to deduce the equation of position angle α about web roll radius R through MATLAB platform, as show in formula (1).

$$
\begin{aligned}
\mathrm{R} = {} & 2.0 * ((145.4 * \cos\beta - 74.53 * \cos\alpha + 17.21 * \sin\alpha \\
& + 33.51 * \sin\beta + 0.006338 * (5.546e8 * \cos\beta^2 \\
& + 5.535e * \sin\beta\hat{\ }2 - 3995.0 * \cos\beta * \sin\beta - 5.51e8)\hat{\ }(1/2))\hat{\ }2 \\
& + (629.9 * \cos\beta - 322.8 * \cos\alpha 74.52 * \sin\alpha + 145.1 * \sin\beta \\
& + 0.02745 * (5.546e8 * \cos\beta\hat{\ }2 + 5.535e8 * \sin\beta^2 \\
& - 3995.0 * \cos\beta * \sin\beta - 5.51e8)\hat{\ }(1/2)\hat{\ }2)\hat{\ }(1/2)
\end{aligned} \tag{1}
$$

According to formula (1), new web roll radius and positioning angle are non-linear, and the equation $\alpha(R)$ is long and difficult to solve, which is not suitable for the actual PLC control. In order to achieve accurate positioning of unwind tower and quickly solved by PLC, non-linear least square method is used to make curve fitting.

3 Nonlinear Least Square Fitting Based on Polynomial Basis Function

The least square algorithm [2] is a classical method for solving optimization problems. The basic idea is to find a set of optimal parameters to minimize the sum $\delta = \sum_{i=0}^{n} (f(x_i) - y_i)^2$ of squares of the difference between the theoretical value and the observed value.

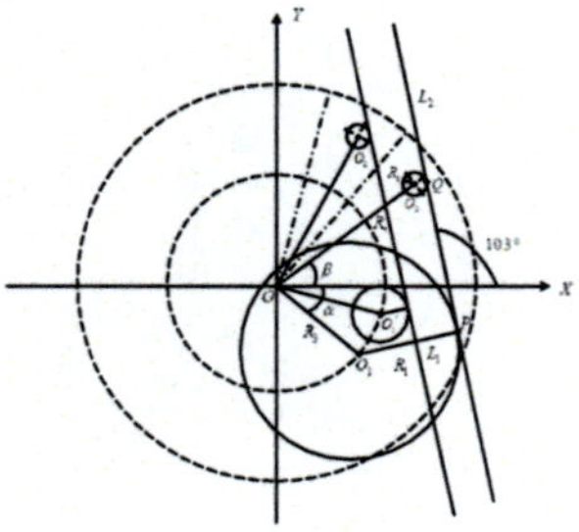

Fig. 4. Unwind tower geometrical relation

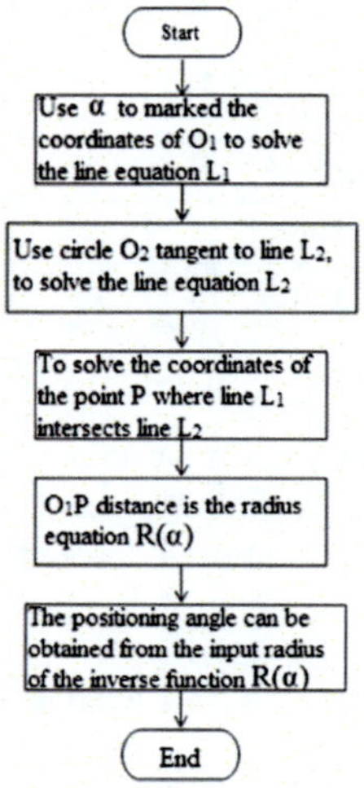

Fig. 5. Unwind tower geometrical derivation flow chart

When adopted the least square method, the basis function should be selected as the fitting model [3] to achieve function approximation of data. If the basis function is too complex, PLC programming will be relatively complex, unable to achieve mechanical equipment in a short time action. If the base function is too simple and fitting accuracy cannot be achieved, the original web cannot be correctly cutter down. Therefore, after comprehensive consideration of the above factors, polynomial is selected as the basis function [4].

$$f(x) = \sum_{i=0}^{n} a_i x^i = a_n x^n + a_{n-1} x^{n-1} + \cdots + a_1 x + a_0 \quad (n < N) \tag{2}$$

According to the least square algorithm, the proposed method was used to fit the simulation data of the rotation position of the unwind tower, then adopted the MATLAB platform written file and draw the fitting curve. This paper compares the fitting accuracy and effect of first-order to fourth-order polynomials [5].

Figure 6 shows the polynomial fitting curves of first to fourth order respectively. Corresponding curve functions and statistical indexes are show in Table 2 and Table 3 respectively Fig. 7 shows the comprehensive comparion of fitting curves of difference orders.

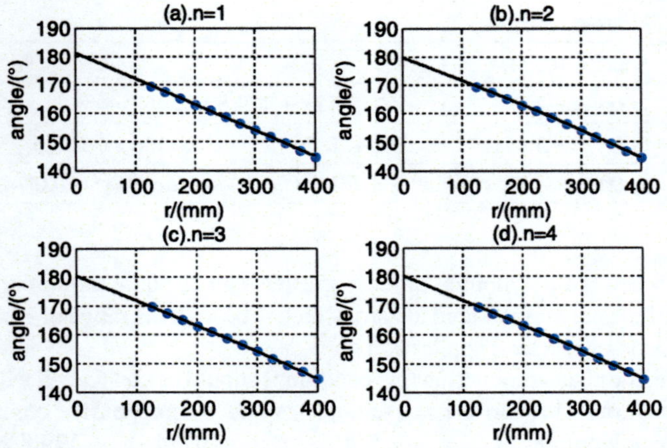

Fig. 6. Based on the first to fourth order polynomial basis function fitting curve dispersion contrast

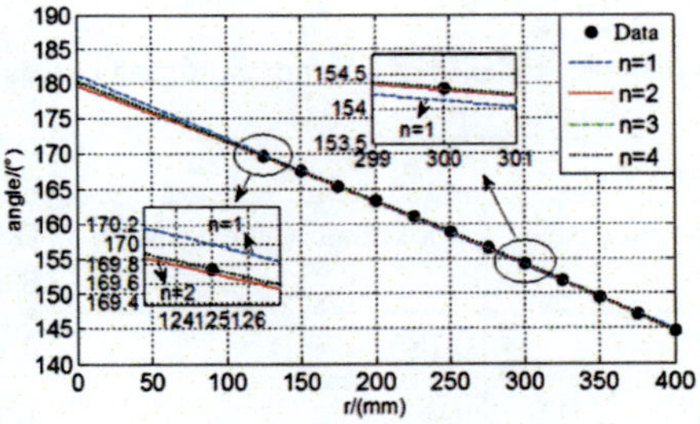

Fig. 7. Comprehen sive comparison of fitting curves based on polynomial basis functions

The comparative analysis of Fig. 6 (a), (b), (c) and (d) shows that the fitting curves in the four figures all show a decreasing trend, which is consistent with the simulation data. Moreover, good fitting effect can be achieved by using difference order polynomials, indicating that the scattered point distribution of data is relatively dense, and polynomial can meet the fitting requirements, which verifies the correctness of the selected basis function.

Figure 7, the first to fourth fitting curves are placed in the same dimension for comprehensive comparison, and selected two data points (125,169.75), (300,154.3) for partial amplification comparison. The four curves overlap highly after the point (125,169.75), but difference before the initial point.

From the two partial amplification images, when the order is greater than or equal to the second order, the fitting curves all pass through data points, and the fitting effect is good. However, the final fitting function cannot be determined only by the figure, and it still needs to be determined by analysis of the fitting function and accuracy.

Table 2. First order to fourth order polynomial fitting functions

Order number	Fitting functions
$n = 1$	$f(x) = 0.09074 \cdot x + 181.3$
$n = 2$	$f(x) = -2.54e^{-5} \cdot x^2 - 0.0774 \cdot x + 179.8$
$n = 3$	$f(x) = -5.031e^{-8} \cdot x^3 + 1.422e^{-5} \cdot x^2 - 0.08713 \cdot x + 180.5$
$n = 4$	$f(x) = -4.755e^{-11} \cdot x^4 - 3.781e^{-10} \cdot x^3 - 4.555e^{-6} \cdot x^2 - 0.08416 \cdot x + 180.4$

Table 2, when the polynomial order is greater than or equal to the third order, the coefficients of the polynomial tend to zero, which is beyond the range of PLC data types and is unrealistic in practical application.

Therefore, the actual production line conditions should be considered when selecting the fitting function. Through the above analysis and comprehensive comparison, the second order polynomial $f(x) = -2.54e^{-5} \cdot x^2 - 0.0774 \cdot x + 179.8$ was initially selected as the fitting function, and the final results still need to be comprehensively analyzed according to the size of each statistical indicator [6].

Table 3. Indexes of polynomial fitting curves

	Indicators		
Order number	Sum variance SSE	Root mean square RMSE	Determinative coefficient R-square
$n = 1$	0.3437	0.1854	0.9995
$n = 2$	0.007245	0.02837	1
$n = 3$	8.82e-05	0.00332	1
$n = 4$	5.572e-05	0.002821	1

Table 3 data shows that SSE and RMSE decrease with the increase of order when the reliability coefficient is 95% [7, 8]. When the order is greater than or equal to 2, r-square is 1, indicating that the curve is highly fitting and most "fits" the data scatter. As the number of polynomials increases, the amount and difficulty of calculation also increase. In the case of similar fitting precision, low-order polynomials should be preferred, which is conducive to the realization of the actual production line.

Above all, selected the quadratic polynomial function $f(x) = -2.54e^{-5} \cdot x^2 - 0.0774 \cdot x + 179.8$ as the fitting curve equation between roll radius and the rotation position of the unwind tower.

4 Practical Application Effect

Figure 8 Tower motor positioning control hardware topology structure.

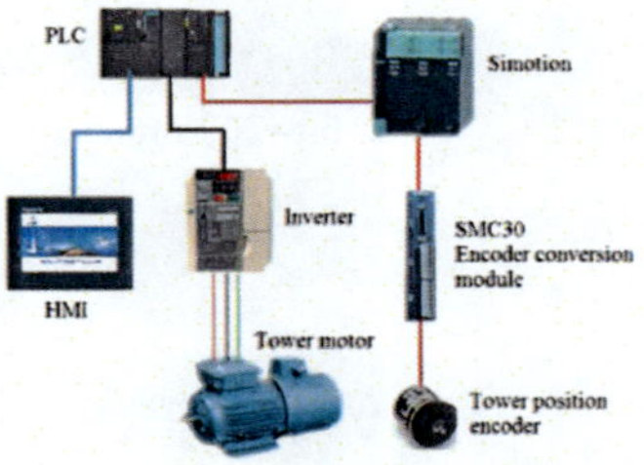

Fig. 8. Tower motor positioning control topology drawing

Input the web roll radius in HMI, PLC calculates the corresponding positioning angle, output to the frequency converter, drive the tower motor to rotate to the positioning point, rotary encoder real-time feedback tower current position.

Production site test machine C.I flexo press model YRC1270-8, which adopts the hardware topology of tower motor positioning control show in Fig. 8. Before the experiment, check the configuration of the production line hardware structure.

1) Installed unwind tower position detection, use SMC30 to read the tower rotation real time position.
2) The tower is controlled by frequency motor and adopted frequency converter Yaskawa CIMR-VB4A004 to control three-phase asynchronous motor Q/320412LDC001.
3) Add accurate positioning program, trace curve and PID control in the main program.

Figure 9 Unwind tower splicing photo, tower axis B new web from the initial position and rotate till positioning angle and stop to prepare well for change web roll.

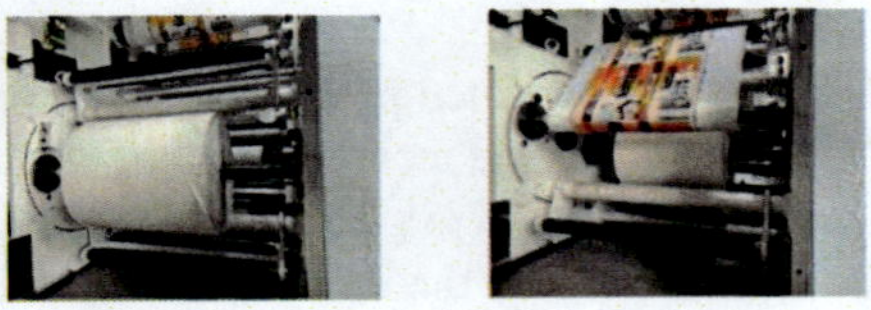

a.Unwind tower splicing initial position **b.**Unwind tower splicing positioning position

Fig. 9. Unwind tower splicing photo

Set the unwind tower converter frequency to 36.94Hz and the maximum speed to 50%. Use Simotion Scout to algorithmic programming. Figure 10 shows the trace curves of positioning anlge for six groups of site tests. The black line is the radius value of web roll and the red curve is the tracking curve of positioning.

a.r=150mm b.r=200mm c.r=350mm

d.r=300mm e.r=350mm f.r=400mm

Fig. 10. 6 Groups new web roll radius corresponding positioning angle trace curve

Figure 10(a), press the splicing button when 24,000 ms, unwind tower start to rotate, when it arrive to 48,000 ms rotate the tower to positioning position and stable at 167.12°. Each group position value and measured at site actual value calculated their error and analyze them.

Table 4. 6 Groups new web roll radius corresponding positioning result and error

New web roll radius r/(mm)	150	200	250	300	350	400
The real value angle/(°)	167.60	163.26	158.84	154.30	149.61	144.72
The fitting value angle/(°)	167.60	163.29	158.85	154.28	149.58	144.76
The positioning value angle/(°)	167.12	162.91	158.99	154.25	149.15	144.62
Error ε	0.0029	0.0040	0.0009	0.0003	0.0031	0.0009

From Table 4, the error between the calculated positioning value and the real value is less than the error index 1%, which meets the technical requirements. Moreover, the nonlinear least square fitting method is adopted to greatly reduce the positioning error, relatively eliminate part of the error caused by the motor encoder, and greatly help to improve the action of the unwind part.

5 Conclusion

This paper from the Flexo press unwind splicing tower motor positioning difficulty, designed a kind of based on nonlinear least squares method of unwind tower precise positioning control algorithm, the method based on least square method, by polynomial approximation, the precise location of the unwind tower is solved, which ensures the speed, accuracy and reliability of the unwind unit.

The site test data show that PLC can quickly calculate the positioning position, and rotation running track well of the unwind tower, and the dynamic performance of the system is improved, which verifies that the propose control strategy is correct and effective. It also saves the manpower of debugging and production operation and the cost of measuring hardware.

References

1. Xu, S., Gao, L., Zhang, P.: Numerical Linear Algebra. Peking University Press (2013)
2. Pan, L.-D., Pan, Y.-D.: System identification and modelling. Ind. Equipment Inf. Eng. Publishing Center (2004)
3. Wang, G., Yang, X., Wang, Z.: On-line deformation monitoring of thin-walled parts based on least square fitting method and lifting wavelet transform. Int. J. Adv. Manuf. Technol. **94**(9–12), 4237–4246 (2017). https://doi.org/10.1007/s00170-017-1145-9
4. Xiao-yong, J.I.A., Chuan-sheng, X.U., Xin, B.A.I.: The invention and way of thinking on least squares. Res. Cent. Hist. Math. Sci. Northwest Univ. **036**(003), 507–511 (2006)
5. Mei, L.: On the least square solutions of incompatible matrix equations. Time Educ. (009), 235–239 (2015)
6. Skala, V.: Least square error method robustness of computation: what is not usually considered and taught. (2018)
7. Shi, Z.: Harbin Engineering University Press (2010)
8. Chen, M., Fang, Y., Chen, J.: Fitting o circular curve based on least square method and iterative method. Sci. Surveying Mapp. **41**(1), 5 (2016)

Study on the Influence of Input Fluctuation on Mixing Effect

Hongwei Xu[1(✉)], Hang Zhang[1], Zhaohua Ma[1], Zhicheng Xue[2], and Darun Xi[2]

[1] Xi'an University of Technology, Xi'an, China
xuhongwei@xaut.edu.cn
[2] Shaanxi Beiren Printing Machinery Co., Ltd., Weinan, China

Abstract. On the basis of expounding the influence of the mixing proportion accuracy of solventless compound A and B on the quality of solventless compound products, this paper analyzes the proportional mixing system of solventless compound mixer. Then the output flow characteristics of the circular arc gear pump used for proportioning are analyzed, and the error of the input flow ratio of the static mixer of the solventless compound mixer caused by the output flow fluctuation of the circular arc gear pump is analyzed. By using FLUENT software, the simulation calculation is carried out for the mixing situation of fluctuating solvent-free composite A and B materials after being input into two different types of static mixers. Through the analysis of the simulation calculation results, it is found that the structure of the static mixer reduces the impact of gear pump output fluctuation on the proportioning accuracy after passing through the static mixer, but the mixing effect of different types of static mixers will be different, so when selecting the static mixer, in order to ensure reliable proportioning accuracy, reasonable selection is required.

Keywords: Solventless compound · Mixing · Proportion accuracy · Flow fluctuation · Simulation

1 Introduction

Solventless compounding technology is one of the three green environmental protection technologies in the printing and packaging industry [1]. However, the current solventless composite technology cannot fully replace the dry composite process with high energy consumption and solvent volatilization. The main reason is that solventless composite products have lower anti stripping performance than dry composite products. The low anti stripping performance of solventless composite products is affected not only by the solventless composite compound itself, but also by the fact that the solventless composite mixer cannot guarantee the high-precision mixing ratio in practical work. The accuracy of mixing ratio has a great impact on the anti-stripping performance of solventless composite products [2].

M. Xu et al. (Eds.): CACPP 2022, LNEE 991, pp. 288–297, 2023.
https://doi.org/10.1007/978-981-19-9024-3_38

Solventless compounding equipment is divided into solventless compounding machine and solventless compounding mixer. Solventless compounding mixer is the auxiliary equipment of solventless compounding machine and provides mixed compound glue for solventless compounding machine [3]. The proportioning and mixing system of solventless compound mixer is the core of solventless compound mixer, which directly affects the performance of solventless compound. Through experiments, it is found that the proportion of solventless compound A and B has an important impact on the anti-stripping performance of solventless composite products [4]. Therefore, the accuracy of the proportion and the mixing effect are very important to the quality of solventless composite products. In the solventless compounding equipment, the mixing effect of the mandrel in the static mixer with different diameters is different. At the same time, different inlet flow rates of materials A and B will also cause different mixing effects [5].

In a static mixer with 180° Kenics elements, the mixing efficiency improved with increasing the number of elements from two to four at a constant number of passes. It was also found that larger values of tube diameter (smaller d/D) lead to a better performance of static mixers [6]. The energy consumption and pressure required to obtain products of equal viscosities is less when using CDDM technology to process plant fibers [6]. For the nitrogen oxide removal processes, the double swirl static mixer resulted in improved mixing performance in a more compact design and energy efficient compared with other SV static mixers [7]. The shear thinning fluids with using lower pressure drop exhibit better mixing quality, lower pressure drop and higher mixing efficiency in a standard SMX static mixer [8]. Therefore, the accuracy of the proportion and the mixing effect are very important to the quality of solventless composite products.

Generally speaking, the solventless compound mixer realizes the ratio of solventless compound materials A and B through the gear pump. The mixing ratio of solventless compound materials A and B is realized by the ratio of gear pump speed for conveying materials A and B. The materials A and B conveyed in proportion enter the static mixer for mixing. Therefore, the gear pump determines the accuracy of solventless compound proportioning, while the static mixer determines the mixing effect. For the proportioning, the proportioning of solventless compound the materials A and B is determined according to the speed of the gear pump, so it is necessary to analyze the relationship between the output flow and speed of the gear pump.

2 Arc Gear Pump Output Flow Analysis

The working principle of arc gear pump is to form an enclosed space through the meshing of two gears. By changing the volume of the closed oil chamber, the oil suction and oil discharge functions can be realized. Through the simulation calculation of the modeling, it can be obtained that the transient outlet flow formula of the circular arc gear pump. The formula is show as Eq. 1 [7].

$$Q_S = w[\tfrac{m^2z^2}{4}\cos^2\alpha \times \left(\theta - \tfrac{(n-1)\pi}{z}\right)^2 - \tfrac{\pi m^2 z}{4}\cos^2\alpha \times \left(\theta - \tfrac{(n-1)\pi}{z}\right) + \tfrac{\pi m^2 z}{4}\cos\alpha\Big]$$
$$\theta \in \tfrac{\mathrm{n}-1\pi}{z}, \tfrac{n\pi}{z} \quad \mathrm{n} \in 1, 2, 3 \ldots \tag{1}$$

where, Q_S is the instantaneous outlet flow of the circular arc gear pump, and ω is the unit of the gear pump rotation speed; m is the module of the gear pump, n is the fluctuation period of the gear pump, and Z is the number of teeth of the gear pump; θ is the turning angle of the gear pump after time t. It can be seen from Eq. 1 that the flow fluctuates periodically with time.

From Eq. 1, it can be seen that the outlet flow of the gear pump fluctuates periodically, so the input of materials A and B in the static mixer also has fluctuating characteristics. We will first analyze what kind of matching error such fluctuations will produce. Table 1 shows the main structural parameters of circular arc gear pump.

Table 1. Arc gear pump main structural parameters

Number of teeth	Modulus	Pressure angle	Effective tooth width	Outlet pressure	Inlet diameter	Center distance
8	5.775	14.5°	52 mm	0.6 MPa	10 mm	24 mm

The specific parameters of gear pump in Table 1 are substituted into Eq. 1. In addition, assuming that when the gear pump feeding speed of material A is 60 r/min. The instantaneous flow parameter formula of material A can be obtained as Eq. 2.

$$Q_A = 32.656 \times (5.04 \times 6.28t - \frac{(n-1)\pi}{8})^2 - 1.98 \times 6.28t - \frac{(n-1)\pi}{8} + 2.04, \quad t \in \frac{n-1}{16}, \frac{n}{16} \quad n \in 1, 2, 3 \ldots \tag{2}$$

Assuming that the volume ratio of solventless compound materials A and B is 1.25:1, the feeding speed of material B is 48/min, and the formula of instantaneous flow parameter of gear pump feeding material B is obtained by substituting into Eq. 1, as shown in Eq. 3.

$$Q_B = 26.1248 \times (5.04 \times (5.024t - \frac{(n-1)\pi}{8})^2 - 1.98 \times (5.024t - \frac{(n-1)\pi}{8}) + 2.04), \quad t \in \frac{\mathrm{n}-1}{12.8}, \frac{n}{12.8} \quad \mathrm{n} \in 1, 2, 3 \ldots \tag{3}$$

The instantaneous flow formula is imaged by MATLAB software, and the flow curve of gear pump with time is obtained. As shown in Fig. 1, the visual instantaneous flow periodic fluctuation diagram after processing Eqs. 2 and 3 respectively.

In order to determine the actual flow ratio of A and B feed liquid input at the mixer inlet, the instantaneous flow curves of the two feed liquids are divided to calculate the actual value of A/B feed liquid flow ratio at different times (i.e. actual flow value = A feed flow value /B feed flow value), and finally the A/B feed liquid flow ratio curve is obtained, as shown in Fig. 2.

It can be seen from Fig. 2 that under the condition that the feed liquid input at the inlet of the mixer is not constant due to the structure of the gear pump, the mixing proportion of the two feed liquids to be mixed in the mixer is inaccurate before the mixing input at the inlet, and there is a certain deviation, and the Fig. 2 shows periodic fluctuations, with A fluctuation period of about 0.347.

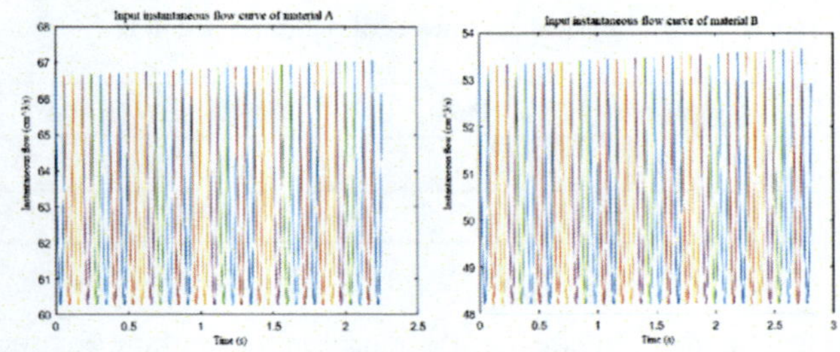

Fig. 1. Visualize instantaneous flow cycle fluctuation diagram

Fig. 2. A/B Curve of material liquid flow ratio

Process the data in Fig. 2, analyze the specific fluctuation, and obtain the flow ratio data of solventless composite A and B, as shown in Table 2.

Table 2. Comparison table of feed liquid input ratio data

Parameter	Minimum value of flow rate	Maximum value of flow rate	Ideal value of flow rate	Range of error
Value	1.1325	1.3794	1.25	0.2469

3 Influence of Input Fluctuation on Mixing Effect

3.1 Pretreatment of Simulation Calculation

The mixer used by the research team is WB8118A/B two-component solventless composite adhesive developed by Shanghai Kangda Chemical Co., Ltd. The specific physical parameters of the adhesive are shown in Table 3.

When considering the unstable input of feed liquid, it is necessary to simulate the fluctuation of inlet velocity with time and introduce time variables, so the transient model

Table 3. Physical parameters of materials A and B

Type	Viscosity 23°C (Pa·s)	Density (Kg/m^3)
Material A	1–1.2	1120
Material B	0.6–0.7	980

is adopted, and the gravity field effect of the model in the working environment needs to be considered.

The reason why the actual fluid flow state presents different patterns is the result of the comparison and confrontation between disturbance factors and viscous stability, that is, the comparison and confrontation between inertial disturbance and viscous stability. Therefore, the fluid flow state is determined by the Reynolds number, and the calculation expression is Eq. 4.

$$Re = \rho v d / \mu \tag{4}$$

where: ρ - Fluid density ($kg/m\hat{}3$); v - flow rate (m/s); d - equivalent diameter (m); μ - Viscosity coefficient ($Pa \cdot s$). Generally, the fluid is laminar flow; At that time, the fluid flow was in a transitional state, and at that time, the fluid flow was in a turbulent state.

Since the maximum feed liquid velocity is about 3.1 m/s, the data is substituted into Eq. 4.

$$Re = \frac{1120 \times 3.1 \times 1.2 \times 10^{-2}}{1.1} \approx 37$$

Therefore, according to *Re* calculation results, the fluid flow pattern of this mixer is laminar flow, and it is a mixture mixing model in two-phase flow. The difference format is set to the second-order upwind format; The solver selects the pressure velocity coupling method as the convergence condition, the time step is 0.001 s, and the iterative calculation is carried out after the setting is completed. Set two inlets as speed inlets, and the outlet adopts free outlet. Materials A and B are respectively input at the inlet A and B, and their flow velocity ratio is 5:4.

3.2 Simulation of Mixing Process in a Static Mixer with Different Rotating Differential Rows

The adopted static mixer with different rotary forks consists of 6 mixing units. The total length of the mixer is 120 mm, the length of the feed liquid inlet input pipe is 8 mm, the inlet pipe diameter is 5 mm, the spacing between the two pipes is 10 mm, the blade thickness of the mixing unit is 1 mm, the diameter is 12 mm, and the screw pitch is 24 mm. After the completion of the mixing unit, there is a stepped contraction transition outlet structure, and the final outlet diameter is 4 mm.

Simulation of mixing process of the static mixer with different rotatory displacement can obtain the volume fraction cloud diagram of material A after mixing through calculation. Figure 3 shows the volume fraction cloud diagram of material A at the outlet of the static mixer at 2.80 s.

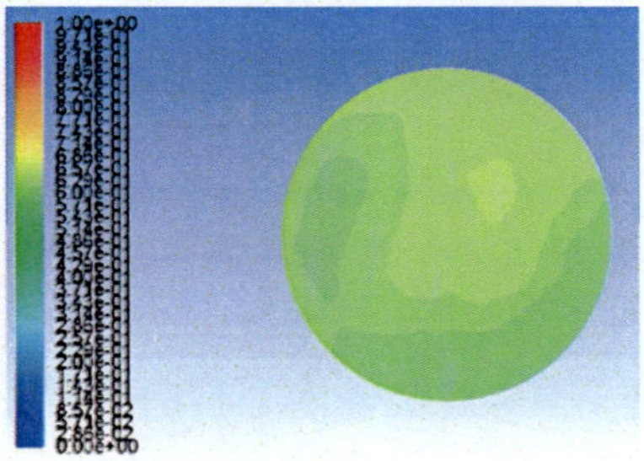

Fig. 3. Cloud diagram of integral number of liquid A at the outlet

Since the ratio of material A and B is 5:4, the ideal mixing effect is that the volume integral value of material A at each node reaches 55.5%. The volume fraction of material A at each node at the mixer outlet at different times is extracted and its mean value is calculated. Take time as the abscissa and the mean volume fraction of material A as the ordinate to draw the curve, as shown in Fig. 4.

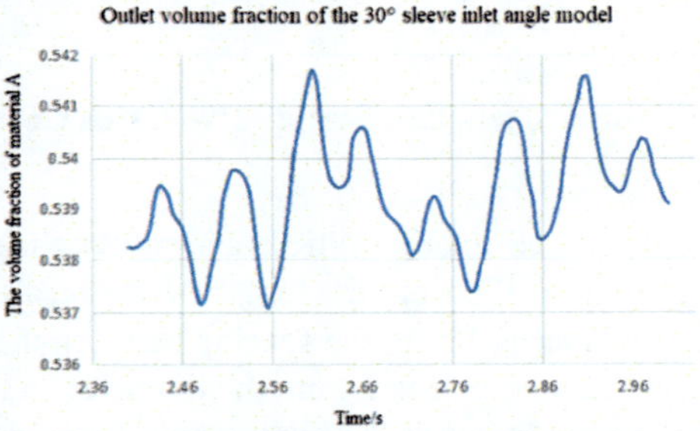

Fig. 4. Outlet volume fraction of the 30°sleeve inlet angle model

It can be seen from Fig. 4 that the mean volume fraction of material A after mixing at the outlet fluctuates with time, and the maximum fluctuation range is 0.00463.

In order to better analyze the mixing effect, the mixing uniformity of material and liquid A at the outlet of the mixer will be studied here. The mixing uniformity will be analyzed by extracting the non-uniformity coefficient of material and liquid A at the outlet of the mixer at different times, and drawing the change curve of the non-uniformity coefficient of material A at the outlet of the mixer with time. As shown in Fig. 5, the change curve of the non-uniformity coefficient of material A at the outlet of the model with time.

It can be seen from Fig. 5 that the maximum mixing non-uniformity coefficient is 0.02473. Based on the above calculation, the relevant parameters of mixing effect of solventless composite materials A and B with fluctuating input flow of the different rotating differential rows static mixer can be obtained. See Table 4.

Fig. 5. Outlet mixing coefficient curve of the 30°sleeve inlet angle model

Table 4. Mixing effect parameters of the different rotating differential rows static mixer

Parameter	Minimum volume fraction of material A	Maximum volume fraction of material A	Volume fraction fluctuation range	Maximum mixing non-uniformity coefficient
Value	0.53707	0.5417	0.00463	0.02473

3.3 Simulation of Mixing Process in a Static Mixer with Co-rotating Differential Rows

The adopted static mixer with co-rotating differential rows consists of 6 mixing units. The total length of the mixer is 120 mm, the length of the feed liquid inlet input pipe is 8 mm, the inlet pipe diameter is 5 mm, the spacing between the two pipes is 10 mm, the blade thickness of the mixing unit is 1 mm, the diameter is 12 mm, and the screw pitch is 24 mm. After the completion of the mixing unit, there is a stepped contraction transition outlet structure, and the final outlet diameter is 4 mm.

Simulation of mixing process of the static mixer with co-rotatory differential rows can obtain the volume fraction cloud diagram of material A after mixing through calculation. Figure 8 shows the volume fraction cloud diagram of material A at the outlet of the static mixer at 2.80 s (Fig. 6).

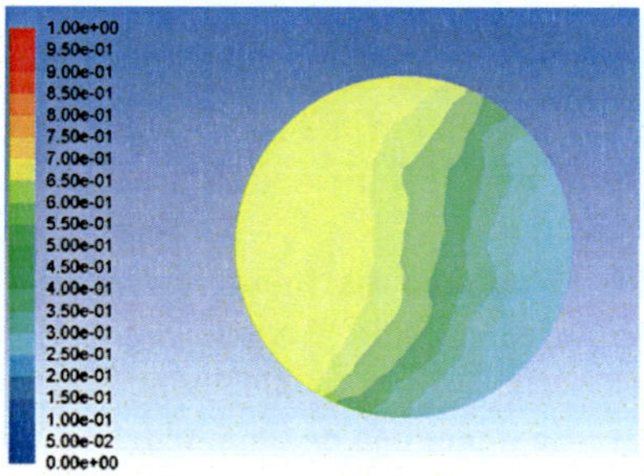

Fig. 6. Cloud diagram of integral number of liquid A at the outlet

The volume fraction of material A at each node at the mixer outlet at different times is extracted and its mean value is calculated. Take time as the abscissa and the mean volume fraction of material A as the ordinate to draw the curve, as shown in Fig. 7.

Fig. 7. Outlet volume fraction of static mixer

It can be seen from Fig. 7 that the mean volume fraction of material A after mixing at the outlet fluctuates with time, and the maximum fluctuation range is 0.004.

In order to better analyze the mixing effect, the mixing uniformity of material and liquid A at the outlet of the mixer will be studied here. The mixing uniformity will be analyzed by extracting the non-uniformity coefficient of material and liquid A at the outlet of the mixer at different times, and drawing the change curve of the non-uniformity coefficient of material A at the outlet of the mixer with time. As shown in Fig. 8, the change curve of the non-uniformity coefficient of material A at the outlet of the model with time.

Fig. 8. Outlet the non-uniformity coefficient of static mixer

It can be seen from Fig. 8 that the maximum mixing non-uniformity coefficient is 0.2605. Based on the above calculation, the relevant parameters of mixing effect of solventless composite materials A and B with fluctuating input flow of the co-rotating differential rows static mixer can be obtained. See Table 5.

Table 5. Mixing effect parameters of the co-rotating differential rows static mixer

Parameter	Minimum volume fraction of material A	Maximum volume fraction of material A	Volume fraction fluctuation range	Maximum mixing non-uniformity coefficient
Value	0.5385	0.5425	0.004	0.2605

4 Conclusion

In this paper, the proportioning system of solventless compounding machine is analyzed and studied. It is found that the output flow of gear pump, which is the key component to ensure the proportioning of solventless compounding machine, fluctuates. Aiming at the fluctuating flow input of solventless compound A and B, the simulation calculation of mixing process through two different static mixers is carried out by using FLUENT software. The calculation results show that both the different rotating differential row static mixer and the co-rotating differential row static mixer can well counteract the impact of flow fluctuation on mixing accuracy, but at the same time, the non-uniformity coefficient of the different rotating differential row static mixer is significantly lower than that of the co-rotating differential row static mixer, which shows that the different rotating differential row static mixer can not only eliminate the impact of input flow fluctuation on mixing effect, At the same time, its mixing effect is also much better than the static mixer with the co-rotating differential row. Therefore, in order to ensure reliable mixing quality, it is best to use the static mixer with different rotating differential rows as the key mixing part of solventless compounding machine.

Acknowledgements. This research project is supported by the Key R&D Plan of Shaanxi Province (approve number: 2021GY-262).

References

1. Xu, H., Wang, X., Lei, R., et al.: Experimental analysis of the effect of vacuum degassing technology on the solventless laminating adhesive performance. In: China Academic Conference on Printing & Packaging and Media Technology, pp. 1123–1129, (2016). https://doi.org/10.1007/978-981-10-3530-2_138
2. Wei, Y., Xu, H., Han, Y., Feng, S.: Tooth profile optimization for mixing proportional pump of solventless laminating mixer. In: Advanced Graphic Communication, Printing and Packaging Technology-Proceedings of 2019 10th China Academic Conference on Printing and Packaging, pp. 523–529 (2019). https://doi.org/10.1007/978-981-15-1864-5_72

3. Xing, B., Xu H., Chen, X., Li, X.: A control algorithm for improving the ratio precision of solventless laminating. In: Proceedings of 2019 IEEE 2nd International Conference on Automation, Electronics and Electrical Engineering, AUTEEE, pp. 542–546 (2019). https://doi.org/10.1109/AUTEEE48671.2019.9033393
4. Liu, L., Xu, H., Li, W., Lu, W.: Study on bubble motion characteristics under the effect of the difference stirring system. In: Advanced Graphic Communication, Printing and Packaging Technology—Proceedings of 2019 10th China Academic Conference on Printing and Packaging, pp. 511–516 (2019). https://doi.org/10.1007/978-981-15-1864-5_70
5. Xu, H., Zhang, H., Liu, D..: Characteristics research of static mixer used in solventless laminator. J. Lect. Notes Electr. Eng. 699–705 (2016).https://doi.org/10.1007/978-981-10-0072-0_87
6. Göbel, F., Golshan, S., Reza Norouzi, H., Zarghami, R., Mostoufi, N.: Simulation of granular mixing in a static mixer by the discrete element method. J. Powder Technol. 171–179 (2019). https://doi.org/10.1016/j.powtec.2019.02.014
7. Harvey, D.H.S., Mothersdale, T., Shchukin, D., et al.: Plant fiber processing using the controlled deformation dynamic mixer. J. Chem. Eng. Technol. 1566–1573 (2019). https://doi.org/10.1002/ceat.201800718
8. Zhuang, Z., Yan, J., Sun, C.: The numerical simulation of a new double swirl static mixer for gas reactants mixing. J. Chin. J. Chem. Eng. 2438–2446 (2020). https://doi.org/10.1016/j.cjche.2020.05.008
9. Michael, V., Dawson, M., Prosser, R., Kowalski, A.: Laminar flow and pressure drop of complex fluids in a Sulzer SMX+TM static mixer. J. Chem. Eng. Res. Des. 157–171 (2022). https://doi.org/10.1016/J.CHERD.2022.03.018
10. Xu, H., Li, W., Wang, X., et al.: Study on the flow mixing results of revolving static mixer used in solventless laminator. In: 2016 China Academic Conference on Printing & Packaging and Media Technology, pp. 783–789 (2016). https://doi.org/10.1007/978-981-10-3530-2_98

Study on the Hot Air Flow Field of TAC Membrane Dehalogenizing Oven Based on Fluent

Ding Wei[1](✉), Xiaojing Xu[1], and Hui Wang[2]

[1] Xi'an Aerospace-Huayang Mechanical & Electrical Equipment Co. Ltd, Xi'an, China
weiding2015@126.com
[2] China Academy of Aerospace Liquid Propulsion Technology, Xi'an, China

Abstract. TAC film (cellulose triacetate) is the primary application material for the flexible OLCD substrate market because of its excellent optical permeability, long bending life and relatively low manufacturing costs.Through the flow field analysis of TAC film dehalogen oven, the intuitive flow field distribution map was obtained. The applicability of the existing TAC oven diversion structure was discussed, and the existing longitudinally opposed air supply, return air outlet layout and uniform flow structure were abandoned. The results show that the flow field in the fluid domain of the dehalogenizing oven is optimized by using a new diagonal hot air layout and a dense hole hot air dredging structure. According to the optimized hot air oven structure, a set of dehalogenizing oven prototype was manufactured, and the actual production test was carried out, and the ideal effect was obtained. From the perspective of the stability of hot air flow field, the probability of scratching of TAC membrane was further reduced, and the qualified rate of products was improved.

Keywords: Flexible OLCD · TAC film · Dehalogenizing oven · Fluent

1 Introduction

Due to its inherent optical and physical properties, TAC film has been applied in electronic and civil fields, especially in the LCD (thin film transistor) industry, and it is irreplaceable as a polarizer support for LCD [1]. There is a certain gap in the performance of PMMA film, PET film and TAC film, and there is still a long way to go in replacing TAC film [2, 3]. TAC film has become an indispensable key raw material in the production of polarizers due to its excellent optical properties and strong mechanical strength. However, the chlorine content of TAC film is above 900 ppm due to the chlorine content in the production formula and the defects in the production process design, which does not meet the EU chlorine standard IEC61249-2-21, and the chlorine content is controlled below 900 ppm [4]. In this paper, several methods to reduce the chlorine content in TAC films are discussed from the aspects of production formulation and production process, which significantly reduces the chlorine content in TAC films.

M. Xu et al. (Eds.): CACPP 2022, LNEE 991, pp. 298–305, 2023.
https://doi.org/10.1007/978-981-19-9024-3_39

Flexible display devices have considerable advantages. Compared with the traditional flat panel display, flexible display has the advantages of light weight, durability, easy to be stored in large quantities, ultra-thin and retractable. With the gradual maturity of processing technology, the application fields of flexible display devices are gradually expanding, such as electronic paper, wearable electronic equipment, electronic posters and electronic and electronic tags [5]. In reference [6], the moisture absorption characteristics of TAC films were studied experimentally. The improvement of moisture absorption of domestic TAC films was taken as the research object, and the foreign competitive TAC films were taken as the improvement goal. Through orthogonal experiment, the drying temperature of TAC membrane was determined to be 130 °C. Due to the previous studies show that: when the temperature is not higher than 130 °C, the apparent quality of TAC film is good, and the chlorine content is close to 900 ppm. When the temperature is 140 °C, the decrease of chlorine content is very small, but at this time, the TAC film has been seriously deformed, which seriously affects the apparent quality of TAC film.

Therefore, in order to further reduce the halogen content of TAC film, it is necessary to find other ways in the drying process. Due to the existing literature research results: with the increase of wind speed, liquid volatilization rate increases, this phenomenon is consistent with the existing theory [7–12]. In the drying process of the coating film, the solubility, viscosity and surface tension of the solvent change with the volatilization of the solvent, which greatly affects the coating structure, appearance and performance [13–15]. New breakthroughs can be found in oven structures that reduce halogen content in TAC films. This paper mainly discusses the method of reducing chlorine in TAC film and protecting the film from being scratched from the perspective of drying air distribution in the oven.

2 Simulation Analysis

2.1 Modeling of Dehalogenizing Oven

The first drying oven of a domestic TAC thin film casting production line is selected as the research object of this analysis. A 3-D model is established. The layout diagram of the box is shown in Fig. 1. The size of the oven is high × length × width: 2700 × 2150 × 2500 mm.

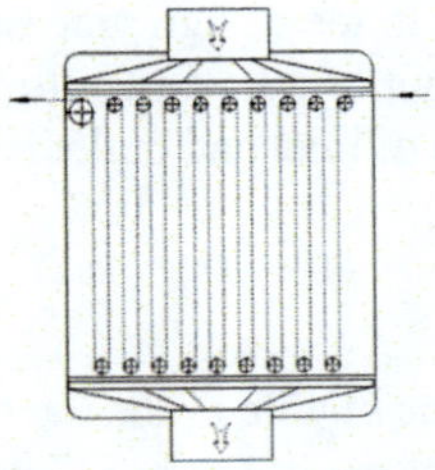

Fig. 1. Schematic diagram of the layout of the dehalogenizing oven

2.2 Fluid Simulation

Firstly, the geometric structure of the dehalogenizing oven is analyzed: as shown in Fig. 1, the main view of the halogen-reducing oven can see dense longitudinally distributed films. Inside the oven, there are multiple sets of longitudinally reentrant TAC films and guide rollers. The hot air entering the oven from the top inlet passes through the oven and discharges the oven from the outlet at the bottom of the oven. The purpose of the airflow path planning in the oven is to make the hot air and the surface of the material film evenly contact, so that the halogen in the TAC film volatilizes from the TAC film under the dual effect of high temperature and airflow, and is carried out by hot air out of the oven to complete the dehalogenizing process.

In order to save computing resources and improve the efficiency of fluid simulation and the whole analysis process, the 1/8 size model of the oven was analyzed to explore the flow field distribution of similar structures.

The geometric model of dehalogenizing oven was established and meshed by DM module of asnsys software. Then the grid file is imported into fluent software, and the turbulence model is set as Standard K-epsilon model; the upper inlet is set as the speed inlet, and the parameter setting is 0.5 m/s; set the outlet of return air taste pressure, parameter setting 0 Pa.

In addition, the parameters related to the flow form in the box are calculated as follows:

In the oven, the Reynolds number and the Nousser number can be calculated using the following equation:

$$Re = \rho v L/\mu$$

wherein: ρ is the density, μ is the dynamic viscosity, L, v is the characteristic velocity and characteristic length of the flow field.

In the design of the oven, the air velocity at the inlet of the oven airflow is the largest, and the Reynolds number at the inlet of the oven is calculated $Re = 2620$, so we can believe that the airflow at the air inlet is in a transitional state. Since the transition flow in the oven can cause harmful fluctuations in the film, further analysis of the flow field conditions in the oven is necessary.

According to the box structure and the distribution of TAC film, the flow field in the box is preliminarily simulated. The simulation goal is that the hot air flows parallelly along the film surface in the box, the hot air film fully contacts, and the halogen in the TAC film is fully volatilized. In the design process, the inlet and outlet of the oven are variables. Therefore, the layout of inlet and outlet of type 3 is set. Combined with practical experience, the location of inlet and outlet is set as shown in Table 1.

2.3 Results of the Simulation

Simulation A inlet is located in the middle of the top of the box, and the return air inlet is located in the middle of the bottom of the box. The simulation results are shown in Fig. 2. When the dry air enters the box from the top of the oven, the airflow is separated by the arc surface of the guide roller when passing through the guide roller. Then, the

Table 1. Simulation settings

Model	Location of inlet	Location of return air inlet
A	Top of the oven	Bottom of the oven
B	G side	M side
C	G side	Top and bottom
D	Top and bottom	Insert-pipe

air flowing through the upper surface of the film can have a good mass transfer and heat transfer with the upper surface of the film. However, on the back of the film, only a small amount of airflow can pass through, and the airflow flowing through the two sides of the film is quite different. On this point, the pressure field distribution in the box can also be seen clearly. In the actual production process, there will also be a rhythmic beating or vibration of the material film caused by the pressure fluctuation on both sides of the film. Obviously, this is unfavorable for the smooth feeding of the material film. The material film will be easy to produce uncontrolled sliding friction on the surface of the guide roller, and eventually lead to the formation of scratches.

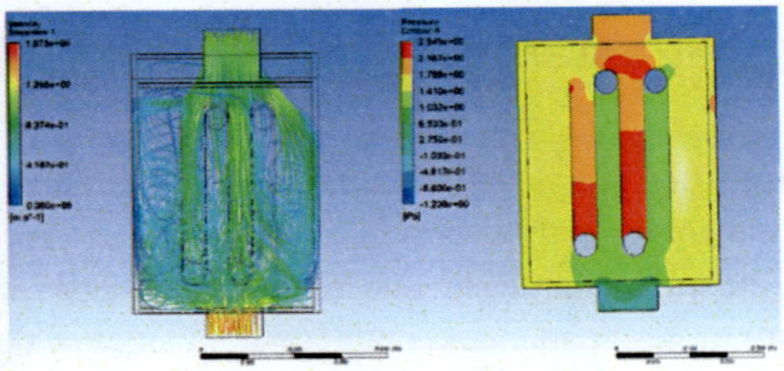

Fig. 2. Simulate A air flow trace diagram and pressure field distribution

The simulation B inlet is located in the middle position of the G side box, and the exhaust port is located in the position of about 100 mm from the bottom of the M side box. The drying airflow passes through the two sides of the film from the G side, and discharges the box from the M side exhaust port. Compared with simulation A, both sides of the film can exchange heat and matter with the dry airflow, but there is obvious large vortex near the G side of the film surface. Large vortex flow field will produce a series of small and random vortices due to the dissipation of vortices, which will cause the beat of the film surface. The maximum air pressure of the oven cross section of this structure can be viewed from the pressure cloud chart 2.545 Pa. Thus, this structure cannot be used either (Fig. 3).

The simulation C inlet is located in the middle position of the box on the G side, and the exhaust outlet is located at the top and bottom of the box, and a section of air guide slope is set for the exhaust outlet to make the airflow smoother. The simulation results show that there are still quite a lot of vortices on the G side of the film surface inside the oven. Although the uniformity of the flow field has been improved compared

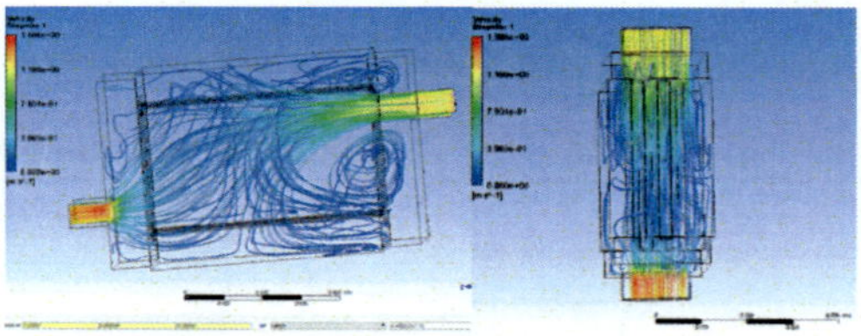

Fig. 3. Simulate B air flow trace diagram

with that of simulation C, the turbulence of the flow field in the box still cannot meet the requirements. The maximum air pressure of the oven cross section of this structure can be viewed from the pressure cloud chart − 5.172 Pa (Fig. 4).

Fig. 4. Simulate C air flow trace diagram and pressure field distribution

Based on the above three groups of simulation, we can simply sum up the flow field operation rule in the dehalogenizing oven: the air flow in the oven is divided into upper and lower sides in the geometric structure due to the barrier between the guide roller and the film. It is necessary to set the inlet and outlet on both sides of the material film at the same time, in order to achieve full contact with the hot air on both sides of the material film, and volatilize the halogen in the film as much as possible, and then discharge from the exhaust outlet.

Under the premise of mastering the above rules, when doing simulation D, the oven size is adjusted to 1:1 size. In order to ensure the accuracy of the calculation results, four groups of guide roller devices and 14 TAC films are set in the oven to simulate the real flow field in the oven. The setting mode of inlet and return air: the top surface of the oven and the ground are set at the same time. The size of the inlet is similar to the spacing of the film. The return pipe is set above the guide roller, and the gradual dense hole is set on the surface of the return pipe. The oven structure is shown in Fig. 5.

For ease of observation, the results of simulation D only output the flow field of the first two inlets and return ducts, as shown in Fig. 6. It can be seen from the airflow nephogram that hot air passes through the surface of the material film uniformly, passes through the dense holes on the inserted return air pipe, enters the waste discharge pipe and discharges the dehalogenizing oven. The hot air runs smoothly in the oven, fully contacts with the material film, and does not produce obvious harmful vortex, reaching the predetermined design goal. It can be seen from the airflow nephogram that the oven with this layout needs to pay some costs while improving the contact efficiency between

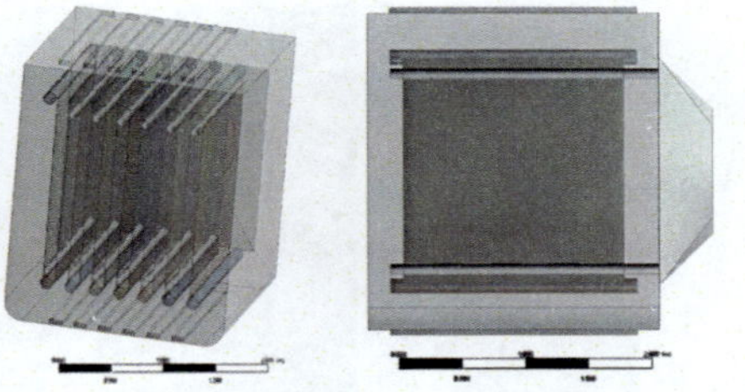

Fig. 5. 3D model diagram of the halogen-reducing oven with a new flow field layout

hot air and material film. That is, there is an air flow with a wind speed of about 13–18 m/s in the waste pipe, and the waste resistance increases. The rated pressure of the waste fan needs to be increased by 150–260 Pa on the current basis.

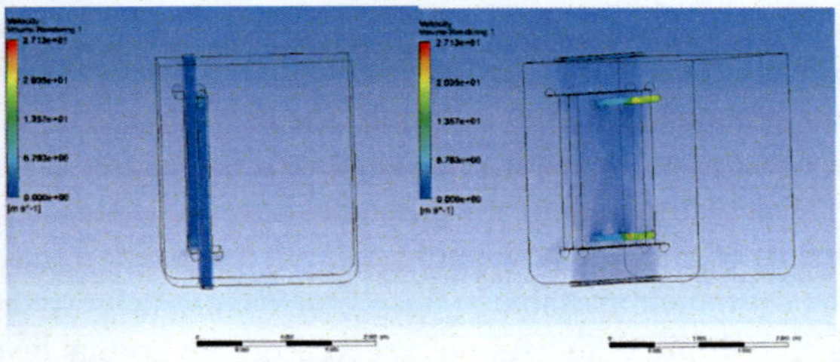

Fig. 6. Simulate D airflow cloud map

2.4 Discussion of the Simulation Results

By comparing the three groups of 1/8 models, the obstruction and dredging effects of the main structures in the TAC oven on the convection field were quickly found. The flow field operation law in the oven was intuitively obtained. Finally, the 1:1 model was used to analyze the flow field distribution in the front area of the oven.

Compared with oven simulation A~D, the airflow trace distribution of simulation A, B and C is uneven in local area. On the basis of summarizing A–C, simulation D made adaptive adjustment to the exhaust layout, increased the intubation exhaust pipe, and adjusted the layout of the inlet to the positive film gap, which avoided the generation of harmful vortex, further reduced the sliding friction between the material film and the guide roller, and provided conditions for improving the yield of products.

3 Installation Test

According to the inlet and outlet structure of the oven in Fig. 5, the specific installation layout of the oven is refined, and Fig. 7 of the manufactured TAC oven is shown. During the stable operation of the equipment, the drying oven guide roller did not scratch the material film. Through 24-h monitoring, The film-forming status of TAC film is greater than 96% in 24 h, the oven of the equipment meets the acceptance standards.

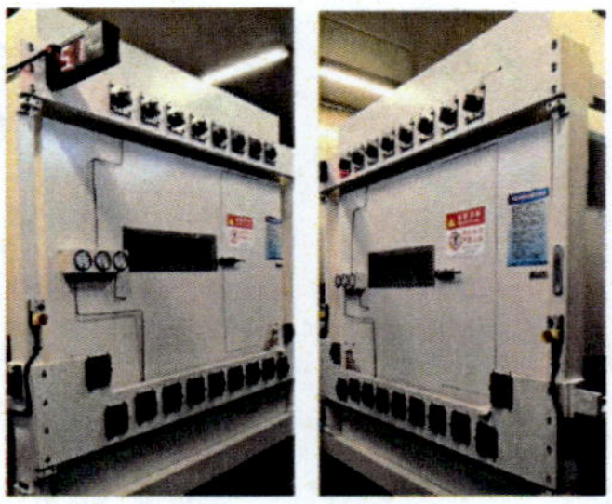

Fig. 7. Installation diagram of a halogen-reducing oven with a new flow field layout

4 Conclusion

In the comparison of the four simulation models, the inlet and outlet structures of the oven are arranged according to experience, and then the fluid simulation software is used to graphically calculate the key calculation results that affect the stable feeding in the flow field. From the graphical simulation results, the flow of airflow in the brine drying oven and the film surface can be intuitively displayed. According to the distribution of eddy current and pressure distribution, the graphical flow field data are screened, and the final simulation results of the oven flow field are applied to the design and production of the equipment. The equipment is stable in operation and can achieve the predetermined technical goals. Through the optimization research on the hot air layout and air duct structure of hot air in the dehalogenation oven, it can provide reference for the structural improvement of TAC dehalogenation oven device in engineering design.

Acknowledgements. This research is supported by Key Lab of Intelligent and Green Flexographic Printing. (No.ZBKT202102).

References

1. Teng L.-Z., Bai, M., Cheng, Z.: Discussion on dope of TAC film for LCD. Inf. Recording Mater. **14**(2), (2013)
2. Xie, Y., Zhang, F., Sun, X.: Research on the competitive situation of optical TAC film. Based on patent analysis. Inf. Recording Mater. 20(6), (2019)
3. Liu, D., Zhong, J., Tang, B.: Research progress of flexible and printed OLED. Chin. J. Liquid Cryst. Displays 36(2), (2021)
4. Xu, X., Song, X., Sun, X.: Research on reducing chlorine content in TAC film. Inf. Recording Mater. **18**(2), (2017)
5. Lan, Z., Wei, J., Yu, Y.: Research progress on materials for flexible display substrate. J. South China Normal Univ. (Nat. Sci. Ed.) **49**(1), (2017)
6. Qiao, S.: The improve of the TAC film hygroscopicity. Inf. Recording Mater. **15**(2), (2014)
7. Wang, Z., Zhu, F., Rong, Z.: Experimental determination and estimation of the evaporation rate of pure solvents. Paint Coat. Ind. **38**(6), (2008)
8. Zhang, Y., Leng, W., Yu, T.: Research on volatilization rate of oil film on ship engine compartment. Ship Sci. Technol. **41**(11), (2019)

9. Liu, W., Bai, C., Liu, Q.: Mechanism and experimental study of high volatile liquid mass transfer rate. Acta Armamentarii **41**(6) (2020)
10. Sun, R., Sun, X.: Effect of non-volatile solutes on boiling point and vapor pressure of mised solvent. CIESC J. **53**(9), (2002)
11. Liu, Y., Liu, B., Xu Z.: Study on halogen reduction by TAC film. Inf. Recording Mater. **41**(11), 9–10 (2019)
12. Wang, C., Yang, Z.-K., Liu, B.: Effect of plasticizer on physical and mechanical properties of TAC film. Anhui Chem. Ind. **21**(7), (2020)
13. Wang, L.: The effect of ink temperature on solvent losses and print quality. China Print (1), 130–133 (2002)
14. Min, W., Hongwei, J.: Factors of influencing fast-drying of moisture-cured polyurethane coating. Paint Coat. Ind. **37**(4), 13–16 (2007)
15. Kuznetsoy, A.V., Xdng, M.: Effect of evaporation on thin film deposition in dip coating. Int. Commun. Heat Mass Transf. **29**(1), 35–44 (2002)

Research on Digital Design Platform of Gravure Printing Press

Pengchao Dou, Peng Liu(✉), Yueyue Xing, and Xianwei Li

School of Printing Packaging and Digital Media, Xi'an University of Technology, Xi'an, China
liupeng@xaut.edu.cn

Abstract. In order to shorten the research and development cycle of gravure printing press, simplify the optimization process of parts and components, and realize the development of product design towards intelligence, serialization and modularization. According to the structural characteristics of gravure printing press and the experience of designers, the SolidWorks software is redeveloped with VB6.0 programming language, and the digital design platform of gravure printing press is developed. The platform can realize parametric modeling of parts, virtual automatic assembly, engineering drawing transformation and structure optimization of key parts. The rapid optimization method and key technologies of the integration of parametric design and finite element analysis are deeply studied. The results show that the digital design platform effectively improves the market response ability of enterprises, shortens the product development cycle, reduces the design cost of enterprises, effectively promotes the standardization, automation and intelligence level of design work, and has good practical application and promotion value.

Keywords: Gravure printing press · SolidWorks · Parametric design · Secondary development

1 Introduction

With the upgrading speed of downstream products accelerating in the flexible packaging industry, the iteration speed of gravure printing press products is also gradually improving. At present, the domestic gravure printing press manufacturing enterprises have problems in product design, such as miscellaneous product models, many repetitive designs and cumbersome optimization process [1]. In order to improve the current situation of gravure printing press manufacturing enterprises, the core problem is to fully implement digital design technology and establish a digital design platform for gravure printing press.

As the core technology of advanced design and manufacturing technology, digital design can effectively improve the whole product development process and shorten the product development cycle [2]. Therefore, scholars at home and abroad have done a lot of research on digital design technology. So far, universities and enterprises have developed some high-level digital design platforms. Bratovanov Nikolay developed a

M. Xu et al. (Eds.): CACPP 2022, LNEE 991, pp. 306–314, 2023.
https://doi.org/10.1007/978-981-19-9024-3_40

universal modeling and motion simulation digital design platform based on SolidWorks API interface [3]; Shaqura Mohammad and Shamma Jeff S developed an automated quadcopter based on SolidWorks API and intelligent dynamic assembly Design platform [4]; Songjian Liu developed a digital design and analysis platform for grenades based on the UG secondary development technology in batch mode, who is from North University of China [5]. But there are few applications for printing equipment manufacturing enterprises [6]. The main reason is that there are many printing machine parts with strong relevance, and it is difficult to develop.

2 Overall Design of the Platform

The platform takes the printing device of a gravure printing press as the research object, adopts the top-down modular design method and uses the Visual Basic 6.0 development tool to carry out the secondary development of SolidWorks, which can realize the parametric three-dimensional modeling of the series parts of the printing device of the gravure printing press. The rapid optimization design of the structural parameters of key parts is realized by the finite element analysis of SolidWorks Simulation. Finally, the digital design platform of the gravure printing press is completed.

The platform contains two functional modules of parametric design and finite element analysis, as shown in Fig. 1. The parametric design module can realize the parametric modeling of key parts, the automatic transformation of engineering drawings and the automatic assembly of parts. Under the premise of ensuring the product performance, the finite element analysis module can determine whether the key structural parameters are optimal, reduce the product quality, reduce the product cost, and quickly obtain the product models with similar parts structure and different sizes.

Fig. 1. Design of the digital platform

3 Parametric Design

There are many parts and components of gravure printing press. According to the different structure and function, this paper divides the six part structure of the printing parts: transmission device, plate cylinder clamping device, lateral adjustment device, longitudinal adjustment device, embossing device and support device. The parametric design of gravure printing press parts is realized by using VB6.0 to develop SolidWorks. The following takes the parameterization process of some parts of gravure printing press as an example to illustrate the implementation of the whole process.

3.1 Parametric Modeling

The macro recording method is used to complete the preliminary programming of the part modeling, then the key structural parameters of the part are extracted, and the key parameters in the program are designed as variables, non-key parameters are determined through the method of dimension chain drive and variable unification, and three-dimensional modeling of the part is performed by invoking a program for parametric modeling of the part. The user can select the type of component to be designed, and then input the main parameters of the part through the component parametric interactive interface, and generate the 3D design model of the required part in SolidWorks. Figure 2 shows the parametric design interface of the transmission device.

Fig. 2. Parametric design interface of transmission device

3.2 Automatic Assembly

The automatic assembly of parts is realized by using assembly features and writing assembly programs. Its essence is to select feature elements on parts, and determine the relative position relationship between parts according to the positioning conditions of components in the actual assembly, and add certain matching constraints [7, 8]. Figure 3 is a three-dimensional model of some parts of the printing device of the gravure press. After automatic assembly is done, the feasibility of the design scheme is preliminarily verified through interference inspection, so as to carry out the next step of product structure analysis and part drawing conversion.

a. Transmission device **b.** Lateral adjustment device

Fig. 3. 3D model of gravure printing press components

3.3 Engineering Drawing Conversion

A two-dimensional part drawing template that conforms to enterprise standards is established. Its fonts, custom attributes, and material attributes are uniformly specified, and its attribute is associated with the part template [9]. The program code is written in combination with the SolidWorks API function, the DrawingDoc object is called, the Create1stAngleViews2 method is used to realize the drawing view layout, and the InsertModelAnnotations3 method is used to realize the automatic dimensioning, and then the automatic generation of the drawing is completed [10]. When converting a drawing, the user should first open the 3D part drawing to be converted.

4 Finite Element Analysis

The design process of gravure printing press mostly relies on the existing experience, which has a certain blindness. The integration of CAD and CAE design method can effectively overcome the defects of part design and improve the level of structural design. In this paper, the rapid optimization design process of wall panel parts of gravure printing press is taken as an example. Under the condition of meeting the product stiffness and strength requirements, the rapid optimization process of lightweight design of wall panel parts is carried out by using SolidWorks Simulation.

4.1 Establish Finite Element Model

80% of the time in the finite element analysis process is spent in the preprocessing. Through the parametric design method, the wall panel can be quickly modeled, which greatly improves the analysis efficiency. The parametric modeling interface design of wall panel parts is shown in Fig. 4.

4.2 Load Solution

The 3D model of the wall panel is imported, and the material properties are added. The material is HT200, the elastic modulus is 145 GPa, the density is 7810.3kg/m^3, and the Poisson's ratio is 0.3. According to the function of wall panel holes, its constraints are

Fig. 4. Interface of the wall panel parameterized design

added. As shown in Fig. 5. A is restrained by bolt fixed connection. B, C, D, E and F are the relationship between the roller bearing and the wall panel hole, and the external load is directly defined to the inner hole surface for loading. G is the installation hole of the lateral adjustment device, and its load is simplified to the inner hole surface of the bolt hole through the principle of equivalence, and the wall panel is fixed to the base through bolts at H, the boss surface of the bolt hole add fixed constraints.

External loads are added to the wall panel. The 3D model of each part is established and the material of the part is added to estimate the mass, and then the external load of the wall panel is determined. The wall panel has its own weight, so its own gravity load of 5280 N should be added. The load on the upper surface of the wall panel is simplified to a uniform load of 5000 N, and the mass of the embossing device is simplified to a concentrated load of 200 N. Printing rollers are installed in the middle of the wall panel on both sides through bearings and the structure is symmetrical. Therefore, the load on one side of the wall panel is half of the weight of the printing roller, which is taken as 250 N. The load on the lateral adjustment device is simplified to an equivalent torque of 250 N.m at the inner hole surface of the bolt hole. After adding constraints and external loads, it is shown in Fig. 6.

The model is meshed that added constraints and loads. The main external load of the wall panel is near the wall panel hole. After automatic meshing, the local meshes can be encrypted. The effect after mesh generation is shown in Fig. 7.

4.3 Result Analysis

After the constraints, external load loading and mesh generation are completed, the finite element static analysis of the wall panel model is carried out, and the stress-strain analysis results of the wallboard parts are shown in Fig. 8.

From the displacement nephogram, it can be seen that the maximum displacement of the installation hole of the lateral adjustment device is 1.395×10^{-3} mm; From the stress-strain nephogram, it can also be seen that the maximum stress of the installation

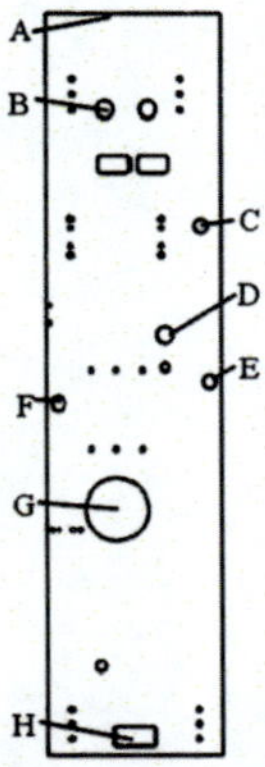

Fig. 5. Wall panel structure drawing

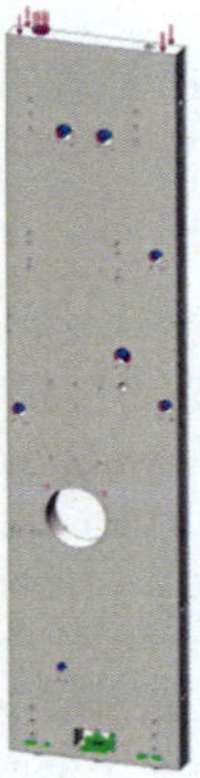

Fig. 6. Add constraints and loads

Fig. 7. Finite element model of the wall panel

a. Equivalent stress of wall panel

b. Total deformation of wall panel

c. Equivalent strain of wall panel

Fig. 8. Finite element static analysis of wall panel

hole of the lateral adjustment device is 1.311 MPa and the strain is 7.315×10^{-6}; The stress and strain in the upper area of the wallboard are small.

Therefore, the thickness of the upper part of the wallboard can be appropriately simplified in the design process. At the same time, in order to meet the stiffness requirements, the stiffener can be designed at the thin wall. The excessive stress near the installation hole of the lateral adjustment device is due to the fact that the lateral adjustment device is suspended outside the wallboard and its self-weight is large. During the operation of the plate cylinder, a certain preload is also required to drive the working device, so the hole structure can be strengthened. At the same time, certain supporting measures shall be taken to reduce the external load on the wall panel.

4.4 Secondary Optimization

The wall panel parts can be re modeled through the parametric design of parts. Under the premise of meeting the use requirements, the upper structure outside the wall panel is simplified. The cutting thickness, cutting area and rib thickness outside the wall panel can be optimized to reduce the wall panel self-weight and improve the structural stiffness. A three-dimensional model can be established and the finite element analysis solution can be performed again. The stress-strain analysis results of the wall panel parts are shown in Fig. 9.

According to the stress-strain distribution of the improved wall panel in the above figure, the maximum stress on the wall panel is 0.7561 MPa and the maximum displacement is 0.7202×10^{-3} mm, the maximum strain is 4.255×10^{-6}. Compared with the original model of the wall panel, each index is significantly reduced. The rigidity and

a. Equivalent stress of improved wall panel

b. Total deformation of improved wall panel

c. Equivalent strain of improved wall panel

Fig. 9. Finite element static analysis of improved wall panel

strength of the improved wall panel meet the use requirements, and the weight is reduced by 28.2%.

5 Conclusion

In order to shorten the design cycle of gravure printing press and improve the design efficiency. In this paper, the rapid optimization method and key technologies of parts parametric design and finite element method of gravure printing press parts are deeply studied. The digital design platform of gravure printing press is built. It can be used for reference for the secondary development of other similar products based on SolidWorks.

Acknowledgements. This research is supported by the Technological Innovation Guidance Plan of Shaanxi Province, China (No. 2020QFY03-02).

References

1. Huang, Q., Chen, F., Bao, N., et al.: Construction of digital design platform for gravure printing press industry. Light Ind. Mach. **27**(04), 6–9 (2009)
2. Duan, Q.: Application of modern digital design in mechanical design and manufacturing technology. Int. Combust. Eng. Part. **04**, 195–196 (2021)
3. Bratovanov, N.: Robot modeling, motion simulation and off-line programming based on SolidWorks API. In: The Third IEEE International Conference on Robotic Computing (IRC). IEEE (2019)

4. Shaqura, M., Shamma, J.S.: An automated quadcopter CAD based design and modeling platform using SolidWorks API and smart dynamic assembly. In: 14th International Conference on Informatics in Control, Automation and Robotics (2017)
5. Liu, S.: Development of Grenade Digital Design and Analysis Platform. North University of China, Shanxi (2019)
6. Zhang, S., Li, Q., Zhu, P.: Development of digital design platform for wheat making equipment based on SolidWorks. Mech. Des. Manuf. **S1**, 164–167 (2017)
7. Zhang, T.: Development of Parametric Design Platform for Regenerative Industrial Furnace Based on SolidWorks. Chongqing University of Technology, Chongqing (2021)
8. Zhang, X.: Valve Digital Design Integration Platform Research and Realization of Parametric Design. Lanzhou University of Technology, Gansu (2020)
9. Yang, Z., Hao, Z., Xu, J., et al.: Research on generation method and common settings of engineering drawings based on SolidWorks. South. Agric. Mach. **48**(16), 119 (2017)
10. Zhang, Q., Zhao, G.: Automatic generation and optimization of engineering drawings in secondary development based on SolidWorks. Agric. Equipment Veh. Eng. **56**(11), 72–75 (2018)

Rigid-Flexible Coupling Modeling and Dynamic Characteristic Analysis of Web Folding Mechanism

Yulong Lin[1], Tao Xue[2], Fanhua Bu[1], and Rui Zhang[1](✉)

[1] College of Mechanical and Electrical Engineering, Beijing Institute of Graphic Communication, Beijing, China
zhangrui@bigc.edu.cn
[2] China North Vehicle Research Institute, Beijing, China

Abstract. Taking the machete arm and the paper as flexible bodies, a rigid-flexible coupling dynamic model of the web paper knife-type folding mechanism was established. The relationship between the deformation of the machete arm and the lateral displacement of the machete head was analyzed to determine the maximum rotation speed of the folding mechanism to ensure the folding accuracy. Then, the dynamic contact excitation between the machete and the paper was analyzed, and the working loads of the machete were determined. Finally, the dynamic analytical model of the rigid folding mechanism was established to compare with the analytical values of the rigid-flexible coupling model, it was found that when the rotation speed of the folding mechanism was greater than 25,000 rph, the deviation of the two support reaction forces was significant, and the influence of the machete arm deformation cannot be ignored. The study has certain reference value for the simulation and experimental research of web folding mechanism.

Keywords: Folding mechanism · Rigid-flexible coupling kinematics · Dynamics

1 Introduction

Multi-body systems can be divided into rigid multi-body system, flexible multi-body system and rigid-flexible coupling multi-body system according to different mechanical properties of their objects [1]. Rigid multi-body system tends to ignore the elastic deformation of objects in the system and treat them as rigid bodies, which is often in low-speed motion. Flexible multi-body system refers to the system in which there is a coupling between the large range of motion of objects and the elastic deformation of objects during motion, so that the objects must be treated as flexible bodies. If a rigid multi-body system has some objects that can be treated as flexible bodies, the system is a rigid-flexible coupled multi-body system, which is the most common model for multi-body systems.

For web knife folding mechanism, the paper is a typical flexible body, the machete arm is a long-arm thin-walled structure with a large span. When the folding mechanism works at high speed, the machete arm will undergo flexible deformation to a certain

M. Xu et al. (Eds.): CACPP 2022, LNEE 991, pp. 315–329, 2023.
https://doi.org/10.1007/978-981-19-9024-3_41

extent under the influence of self-weight, load and moment of inertia. Therefore, it is an inevitable choice to study the web folding mechanism as a rigid-flexible coupled multi-body system. However, the current studies [2, 3] on the web knife folding mechanism usually simplify them to rigid systems, ignoring the deformation of weak parts, and the working load between the machete and the paper cannot be determined. In view of the above problems, based on the theory of rigid-flexible coupling, this paper establishes a multiple rigid-flexible coupling dynamics model based on rigid-flexible coupling theory with the machete arm and paper as flexible bodies and other components as rigid bodies, and analyzes the dynamic characteristics of the rigid-flexible coupling of the folding mechanism and the dynamic contact excitation between the machete and the flexible paper.

2 Rigid-Flexible Coupling Dynamic Modeling of Folding Mechanism

The analysis of rigid-flexible coupling model usually takes one or more rigid body parts for the flexible treatment, and then forms a new stable system with multiple rigid bodies through different constraints [4]. For web knife folding mechanism, the paper is a typical flexible body. The machete arm is a long-arm thin-walled structure with a large span. When the folding mechanism works at high speed, the load on the machete arm is large and its deformation should not be ignored. Therefore, the paper and the machete arm should be treated as flexible bodies while other components are still rigid bodies during modeling. Then a rigid-flexible coupling dynamic model is established, which can more accurately reflect the dynamic characteristics of the folding mechanism and the dynamic contact excitation between the machete and the paper.

2.1 Rigid Body Modeling of Folding Mechanism

The structure of the web folding mechanism is shown in Fig. 1, which is mainly composed of crank disc, connecting rod, machete arm, rollers, wallboard and other components. The machete arm is driven by the crank and the connecting rod to do reciprocal swing, the machete is installed on the machete arm. The paper is delivered to the bottom of machete by the conveyor belt of paper feeding table. The machete pushes the paper between the two rollers, and then the rollers squeeze the paper to complete a horizontal folding action.

The 3D rigid model of each component of the folding mechanism is established in the 3D modeling software Creo, and then the components are assembled to obtain a 3D multi-rigid body model, as shown in Fig. 2. It was saved in parasolid format and imported into the multi-body dynamic analysis software ADAMS.

2.2 Establishment of Flexible Parts for Folding Mechanism

There are three main methods to obtain the flexible body of machete arm and paper required in this paper [5, 6]: (1) the flexible body is established by flexible connection of discrete rigid parts, but it may produce large errors; (2) the flexible body is created

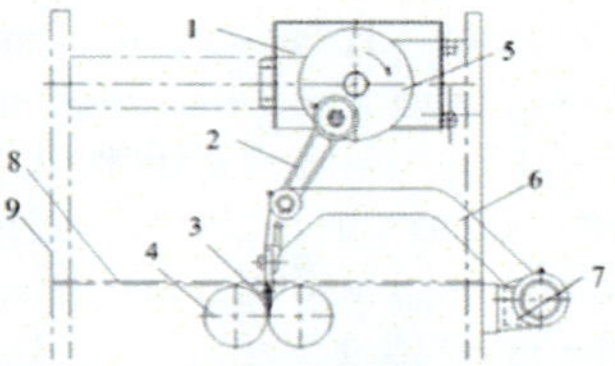

1. Transmission 2. Connecting rod
3. Machete 4. Roller 5. Crank disc 6. Machete arm
7. Knife body base 8. Paper feeding table 9. Wallboard

Fig. 1. Schematic structure of web webfolding mechanism

Fig. 2. 3D rigid body model of folding mechanism

by stretching method or geometric shape description method, but it is only suitable for simple models; (3) the flexible body is completed by finite element software and then generate mnf format modal neutral file, this method has a wide range of applications and high accuracy. Considering the complex structure of the machete arm and the material of the paper, the third method is used to build the flexible bodies using the rigid region method of ANSYS to generate the modal neutral files for importing into ADAMS [7, 8].

Modeling of Flexible Machete Arm

(1) The 3D solid model of the machete arm is imported into ANSYS software, the element types and material properties are set and the division of structured mesh is completed. The material of machete arm is 45# steel, the Yang's modulus is 209 GPa, the Poisson's ratio is 0.269, the material density is 7850 kg/m^3, the element type is set to solid186, the number of mesh elements after division is 64,666 and the number of nodes is 18,199. The mesh division effect is shown in Fig. 3.
(2) The central nodes at the two hinges of machete arm are selected to generate two tiny mass units, the unit type is selected as mass21 and the unit attribute is mass21p. The attribute value of mass21 quality unit is taken as a very small value in order that the quality does not affect the mass distribution of machete arm. Then the center nodes are used as the master nodes with the respective surrounding slave nodes to form the rigid region, which is the undeformed region of flexible body connected with the outside world in ADAMS. Through the GUI command "Preprocessor > Coupling/ Ceqn > Rigid Region", the central master nodes are selected first, and then the nodes around master nodes are connected to them. The generated rigid areas are shown in Fig. 4.

(3) The ANSYS software's macro command of adams.mac can generate modal neutral files for the desired flexible machete arm, and generate *.mnf file through the GUI command "Solution > ADAMS Connection > Export to ADAMS", which creates the master nodes of the rigid area as the interface nodes. In ADAMS, when it is required to associate kinematic pairs with the machete arm, the corresponding constraints will be added to the interface nodes.

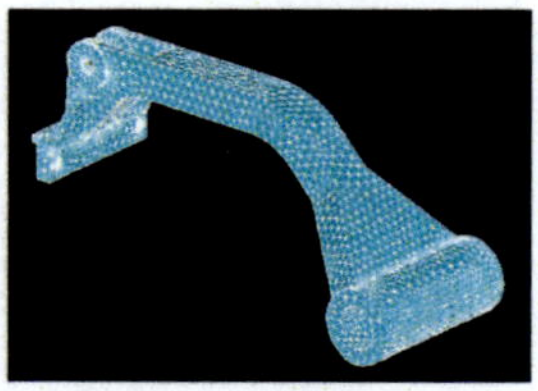

Fig. 3. Meshing of machete arm

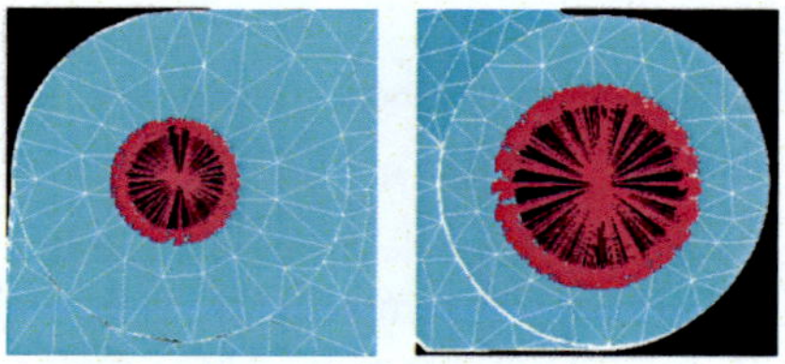

Fig. 4. Rigid area

Modeling of Flexible Paper

(1) The research object is No. 787 paper of sextodecimo. Since the paper is in the stage of elastic deformation when the folding mechanism machete pushes the paper into rollers, and relevant studies have shown that the paper is approximately orthotropic [9], so the paper material type is selected as linear elastic orthotropic material. With reference to the relevant parameters in the literature [10–12], the material properties of the paper are determined as shown in Table 1.

The paper is a thin sheet structure and has a small bending stiffness. The shell181 shell element is used for grid division. The edge of the paper element is 1mm long and the shell thickness is 0.1 mm. After division, the number of grid elements is 2491 and the number of nodes is 2477.

(2) In the same way as the rigid region generation method for the machete arm described above, four tiny mass units are generated at the nodes of the four corners of the paper model, which are used as master nodes with their respective surrounding slave nodes to form rigid regions.

(3) Through ANSYS software's macro command of adams.mac to generate the modal neutral file of the flexible paper model, and the master nodes of the paper model's rigid region are created as interface nodes. In ADAMS, force vectors are added

to the master nodes of the four rigid regions to simulate the friction of the paper feeding table belt on the paper. The master nodes of the rigid regions are shown in Fig. 5.

Table 1. Material properties of paper

Paper size (mm)	Density (kg/m^3)	Elastic modulus/MPa			Poisson's ratio			Shear modulus (MPa)		
		E_x	E_y	E_z	PR_{xy}	PR_{yz}	PR_{xz}	G_{xy}	G_{yz}	G_{zx}
260 × 185 × 0.1	960	2940	1500	15	0.34	0.01	0.01	807	14.85	14.63

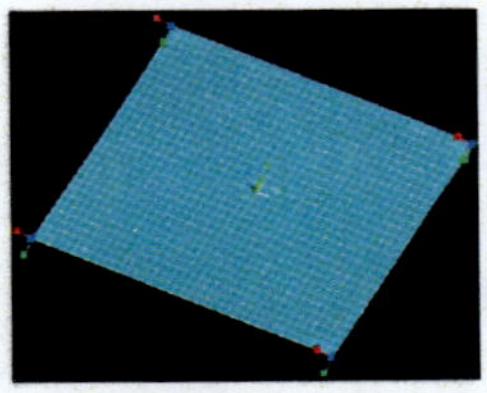

Fig. 5. Master nodes of paper's rigid regions

2.3 Establishment of Rigid-Flexible Coupling Model for Folding Mechanism

The ADAMS/Flex module provides a two-way data exchange interface between ANSYS and ADAMS [13, 14], through which the modal neutral files are imported into the 3 D rigid body model of the folding mechanism in Sect. 1.1. The rigid machete arm in the rigid body model is replaced by the flexible machete arm, and the flexible paper is arranged on the correct position of the paper feeding table. Then constraints are added to each component, and the components that do not affect the transmission are hidden in order to improve the analysis efficiency. The constraint relationships of components of the simplified folding mechanism are shown in Table 2.

Ultimately, the generated rigid-flexible coupling dynamic model of the folding mechanism is shown in Fig. 6.

3 Dynamic Characteristic Analysis of Folding Mechanism

The kinematic equation of the folding four-bar mechanism is established, and the kinematic analytical value is compared with the calculated value of the rigid-flexible coupling model to verify the rationality of the rigid-flexible coupling model and the constraint

Table 2. Constraint relationships between components

Number	Component	Constraint
1	Crank disc & ground	Revolute joint
2	Crank disc & connecting rod	Revolute joint
3	Connecting rod & machete arm	Revolute joint
4	Base & machete arm	Revolute joint
5	Machete & machete arm	Fixed joint
6	Base & ground	Fixed joint
7	Machete & paper	Contact
8	Paper & paper feeding table	Contact
9	Paper	Force vector
10	Paper feeding table & ground	Fixed joint
11	Roller & ground	Revolute joint

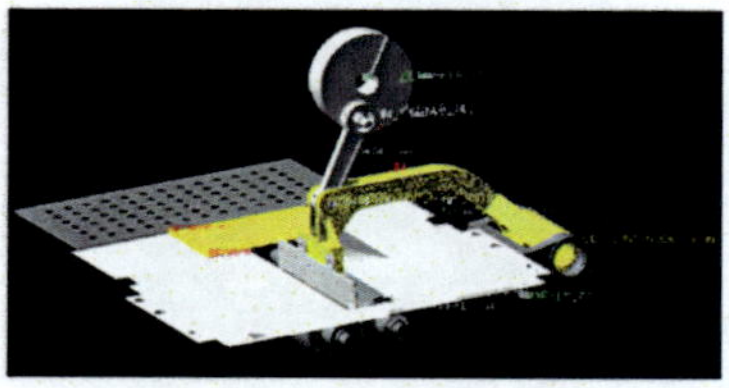

Fig. 6. Rigid-flexible coupling dynamics model of folding mechanism

relationship of the folding mechanism. The web knife folding mechanism is essentially a crank rocker mechanism, whose motion diagram and coordinate system are shown in Fig. 7.

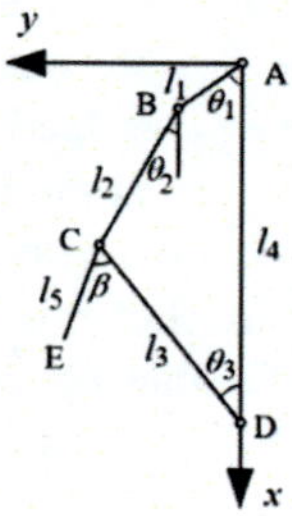

Fig. 7. Motion diagram of folding mechanism

3.1 Kinematic Analysis of Folding Four-Bar Mechanism

Through the closed vector equation of the plane four-bar mechanism, the motion equation of this four-bar mechanism is obtained as follows:

$$\begin{cases} l_1 \cdot \cos\theta_1 + l_2 \cdot \cos\theta_2 + l_3 \cdot \cos\theta_3 - l_4 = 0 \\ l_1 \cdot \sin\theta_1 + l_2 \cdot \sin\theta_2 - l_3 \cdot \sin\theta_3 = 0 \end{cases} \tag{1}$$

The derivative of Eq. (1) with respect to time yields the angular velocity relationship as follows:

$$\begin{cases} l_1 \cdot \dot{\theta}_1 \cdot \sin\theta_1 + l_2 \cdot \dot{\theta}_2 \cdot \sin\theta_2 + l_3 \cdot \dot{\theta}_3 \cdot \sin\theta_3 = 0 \\ l_1 \cdot \dot{\theta}_1 \cdot \cos\theta_1 + l_2 \cdot \dot{\theta}_2 \cdot \cos\theta_2 - l_3 \cdot \dot{\theta}_3 \cdot \cos\theta_3 = 0 \end{cases} \tag{2}$$

The derivative of Eq. (2) with respect to time yields the angular acceleration relation as follows:

$$\begin{cases} l_1 \cdot \ddot{\theta}_1 \cdot \sin\theta_1 + l_1 \cdot \dot{\theta}_1^2 \cdot \cos\theta_1 + l_2 \cdot \ddot{\theta}_2 \cdot \sin\theta_2 \\ \quad + l_2 \cdot \dot{\theta}_2^2 \cdot \cos\theta_2 + l_3 \cdot \ddot{\theta}_3 \cdot \sin\theta_3 + l_3 \cdot \dot{\theta}_3^2 \cdot \cos\theta_3 = 0 \\ l_1 \cdot \ddot{\theta}_1 \cdot \cos\theta_1 - l_1 \cdot \dot{\theta}_1^2 \cdot \sin\theta_1 + l_2 \cdot \ddot{\theta}_2 \cdot \cos\theta_2 \\ \quad - l_2 \cdot \dot{\theta}_2^2 \cdot \sin\theta_2 - l_3 \cdot \ddot{\theta}_3 \cdot \cos\theta_3 + l_3 \cdot \dot{\theta}_3^2 \sin\theta_3 = 0 \end{cases} \tag{3}$$

where, l_1 is the length of crank, $l_1 = 63.5$ mm; θ_1 is the angle between crank and frame; l_2 is the length of link, $l_2 = 160.5$ mm; θ_2 is the angle between link and frame; l_3 is the length of rocker, $l_3 = 432$ mm; l_4 is the length of frame, $l_4 = 455.5$ mm; θ_3 is the angle between rocker and frame.

The initial conditions are as $\theta_1 = 70°$, $\theta_2 = 0°$, $\theta_3 = 29°$. The crank rotates at a constant speed, and when the rotation speed is 36,000 rph, substituting the above data into Eqs. (1)–(3) to obtain the analytical value of the motion characteristics of the four-bar mechanism.

The rigid-flexible coupling model of the folding mechanism is solved and the simulated values of the motion characteristics of followers are measured in ADAMS software, which is compared with the above analytical values and the comparison results are shown in Fig. 8, the time steps for both simulated and analytical values are 1000 steps. It can be seen from Fig. 8 that the displacement and velocity of point C at the hinge between the connecting rod and the machete arm, and the angular velocity of the connecting rod and the machete arm are in good agreement with the simulation value, which verifies the rationality of the rigid-flexible coupling modeling and constraint relationship of the folding mechanism. However, in Fig. 8d, the average deviation between the analytical value of the angular acceleration of the connecting rod and the solution result of the rigid-flexible coupling model is 9.6 rad/s^2, and the average deviation rate is 1.2%. The average deviation of the angular acceleration of the machete arm is 0.38 rad/s^2 and the average deviation rate is 1.56% when comparing the analytical value of the angular acceleration of the machete arm with the solution result of the rigid-flexible coupling model. The deformation of the machete arm has a definite effect on the angular acceleration.

a.Comparison of displacement b.Comparison of velocity

(c) Comparison of angular velocity (d) Comparison of angular acceleration

Fig. 8. Comparison between simulation value and analytical value

3.2 Relationship Determination of Lateral Displacement Deviation of Machete Head and Rotational Speed

The deformation of the machete arm will cause small changes in the movement trajectory of the machete head, which will directly affect the folding accuracy. The longitudinal displacement deviation of the machete head affects the depth of the machete folding and does not produce folding deviation, which can be adjusted and corrected under working state. However, the deviation of the lateral displacement of the machete head directly affects the accuracy of folding and is difficult to adjust. Therefore, it is necessary to analyze the relationship between the deformation of the machete arm and the lateral displacement of the machete head.

The relationship between the coordinate position (x_E, y_E) of point E of the machete head and the geometric position of each link can be expressed as

$$\begin{cases} x_E = l_4 - l_3 \cdot \cos\theta_3 + l_5 \cdot \cos(\beta - \theta_3) \\ y_E = l_3 \cdot \sin\theta_3 + l_5 \cdot \sin(\beta - \theta_3) \end{cases} \tag{4}$$

where, β is the included angle between the machete arm and the machete with a size of 80.25°.

When the crank speed ω_1 is 36,000 rph, the machete head's lateral displacement of the rigid body folding mechanism and the rigid-flexible coupling folding mechanism is shown in Fig. 9.

It can be seen from the above figure that when the crank angle is 36°, the lateral displacement deviation of the machete head is the largest with a maximum deviation value is 0.049 mm. The average deviation is 0.040 mm and the minimum deviation is 0mm during paper cutting (crank angle is 0–67.9° and 337.3–360°). When the rotational speed of the folding mechanism is increasing, the relationship between the lateral displacement deviation of the machete head and the rotational speed is shown in Table 3.

The practical experience of folding mechanism shows that in order to ensure a high folding accuracy, the lateral deformation of the machete head must be controlled within

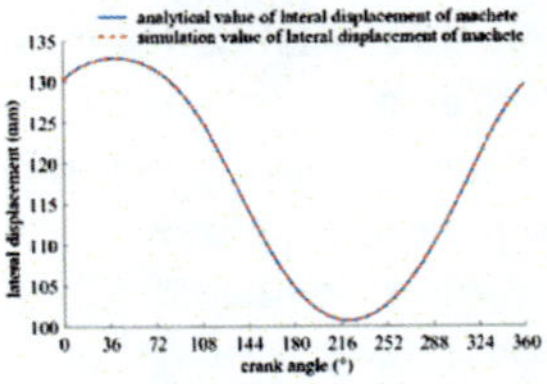

Fig. 9. Lateral displacement of machete

Table 3. Relationship between lateral displacement deviation of machete head and rotational speed

Rotational speed (rph)	Maximum deviation (mm)	Minimum deviation (mm)	Average deviation (mm)
36,000	0.049	0	0.040
37,000	0.056	0	0.045
38,000	0.061	0	0.051
39,000	0.076	0	0.062
40,000	0.081	0	0.071
41,000	0.097	0	0.078
42,000	0.107	0	0.089
43,000	0.117	0	0.102

0.1 mm [15]. According to the above table, when the crank speed is less than or equal to 41,000 rph, the lateral deformation of the machete head is within the allowable range and the folding mechanism can meet the working requirements. When the crank speed reaches 42,000 rph, the maximum lateral deformation deviation of the machete arm exceeds the permissible value. When the crank speed is increased to 43,000 rph, the maximum deviation and average deviation of the machete arm exceeds the allowable value, and the folding accuracy cannot be guaranteed. Considering the possible clearance and wear of the kinematic pair will somewhat increases the deformation of the machete head, it is recommended that the limit working speed of the N160 type knife folding mechanism should not exceed 41000 rph.

3.3 Dynamic Analysis of Folding Four-Bar Mechanism

Dynamic Contact Excitation Calculation of Machete and Paper. Since the dynamic contact force between the machete and the paper is difficult to obtain by experiments or other means, it is a common practice to ignore the working load in the dynamic analysis of the folding mechanism. In order to solve this problem, the rigid-flexible coupling model of the folding mechanism is used to calculate the dynamic contact excitation between the machete and the paper and to determine the working load of the machete.

Referring to table of various materials' collision parameters of ADAMS contact force [16], the contact parameters between the machete and the paper, and between the paper feeding table and paper for the rigid-flexible coupling model are shown in Table 4. The step function is used to add point wise tension to the master nodes of the four rigid regions of the flexible paper model to simulate the friction between the paper and the paper conveyor belt. The function formula is step (time, 0, 0, 0.01, − 2.15) + step (time, 0.01, 0, 0.055, 0) + step(time, 0.055, 0, 0.072, 2.15) The crank speed is set to 36,000 rph, the solution time is 0.1s, and the number of steps is 1000. In this case, the dynamic contact force between the machete and the paper during one rotation of the folding mechanism is calculated.

Table 4. Contact parameter

Parameter	Value
Stiffness/($N{\cdot}mm^{-1}$)	3800
Damping/($N{\cdot}s{\cdot}mm^{-1}$)	1.52
Force exponent	2
Penetration depth/(mm)	0.1
Dynamic coefficient	0.05
Static coefficient	0.08

Within time of 0–0.072 s, the paper moves along the paper feeding table under the action of point wise tension. At time of 0.072 s, the paper moves directly below the machete and comes into contact with it. During the time of 0.072–0.092 s, the paper is in dynamic contact with the machete, and the paper is deformed elastically under the action of the machete which is as shown in Fig. 10. At time of 0.092 s, the machete reaches the bottom, then rises and separates from the paper, and the paper finally completes the folding action under the roller's crushing action. The contact force between the machete and the paper is extracted in the post-processor as shown in Fig. 11. At the crank angle 336.82(0.072 s), the machete and the paper impact force is 120.78 N, then the machete contacts the paper dynamically and the force changes constantly. When the crank angle is 336.82° (at time of 0.072 s), the machete is in impact contact with the paper, and the impact force in the x direction is 120.78 N. Then the machete is in dynamic contact with the paper, and the force is constantly changing. When the crank angle is 67.9° (at time of 0.092 s), the machete is separated from the paper and the contact force goes to 0. The average amplitude of the x-directional force between the machete and the flexible paper is 39.6 N, and the average amplitude of the y-direction force is 3.2 N. The determination of this force can provide mechanical boundary data for the kinematic analytical model of the folding mechanism.

Dynamic Equation of Folding Four-Bar Mechanism. For the folding mechanism, the force analysis sketch is shown in Fig. 12. The structural parameters of each component of the folding mechanism are as follows: the crank mass is m_1, the moment of inertia is

Fig. 10. Dynamic contact of paper

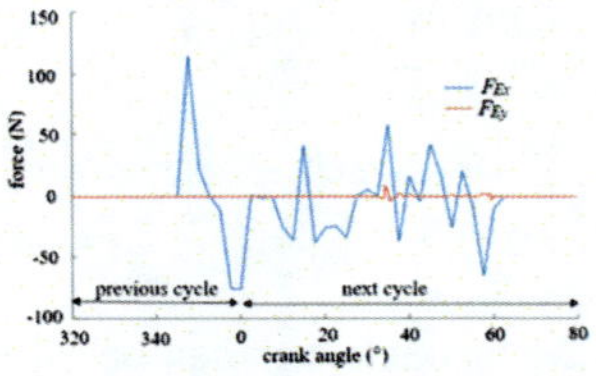

Fig. 11. Contact force of machete and machete and paper

J_{s1}, the coordinates of the mass center S_1 are (x_{s1}, y_{s1}), and its distance from point A is l_{s1}. The mass of the connecting rod is m_2, the moment of inertia is J_{s2}, the coordinates of the mass center S_2 are (x_{s2}, y_{s2}), and its distance from point B is l_{s2}. The mass of the machete arm is m_3, the moment of inertia is J_{s3}, the coordinates of the mass center S_3 are (x_{s3}, y_{s3}), and its distance from point D is l_{s3}. F_{Ax} and F_{Ay} are the components of the restraint reaction force at hinge A in the x and y directions. F_{Bx} and F_{By} are the components of the restraint reaction at hinge B in the x and y directions. F_{Cx} and F_{Cy} are the components of the restraint reaction at hinge C in the x and y directions. F_{Dx} and F_{Dy} are the components of the restraint reaction at hinge D in the x and y directions. F_{Ex} and F_{Ey} are the components of the resistance of the paper to the machete in the x and y directions. M is the torque acting on the crank.

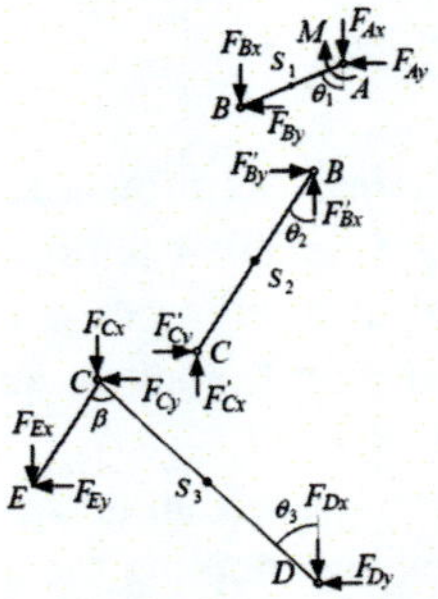

Fig. 12. Stress analysis diagram of folding mechanism

The balance equation for the crank can be formulated as:

$$\begin{cases} F_{Ax} + F_{Bx} = m_1 \cdot \ddot{x}_{s1} - m_1 \cdot g \\ F_{Ay} + F_{By} = m_3 \cdot \ddot{y}_{s1} \\ F_{Bx} \cdot l_{s1} \cdot \sin\theta_1 + F_{Ay} \cdot (l_1 - l_{s1}) \cdot \cos\theta_1 \\ \quad -F_{Ax} \cdot (l_1 - l_{s1}) \cdot \sin\theta_1 - F_{By} \cdot l_{s1} \cdot \cos\theta_1 - M = J_{s1} \cdot \ddot{\theta}_1 \end{cases} \tag{5}$$

The balance equation for the connecting rod can be formulated as:

$$\begin{cases} -F'_{Bx} - F'_{Cx} = m_2 \cdot \ddot{x}_{s2} - m_2 \cdot g \\ -F'_{Cy} - F'_{By} = m_2 \cdot \ddot{y}_{s2} \\ F'_{Bx} \cdot (l_2 - l_{s2}) \cdot \sin\theta_2 + F'_{Cy} \cdot l_{s2} \cdot \cos\theta_2 \\ \quad -F'_{Cx} \cdot l_{s2} \cdot \sin\theta_2 - F'_{By} \cdot (l_2 - l_{s2}) \cdot \cos\theta_2 = J_{s2} \cdot \ddot{\theta}_2 \end{cases} \tag{6}$$

The balance equation for the machete arm can be formulated as:

$$\begin{cases} F_{Cx} + F_{Dx} = m_3 \cdot \ddot{x}_{s3} - m_3 \cdot g - F_{Ex} \\ F_{Cy} + F_{Dy} = m_3 \cdot \ddot{y}_{s3} - F_{Ey} \\ F_{Cx} \cdot (l_3 - l_{s3}) \cdot \sin\theta_3 + F_{Cy} \cdot (l_3 - l_{s3}) \cdot \cos\theta_3 \\ \quad -F_{Dx} \cdot l_{s3} \cdot \sin\theta_3 - F_{Dy} \cdot l_{s3} \cdot \cos\theta_3 = J_{s3} \cdot \ddot{\theta}_3 \end{cases} \tag{7}$$

The structural parameters of the components are shown in Table 5.

Table 5. Structural parameters of components

Property	Crank	Connecting rod	Machete arm
Position of mass center/mm	3.17	80.25	177.12
Mass/kg	10.72	1.19	9.61
Moment of inertia/kg·mm^2	2.90×10^4	2571	5.43×10^5

Comparison of Dynamic Analytical Solution and Simulated Solution. When the crank speed is 36,000 rph, the rigid-flexible coupling model of the folding mechanism is solved to measure the support reaction forces at the hinges of machete arm in the ADAMS software and compared with the above analytical values. The comparison results are shown in Figs. 13 and 14.

For the support reaction in the x direction at the hinge of point C, comparing the simulation results of the rigid-flexible coupling model with the analytical values, it is found that the maximum deviation of the two is 37.95 N which occurred at the crank angle of 36°, and the average deviation is 16.49 N. The maximum deviation rate of the two is 8.25% which occurred at the crank angle of 116.36°, and the average deviation rate is 3.87%. For the support reaction in the y direction at the hinge of point C, comparing the simulation results of the rigid-flexible coupling model with the analytical values, it is found that the maximum deviation of the two is 13.79 N which occurred at the crank

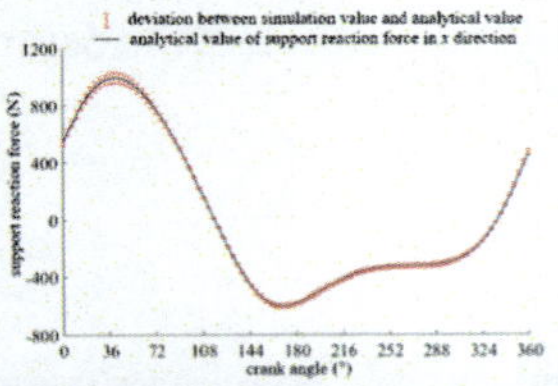

a. Support reaction in x direction

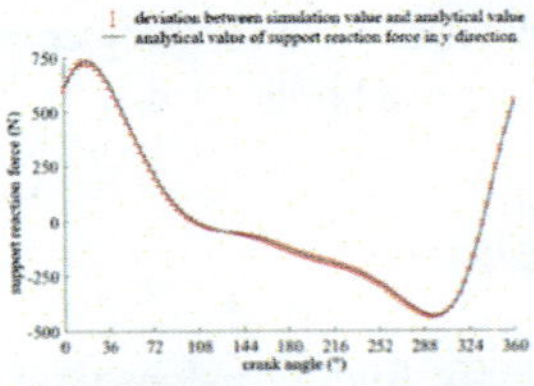

b. Support reaction in y direction

Fig. 13. Error bar of support reaction at point C

angle of 83.63°, and the average deviation is 10.11 N. The maximum deviation rate of the two is 10.01% which occurred at the crank angle of 109.09°, and the average deviation rate is 3.34%.

(a) Support reaction in x direction

(b) Support reaction in y direction

Fig. 14. Error bar of support reaction at point D

For the support reaction in the x direction at the hinge of point D, comparing the simulation results of the rigid-flexible coupling model with the analytical values, it is found that the maximum deviation of the two is 3.38 N which occurred at the crank angle of 330.91°, and the average deviation is 1.24 N. The maximum deviation rate of the two is 6.16% which occurred at the crank angle of 341.82°, and the average deviation rate is 0.78%. For the support reaction in the y direction at the hinge of point D, comparing the simulation results of the rigid-flexible coupling model with the analytical values, it is found that the maximum deviation of the two is 4.79 N which occurred at the crank angle of 334.54°, and the average deviation is 1.94 N. The maximum deviation rate of the two is 6.94% which occurred at the crank angle of 272.73°, and the average deviation rate is 0.85%.

According to the above analysis, it can be seen that the deformation of the machete arm has a certain effect on the angular acceleration and a larger effect on the restrained reaction force of each kinematic pair. Further comparative analysis found that the deviation between the results of the rigid-flexible coupling model and the analytical values is more significant as the speed of the folding mechanism increases. However, when the rotational speed of the folding mechanism is no more than 25,000 rph, the average deviation rate of both x-directional and y-directional support reaction force at point C is less than 1%, and the average deviation rate of both x-directional and y-directional support reaction force at point D is less than 0.2%. The solution results of the rigid-flexible coupling model are approximately equal to the analytical value, and the deformation of the machete arm can be ignored.

4 Conclusions

For the web knife folding mechanism, a multi rigid-flexible coupling dynamic model is established, in which the machete arm and the paper are flexible bodies and other components are rigid bodies. The rigid-flexible coupling dynamic characteristics of the folding mechanism are studied. The main conclusions are as follows:

(1) The analytical values of the displacement and velocity of point C at the hinge between the connecting rod and the machete arm and the angular velocity of followers are in good agreement with the simulation values, which verifies the rationality of the rigid-flexible coupling modeling and constraint relationship of the web folding mechanism.
(2) The influence of the flexible machete arm's deformation on the folding accuracy is studied, and the relationship between the lateral deformation of the machete and the crank speed is determined. According to the folding accuracy requirements, the theoretical limit working speed of the folding mechanism is analyzed to be 41,000 rph.
(3) The dynamic contact excitation between the machete and the paper is analyzed to determine the working load of the machete. The average amplitude of the x-directional force between the machete and the flexible paper is 39.6 N and the average amplitude of the y-directional force is 3.2 N at the crank speed of 36,000 rph.
(4) When the rotational speed of the folding mechanism is within 25,000 rph, the deformation of the machete arm has little influence on the kinematic and dynamic characteristics of the folding mechanism, which can be analyzed as a rigid system. When the rotational speed of the folding mechanism is greater than 25,000 rph, the deformation of the machete arm has negligible effect on the trajectory and speed of the folding mechanism, but it has a certain effect on the angular acceleration and a significant effect on the machete arm support reaction force.

Acknowledgements. The research was supported by basic research project of BIGC (Eb202003), science and technology project of BMEC (KM202110015005).

References

1. Zhao, L.J., Ma, Y.Z.: Research on key technologies of modeling and simulation of rigid-flexible coupling system. Comput. Eng. Appl. **46**(02), 243–248 (2010)
2. Xiao, J.J., Lin, D.L., Cheng, G.Y., et al.: Motion simulation of knife folding machine based on SolidWorks. Mech. Eng. (07), 11–12+15 (2019)
3. Shi, X.D., Lu, N.C.: Optimization design of the 16K folders based on ADAMS. Packag. Eng. (06), 175–178 (2006)

4. Hill, D.E.: Dynamics and control of a rigid and flexible four bar coupler. J. Vib. Control **20**(1), 131–145 (2014)
5. Tan, N.: Calibration for accuracy improvement of serial manipulators based on compressed sensing. Electron. Lett. **51**(11), 820–822 (2015)
6. Hou, Y., Xiong, X.Y., Wang, X.: Dynamic co-simulation based on ANSYS and ADAMS and its application. Mining Machinery **42**(1), 111–115 (2014)
7. Madenci, E., Guven, I.: The Finite Element Method and Applications in Engineering Using ANSYS. Springer, Boston, MA (2006)
8. Musa, S.M., Kulkarni, A.V., Havanur, V.K.: Finite Element Analysis. Mercury Learning and Information (2013)
9. Szewczyk, W., Marynowski, K., Tarnawski, W.: An analysis of young's modulus distribution in the paper plane. Fibres Text. Eastern Eur. **14**(4), 91–94 (2006)
10. Xie, Y.H.: Relationship between elastic modulus and shear modulus of paper materials. Packag. Eng. **33**(21), 37–40 (2012)
11. Zeng, F.L.: Mechanical analysis of how many times a piece of paper can be folded in half at most. Mech. Pract. **41**(04), 483–487 (2019)
12. Hua, G.J., Luo, D.T.: Simulation analysis of corrugated box strength based on yield criterion. J. Packag. **2**(1), 18–21 (2010)
13. Shabana.: Dynamics of Multibody Systems. Cambridge University Press (1998)
14. MSC.ADAMS/Flex and AUTOFlex training materials. Science Press, Beijing (2006)
15. Yuan, Y.C.: Dynamic research and robust desing of web offset press's folding mechanism with clearance. Central South University (2011)
16. Guo, W.D., Li, S.Z.: Virtual Prototype Technology and Adams Application Example Tutorial, 2nd edn. Beijing University of Aeronautics and Astronautics Press (2018)

Residual Volume of Entrained Air in Wound Roll

Li'e Ma(✉), Zhengyang Guo, Jimei Wu, Donghao Ma, and Haiyang Ji

Faculty of Printing, Packing and Digital Media Engineering, Xi'an University of Technology, Xi'an, China

malie@xaut.edu.cn

Abstract. In the roll-to-roll printing process, air entrainment is an important factor affecting the quality of wound roll when the winding speed is high and the air permeability of the film is poor. In the winding process, a spiral air layer is formed in wound roll. At the same time, air leakage will occur with the increase of radial stress. The residual air inside the web will affect the magnitude and distribution of the stress. Therefore, it is important to study air residual volume. In this study, we get the air layer thickness equation from the relevant theory of air leakage. Secondly, incremental equation of the radial stress in wound roll is derived. Finally, the increment of the internal stress for each layer is substituted into the air layer thickness expression to obtain the air layer thickness which is the air residual volume. This study predicts the distribution of entrained air in the wound roll, which can be used to improve the analysis of the effect of entrained air on the internal mechanical properties of the wound roll.

Keywords: Roll-to-Roll · Air entrainment · Squeeze flow · Residual volume

1 Introduction

In order to meet the needs of the development of the flexible electronics industry, roll-to-roll flexible electronics printing systems have been applied to the manufacture of thin flexible materials [1]. The roll-to-roll printing system has the advantages of high automation, high production efficiency and low cost. In this process, winding is an important process that determines the quality of the final product. Too much or too little air entrainment is not conducive to improving winding quality. Therefore, it is important to study the residual volume of air entrainment according to the material characteristics of flexible electronic rolls [2].

In the winding process, as air is entrained between the films, a spiral layer of entrained air is formed in the web, and this entrained air layer affects the properties of the entire wound roll. In this regard, Good [3, 4] investigated the effect of air entrainment layer on the in-roll tension and the radial elastic modulus of the web for the center wound rolls. It was assumed that the air was kept in wound roll and the air layers were treated as equivalent layers, so this theoretical model is applicable to non-permeable film winding. Hashimoto [5] used Newton-Raphson iterative method to solve the two-dimensional

M. Xu et al. (Eds.): CACPP 2022, LNEE 991, pp. 330–336, 2023.
https://doi.org/10.1007/978-981-19-9024-3_42

Reynolds equation and the film balance equation based on the finite width compressible foil bearing theory, and the pressure and film thickness were obtained. Lei [6] developed a nonlinear model for stresses within a winding roll considering contact forces based on entrained air theory, and then used a thermal stress model to investigate the effect of temperature on the stresses in the wound roll. Kanda [7, 8] developed a theoretical prediction model for the stress state in the wound roll with air escaped from the wound roll. A theoretical prediction model for the unsteady in-roll stresses in winding considering the effect of entrained air on heat transfer was also proposed and verified experimentally. Kiribe [9] used a model with air entrainment effect for studying the effect of thickness inhomogeneity of the web in the width direction on the in-roll stresses in winding and verified it experimentally. Hashimoto [10] predicted the thickness of the entrained air layer in wound roll which used the Hakiel's nonlinear model with air entrainment effects.

In this study, we get the air layer thickness in wound roll which is based on the air layer thickness equation and derive the radial stress increment equation.

2 Analysis of Air Residual Volume

2.1 Model of Air Entrainment Squeeze Film

For the air entrainment residual volume, the thickness of the initial air layer should be clarified. The initial air layer thickness can be obtained from Eq. 1 [5].

$$h_0 = r_{out}\left(\frac{12\eta V}{T}\right)^{2/3}\left(0.589 - \frac{1.614}{\lambda} + \frac{1.764}{\lambda^2}\right) \tag{1}$$

where $\lambda = \frac{L}{2r_{out}}\left(\frac{12\eta V}{T}\right)^{-1/3}$, η is air viscosity, T is winding tension, V is winding velocity, r_{out} is outside diameter of wound roll and L is the width of the film.

The air layer is compressed by the radial stress and the air flows axially between the film layers, gradually flowing out from the wound roll. This phenomenon can be seen as squeeze film flow of air between parallel plates, as shown in the Fig. 1.

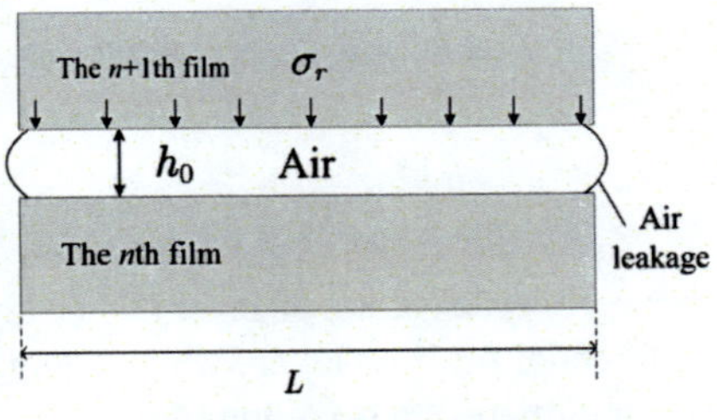

Fig. 1. Model of squeeze film

The relationship between air layer thickness h, radial stress σ_r and time t is obtained according to Fig. 1, as shown in Eq. 2 [7].

$$h = \left(\frac{12t}{\eta L^2}\sigma_r + \frac{1}{h_0^2}\right)^{-\frac{1}{2}} \tag{2}$$

2.2 Incremental Solution Equation of the Radial Stress

It was assumed that the wound roll was under the following condition:

a. The cross-section is concentric circles stacked layer by layer.
b. The wound roll without the axial stress is in an axisymmetric stress state.
c. The interlayer slips are not considered.

Based on these assumptions, film winding mechanical model is as shown in Fig. 2:

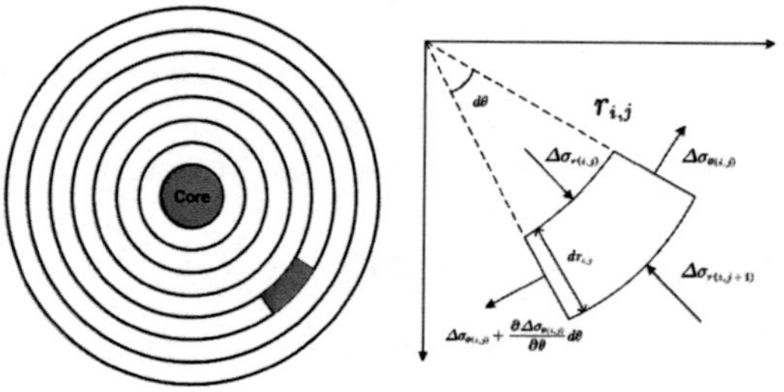

Fig. 2. Film winding mechanical model

The equilibrium equation can be obtained as follows:

$$\Delta\sigma_{\theta(i,j)} + \Delta\sigma_{r(i,j+1)} + r_{i,j}\frac{\Delta\sigma_{r(i,j+1)} - \Delta\sigma_{r(i,j)}}{H} = 0 \tag{3}$$

where $\Delta\sigma_{r(i,j)}$ is the increment of radial stress on layer j when layer i is winding, $\Delta\sigma_{\theta(i,j)}$ is the increment of circumferential stress on layer j when layer i is winding, $r_{i,j}$ is the radius of layer j when layer i is winding and H is the film thickness.

According to Hooke's law:

$$\Delta\varepsilon_{\theta(i,j)} = \frac{1}{E}(\Delta\sigma_{\theta(i,j)} + v\Delta\sigma_{r(i,j)}) \tag{4}$$

$$\Delta\varepsilon_{r(i,j)} = -\frac{1}{E}(\Delta\sigma_{r(i,j)} + v\Delta\sigma_{\theta(i,j)}) \tag{5}$$

where $\Delta\varepsilon_{r(i,j)}$ is the increment of radial strain on layer j when layer i is winding, $\Delta\varepsilon_{\theta(i,j)}$ is the increment of circumferential strain on layer j when layer i is winding, E is the film modulus of elasticity, v is the Poisson's ratio of the film.

Circumferential and radial strain are expressed by displacement.

$$\Delta\varepsilon_{\theta(i,j)} = \frac{\Delta u_{i,j} - \Delta u_{i,j+1}}{2r_{i,j}} \tag{6}$$

$$\Delta\varepsilon_{r(i,j)} = \frac{\Delta u_{i,j+1} - \Delta u_{i,j}}{H} \tag{7}$$

where $\Delta u_{i,j}$ is the increase of radial displacement of layer j when layer i is winding.

Substituting Eqs. 6 and 7 into Eqs. 4 and 5 respectively, we get,

$$\frac{\Delta u_{i,j} - \Delta u_{i,j+1}}{2r_{i,j}} = \frac{1}{E}\left[\Delta\sigma_{\theta(i,j)} + \frac{1}{2}\nu(\Delta\sigma_{r(i,j)} + \Delta\sigma_{r(i,j+1)})\right] \tag{8}$$

$$\frac{\Delta u_{i,j+1} - \Delta u_{i,j}}{H} = -\frac{1}{E}\left[\frac{1}{2}(\Delta\sigma_{r(i,j)} + \Delta\sigma_{r(i,j+1)}) + \nu\Delta\sigma_{\theta(i,j)}\right] \tag{9}$$

Substituting Eq. 8 into Eqs. 9 and 3, we get,

$$\begin{cases} \left(\frac{\nu}{2r_{i,j}} - \frac{1}{H}\right)\Delta u_{i,j} + \left(\frac{\nu}{2r_{i,j}} + \frac{1}{H}\right)\Delta u_{i,j+1} = (\Delta\sigma_{i,j+1} + \Delta\sigma_{r(i,j+1)})\frac{\nu^2-1}{2E} \\ \frac{E}{2r_{r,j}}(\Delta u_{i,j} + \Delta u_{i,j+1}) = \left(\frac{\nu}{2} + \frac{r_{i,j}}{H} - 1\right)\Delta\sigma_{r(i,j)} + \left(\frac{\nu}{2} - \frac{r_{i,j}}{H}\right)\Delta\sigma_{r(i,j+1)} \end{cases} \tag{10}$$

The first boundary condition is the displacement of the outer layer of film.

$$\Delta u_{i,j} = -\frac{r_{out}}{E}\left(\frac{r_{out}^2 + r_c^2}{r_{out}^2 - r_c^2} - \mu\right)\Delta\sigma_{r(i,i)} \tag{11}$$

where r_c is the radius of the core.

The second boundary condition is the stress of the outer layer of film.

$$\Delta\sigma_{r(i,j)} = \frac{T}{Lr_{i,i}} \tag{12}$$

3 Numerical Analysis

3.1 Increment of Radial Stress

From Eqs. 10, 11, and 12, the increment of radial stress in the film be calculated for the 3000-layer roll by MATLAB, and the specific parameters are listed in Table 1.

When $j = 1$, 1000, 2000, radial stress as shown in Fig. 3(a). As can be seen from Fig. 3(a), under the condition of certain winding tension, the initial radial stress increases rapidly, but with the increasing number of winding film layers, the incremental radial stress of film winding will gradually decrease.

Similarly, j = 2500, 2800, 2900 and 2950 are taken to plot the radial stress increase curve. As we know from Fig. 3(b), the more layers are wound, the more obvious the linear relationship of stress increment. Secondly, the closer the film layer is to the core, the higher the radial stress is.

3.2 Residual Volume of Entrained Air

The data of the above plots are fitted as a function of the number of winding layers and the radial stress. Since the variable of Eq. 2 is time, so the relationship between the number of winding layers and time is obtained according to Eq. 13.

$$(n \cdot H + r_c)^2\pi - r_c^2\pi = V \cdot t \cdot H \tag{13}$$

Table 1. The winding parameters

Parameter	Symbol	Numerical value	Unit
Wound roll radius	r_c	0.1	m
Winding velocity	V	5	m/s
Winding tension	T	50	N
Film Poisson's ratio	v	0.3	
Film thickness	H	6×10^{-5}	m
Film elastic modulus		2500	Mpa
Film Width	L	1.25	m

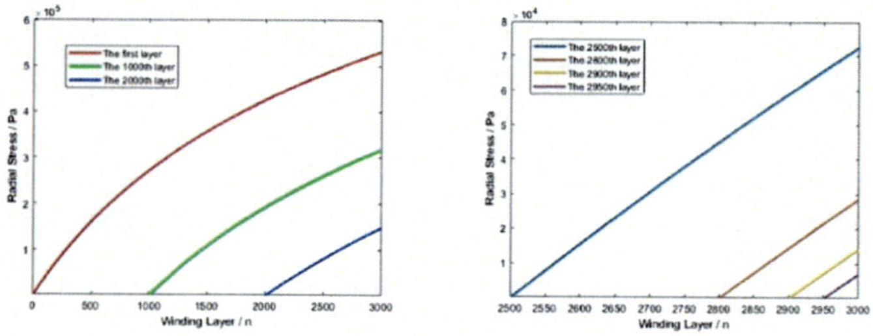

(a) first, 1000th, 2000th layer (b) 2500th, 2800th, 2900th, 2950th layer

Fig. 3. Radial stress of different wingding layers

The fitted function and Eq. 13 are substituted into Eq. 2 to obtain the expression for the air leakage with respect to time. We analyze the thickness of the first, 1000th, 2000th 2500th, 2800th, 2900th, 2950th entrained air layer, and gave the plots of the representative first layer and the 2950th layer.

It can be seen in Fig. 4(a) that the horizontal coordinate 716 s is the calculated time used to wind 3000 layers, and the air layer thickness decreases rapidly with the increase of the number of winding layers and gradually stabilizes at about 0.2 μm. In Fig. 4(b), it can be seen that the air layer thickness decreases more slowly than that shown in Fig. 4(a). This is due to less time and less stress in Fig. 4(b) than in Fig. 4(a).

Summarizing the final air film thickness of the 1th, 1000th, 2000th 2500th, 2800th, 2900th, 2950th layer, the air film thickness of each layer can be obtained when 3000 layers are wound. The relationship between the winding layer and the air layer thickness is shown as in Fig. 5.

It can be seen more visually in Fig. 5 that the thickness of the air layer is small before the 2500th layer and increases rapidly after the 2500th layer. This result is the same as the trend of the thickness of the entrained air layer predicted by the Hakiel's nonlinear model with air entrainment effects in Hashimoto's paper [10], which supports the correctness of the incremental radial stress for calculating the thickness of the entrained air layer in this paper.

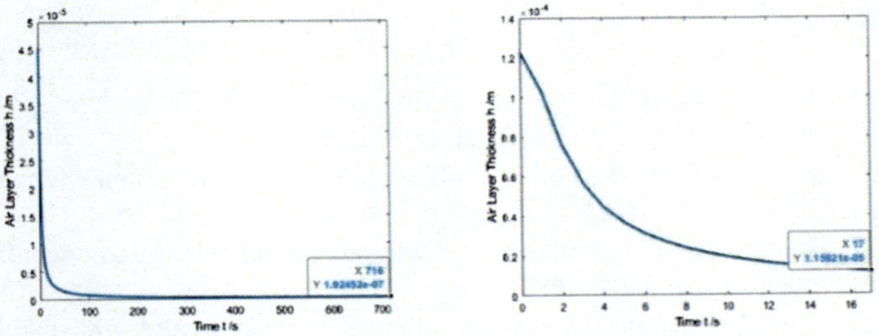

(a) first layer (b) 2950th layer

Fig. 4. Thickness of entrained air layer

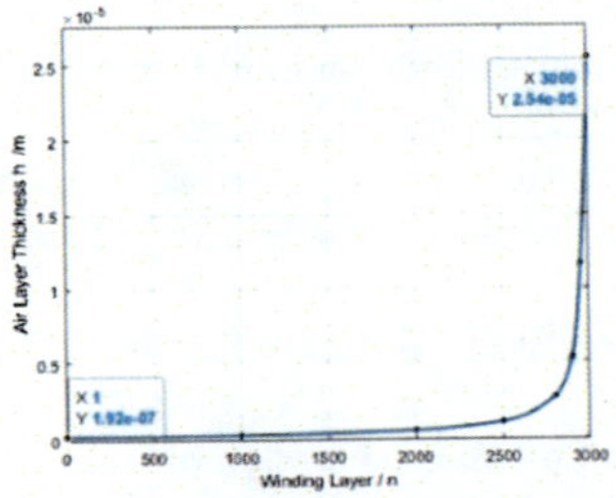

Fig. 5. The thickness of each layer of air film in wound roll

4 Conclusions

In this study, the air layer thickness expression containing stress and time variables is obtained by considering the air leakage as the squeeze flow of air between parallel plates, The expression was a theoretical method to predict the air layer thickness of entrained air in the wound roll. Secondly, incremental solution equation of radial stress in the winding process is derived according to the plane axisymmetric elasticity theory, so that the increase of radial stress is analyzed numerically and fitted as a function, and finally the fitted function is substituted into the air layer thickness expression to obtain the air layer thickness. The results show that the entrained air is mostly distributed between the outer film layers of the roll, while the inner film layers of the wound have a small entrained air layer thickness due to both stress and time. This study provides the distribution of entrained air in the wound roll, which can be used to improve the analysis of the effect of entrained air on the internal mechanical properties of the wound roll.

Acknowledgements. This research is supported by the National Key Research and Development Program of China (2019YFB1707200), the Technology Innovation Leading Program of Shaanxi Province (No. 2020QFY03-04 and No. 2020QFY03-08) and the Key Research and Development Program of Shaanxi Province (No. 2020ZDLGY14-06).

References

1. Palavesam, N.: Roll-to-roll processing of film substrates for hybrid integrated flexible electronics. Flex. Printed Electron. **3**(1), 014002 (2018)
2. Chen, J.K., Jin, Y.W.: Review of wound roll stress in roll-to-roll manufacturing of flexible electronics. Chin. Sci. Bull. **64**(5–6), 555–565 (2019)
3. Good, J.K.: The effect of air entrainment in center-wound rolls. In: International Conference on Web Handling, pp. 246–264 (1993)
4. Good, J.K.: Air entrainment and residual stresses in rolls wound with a rider roll. In: International Conference on Web Handling, pp. 95–112 (1995)
5. Hashimoto, H.: Air film thickness estimation in web handling processes. ASME J. Tribol. **121**(1), 50–55 (1999)
6. Lei, H., Cole, K., Weinstein, S.: Modeling air entrainment and temperature effects in winding. J. Appl. Mech. **70**(6), 902–914 (2003)
7. Kanda, T.: Air leaking effect in winding on in-roll stress analysis of wound roll. Trans. Jpn. Soc. Mech. Eng. Ser. C **76**(772), 3736–3743 (2010)
8. Kanda, T.: A winding model for unsteady thermal stress within wound roll considering entrained air effect on heat conduction. Trans. Jpn. Soc. Mech. Eng. Ser. C **77**(780), 3161–3174 (2011)
9. Kiribe, S.: Internal stress analysis of a wound roll across widthwise direction. Trans. Jpn. Soc. Mech. Eng. 14–00143 (2014)
10. Hashimoto, H.: Optimization of wind-up tension of webs preventing wrinkles and slippage with experimental verification. In: International Conference on Web Handling, pp. 83–104 (2009)

Cooling Water Monitoring and Early Warning Device for Gravure Printing Machine Based on 51 Single Chip Microcomputer

Yishen Wang[1], Yuansheng Qi[2](✉), Yongbin Zhang[1], Yingzhe Ma[1], and Wenjing Ma[1]

[1] School of Mechanical and Electrical Engineering, Beijing Institute of Graphic Communication, Beijing, China
[2] School of Printing and Packaging Engineering, Beijing Institute of Graphic Communication, Beijing, China
yuansheng-qi@bigc.edu.cn

Abstract. For the gravure printing machine in the process of high-speed operation, the cooling water delivered by the cooling system before reaching the cooling roller by the length of the pipe and the environment, the temperature will rise, resulting in the printing paper can not be cooled in time after drying, affecting the quality of printing. This paper designs a STC microcontroller based gravure printing machine cooling water monitoring and early warning device, using DS18B20 temperature sensor and YF-B9 Hall water flow sensor to monitor the cooling water inlet of the cooling roller, to warn when the cooling water temperature is too high or before the leakage or rupture of the piping system, and to display the cooling water temperature and flow rate in real time on the LCD display, realizing the real-time monitoring of the cooling water temperature and flow rate and fault warning function, which has certain guiding significance for the operation and maintenance of domestic gravure printing machines.

Keywords: Microcomputer · Sensor · Real-Time monitoring · Gravure printing machine

1 Introduction

With the development of China's economy, the printing industry is transforming and upgrading towards digitalization and informatization, and the equipment structure and workshop layout of domestic printing factories are also changing [1]. Previously, during the high-speed operation of the gravure printing machine, influenced by the length of the pipe and the ambient temperature, the cooling water delivered by the cooling system will rise in temperature before reaching the cooling rollers, resulting in the printing paper not being cooled in time after printing and drying, resulting in paper deformation and ink "crystallization" and other problems [2]. Gravure printing machine normal operation of the temperature should be controlled between 21 and 25 °C, now the application of printing chillers to better solve this problem. It can transport the cooling water to the cooling roller through the pipeline, and absorb the heat energy generated in the printing

M. Xu et al. (Eds.): CACPP 2022, LNEE 991, pp. 337–344, 2023.
https://doi.org/10.1007/978-981-19-9024-3_43

process through the process of water circulation cooling - heat absorption - cooling - cooling - cooling again - heat absorption again [3]. Since the cooling system's control of the cooling water temperature is limited to the chilled water tower, which is installed outside the workshop, the cooling water delivered by the cooling system needs to pass through a long pipe to reach the cooling rolls. Due to the influence of the environment and the length of the pipes, the temperature of the cooling water delivered in the pipes will increase when it reaches the cooling rolls, no longer 17 °C just from the chiller, but even up to 25 °C in summer. The cooling effect will be significantly reduced, and in winter it will also be reduced to less than 10 °C, which will affect the service life of the gravure printing machine. Therefore, it is necessary to monitor the temperature of the cooling water at the inlet of the cooling roller [4]. In addition, in order to prevent the cooling system from failing due to, for example, water leakage and rupture of water pipes, damage to water pumps and damage to motors, the cooling water flow rate needs to be monitored. It is planned to install the monitoring device at the cooling roller inlet, as shown in Fig. 1, which shows the cooling roller inlet of the printing machine [5].

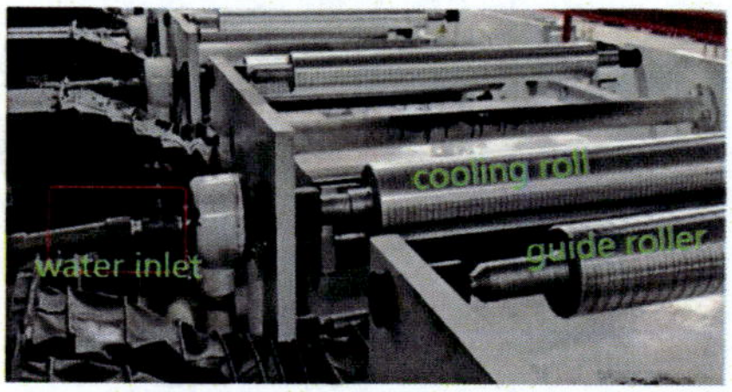

Fig. 1. Printing machine cooling roller inlet

To this end, this paper designs an early warning device for monitoring the temperature and flow rate of the cooling water pipeline of the gravure printing machine, which issues an alert when the cooling water temperature is higher than 18 °C and an alert when the flow rate is lower than 50 L/min (the temperature and flow rate data here are from the actual factory research, the cooling system water temperature is usually set at 17 °C, and the cooling effect is significantly reduced if the temperature is higher than this). The cooling water temperature and flow rate are displayed in real time on the LCD screen, so that the operator can monitor the operation of the cooling system by observing the values to ensure that the gravure printing machine is working in normal condition.

2 Monitoring and Early Warning System Program Design

The monitoring system designed in this paper consists of a hardware part and a software part, the specific implementation process as follows: flow sensor YF-B9 and temperature sensor DS18B20 to monitor the flow and temperature of cooling water, the data is uploaded to the microcontroller through the interface [6]. The microcontroller analyzes and processes the received flow data and temperature data, and displays the current cooling water temperature and flow rate in real time through the LCD1602 liquid crystal display. Through the threshold judgment of temperature and flow data, if the temperature is higher than 14°C and the flow rate is lower than 50 L/min, the buzzer will alarm and make over-limit warning, high will display "H" and low will display "L" to warn the staff of abnormal cooling system [7].

The system adopts STM89C51 microcontroller as the core, mainly composed of cooling water flow collection module, cooling water temperature collection module, LCD display module, warning module and control program. The system composition is shown in Fig. 2.

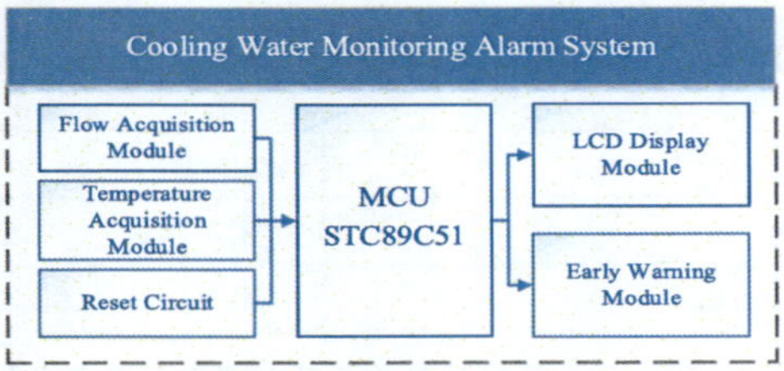

Fig. 2. System composition diagram

2.1 Hardware Design of the System

Microcontroller and Reset Circuit. Microcontroller has a series of common computer hardware such as CPU, RAM, ROM and communication interface integrated inside, which has the advantages of small size, easy manipulation, simple structure and low cost compared to computer. For industrial control and artificial intelligence and other areas of great use. The chip used in this paper is STC89C51, which is an eight-bit microcontroller based on the 51 core board and made by CMOS process technology. Its main function is to rely on sensors for information collection, CPU for information data processing and control of the display and buzzer.

The reset circuit plays a huge role in the operation of the program as the smallest application system of the microcontroller. The reset circuit resumes the execution of the program by resetting the program when it is running. The reset circuit is relatively simple compared to other circuits, and its operation methods include power-on reset, automatic reset according to the program, and manual reset. This design uses the keypad to reset manually. In order to improve the real-time monitoring of the microcontroller data, an interrupt system is added so that the microcontroller CPU can work synchronously with the external devices, and the Hall sensor frequency flow conversion calculation is placed in the interrupt program to eliminate the waiting phenomenon that the CPU needs to query the status of external devices several times when running.

Cooling Water Flow Collection Module. This module is to collect the cooling water flow data through the sensor, the cooling water flow collection using the YF-B9 water flow Hall sensor, using Hall effect of Hall elements to detect the flow rate of cooling water, Hall sensor for real-time monitoring of cooling water flow, when the water through the turbine to push the magnetic rotor rotation, it will generate a rotating magnetic field cutting the magnetic induction line to generate a pulse signal [8]. According to the empirical formula for the conversion of the frequency of the pulse signal to the flow rate:

$$Q = (f + 3) \div 8.1$$

In the formula: f—pulse signal frequency. Q—the flow of water, The unit of flow is L/min. The YF-B9 water flow Hall sensor has the characteristics of long service life and high stability, and its parameters are shown in Table 1 [9].

Table 1. YF-B9 water flow hall sensor parameters

Parameter name	Parameter range
Rated working voltage	DC3 ~ 24V
Temperature applicable range	– 20 °C to + 100 °C
Withstand voltage range	1.75 MPa

Cooling Water Temperature Data Acquisition Module. This module is to collect the cooling water temperature data through the sensor, this paper uses DS18B20 temperature sensor probe as the cooling water temperature collection sensor. DS18B20 is a digital temperature sensor produced by DALLAS, USA, with independent 64-bit read-only memory, and the control command and collection data are input and output in the form of digital signals. Compared with the analog temperature sensor, it has strong functions, simple hardware facilities and strong resistance to infection, etc. The DS18B20's communication interface uses a single bus, only one communication line to achieve bi-directional data transmission, saving the occupation of the microcontroller I/O port, with a small size and simple interface characteristics [10].

LCD Display Module. This module is to be able to reflect the data after the analysis and processing of the microcontroller to the display in real time, giving the staff an intuitive way of feeling. This paper uses LCD1602 character LCD as the display module, which has 160 kinds of character graphics stored inside, and can display ASCII code standard characters and some other special characters. It can also have 8 custom characters, and can also display 32 characters at the same time, the characters are 5 × 7 dot matrix. When programming in Keil software, you can also use variables to assign values to it. In addition, the LCD1602 LCD also comes with its own control program and scanning circuit function, just send the content through the microcontroller to the LCD display can display the corresponding content, has the advantages of convenient operation, low cost and simple structure.

Early Warning Module. This module is designed to be able to sound an alarm to warn the operator in case of abnormalities in the cooling system. The buzzer is an electronic interrogator that converts electrical signals into sound signals and can be used to generate warning signals such as key tones and alarm tones by microcontroller. In this design, the active buzzer is used as the sound source of the warning module. The source buzzer has an internal oscillation source, and it only needs to connect the positive and negative terminals to the DC regulated power supply to make a continuous sound during operation. In this design, when the cooling water flow rate and temperature are lower or higher than the limit, the buzzer will sound to warn until the flow rate and temperature return to the normal value [11].

Hardware Diagram of Early Warning System. For the physical connection of the device, the flow sensor, temperature sensor probe and LCD are powered by 5V DC regulated power supply, the buzzer is powered by 3V DC regulated power supply, the flow sensor is connected to microcontroller pin P34, the temperature sensor probe is connected to microcontroller pin P37, the buzzer port is connected to microcontroller pin P15, the temperature sensor probe and the flow sensor can be combined together, the LCD display shows real-time temperature and flow, the system circuit diagram is shown in Fig. 3, the physical hardware diagram is shown in Fig. 4.

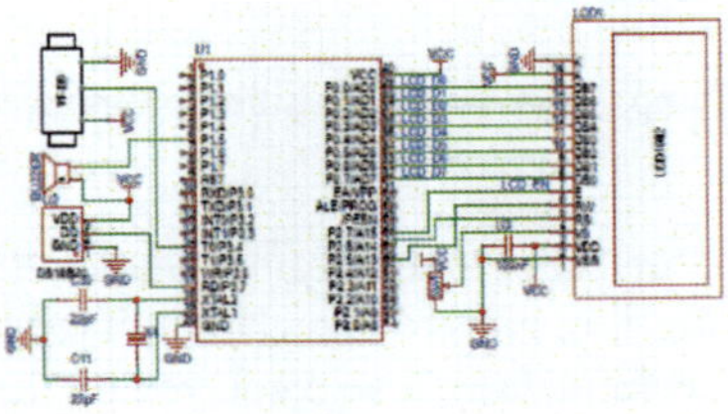

Fig. 3. System circuit diagram

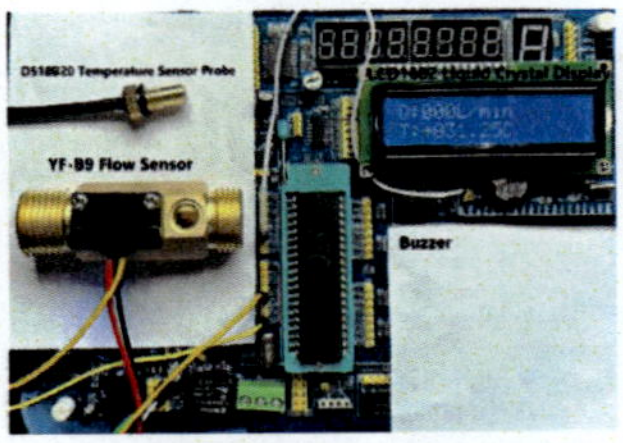

Fig. 4. Hardware physical map

2.2 System Software Design

The program designed in this paper is programmed in C language, and the programming software is Keil. The design program includes the main program, threshold judgment program, data acquisition program, warning program and data display program, and the main functions of the system software are shown in the Figs. 5 and 6.

This warning device is installed at the water inlet of the cooling roller, and the cooling system will start at the same time when the gravure printing machine starts. The cooled cooling water will be piped through the warning device into the cooling rolls, and the data monitored by the flow and temperature sensors will be compared with the pre-set

```
void T1_time()interrupt 3
{
 TH1=(65536-45872)/256;//50毫秒
 TL1=(65536-45872)%256;
 num++;
 if(num==20)        //1s更新一次数据，送至数码管显示
 {
  num=0;
  F=read();         //每隔1s读一次计数器0，该值则为频率，计算出Q后立马把计数器0清零重新计数
  if(F>0)
  {
   Q=(F+3)/8.1;     //流量传感器经验公式
   TH0=0;
   TL0=0;
  }
  else
  {
   Q=0;             //如果不加这句，当F=0，由公式知道Q! =0。
   TH0=0;
   TL0=0;
  }
 }
```

Fig. 5. Flow data collection program segment

```
float DS18B20_ReadT(void)
{
  unsigned char TLSB,TMSB;
  int Temp;
  float T;
  OneWire_Init();
  OneWire_SendByte(DS18B20_SKIP_ROM);
  OneWire_SendByte(DS18B20_READ_SCRATCHPAD);
  TLSB=OneWire_ReceiveByte();
  TMSB=OneWire_ReceiveByte();
  Temp=(TMSB<<8)|TLSB;
  T=Temp/16.0;
  return T;
```

Fig. 6. Temperature data acquisition program segment

threshold value in the microcontroller to determine whether the limit is exceeded. If the limit is not exceeded, the value is displayed in real time. If the limit is exceeded, a buzzer will be triggered to warn and the data will be displayed on the LCD screen in real time until the cooling water temperature and flow rate return to the normal range, achieving the functions of monitoring and warning at the same time. The system operation flow chart is shown in Fig. 7. After testing, the device can display the temperature and flow rate of cooling water in real time during the normal operation of the gravure printing machine equipment. By manually adjusting the chiller water temperature over the limit can normally trigger the buzzer alarm, and display "T: H" as shown in the Fig. 8.

3 Conclusion

This paper designs an early warning device for cooling water monitoring of gravure printing machine based on STC microcontroller. The temperature and flow rate of cooling water in the cooling system pipeline are monitored in real time using sensors and displayed in real time on the LCD screen. The flow rate monitoring can provide early warning before leakage or rupture of the cooling water pipeline occurs, and the temperature monitoring can provide timely warning when the cooling water temperature is abnormal. The cooling rollers of gravure printing machine can be kept at normal temperature to ensure timely cooling of printing paper after drying and avoid deformation of printing paper and "crystallization" of ink, which has certain guiding significance to the operation and maintenance of domestic gravure printing machine.

The STC microcontroller-based cooling water monitoring and warning device for gravure printing machines designed in this paper still needs to be improved in the following aspects.

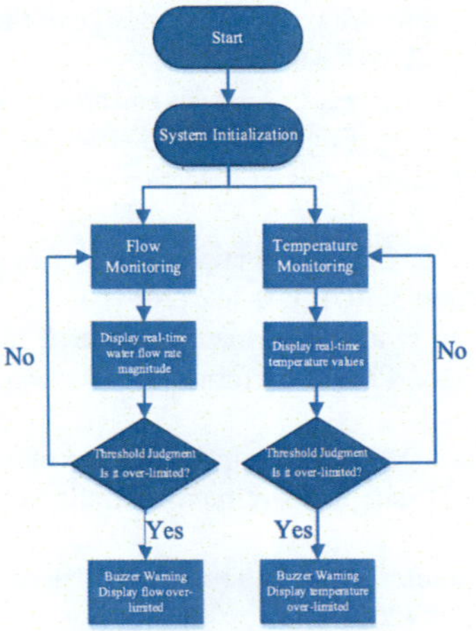

Fig. 7. Main program flowchart

Fig. 8. System circuit diagram

1. Monitoring PH value can also be added to the monitoring data type to prevent corrosion and rusting of pipes, which will be studied in depth subsequently.
2. New technologies such as IoT and cloud computing can also be used to upload the monitored data to the cloud to achieve cloud visualization of the whole monitoring process, which will be expanded in the subsequent research.

Acknowledgements. This research is supported by National Key R&D Program Project (No. 2019YFB1707202) Research and Demonstration of Intelligent Packaging Printing and Sewing Machinery Manufacturing Industry Network Collaborative Manufacturing Integration Technology.

References

1. Qi, Y.-S., Gao, S., Wu, M., et al.: Research progress on key technologies of printing intelligent manufacturing. Digital Printing (3), 1–13 (2021)
2. Hui, W.: Factors affecting printing quality and adjustment. Printing Mag. **03**, 48–52 (2016)
3. Wu, Q.M., Wu, J.-M., Wang, Y., et al.: Study on the servo performance of cooling roller in gravure press based on CFD. J. Xi'an Univ. Technol. **34**(03): 253–256+250 (2018)

4. Chen, B.-R.: Fault analysis and partial innovation design of the cooling roll of gravure press. J. Changzhou Inst. Technol. **29**(01), 31–33 (2016)
5. Geng, Y.-J., Liu, Y.-H., Zhao, T.-P., et al.: Analysis and optimization of offset ink roller cooling structure based on fluent. Packag. Eng. **37**(03), 165–169 (2016)
6. Zhang, Z.-J.: Research on printing pressure detection method based on single-chip. Electron. Test. (15), 46–48 (2018)
7. Xia, X.-Y., Xu, F.-F., Yu, Y.-T.: Design and implementation of a printing press ink adjustment control panel. Manuf. Autom. **35**(07), 110-112 (2013)
8. Zhang, B., Li, Z., Li, Y.-J., et al.: Experimental research on the influence of switch hall installation phase deviation on PMSM performance. Electron. Meas. Technol. **45**(01), 8–14 (2022)
9. Li, P., Liu, H., Li, M., et al.: Design and Implementation of an intelligent ward temperature control system based on STC simple chip microcontroller. Comput. Meas. Control **30**(03), 126-132 (2022)
10. Gu, Y.-L.: Temperature monitoring and alarm system based on AT89S52 microcontroller. Electron. Manuf. (12), 76–78 (2021)
11. Li, J., Wang, J., Li, S., et al.: Design of a smoke alarm based on 51 single chip microcomputer. South. Agric. Mach. **53**(08), 129–131+140 (2022)

Research on Control Algorithm of Solvent-Free Compound Mixing Ratio Based on Feedforward Control

Hongwei Xu[1](✉), Jiacheng Huang[1], Xiao Xu[1], Wenbin Ye[1], Zhicheng Xue[2], and Darun Xi[2]

[1] Xi'an University of Technology, Xi'an, China
xuhongwei@xaut.edu.cn
[2] Shaanxi Beiren Printing Machinery Co., Ltd., Weinan, China

Abstract. The output flow of the circular arc gear pump, the core component of the proportioning system of the solvent-free compound mixer, fluctuates, which affects the proportioning accuracy of the solvent-free compound mixer. In this paper, through the theoretical analysis of the output flow of the circular arc gear pump, the basic characteristics of the output flow of the circular arc gear pump are revealed. By decomposing the output flow into stable output flow and disturbance flow, the feedforward control algorithm is adopted to suppress the output flow fluctuation of circular arc gear pump. By decomposing and linearizing the output flow signal, a feedforward control system for the output flow of circular arc gear pump is established. The system modeling and simulation calculation with MATLAB verify that the feedforward control algorithm can better solve the problem of output flow fluctuation of circular arc gear pump.

Keywords: Solvent-free compound · Arc gear pump · Proportion accuracy · Feedforward control

1 Introduction

Solvent-free compound mixer is a key component of solvent-free compound equipment, providing mixed solvent-free compound glue for solvent-free compound mixer. Solvent-free compound glue is composed of solvent-free compound A and B materials. When mixing is required, it is mixed by solvent-free compound mixer [1]. The proportion of solvent-free compound A and B has a great influence on the performance of solvent-free compound, so the proportion accuracy of solvent-free compound A and B plays an important role in ensuring the performance of solvent-free compound.

For the solvent-free compound mixer, the key component of its proportioning system is the circular arc gear pump. In practical application, by adjusting the speed ratio of solvent-free compound A and B materials to the circular arc gear pump, the mixing ratio of solvent-free compound A and B materials can be realized. Some researchers analyzed the working principle of the circular arc gear pump. Because the theory error, there are fluctuation in the output flow of gear pump [2, 3]. The fluctuating output flow cannot

M. Xu et al. (Eds.): CACPP 2022, LNEE 991, pp. 345–352, 2023.
https://doi.org/10.1007/978-981-19-9024-3_44

guarantee the reliable solvent-free compounding accuracy, so this paper will explore a solution to this problem from the control aspect.

At present, there are many control algorithms, and some researchers have also studied the control of restraining the output flow fluctuation of circular arc gear pump [4]. If we want to suppress the output flow fluctuation of the circular arc gear pump through the control algorithm, we first need to understand the output flow characteristics of the circular arc gear pump. By analyzing the output flow characteristics of circular arc gear pump, we can better find out the appropriate control algorithm to achieve the goal of restraining the output flow fluctuation. This paper studies the control algorithm based on the analysis of the output flow characteristics of the circular arc gear pump.

2 Analysis of Output Fluctuation of Arc Gear Pump

Due to the working principle of the arc gear pump, the output flow of the gear pump is fluctuating. Through the theoretical analysis of the flow of the gear pump, the instantaneous flow pulsation formula of the circular arc gear pump can be obtained as follows [2]:

$$Q_s = \omega B\left(\frac{m^2Z^2}{4}cos^2\alpha\cdot\left(\theta-\frac{(n-1)\pi}{Z}\right)^2-\frac{\pi m^2Z}{4}cos^2\alpha\left(\theta-\frac{(n-1)\pi}{Z}\right)+\frac{\pi m^2Z}{4}cos\alpha\right)\theta\in\left[\frac{(n-1)\pi}{Z},\frac{n\pi}{Z}\right],(n=1,2,\ldots,\infty) \tag{1}$$

where, ω is the rotation angle speed of gear pump; *B* is the tooth width; *m* is modulus; *Z* is the number of teeth; *α* is the pressure angle.

In the proportional control system of the solvent-free compound mixer, the basic parameters of the circular arc gear pump used are shown in Table 1.

Table 1. Gear pump parameters

Teeth number Z	Modulus m/mm	Teeth width B/mm	Pressure angle α
8	5.775	52	14.5°

Through the rotation characteristics of the gear pump, the relationship between the rotation angle, angular velocity and time of the gear pump can be obtained.

$$\theta = \omega * t \tag{2}$$

By introducing Eq. (2) and known parameters into Eq. (1), the flow pulsation formula of circular arc gear pump can be obtained.

$$Q_s = 5.2\omega*\left(5.04*\left(\omega t-\frac{(n-1)\pi}{16}\right)^2-1.98*\left(\omega t-\frac{(n-1)\pi}{16}\right)+2.04\right)$$

$$t\epsilon\left[\frac{(n-1)\pi}{16\omega},\frac{n\pi}{16\omega}\right],(n=1,2,\ldots,\infty) \tag{3}$$

Set angular velocity ω = 3.14 rad/s, the output flow fluctuation curve of gear pump can be obtained, as shown in Fig. 1:

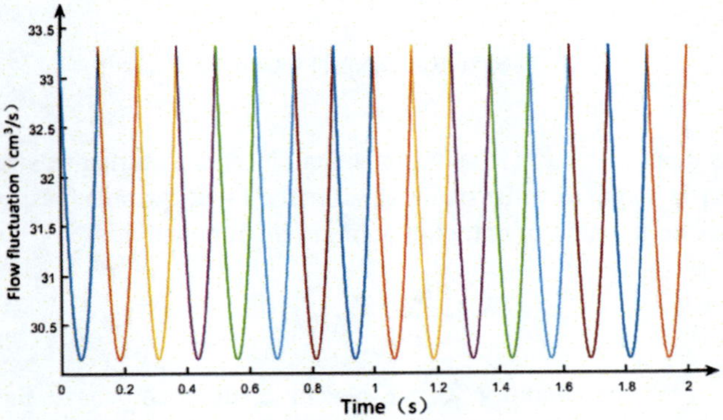

Fig. 1. Flow pulsation simulation curve of gear pump

It can be seen from Fig. 1 that the pulsating effect of the instantaneous flow of the gear pump, and the flow value changes periodically with the change of the gear angle. It can be seen that it is a parabola change law, and it fluctuates every time the two gears with the same parameters mesh.

3 Feedforward Control Algorithm

According to the output flow characteristics of circular arc gear pump, it can be seen that its output flow has the characteristics of periodic fluctuation. Because the output flow fluctuation is regular, the feedforward control algorithm is a better method to suppress the flow fluctuation. Feedforward control is an open-loop regulation system that compensates according to the disturbance, that is, after the system disturbance occurs but before the controlled variable changes, the feedforward controller will correct the control variable according to the disturbance amplitude and change trend to compensate the influence of the disturbance on the controlled variable and keep the controlled variable unchanged [5]. The structure diagram of feedforward control is shown in Fig. 2.

It can be seen from Fig. 2 that the feedforward control algorithm mainly includes the feedforward control link $G_{ff}(s)$, forward channel link $G_{PC}(s)$ and jamming signal link $G_{PD}(s)$. When an input quantity x is given, it passes through the control channel. When the system generates a disturbance quantity n and passes through the disturbance channel, the transfer function of feedforward control is obtained to eliminate the disturbance signal, so that the output value y can maintain the ideal effect and achieve the purpose of control. According to the obtained feedforward control block diagram, the relationship of the system transfer function can be obtained as follows [6]:

$$Y(s) = [X(s) + N(s)G_{ff}(s)]G_{Pc}(s) + N(s)G_{Pd}(s) \tag{4}$$

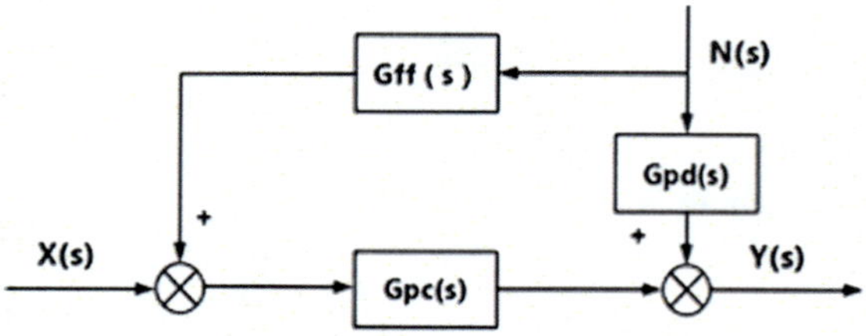

Fig. 2. Feed-forward control schematic diagram

where, X(s) is input signal, Y(s) is output signal, N(s) is disturbance signal. The controlled quantity is required to be obtained after the input quantity passes through the control channel, so the ratio of output quantity to input quantity is:

$$\frac{Y(s)}{X(s)} = G_{PC}(s) \tag{5}$$

If $N(s) = X(s)$, the condition for the system to fully compensate the interference signal is:

$$G_{ff}(s) = -\frac{G_{PD}(s)}{G_{PC}(s)} \tag{6}$$

The feed-forward compensation device satisfying Eq. (6) makes the controlled variable unaffected by the disturbance change. In case of interference, the adjustment effect generated by the feed-forward compensation device is opposite to the direction of interference formation and has the same amplitude, so as to realize full compensation for the interference. Through this adjustment effect, the final synthetic result can achieve ideal control and be continuously maintained at a constant set value. Because the output fluctuation of the gear pump is known, the output flow of the gear pump can be effectively stabilized by using the feedforward control algorithm to control the flow fluctuation part.

4 Feedforward Control Algorithm for Stable Output Flow of Arc Gear Pump of Solvent-Free Compound Mixer

4.1 Decomposition and Linearization of Flow Fluctuation Signal

Expand the flow output formula (3), and obtain:

$$Q_s = 5.2\omega * 5.04\left(\omega t - \frac{(n-1)\pi}{Z}\right)^2 - 5.2\omega * 1.98\left(\omega t - \frac{(n-1)\pi}{Z}\right) + 5.2\omega * 2.04 * 2.04 \tag{7}$$

Make $n = 1$, and obtain.

$$Q_s = 5.2\omega * 5.04 * \omega^2 t^2 - 5.2\omega * 1.98 * \omega t + 5.2\omega * 2.04 \tag{8}$$

The mean value of fluctuation is:

$$Q_s = 10.103\omega \tag{9}$$

The decomposable interference signal is:

$$Q_r = 26.208 * \omega\left(\omega t - \frac{(n-1)\pi}{Z}\right)^2 - 10.296 * \omega\left(\omega t - \frac{(n-1)\pi}{Z}\right) + 0.505 * \omega \tag{10}$$

Set ω =3.14 rad/s, $Z = 8$, then by linearizing Eq. (10) with Fourier series, the first three Fourier series can be obtained as follows:

$$\begin{aligned} Q_f = & 1.3 * cos(16\pi * t) + 0.32 * cos(32\pi * t) \\ & + 0.14\,cos(48\pi * t) + 1.6e^{-3} * sin(16\pi * t) \\ & + 8.1e^{-4} sin(32\pi * t) + 5.4e^{-4} sin(48\pi * t) - 0.55 \end{aligned} \tag{11}$$

After the linearization of the interference signal is obtained by Fourier series, one of the waveforms is compared with one of the cycles of the original signal. The comparison diagram is shown in Fig. 3:

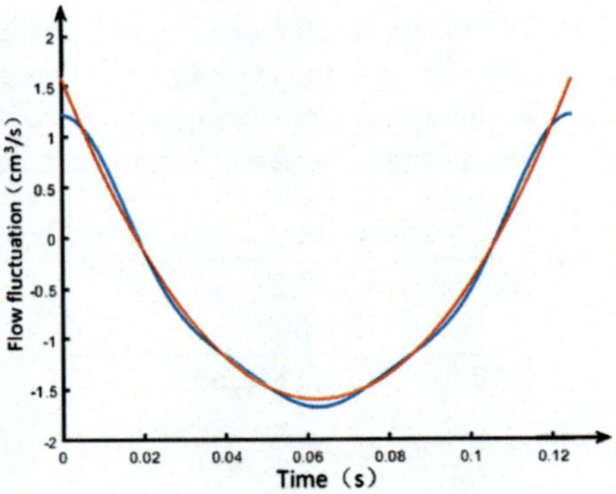

Fig. 3. Contrast diagram of interference signal and linearization signal

The red line in the figure indicates one cycle of the original interference signal, and the blue line indicates a section of waveform after linearization. The variation range of the original interference signal is [– 1.585 cm^3/s, 1.587 cm^3/s], and the variation range after linearization is [– 1.66 cm^3/s, 1.21 cm^3/s]. By comparing the two cases, the error between the linearized signal and the original signal can be obtained, as shown in Fig. 4.

Through the obtained error value image, it can be seen that the maximum value in the flow error diagram is 0.3765 and the minimum value is – 0.1225. The maximum value of the error obtained by subtracting the two is 0.5. Therefore, the error range between the linearized signal and the original interference signal is controlled within 0.5, while the variation range of the interference signal is 3.172. The relative error of

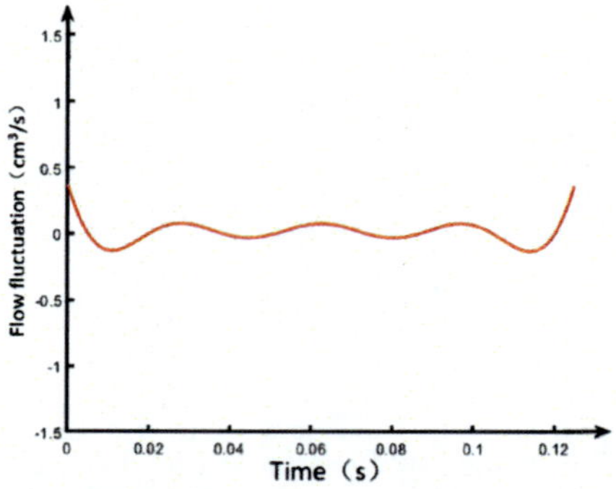

Fig. 4. Error diagram of interference signal and linearization signal

linearization is 15%, and the error range is relatively small. The obtained linearized signal is obtained through the expansion of the three terms of Fourier series, and the error range is acceptable. When a smaller error range is required, the waveform of the interference signal can be more approximate by increasing the Fourier coefficient, so as to reduce the error.

4.2 Simulation Calculation

The linearization formula (11) of the interference signal is the time domain signal. In order to establish the transfer function of the system, it is necessary to convert the time domain signal into a complex frequency domain signal, which is changed by Laplace, and the frequency domain signal of the interference signal can be obtained as follows:

$$N(s)Y_{PD}(s) = \frac{1625s + 100}{1250s^2 + 3{,}158{,}273} + \frac{1000s + 254}{3125s^2 + 31{,}582{,}734} + \frac{875s + 509}{6250s^2 + 142{,}122{,}303} - \frac{0.55}{s} \quad (12)$$

Laplace transforms Eq. (11), and obtain.

$$\mathrm{Y}(s) = \frac{32}{s} \quad (13)$$

Since $\mathrm{N}(s) = \mathrm{X}(s)$, then, the Fig. 2 will be changed to Fig. 5.

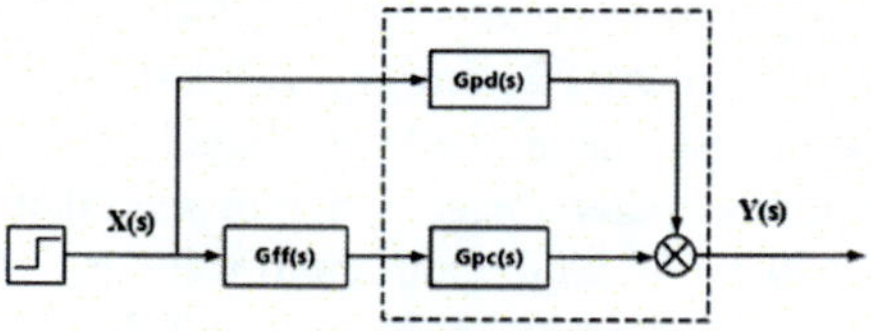

Fig. 5. Feedforward control block diagram

The input of the system is unit step signal, i.e.

$$X(s) = \frac{1}{s} \tag{14}$$

Then, the transfer function of disturbance link is:

$$\begin{aligned} G_{PD}(s) &= \frac{\frac{1625s+100}{1250s^2+3{,}158{,}273} + \frac{1000s+254}{3125s^2+31{,}582{,}734} + \frac{875s+509}{6250s^2+142{,}122{,}303} - \frac{0.55}{s}}{\frac{1}{s}} \\ &= \frac{1625s^2+100s}{1250s^2+3{,}158{,}273} + \frac{1000s^2+254s}{3125s^2+31{,}582{,}734} \\ &\quad + \frac{875s^2+509s}{6250s^2+142{,}122{,}303} - \frac{11}{20} \end{aligned} \tag{15}$$

The transfer function of control link is:

$$G_{PC}(s) = \frac{Y(s)}{X(s)} = \frac{32}{s} / \frac{1}{s} = 32 \tag{16}$$

Then, the transfer function of feedforward control link is:

$$\begin{aligned} G_{ff}(s) &= \frac{11}{640} - \frac{1625s^2+100s}{40,000s^2+101,064,736} - \frac{1000s^2+254s}{100,000s^2+1,009,8647,488} \\ &\quad - \frac{875s^2+509s}{200,000s^2+4,547,913,696} \end{aligned} \tag{17}$$

Then, using the software Matlab to found the model of this system as shown as (Fig. 6).

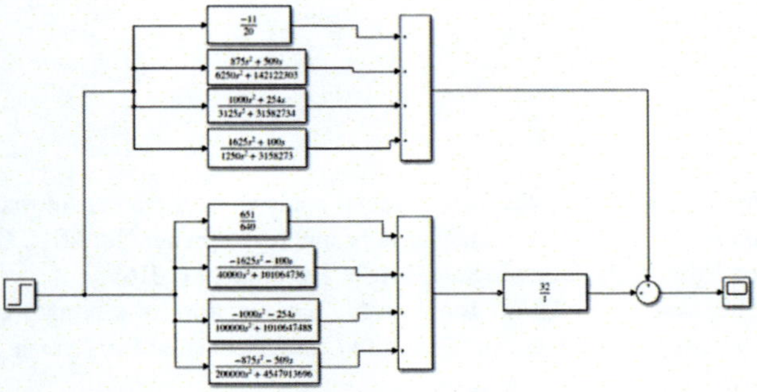

Fig. 6. Feedforward control simulation diagram

Through the simulation computation, the result is shown in Fig. 7.

It can be seen from Fig. 7 that the peak value of the fluctuation signal is 3 cm^3/s. When the fluctuation signal is compensated by the feedforward control, the output signal is shown as the black line which has no fluctuation. The result shows that the feedforward control can play a good role in compensating the flow interference and reduce the output fluctuation of the flow.

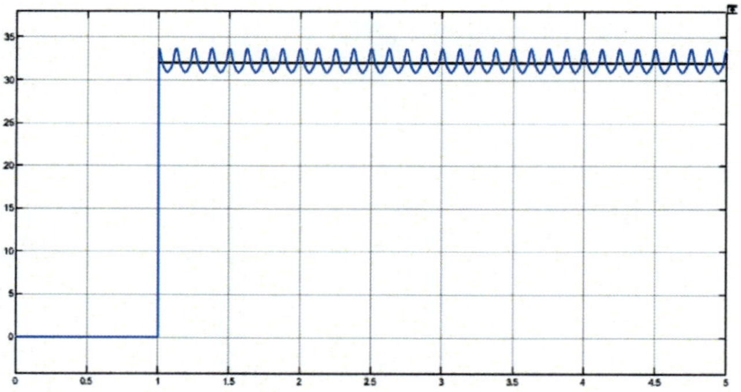

Fig. 7. Feedforward control response signal

5 Conclusion

As the core component of the solvent-free compound proportioning system, the output of the circular arc gear pump fluctuates. Through the analysis of the output waveform of the gear pump, it can be seen that its characteristics are suitable to use the feedforward control algorithm to reduce the impact of the fluctuation on the proportioning. By decomposing and linearizing the output signal of circular arc gear pump, the corresponding feedforward control system is established. Matlab software is used for simulation calculation. The calculation results show that the output fluctuation can be effectively reduced through the feedforward control algorithm, so as to effectively ensure the solvent-free compound proportioning accuracy.

Acknowledgements. This research project is supported by the Key R&D Plan of Shaanxi Province (approve number: 2021GY-262).

References

1. Xu, H., Wang, X., Lei, R., et al.: Experimental analysis of the effect of vacuum degassing technology on the solvent-free laminating adhesive performance. In: 2016 China Academic Conference on Printing & Packaging and Media Technology (2016)
2. Wei, Y., Xu, H., Han, H., Feng, S.: Tooth profile optimization for mixing proportional pump of solvent-free laminating mixer. In: 2019 10th China Academic Conference on Printing and Packaging (2019)
3. Zardin, B., Natali, E.: Evaluation of the hydro—mechanical efficiency of external gear pumps. Energies **13**(12), 2468–2471 (2019)
4. Xing, B., Xu, H., Chen, X., Li, X.: A control algorithm for improving the ratio precision of solvent-free laminating. In: 2019 IEEE 2nd International Conference on Automation, Electronics and Electrical Engineering, AUTEEE (2019)
5. Landau, I.D., Alma, M., Airimitoaie, T.-B.: Adaptive feedforward compensation algorithms for active vibration control with mechanical coupling. Automatic **47**, 2185–2196 (2011)
6. Jiang, P.: Multiple-targets tracking control algorithm for a class of nonlinear systems with feedforward compensations. Neurocomputer **196**, 210–213 (2016)

Performance Analysis and Structure Optimization of Knife Folding Mechanism's Machete Arm

Yulong Lin, Taotao Chen, Han Jiang, and Yuansheng Qi(✉)

College of Mechanical and Electrical Engineering, Beijing Institute of Graphic Communication, Beijing, China
yuansheng-qi@bigc.edu.cn

Abstract. When the speed of knife folding mechanism of N160 web press is ramped up from 25,000 to 36,000 rph, the root of machete arm will be damaged in a short period of time. This paper explores the causes of damage to the machete arm in terms of strength and stiffness, and the corresponding structural improvement methods are provided. After optimization, the new machete arm ensures the strength while avoiding the resonant regions of natural frequencies for each working dominant frequency. A preliminary workshop test verifies the rationality of the new machete arm structure. The research results provide a basis for fault analysis and structural optimization for the knife folding mechanism of web press.

Keywords: Folding mechanism · Strength calculation · Constrained mode · Vibration test · Structural optimization

1 Introduction

In recent years, with the rising cost of the raw materials and labor, higher printing speed is demanded by printers to ensure profits. However, the increase of printing speed brings problems of reliability and stability. For example, the knife folding mechanism of N160 printing unit works stably at the speed of 25,000 rph, but its folding mechanism becomes problematic when the speed is increased to 36,000 rph. The root area of machete arm will crack and damage in a very short time. Many scholars have carried out relevant researches on the knife folding four-bar mechanism.

In order to improve the working stability of the web folding mechanism, Yuan Yingcai et al. [1, 2] carried out robust design of the folding mechanism. Li Zhuangju et al. [3] established a dynamic model of the web folding mechanism, which was solved by the direct integration method of nonlinear system. Wang et al. [4] used geometric parameters as design variables and adopted robust optimization design for a planar four-bar mechanism. Selcuk et al. [5] used rigid rod to represent joint clearance and carried out a dynamic calculation for four-bar mechanism with clearance.

M. Xu et al. (Eds.): CACPP 2022, LNEE 991, pp. 353–364, 2023.
https://doi.org/10.1007/978-981-19-9024-3_45

The above researches mainly focus on theoretical analysis and design, there are few reports concerning on the actual engineering problems of the damage to key parts of the knife folding mechanism after speeding up. The knife folding mechanism's machete arm of N160 printing unit is prone to breakage after speeding up, which seriously affects the production efficiency of printing enterprises. In this paper, the causes of damage to the machete arm after speeding up are explored and the corresponding structural improvement methods are provided, with the view to solve this key technical problem restricting the development of printing enterprises.

2 Strength Analysis of Folding Mechanism's Machete Arm

The structure of web folding mechanism is shown in Fig. 1, which is mainly composed of flywheels, connecting rod, machete arm, rollers, wall plates, and so on. The machete arm is driven by the crank and connecting rod to do reciprocal swing which is loaded with machetes. The conveyor belt of paper feeding table transports the paper under machete which pushes the paper between the two rollers, and then the rollers squeeze the paper to complete a cross-folding action.

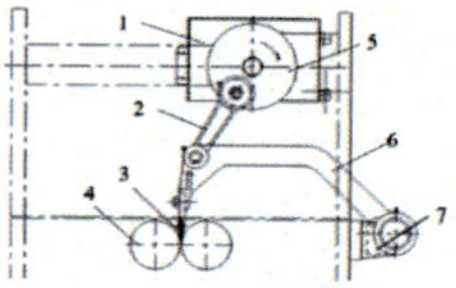

1. Gearbox 2.Connecting rod 3.Machete
4.Roller 5.Flywheel 6.Machete arm 7.Base

Fig. 1. Knife folding mechanism of web printing machine

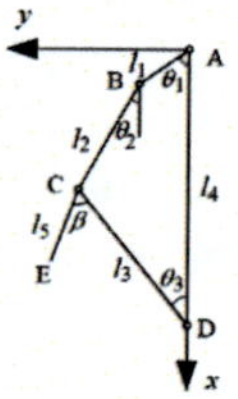

Fig. 2. Kinematic diagram of folding mechanism

The kinematic and dynamic equations of the folding mechanism are established to calculate the strength of the machete arm theoretically. The folding four-bar mechanism is essentially a crank rocker mechanism, whose kinematic diagram is shown in Fig. 2.

2.1 Kinematic Analysis of Folding Four-Bar Mechanism

Through the closed vector equation of the four-bar mechanism, its angular acceleration relation is obtained as

$$\begin{cases} l_1 \cdot \ddot{\theta}_1 \cdot \sin\theta_1 + l_1 \cdot \dot{\theta}_1^2 \cdot \cos\theta_1 + l_2 \cdot \ddot{\theta}_2 \cdot \sin\theta_2 + l_2 \cdot \dot{\theta}_2^2 \cdot \cos\theta_2 \\ \quad + l_3 \cdot \ddot{\theta}_3 \cdot \sin\theta_3 + l_3 \cdot \dot{\theta}_3^2 \cdot \cos\theta_3 = 0 \\ l_1 \cdot \ddot{\theta}_1 \cdot \cos\theta_1 - l_1 \cdot \dot{\theta}_1^2 \cdot \sin\theta_1 + l_2 \cdot \ddot{\theta}_2 \cdot \cos\theta_2 - l_2 \cdot \dot{\theta}_2^2 \cdot \sin\theta_2 \\ \quad - l_3 \cdot \ddot{\theta}_3 \cdot \cos\theta_3 + l_3 \cdot \dot{\theta}_3^2 \cdot \sin\theta_3 = 0 \end{cases} \tag{1}$$

where, l_1 is the length of the crank, $l_1 = 63.5$ mm; θ_1 is the included angle between the crank and the frame; l_2 is the length of the connecting rod, $l_2 = 160.5$ mm; θ_2 is the included angle between the connecting rod and the frame; l_3 is the length of the rocker, $l_3 = 432$ mm; l_4 is the length of the frame, $l_4 = 455.5$ mm; θ_3 is the included angle between the rocker and the frame.

The initial conditions are as $\theta_1 = 70°$, $\theta_2 = 0°$, $\theta_3 = 29°$. The crank rotated at a constant speed of 25,000 rph and 36,000 rph. By substituting the above data into Eq. (1), the angular acceleration of followers are obtained as shown in Fig. 3.

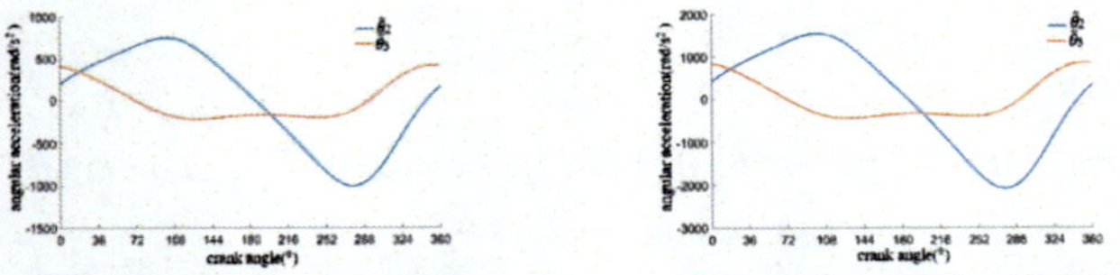

a. rank speed is 25000rph b Crank speed is 36000rph

Fig. 3. Angular acceleration of followers

2.2 Dynamic Equation of Folding Four-Bar Mechanism

For this folding mechanism, the force analysis diagram is as shown in Fig. 4. The structural parameters of each component of the folding mechanism are as follows: the crank mass is m_1, the moment of inertia is J_{s1}, the coordinates of the mass center S_1 are (x_{s1}, y_{s1}) whose distance from point *A* is l_{s1}; the mass of the connecting rod is m_2, the moment of inertia is J_{s2}, the coordinates of the mass center S_2 are (x_{s2}, y_{s2}) whose distance from point B is l_{s2}; the mass of the machete arm is m_3, the moment of inertia is J_{s3}, the coordinates of the mass center S_3 are (x_{s3}, y_{s3}) whose distance from point *D* is l_{s3}; F_{Ax}, F_{Ay} are the components of the restraint reaction force at hinge *A* in the *x* and *y* directions; F_{Bx}, F_{By} are the components of the restraint reaction force at hinge *B* in the *x* and *y* directions; F_{Cx}, F_{Cy} are the components of the restraint reaction force at hinge *C* in the *x* and *y* directions; F_{Dx}, F_{Dy} are the components of the restraint reaction force at hinge *D* in the *x* and *y* directions; F_{Ex}, F_{Ey} are the components of the resistance of paper to machete in the *x* and *y* directions when folding; *M* is the torque acting on the crank.

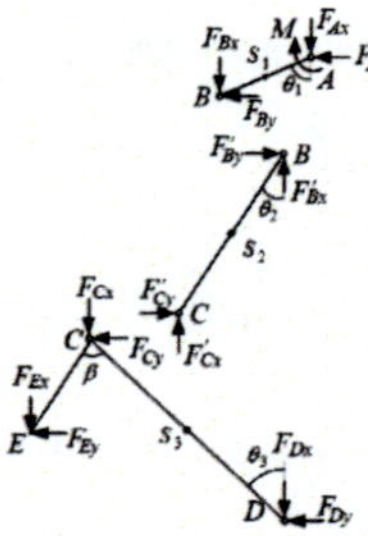

Fig. 4. Force analysis diagram of folding mechanism

The force equilibrium equation for the crank can be listed as follows:

$$\begin{cases} F_{Ax} + F_{Bx} = m_1 \cdot \ddot{x}_{s1} - m_1 \cdot g \\ F_{Ay} + F_{By} = m_3 \cdot \ddot{y}_{s1} \\ F_{Bx} \cdot l_{s1} \cdot \sin\theta_1 + F_{Ay} \cdot (l_1 - l_{s1}) \cdot \cos\theta_1 - F_{Ax} \cdot (l_1 - l_{s1}) \cdot \sin\theta_1 \\ \quad -F_{By} \cdot l_{s1} \cdot \cos\theta_1 - M = J_{s1} \cdot \ddot{\theta}_1 \end{cases} \tag{2}$$

The force equilibrium equation for the connecting rod can be listed as follows:

$$\begin{cases} -F'_{Bx} - F'_{Cx} = m_2 \cdot \ddot{x}_{s2} - m_2 \cdot g \\ -F'_{Cy} - F'_{By} = m_2 \cdot \ddot{y}_{s2} \\ F'_{Bx} \cdot (l_2 - l_{s2}) \cdot \sin\theta_2 + F'_{Cy} \cdot l_{s2} \cdot \cos\theta_2 - F'_{Cx} \cdot l_{s2} \cdot \sin\theta_2 \\ \quad -F'_{By} \cdot (l_2 - l_{s2}) \cdot \cos\theta_2 = J_{s2} \cdot \ddot{\theta}_2 \end{cases} \tag{3}$$

The force equilibrium equation for the machete arm can be listed as follows:

$$\begin{cases} F_{Cy} + F_{Dy} = m_3 \cdot \ddot{y}_{s3} - F_{Ey} \\ F_{Cx} + F_{Dx} = m_3 \cdot \ddot{x}_{s3} - m_3 \cdot g - F_{Ex} \\ F_{Cx} \cdot (l_3 - l_{s3}) \cdot \sin\theta_3 + F_{Cy} \cdot (l_3 - l_{s3}) \cdot \cos\theta_3 \\ \quad -F_{Dx} \cdot l_{s3} \cdot \sin\theta_3 - F_{Dy} \cdot l_{s3} \cdot \cos\theta_3 = J_{s3} \cdot \ddot{\theta}_3 \end{cases} \tag{4}$$

The structural parameters of main components of the folding four-bar mechanism are as shown in Table 1.

Table 1. Structural parameters of main components of folding mechanism

Property	Crank	Connecting rod	Machete arm
Position of mass center/mm	3.17	80.25	177.12
Mass/kg	10.72	1.19	9.61
Moment of inertia/kg·mm^2	2.90×10^4	2571	5.43×10^5

The crank rotates at a constant speed of 25,000 rph and 36,000 rph, by substituting the above data into Eqs. (2) to (4), the support reaction force of the machete arm is calculated as shown in Figs. 5 and 6.

(a) force of connecting rod on machete arm (b) force of base on machete arm

Fig. 5. Support reaction force of machete arm at crank speed of 25,000 rph

a. Force of connecting rod on machete arm b. Force of base on machete arm

Fig. 6. Support reaction force of machete arm at crank speed of 36,000 rph

2.3 Stress Calculation of Machete Arm

The stress of a section of the machete arm is calculated as:

$$\begin{cases} \sigma = F_n/A + F_{t_1} \cdot L \cdot Y/I + F_{t_2} \cdot (L - L_m) \cdot Y/I \\ F_{t_1} = -F_{Dx} \cdot \sin\phi_3 + F_{Dy} \cdot \cos\phi_3 \\ F_{t_2} = \alpha_3 \cdot L_m \cdot m \\ F_n = -F_{Dx} \cdot \cos\phi_3 - F_{Dy} \cdot \sin\phi_3 + \omega_3^2 \cdot L_m \cdot m \end{cases} \tag{5}$$

where, σ is the stress of the machete arm section; A is the sectional area of the machete arm; L is the distance from the section to the base; Y is centroid height; I is the moment of inertia of the section to the centroid axis; L_m is the distance from the mass center of the lower part of machete arm section to the base.

The minimum sectional area of the machete arm is 9.3×10^{-4} m^2 and the maximum sectional area is 3.65×10^{-3} m^2. For this kind of variable section structure, the position of mass center, the area of section, the height of centroid and the moment of inertia can be easily obtained by using the cross-section mass attribute function of Creo software. Using Eq. (5), the maximum stress of the machete arm at 25,000 rph is calculated to be 5.59 MPa and the maximum stress of the machete arm at 36,000 rph is calculated to be 11.90 MPa, both of which are located at the bending of the machete arm. The maximum stress in the root area of the machete arm under two working conditions is 3.65 Mpa and 8.03 Mpa respectively. The theoretical calculation results show that the stress on the machete arm is much less than the yield strength of the material which means the machete arm structure has sufficient strength, and it can be judged that the damage to the root of the machete arm is not caused by force problem.

3 Model Analysis of Folding Mechanism

The machete arm not only has strength requirements which can bear a certain force, but also need to meet the requirements of stiffness. According to the vibration theory, when the external excitation frequency of the structure is close to the natural frequency of a certain mode, the system amplitude will increase significantly, that is, the resonance occurs. Resonance will cause large deformation and dynamic stress of mechanical structure and result in serious damage [6]. Modal analysis is helpful to understand the vibration, natural frequency, vibration mode and other characteristics of the folding mechanism. It is convenient to investigate the relationship between the natural frequency and working frequency of the folding mechanism, so as to avoid resonance caused by the natural frequency of the structure within the range of the working frequency.

3.1 Boundary Conditions Setting

In order to accurately predict the dynamic characteristics of the structure, it is necessary to impose correct boundary conditions on the analysis model according to the actual structure and stress form, that is, constrained modal analysis [7]. For any system, with boundary conditions are changed, the dynamic characteristics will also change. Different constraints will significantly affect the modal frequency and vibration mode of the system [8]. Constrained modal analysis of the folding mechanism must strictly follow the actual constraints.

A virtual prototype of the folding mechanism is built in ADAMS software to extract the load on the machete arm (see Fig. 7). Import the 3D model of the folding mechanism into ADAMS in *x-t* format, set the material properties (steel) of each component, add constraints (a total of 21 fixed pairs and 4 rotating pairs), friction of the moving pairs and gravity. The speed of flywheel is set to be 36000rph, the simulation time is 0.1 s which the machete arm moves for one cycle. The top of the machete arm is hinged to the connecting rod and the bottom of the machete arm is hinged to the base. The loads at the machete arm hinges are extracted in the post-processor of the virtual prototype, which are shown in Figs. 8 and 9.

Fig. 7. Virtual prototype of folding mechanism

From Figs. 8 and 9, it can be seen that both the top and bottom of the machete arm are subject to forces in the x, y and z axes and torques along the x and y axes, with no torque along the z axis. Therefore, the machete arm is subject to movement constraints along the x, y and z axes, rotation constraints along the x, y axes, and there is only rotational degree of freedom along the z axis (the direction of the coordinate axis is as shown in Fig. 7.

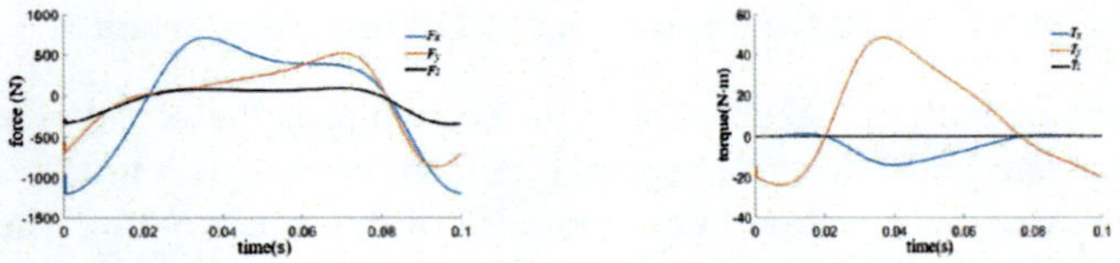

a.Force b.Moment of force

Fig. 8. Loads on the top of machete arm

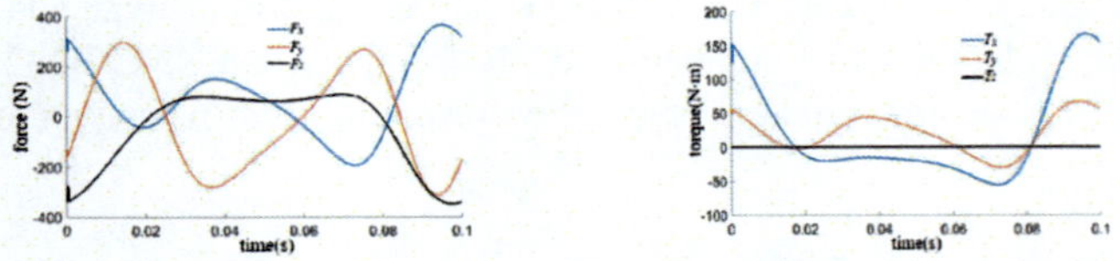

a.Force b.Moment of force

Fig. 9. Loads on the bottom of the machete arm

3.2 Modal Analysis Results

The Block Lanczos algorithm [9] is chosen to extract the first 6th natural frequencies and vibration modes of the machete arm, which is shown in Table 2.

Table 2. Description of the first 6th natural frequencies and vibration modes of the machete arm

Modal order	Natural frequencyvalue (Hz)	Main vibration mode
1	142.12	The vibration mode is mainly manifested as the left and right torsion of the machete arm
2	331.09	The vibration mode is mainly manifested as the front end of the machete arm bending up and down
3	591.91	The vibration mode is mainly manifested as the front end of the machete arm twisting around
4	956.83	The vibration mode is mainly manifested as the upper and lower bending of the machete arm
5	1001.6	The vibration mode is mainly manifested as the front end of the machete arm bending and twisting
6	1125.9	The vibration mode is mainly manifested as bending and twisting of the middle end of the machete arm

3.3 Analysis on Vibration Characteristics of Folding Mechanism

Vibration Test of Folding Mechanism. This section tests the actual vibration of the folding mechanism under workshop operation, and builds a vibration test system to measure the vertical acceleration in the root area of the machete arm. The test instruments used for vibration test include an INV3020 dynamic signal collector, which has a built-in embedded computer module and hard disk storage and a fully shielded chassis with good anti-interference capability; an INV9824 acceleration sensor with frequency range of 1~15kHz and sensitivity of 5 mV/g; a set of multi-channel signal acquisition and real-time analysis software DASP V11, which can easily analyze and process the collected data; an laptop and a number of M5-BNC cables and BNC-BNC connectors. The hardware device and software acquisition program used in the test are shown in Fig. 10.

Fig. 10. Vibration test

The test was carried out in September 2020 on the folding mechanism of N160 printing unit under the following conditions: (1) the rotation speed of the folding mechanism is 25,000 rph; (2) the rotation speed of the folding mechanism is 36,000 rph. Once the folding mechanism was running smoothly, data acquisition began with a sampling frequency of 10 kHz and 30 s of data was acquired at a time.

Vibration Signal Analysis of Folding Mechanism. The time domain vibration signals at the root of the machete arm for the two operating conditions collected from the test are shown in Fig. 11.

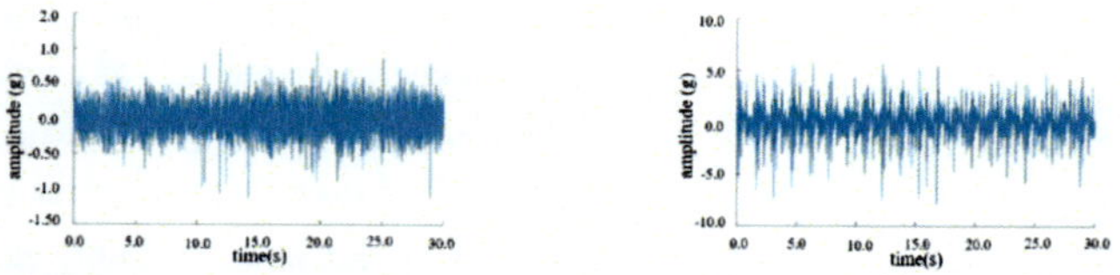

a.Rotation speed is 25000 rph b.Rotation speed is 36000 rph

Fig. 11. Vibration signals at the root of machete arm

It can be seen from the above figures that the overall vibration signal is relatively stable and the overall amplitude is small under the first working condition, the absolute values of the maximum and minimum vertical acceleration are within 1.5 g and the effective value and standard deviation are within 0.3 g. Under the second working condition, the vibration is become severer, whose absolute values of the maximum and minimum vertical acceleration are more than 5g and the effective value and standard deviation are more than 3 g. The waveform amplitude is large and discrete, indicating that the eccentric mechanism (flywheel) has a significant impact on the machete arm.

For the above vibration signals, the power spectrum analysis method can better highlight the main frequency components of the spectrum [10]. To study the low frequency vibration in the vibration signal which has a greater influence on the vibration characteristics of the machete arm, the high frequency interference signal above 200 Hz was filtered and the power spectrum density of the vibration signal was obtained as shown in Figs. 12.

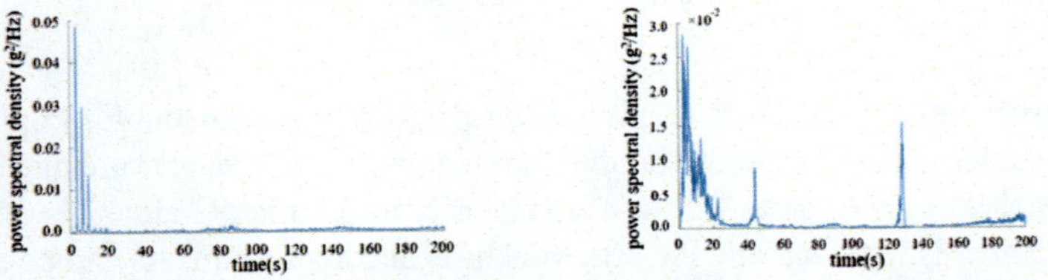

a.Rotation speed is 25000 rph b.Rotation speed is 36000 rph

Fig. 12. Power spectral density of vibration signal at the root of machete arm

In engineering, the interval between 0.75 and 1.25 times of the natural frequency of the system is usually taken as the resonance region [11]. Under the first working condition, the working dominant frequencies at the root of the machete arm are all less than 20 Hz. Under the second working condition, most of the working dominant frequencies of the machete arm are below 50 Hz, but there is a working dominant frequency of 135.83 Hz which is close to the first-order natural frequency of the machete arm and resonance will occur. This may be the cause of damage to the root of the machete arm.

4 Structural Improvement

4.1 Optimization Model Setting

The Shape Optimization module of ANSYS software was used for topology optimization of the machete arm. The grid selected in this part is solid95. As the structure of the machete arm is relatively regular, the grids are divided by hexahedral meshing predominant with tetrahedral meshing filling in. The setting of boundary conditions is an extremely important aspect. The top of the machete arm is hinged to the connecting rod and the bottom of the machete arm is hinged to the knife body holder.

According to the multiple parameter iterative analysis of topology optimization, the convergence is good when the percentage of material removed is greater than 70%, and the percentage is too large for convergence when it exceeds 90%. The flexibility is taken as objective function and the volume is taken as constraint function for optimization. The percentage of material reduction is set to be 80%. For the loadings of the machete arm in Figs. 8 and 9, a set of data is taken every 0.01s as the boundary condition for the optimization analysis. A total of 10 optimization calculations are carried out, the overlap of the removed material from the 10 calculations is shown in Fig. 13.

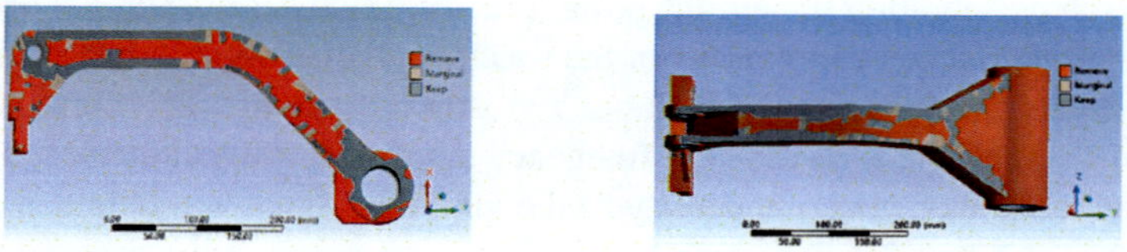

a.Side of machete arm b.Back of machete arm

Fig. 13. Topology optimization results

The structure of the machete arm is improved according to the topology optimization results. It can be seen from Fig. 13(a) that the side of the machete arm are mainly loaded by the upper and lower parts with excess material in the middle area. The material in the middle area is removed and the arm thickness becomes 8mm. As can be seen from Fig. 13(b), the back of the machete arm is mainly loaded by the left and right parts with excess material in the middle area. The machete arm is changed into a double arm structure as the material in the middle area is removed. The width of each arm becomes one third of the original structure, which is 16 mm. The top of the machete arm needs to retain its existing structure to mount the machete. The cylindrical structure at the bottom is connected with the long shaft of the body base through bearings, so the material in this section shall be retained to ensure the safety and cleanliness during operation. The improved double arm hollow machete arm is shown in Fig. 14, the mass of the new arm is 5.21 kg, which is 45.78% lower than the original structure.

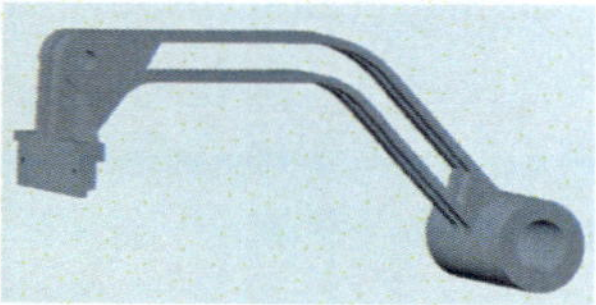

Fig. 14. Double arm hollow structure of machete arm

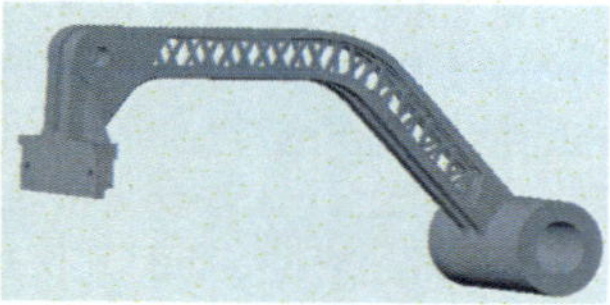

Fig. 15. Full diagonal web member truss structure of machete arm

4.2 Model Analysis of Machete Arm's New Structure

Double Arm Hollow Structure. The same method as in Sect. 2 is used to calculate the constrained modalities of the double arm hollow structure machete arm. The first 6th natural frequencies are 49.88 Hz, 169.25 Hz, 318.4 Hz, 464.8 Hz, 493.44 Hz, and 521.31 Hz.

It can be seen that the natural frequencies of each mode of the improved double arm hollow structure machete arm is reduced, but they are still much larger than the working dominant frequencies under the first condition. The first-order natural frequency of the double arm hollow structure machete arm is close to the working dominant frequency of 45.27 Hz under condition 2, the second-order natural frequency is close to the working dominant frequency of 135.83 Hz, and the structure will still resonate. For the machete arm which is long arm structure with large span, truss structure is adopted to increase the stiffness appropriately to avoid the working main frequencies.

Truss Structure Machete Arm. According to the different arrangement of vertical and web members, the truss can be divided into "diagonal web member symmetrical type", "diagonal web member staggered type", "full diagonal web member type" and "open web member type". In the literature [12], a comparative analysis of the internal force distribution, node deflection variation and steel consumption of the above-mentioned truss structure forms revealed that the inverted triangle full diagonal web member type structure has the most obvious advantages. The full diagonal web truss structure is chosen to change the structural stiffness of the machete arm, which is placed on the symmetrical center surface of the hollow part of the machete arm in the vertical direction. The shape of the inverted triangular unit of the full diagonal web member truss is chosen as equilateral triangle, and the height of the equilateral triangle is equal to the height of the hollow part of the topologically optimized double arm structure, with a value of 26mm. The width of each arm is 16mm, so the rod width of the triangle unit is no greater than 16mm. The rod width is divided into 4 levels, the rod thickness is half of the rod width which is also divided into 4 levels. Taking the first 6th order modal natural frequencies of the machete arm as evaluation index, 2-factor and 4-level tests are carried out to determine the optimal parameters for the triangular unit rod width and thickness. Factor A is the rod thickness of the triangular unit, the 4 levels is 2 mm, 4 mm, 6 mm, and 8 mm. Factor B is the rod width of the triangular unit, the 4 levels is 4 mm, 8 mm, 12 mm, and 16 mm.

A total of 16 sets of constrained modal analysis tests are conducted. The comparison results show that when the rod thickness is 4mm and the rod width is 8 mm, the first-order natural frequency of the full diagonal web member truss structure machete arm is 87.481 Hz, and its resonance zone is 65.61–109.35 Hz which is avoiding the working dominant frequencies of 45.27 Hz and 135.83 Hz under condition 2. The second-order natural frequency is 424.29 Hz, which is far greater than the working dominant frequencies, and the structure will not resonate. The final three-dimensional model of the full diagonal web member truss structure machete arm is as shown in Fig. 15.

A preliminary workshop test of the full diagonal web member truss structure machete arm was carried out, which was found that the machete arm of the folding mechanism working smoothly under the two working conditions of 24,000 rph and 36,000 rph.

However, the new machete arm still needs to be further investigated and studied for performance in future processing and production.

5 Conclusions

Aiming at the problem that the knife folding mechanism of the N160 printing unit is prone to damage at the root of the machete arm under the speed of 36,000 rph, the causes of damage to the machete arm are explored in terms of strength and stiffness, which is found that resonance is the main reason. Topology optimization is adopted to improve the structure of the machete arm by changing the structure from a single-arm thin-walled structure to a double-arm full diagonal web member truss structure, and 2-factor and 4-level tests are carried out to determine the optimal rod thickness and rod width of the truss triangular units. The new machete arm structure ensures the strength without resonance. A preliminary workshop test shows that the machete arm operates smoothly without abnormalities.

Acknowledgements. The research was supported by Basic research project of BIGC (Eb202003) and Science and technology projects of Beijing Municipal Commission of Education (KM202110015005).

References

1. Yuan, Y.C., Liu, Y.L., Li, Y.: Optimization design of the web press's fold mechanism based on robustness. Adv. Mater. Res. **174**, 277–281 (2011)
2. Yuan, Y.C., Liu, Y.L., Li, Y.: The robust design of four-bar linkage with clearance based on sensitivity analysis. Appl. Mech. Mater. **34–35**, 1656–1660 (2010)
3. Li, Z.-J., Tian, S.-Y., Zhao, W., Cao, S.-Z.: Direct-Integrating approach for solving state equation of the mechanics model of web offset press's folding mechanism with clearance. Acta Electronica Sinica, **46**(12), 3037–3043 (2018)
4. Wang, T.: Robust Optimization Design of Plane Four-Bar Linkage Based on Six-Sigma Theory. Hoisting and Conveying Machinery (2009)
5. Selcuk, E., Uzmay, I.: Investigation on Effect of Joint Clearance on Dynamics of Four-Bar Mechanism. Nonlinear Dynamics (2009)
6. You-tang, L.I.: Theory and Application of Mechanical Vibration. Science Press, Beijing (2020)
7. Zhi-fang, F.U., Hong-xing, H.U.A.: Theory and Application of Modal Analysis. Shanghai Jiaotong University Press, Shanghai (2000)
8. Lin, J.-L.: Modal Analysis and Experiment. Tsinghua (2011)
9. Zhang, H.-C.: ANSYS 14.0 Theoretical Analysis and Engineering Application Examples. Machinery Industry Press, Beijing (2013)
10. Gao, X.-Q., Ding, M.-Y.: Digital Signal Processing. Xi'an University of Electronic Science and Technology Press, Xi'an (2016)
11. Zhang, Y.-M.: Mechanical Vibration. Tsinghua University Press, Beijing (2011)
12. Wang, Z., Xu, F., Xu, X.-X.: Comparative study on web member arrangement of truss pipe rack. In: Shenyang Annual Conference on Science (2015)

Study on the Structure Optimization and Simulation Analysis of Oven System in Gravure Press

Yueyue Xing, Peng Liu(✉), Boqi Deng, and Yinhua Zhai

School of Printing Packaging and Digital Media, Xi'an University of Technology, Xi'an, China
liupeng@xaut.edu.cn

Abstract. Based on traditional gravure press drying system in the uneven distribution of internal hot air oven, drying problem such as low efficiency and large energy consumption, refer to the original structure of the wind drying oven design experience, design a kind of adjustable air nozzle, optimize the structure of original oven, at the same time in the middle of the oven and wind mouth entrance increase wind plate uniform, chamfering is set in the oven outlet. In this way, the hot air reaching the inlet of the air nozzle, the outlet of the air nozzle and the surface of the base material is more uniform, improving the drying efficiency of the oven, saving energy and reducing consumption. CFD simulation software was used to establish the fluid domain model of air nozzle and oven. Simulation analysis was carried out on the adjustable air nozzle and the whole oven before and after optimization, and the velocity values of key parts were extracted for numerical analysis. Finally, the oven structure with the highest drying efficiency was obtained. The experimental platform of oven system was built, and the results of the optimized oven were verified experimentally. The simulation results are basically consistent with the experimental data analysis results.

Keywords: CFD · Gravure press · Oven system · Uniformity

1 Introduction

According to the theory of hydrodynamics and aerodynamics, this paper simulates and analyzes the oven of gravure press, establishes a three-dimensional model, and uses the hydrodynamic analysis software Fluent to simulate the local and the whole oven system. According to the simulation results, the velocity value of hot air is extracted on the surface of base material [1], so as to simulate the movement state of hot air in the local and the whole oven system. Data processing and numerical analysis are carried out on the extracted data, and the average velocity is used as an index to evaluate the uniformity of the corresponding wind field [2]. Finally, the accuracy of simulation is analyzed by experiment [3]. It can provide a basis for the structural optimization of the subsequent oven system [4, 5].

M. Xu et al. (Eds.): CACPP 2022, LNEE 991, pp. 365–374, 2023.
https://doi.org/10.1007/978-981-19-9024-3_46

In recent years, theoretical and practical research on the structure of air nozzle in drying system has been carried out in China, and some research results and literatures have been obtained. Document [6], the oven of gravure press is analyzed and optimized, and the hot air inside the oven is analyzed by fluid mechanics, and the optimal design scheme of the oven is put forward. Document [7], according to the oven of gravure press, the three-dimensional flow model of air nozzle in drying device is established, and the distribution of wind field from air nozzle is obtained. Document [8], the hot air flow field in the air nozzle of the drying oven of the coating machine is simulated and calculated, and the distribution law of the hot air velocity field in the air nozzle is obtained. Document [9], the structure of the air nozzle of the drying device of gravure printing machine was optimized, and the air nozzle model with cross air plates was established. The wind field distribution of air nozzles with different widths of cross air plates was analyzed. Document [10], the impact flow field of the oven nozzle of printing and coating equipment was analyzed, and the relationship between the height and width of the nozzle and the jet velocity near the wall was studied.

2 Oven Structure and Simulation Analysis

2.1 Oven Structure

This paper takes the structure of a certain oven of a certain company as the research object. Use SolidWorks software to establish the three-dimensional model of the oven, as shown in Fig. 1.

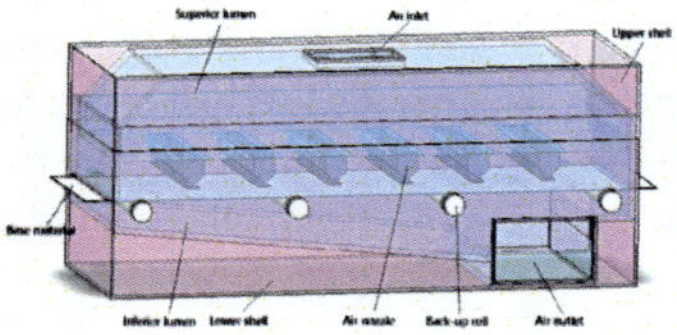

Fig. 1. Three-dimensional diagram of oven

It can be seen from the Figure that the oven structure is made up of two shells, and there is a rectangular opening in the center of the upper shell, it is the air inlet of the oven. There is a rectangular opening at the lower right position in front of the lower shell, it is the air outlet of the oven. The inner layer of the shell is composed of upper and lower inner cavities. A row of air nozzles with the same style and size are installed on the bottom surface of the upper inner cavity, which runs through the inner part of the lower inner cavity. A row of backup rollers are installed above the outlet of the oven, and the base material is between the backup roller and the air nozzle [11]. The air is sent to the air inlet of the oven through the fan, it flows downward, enters the air nozzle through the inlet of the air nozzle, and finally ejects at a high speed from the narrow outlet of the air nozzle to the surface of the base material to dry the base material. The dried exhaust gas continues to flow down and finally is discharged from the air outlet [12].

There are three research objects in this paper, respectively, to study the velocity uniformity of air nozzle inlet, air nozzle outlet and the surface of the base material of the oven. Therefore, when determining the computational fluid domain, it is necessary to combine the research object and the analysis target, and simplify the structure as much as possible on the premise of ensuring the computational accuracy. Through Boolean operation, the fluid domain model finally established in this paper is shown in Fig. 2. For the convenience of the description in the numerical analysis process, the nozzles are numbered, as shown in Fig. 2, and numbered from left to right as nozzles 1–6.

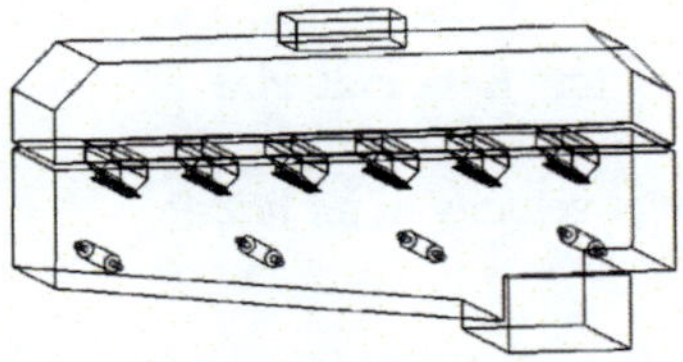

Fig. 2. Oven fluid domain

2.2 Simulation Analysis

Analysis of simulation results. Figure 3 shows the hot air trace diagram of the oven. From the trace diagram, you can clearly see the movement state of hot air inside the oven. The color of the line represents the speed, and the colder the color tone, the smaller the speed, and the warmer the color tone, the larger the speed. The line with arrow indicates the direction, and the density of the line also represents the air volume.

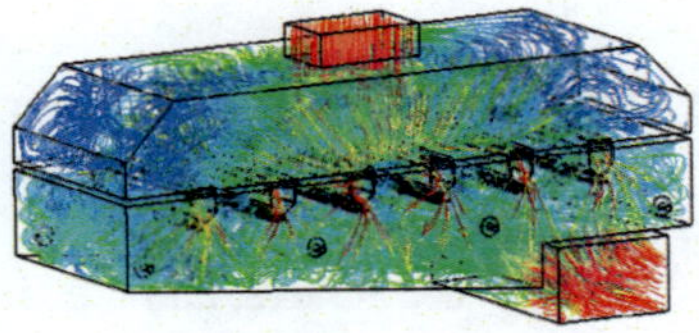

Fig. 3. Trace diagram of oven

Slice the velocity cloud image of the nozzle inlet, nozzle outlet and base material. The slice of velocity cloud image is shown in Fig. 4.

From Fig. 4 (a), we can clearly see the speed of the inlets of air nozzle. The color uniformity represents the speed uniformity, and different colors represent different speeds. Therefore, it can be seen that the inlet speed from the middle to both ends of the tuyere is getting smaller and smaller. The six thin bars in Fig. 4(b) represent the velocity distribution at the outlet of the air nozzle. It can be known that the velocity of No. 3 and No. 4 nozzle is the highest, that of No. 1 and No. 6 nozzle is the lowest, and that of No. 2 and No. 5 nozzle is in the middle. Figure 4(c) shows the cloud image of the surface velocity of base material, which can clearly show the unevenness.

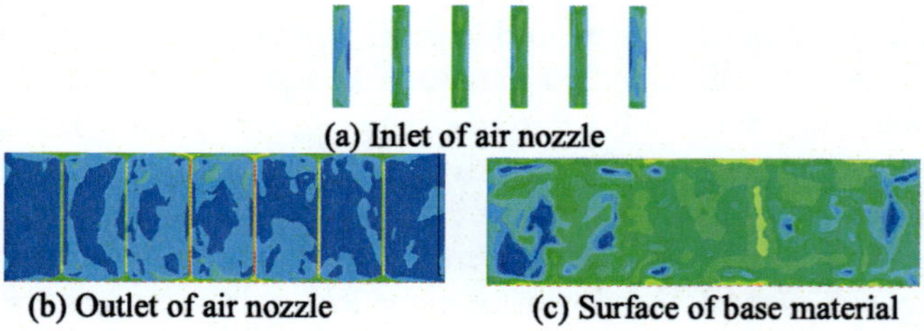

Fig. 4. Slice of velocity cloud picture

Numerical analysis of nozzle and base material

1) Numerical analysis of inlet velocity of air nozzle.

The numerical extraction method of the nozzle inlet is: divide the nozzle width by 4 and the nozzle length by 10, and average the velocity values of each nozzle width and length to obtain the average velocity curve, as shown in Fig. 5.

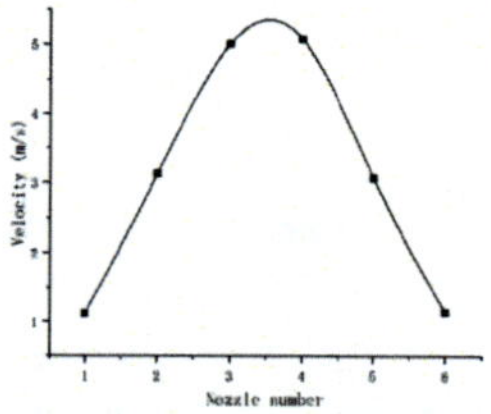

Fig. 5. Graph of average speed

As can be seen from Fig. 5, the average value of each air nozzle inlet of the oven is quite different. No. 3 and No. 4 nozzles have the largest average value, which is about 5 m/s; Second, fifth, around 3 m/s; No. 1 and No. 6 nozzles have the smallest average value, which is about 1m/s. The range is 3.95 m/s, and the air inlet uniformity of the six air nozzles of the oven is very poor.

2) Numerical analysis of outlet velocity of air nozzle.

The numerical extraction method of the nozzle outlet is: Divide the nozzle width by 4 and the nozzle length by 10, and average the velocity values of each nozzle width and length to obtain the average velocity curve, as shown in Fig. 6.

As can be seen from Fig. 6, there is a big difference in the average value of air outlet of each air nozzle of the oven, and the average value from No.1 air nozzle to No. 6 air nozzle increases first and then decreases. The range is 13.26 m/s.

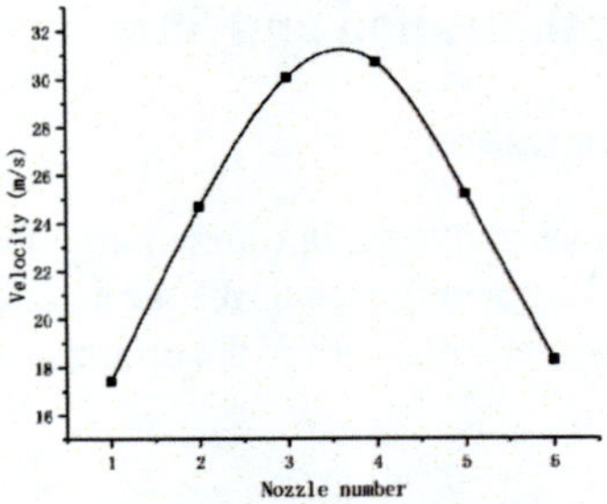

Fig. 6. Graph of average speed

3) Numerical analysis of surface velocity of base material.

In this paper, the velocity values of the surface of base material on each slice were extracted by dividing 10 parts in the longitudinal direction and 6 parts in the transverse direction of the oven. Calculate the average velocity of each point, and finally integrate the values of the six parts to produce the average velocity curve of the substrate surface, as shown in Fig. 7.

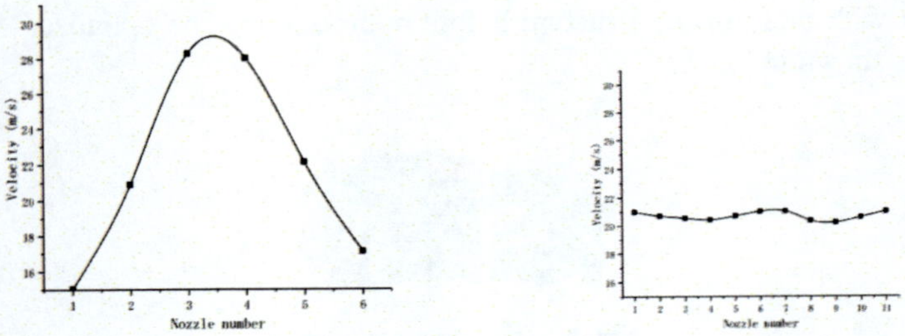

(a) Horizontal (b) Vertical

Fig. 7. Graph of average speed

As can be seen from Fig. 7, the hot air sprayed from the air nozzle acts on the surface of the base material, and the speed of different areas of the base material varies greatly, ranging from 15 to 28.46 m/s, and the longitudinal average speed indicates the average wind speed of the areas opposite to different air nozzles at the same slice position. The average transverse speed indicates the average wind speed at different slice positions in the same nozzle area. It can be seen from the Fig. 7 that the longitudinal average difference is small, basically within 0.5 m/s; The average difference in the transverse direction is large, and the difference between the maximum value and the minimum value is 13.24 m/s, which indicates that the uniformity of the surface velocity of the substrate in the longitudinal direction is good, and the uniformity in the transverse direction needs to be improved.

3 Oven Structure Optimization and Simulation Analysis

3.1 Local Structure Optimization

In this paper, the local structure of oven is optimized. Including adding an air equalizing plate at the inlet, adjusting the opening size of the air nozzle, and chamfering the outlet position (diversion). The schematic diagram of the optimized structure is shown in Fig. 8.

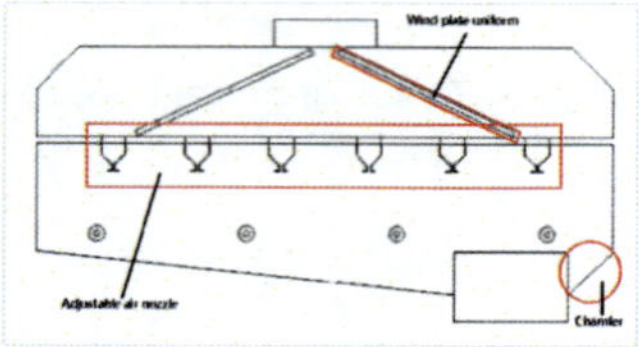

Fig. 8. Schematic diagram of optimized structure

3.2 Comparative Analysis Before and After Optimization

The pretreatment of the optimized model and the setting of boundary conditions are consistent with those of the structure before optimization. The optimized oven trace diagram is shown in Fig. 9.

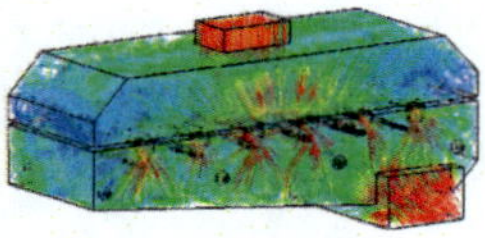

Fig. 9. Trace diagram of oven

Wind speed at the inlet of the nozzle. Figure 10 shows the inlet velocity of air nozzle of the optimized oven.

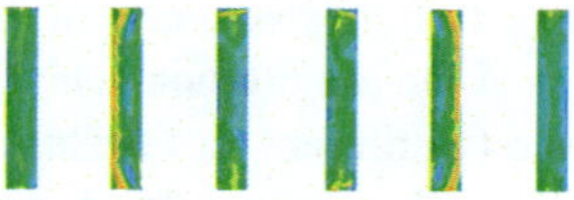

Fig. 10. Image of inlet velocity of air nozzle

It can be seen by comparing Fig. 10 and 4 (a): after optimization, the inlet speed of No. 3 and No. 4 nozzles has decreased, that of No. 1 and No. 6 nozzles has increased, and that of No. 2 and No. 5 nozzles has no obvious change. Compared with the cloud images, it can be clearly and intuitively seen that the optimized cloud images are more uniform

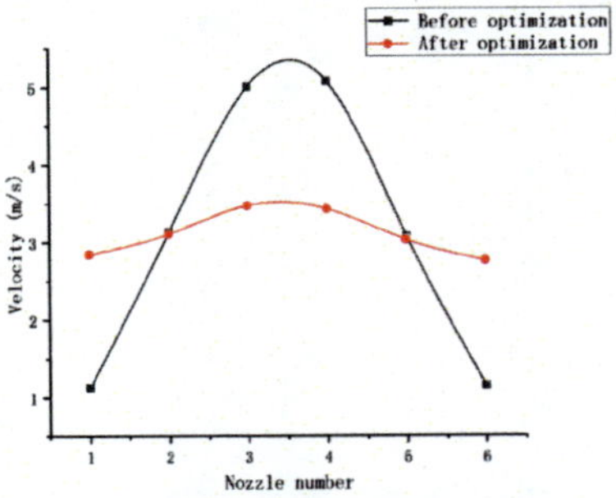

Fig. 11. Comparison of average velocity at inlet of air nozzle

in color, that is, the speed uniformity is greatly improved. The numerical analysis is carried out. The comparison of the average velocity of air inlet is shown in Fig. 11.

It can be seen from the graph that the graph before optimization fluctuates greatly, the middle position is higher, the two ends are lower, and the difference between the highest point and the lowest point is larger; After optimization, the value of the graph is kept at about 3 m/s, and the graph is smoother and more stable, which indicates that the velocity uniformity of the whole air nozzle inlet in the oven is greatly improved after adding the air distributor to the air inlet.

Wind speed at the outlet of the nozzle. Figure 12 shows the outlet velocity of air nozzle of the optimized oven.

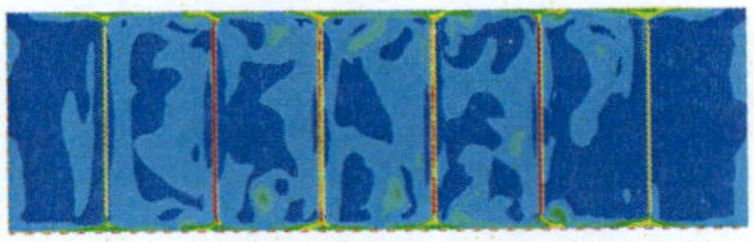

Fig. 12. Image of outlet velocity of air nozzle

It can be seen by comparing Figs. 12 and 4 (b): after optimization, the inlet speed of No. 3 and No. 4 nozzles has decreased, that of No. 1 and No. 6 nozzles has increased, and that of No. 2 and No. 5 nozzles has no obvious change. Compared with the cloud images, it can be clearly and intuitively seen that the optimized cloud images are more uniform in color, that is, the speed uniformity is greatly improved. Through numerical analysis, the comparison of the average velocity at the outlet of the nozzle is shown in Fig. 13.

As shown in Fig. 13. It can be clearly seen from the Fig that the outlet velocity curve of the air nozzles tends to be stable after the optimized structure of the oven, that is, the velocity uniformity is improved, and finally, the outlet velocity of the six air nozzles is maintained at about 22.5 m/s, with moderate velocity and good uniformity.

Surface wind speed of base material. Figure 14 hows the surface velocity of base material of the optimized oven.

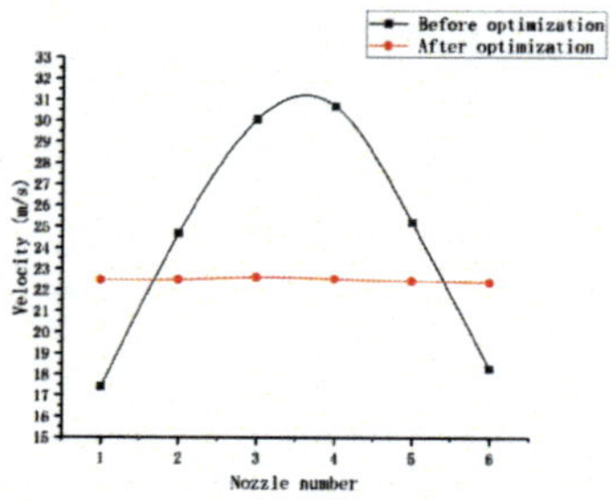

Fig. 13. Comparison of average velocity at outlet of air nozzle

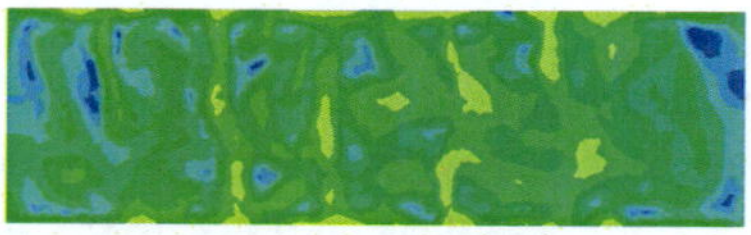

Fig. 14. Images of surface velocity of base material

It can be seen by comparing Fig. 14 and 4 (c): it is found that the surface uniformity of the base material is obviously improved after the optimization of adding the air distributor and adjusting the air nozzle angle. The numerical analysis is carried out. Comparison of average surface velocity of base material is shown in Fig. 15.

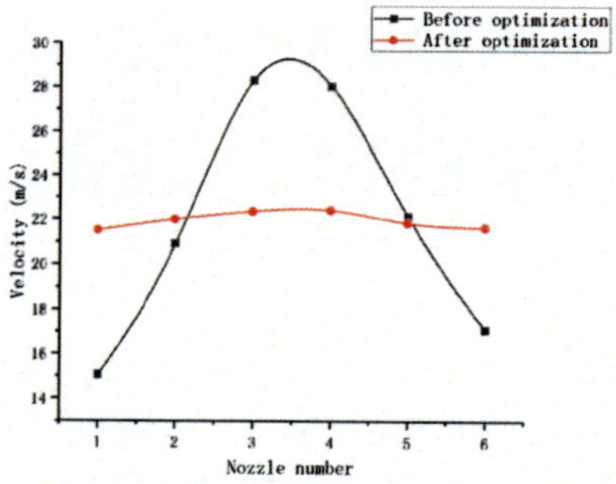

Fig. 15. Comparison of average surface velocity of base material

It can be seen from the Fig. 15 that the velocities in different areas are all maintained at about 21 m/s, and the lateral uniformity of the substrate surface is significantly improved after adding the air distributor to the air inlet and adjusting the angle of the air nozzle.

4 Experimental Verification

In order to verify the correctness of the analysis conclusion of the optimized oven, according to the optimized three-dimensional model of the oven, a small oven experimental device was set up by scaling it down. According to the optimized three-dimensional model of the oven, a small-scale oven experimental device was built by reducing the size of the oven. As shown in Fig. 16. Use the anemometer to measure the velocity distribution at the outlet of the air nozzle and the surface of the base material, collect the data of the research location through the wind speed experiment, as shown in Fig. 17, compare and analyze the experimental measured velocity value with the theoretical extracted velocity value, summarize the experimental data distribution law, and verify that the uniformity of the hot air velocity on the surface of base material of the optimized oven structure has been greatly improved. The distribution law of the experimental velocity field is consistent with that of the numerical simulation.

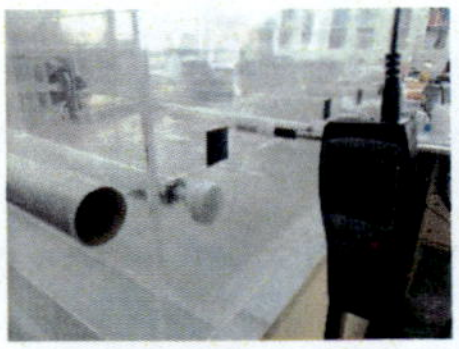

Fig. 16. Oven experimental device

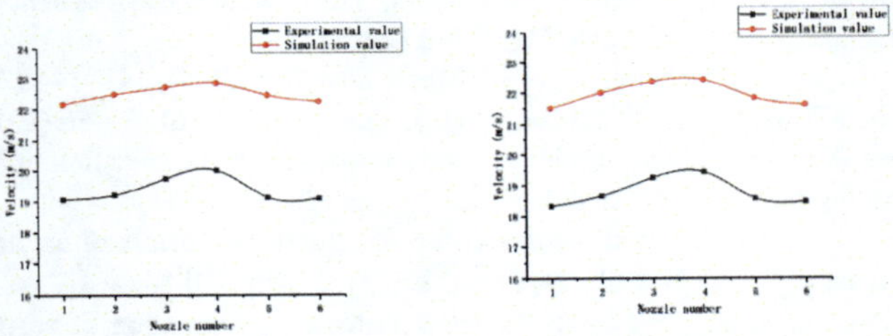

(a) Outlet of air nozzle (b) Substrate surface

Fig. 17. Comparison of -average velocity at outlet of air nozzle and surface of substrate

5 Conclusion

In this paper, the nozzle inlet, nozzle outlet and substrate surface of the oven are taken as the research objects, and the velocity uniformity of these three positions is taken as the research means. The oven is simulated and optimized by means of simulation numerical analysis, and the velocity uniformity of each position is greatly improved by structural optimization. Finally, the hot air reaches the surface of the substrate uniformly, and the average velocity of six areas corresponding to six air nozzles is kept at about 21 m/s, with moderate velocity and good uniformity. In order to improve the drying performance of the oven, the simulation is proved to be accurate and reliable by experiments. The research process of velocity uniformity and the idea of structure optimization in this paper provide a reference for the analysis and structure design of the oven in the future. In this paper, the adjustable air nozzle structure is designed for the first time. According to the state of the wind field in the oven, the openings of the air nozzles at different positions are adjusted to different widths, so as to achieve the purpose of wind field uniformity, which opens up a new idea for the subsequent air nozzle structure design.

Acknowledgements. This research is supported by the Technological Innovation Guidance Plan of Shaanxi Province, China (No. 2020QFY03-02).

References

1. Yanfei, G.: Design of energy-saving air nozzle and oven for gravure printing press. Printing J. **12**, 45–46 (2014)
2. Elsayad, S., Morsy, F., El-Sherbiny, S., et al.: Some factors affecting ink transfer in gravure printing. Pigm. Resin Technol. **31**(4), 234–240 (2002)
3. Wenhua, P.: Working principle and core technology of gravure printing machine. Light Ind. Mach. **28**(3), 10–11 (2006)
4. Rahman, S.M.A., Saidur, R., Hawlader, M.N.A.: An economic optimization of evaporator and air collector area in a solar assisted heat pump drying system. Energy Convers. Manage. **76**, 376–384 (2013)
5. Jian, L., Haiyan, Z., Linlin, L.: Fluid dynamics analysis of wind speed loss in nozzle of gravure printing machine. Light Ind. Mach. **29**(3), 1–3 (2011)
6. Haiyan, Z., Jinjin, L., Zhicheng, X., et al.: Mech. Des. Res. **31**(4), 143–144 (2015)
7. Wu Jimei, XU Zonglei, Chen Yunchun, et al. J. Vibr. Shock **31**(6), 53–56 (2012)
8. Wu, J., Shen, X., Liu, L., et al.: Fluid analysis and parameter optimization of YF93 oven for gravure printing press. J. Vibr. Shock **32**(22), 63–66 (2013)
9. Zhang, H., Liu, J., Xu, Z., et al.: Research on structural parameters of air nozzle cross air plate in gravure drying device. Packag. Eng. **35**(23), 94–95 (2014)
10. Wei, Y., Hou, H., Zhang, H.: Study on oven structure optimization of suspension drying system. J. Xi'an Univ. Technol. **31**(2), 171–172 (2015)
11. Weili, W., Wenge, C.: Research on drying system of gravure printing machine. Packag. Eng. **6**, 98–100 (2008)
12. Huang, Q., Chen, F., Xu, P., et al.: Construction of energy saving and emission reduction efficiency research system of gravure printing press drying system. Packag. Eng. **31**(3), 26–28 (2010)

Environment Detection System of Printing and Packaging Workshop Based on NB-IOT 4GCAT1

Boqi Deng, Peng Liu(✉), Yueyue Xing, and Hao Wan

School of Printing Packaging and Digital Media, Xi'an University of Technology, Xi'an, Shaanxi Province, China
liupeng@xaut.edu.cn

Abstract. According to the new demand of printing standardization and intelligent development, the printing and packaging workshop needs to monitor the temperature and humidity and VOC gas concentration of the printing workshop in time, while the traditional monitoring products based on PLC technology are expensive and difficult to be popularized. This paper takes advantage of NB-IoT technology's wide coverage, low power consumption and low cost, and integrates NB-IOT communication technology, temperature/humidity/volatile toxic gas sensors, cloud services and other industrial IoT technologies to develop a low-power environment monitoring system based on NB-IoT, which greatly reduces the cost while ensuring monitoring accuracy. It can upload the environmental status of the production workshop to the cloud platform for data storage, analysis and processing through NB-IOT or 4G network. Through the mobile phone or system management platform, the temperature, humidity and VOC gas concentration of the printing and packaging workshop can be monitored and alarmed in a timely manner.

Keywords: NB-IOT · 4GCAT · The Internet of things · Printing and packaging · Environmental testing

1 Introduction

The printing and packaging workshop has high requirements for temperature and humidity. No matter whether the temperature and humidity are too high or too low, it will lead to printing problems, more waste products, increased costs. Low humidity will also make equipment wear and tear, and prone to failure. In addition, due to the use of ink and volatile toxic solvents in the production process, it will inevitably produce some toxic and harmful gases, and the concentration exceeding the standard will affect the life safety of personnel. Therefore, it is of great significance to monitor the temperature, humidity and VOC concentration in the printing and packaging workshop. However, the high cost of traditional monitoring tools makes it difficult to spread them.

In order to solve the appeal problem, this paper constructs an environmental detection technology system for the printing and packaging workshop based on NB-IOT

M. Xu et al. (Eds.): CACPP 2022, LNEE 991, pp. 375–383, 2023.
https://doi.org/10.1007/978-981-19-9024-3_47

and 4G CAT1, which is composed of six parts: sensor part, data acquisition part, wireless communication part, cloud processing program, mobile app and client application. Using temperature sensor, humidity sensor, photo ionization gas sensor and other sensing devices, the temperature, humidity and VOC gas concentration in the printing and packaging workshop are detected and alarmed. The data acquisition and control system is composed of MCU circuit and its embedded program. The collected data is packaged and sent to the cloud server through the wireless communication module; The control of wireless communication part is completed by MCU through command word and relevant Internet of things protocol. The cloud processing program and database are designed based on the combination of Swoole and SQL to complete the functions of packet parsing, instruction generation, data storage, etc., and realize the real-time monitoring of the environmental status of the printing and packaging workshop [1, 2].

2 Overall System Design

2.1 System Composition

The system mainly includes the following modules, system power supply, MCU circuit, communication circuit, SIM card circuit, temperature sensor circuit, humidity sensor circuit, optical ionization detector circuit, battery power detection circuit, cloud service unit and power supply system, The overall design scheme is shown in Fig. 1. Through the above circuit combination and software design, the following functions are finally realized.

(1) Temperature and humidity, VOC gas concentration real-time monitoring and abnormal alarm.
(2) Support data transmission of Internet of things protocols such as TCP/MQTT/COAP.
(3) The system has the functions of Automatic wake-up and sleep, and automatically wakes up the uploaded data.

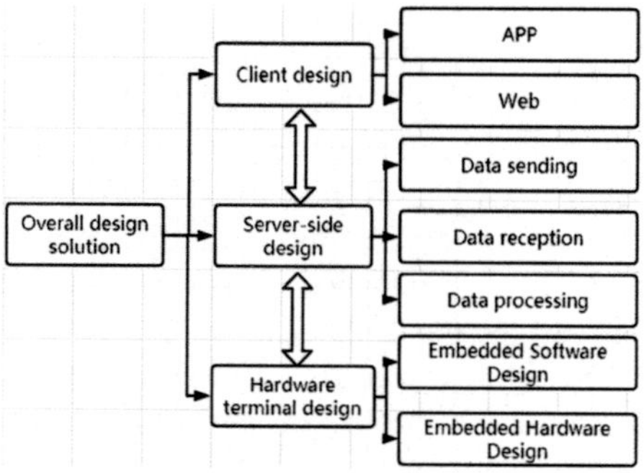

Fig. 1. System framework

2.2 Measuring Principle

Principle of Temperature and Humidity Measurement. The DHT11 is an integrated wet and temperature digital sensor. The sensor consists of a resistive humidity measurement element and an NTC temperature measurement element, connected to a high performance 8-bit microcontroller. The DHT11 can communicate with the microcontroller using a simple single bus, requiring only one I/O port. The sensor's internal humidity and temperature data is transmitted to the microcontroller in a single pass at 40 Bit, and the data is check summed to ensure the accuracy of the data transmission [3]. The sensor has integrated signal acquisition and processing circuitry, providing three pins GND, VCC and DATA, connected to the ground and power supply lines, only the DATA pins need to be connected to the sensor's IO port with a Dupont wire, the microcontroller can receive data in the serial port when working [4].

Principle of VOC Gas Measurement. The system use PID-TECH photo ionization detection method to detect VOC gas. The ultraviolet lamp is used as its energy source to ionize the incoming gas inside the photoionization gas sensor. The incoming VOC gas will be ionized into corresponding ions and electrons after being irradiated by the ultraviolet lamp. After the ionized ions and electrons pass through the electric field generated by the internal electrode, the electrons move towards the signal electrode, and the ions move towards the polarization electrode connected to the high-voltage DC source. During the movement, a current will be formed, and the generated current will be amplified by the subsequent amplifier. The output analog signal is the concentration of the incoming VOC gas [5].

3 Circuit Design

3.1 System Power Circuit Design

As shown in Fig. 2, the VBAT in this system is the lithium battery access voltage, which is a direct power supply to the MCU. Based on the sensor's need for a stable operating voltage, the RT9013-33GB model regulator is used to design the step-down circuit shown in the figure to generate a stable 3.3 V voltage for the sensor.

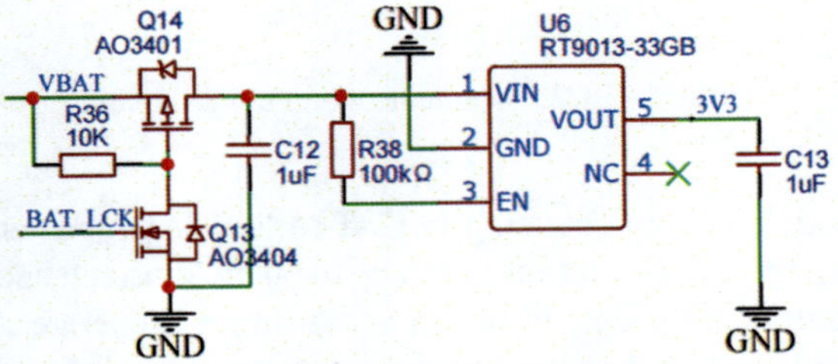

Fig. 2. Power buck circuit

3.2 MCU Circuit Design

ATMEGA328P-AU, the core processor is AVR, which adopts RISC architecture, with a clock frequency of 20 MHz and a power supply voltage input range of 1.8~5.5 V. As shown in Fig. 3, its internal 8-channel analog-to-digital conversion input is very suitable for the detection of temperature, humidity and VOC concentration in this design [6].

Fig. 3. MCU circuit design

3.3 Communication Circuit Design

Communication Module Circuit. The NB-IOT module adopts the BC25 model module, which uses a single power supply with a power supply range of 3.0–4.3 V, and has two pins as power input. Power is supplied to the internal RF and baseband circuits via these two pins (Fig. 4).

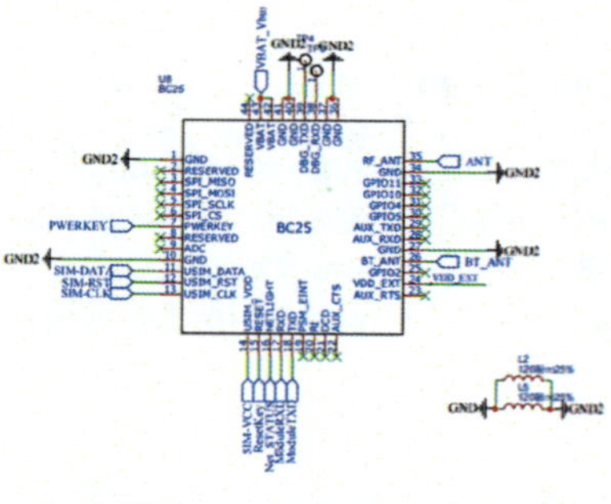

Fig. 4. BC25 communication module circuit

When the module is only operating in CAT-M, NB-IoT mode and transmitting at maximum power, the peak current can reach up to 600mA instantaneously. The module has low power consumption, high sensitivity and is simple to operate. All that is required is to set the address and port number of the server to be uploaded, and the VOC gas concentration values monitored by the VOC monitor can be sent to the module via the serial port, where the data is processed and uploaded [7].

SIM Card Circuit Design. The data shown in Fig. 5, where SIM_DATA: SIM bus data; SIM_RST: SIM bus reset output. SIM_CLK: SIM bus clock output; SIM_VCC: SIM card circuit power supply [8].

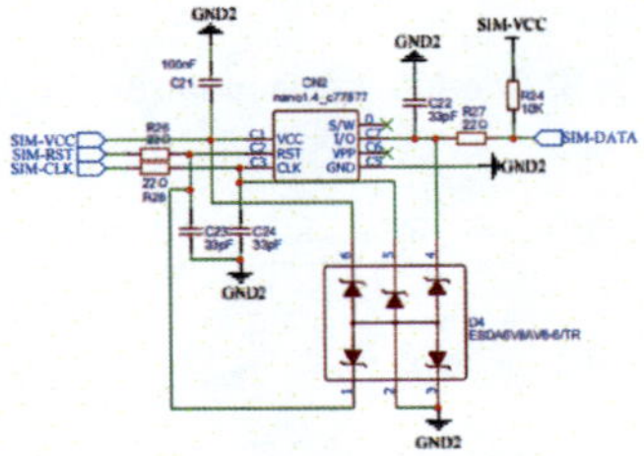

Fig. 5. Sim card circuit design

Module Power-Up and Module Restart. PWRKEY: Switching control input, Active low, the maximum value of the active low level of the input is 0.5 V. Used to control the power on/off of the module (Figs. 6 and 7).

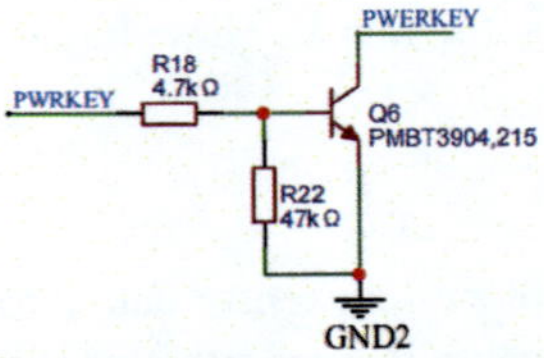

Fig. 6. Module power-up circuit

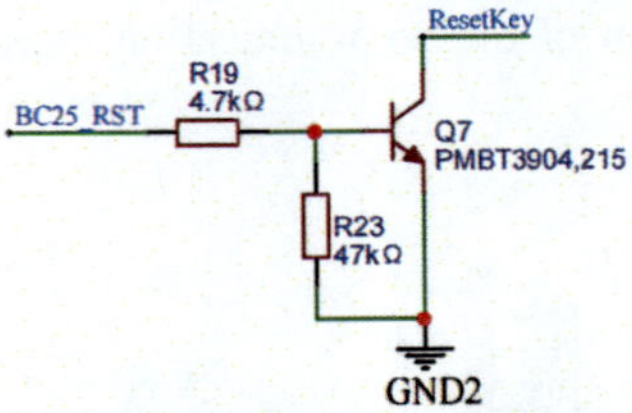

Fig. 7. Module restart circuit

Module Antenna, Network Indicator and Power Control Circuit. In order toreduce the power consumption of the system and extend the usage time of the device, the module power control circuit is designed so that it will only power up and send data when the module wakes up. In order to clearly determine whether the module is online and powered on, the network indicator circuit is designed so that when the power is on and waiting for the module to be online, the indicator status is green flashing. When the module is successfully online, the indicator light will be green and always on. The antenna circuit is an airtight room with a certain shielding effect on the signal (Figs. 8, 9 and 10).

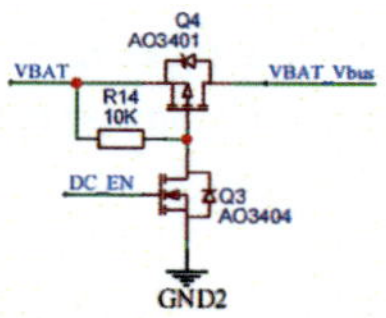

Fig. 8. Module power control circuit

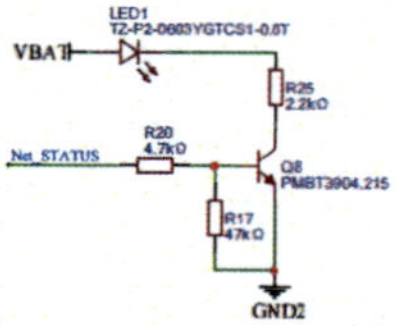

Fig. 9. Indicator light circuit

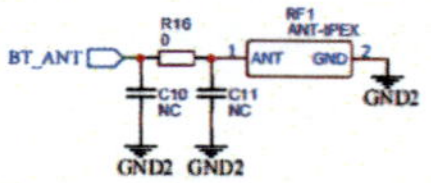

Fig. 10. Module power-up circuit

3.4 Sensor Interface Circuit

The system collects LM335 temperature sensor data through SPI communication interface. The data of PID-TECH photo ionized gas sensor is collected through analog-to-digital conversion interface [9, 10]. For the 4GCAT1 or NB-IOT communication module, a unified interface circuit unit is used to send at commands through the serial port to control the data transmission of the communication module. This series of interface circuits is shown in Fig. 11.

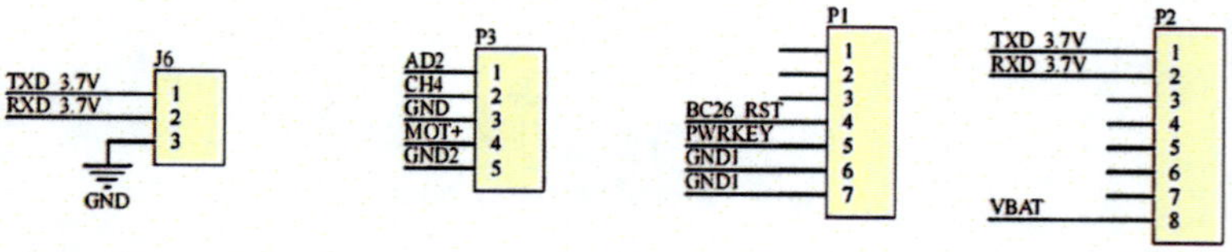

Fig. 11. Sensor interface circuit

4 System Software Design

The main functions of the software are equipment communication, registration, management and real-time data display. Specific to the software Support for two-way communication of IoT devices, including uplink of device data and downlink of cloud device control commands. Provide device side SDK, fast connection equipment to the cloud, while supporting remote equipment access, heterogeneous network equipment access, multi-environment equipment access, multi-protocol access. Supports device ID input, device installation latitude and longitude automatic input. The communication between the data on the hardware and the server is implemented in Fig. 12, where the data is uploaded to the server, then stored in the RDS database and forwarded to the MQ queue service. Grasp the real-time temperature, humidity and VOC gas concentration data in the APP interface shown in Fig. 14. When the temperature, humidity or VOC gas concentration exceeds the specified value, the alarm information is pushed and can be viewed in the records in the navigation bar below the APP (Fig. 13).

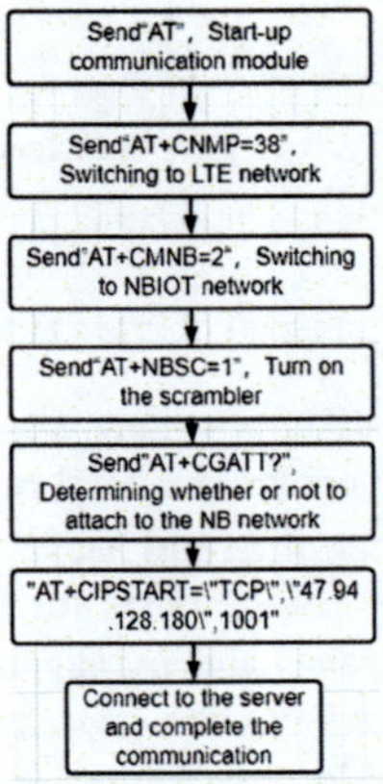

Fig. 12. Communication between modules and servers

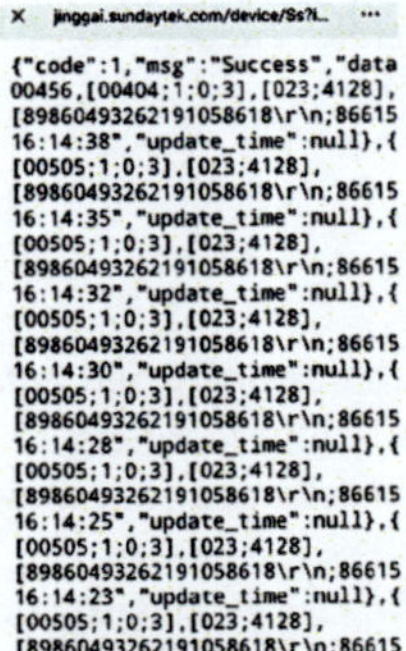

Fig. 13. Data of server

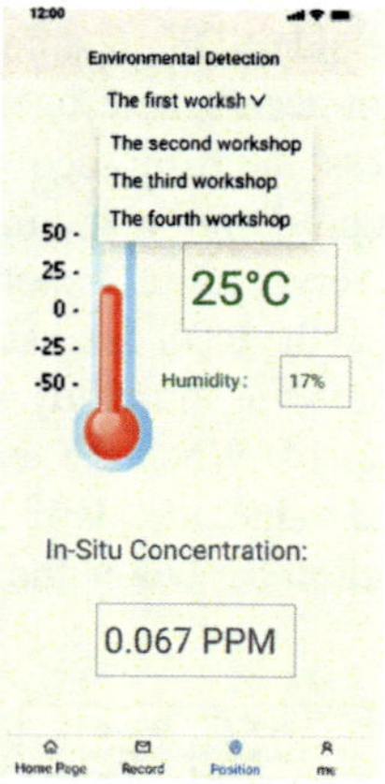

Fig. 14. APP interface

5 Conclusion

This paper designs a set of printing and packaging workshop based on NB-IOT, 4GCAT1 technology and cloud platform. The environmental monitoring system realizes the real-time monitoring of temperature, humidity and VOC gas concentration in the workshop by mobile terminal, and also provides the basis for large-scale multi-table integration in the workshop. It not only improves the management efficiency of enterprises for workshop environment, but also reduces the probability of accidents. Therefore, this technical scheme provides a feasible way for the informatization and intelligence of environmental monitoring technology in printing and packaging workshop.

Acknowledgements. This research is supported by the Technological Innovation Guidance Plan of Shaanxi Province, China (No. 2020QFY03-02).

References

1. Wang, Z.: A brief discussion on temperature and humidity control in printing workshops. Shanghai Packag. **08**, 44–46 (2019). https://doi.org/10.19446/j.cnki.1005-9423.2019.04.011
2. Su, P.: The influence of ambient temperature and humidity on printing. Printing Technol. **12**, 52–53 (2008)
3. Wang, S., Liu, Q., Yang, J., Ge, X.: Design and experimental study of a ground-based meteorological measurement temperature sensor. Sens. Microsyst. **41**(06), 25–28 (2022). https://doi.org/10.13873/J.1000-9787(2022)06-0025-04
4. Xu, J.Z., Sun, H.: Design and implementation of an Arduno-based system for automatic soil moisture regulation of flowers. Electron. Prod. **30**(11), 20–22+26 (2022). https://doi.org/10.16589/j.cnki.cn11-3571/tn.2022.11.003
5. Lu, H.J.: Research on the online monitoring system of volatile organic gas based on NB-IoT. Shandong Institute of Industry and Commerce (2019)

6. Hao, J.: Research and design of multi-mode current-mode controlled boost converter. Xi'an University of Electronic Science and Technology (2020).https://doi.org/10.27389/d.cnki.gxadu.2020.003499
7. Kai, Z., Yang, O., Liu, J., Tian, Z.: Design of smart meter monitoring system based on NB-IoT wireless communication. Electron. Des. Eng. **30**(11), 131–134+139 2022. https://doi.org/10.14022/j.issn1674-6236.2022.11.028
8. Wang, G., Wang, H. Tang, Y.: Temperature monitoring system for data centre based on NB-IoT technology. Ind. Control Comput. **35**(05), 42–43 (2022)
9. Li, Y., Chen, K.: A design of an improved monitoring system based on Internet of Things. J. Shaoguan College **43**(03), 39–42 (2022)
10. Xingping, S.: NB-IoT-based manhole cover missing warning and monitoring system. Integr. Circ. Appl. **38**(12), 64–65 (2021). https://doi.org/10.19339/j.issn.1674-2583.2021.12.023

Heat Transfer Analysis of Regenerative of VOCs Treatment Equipment in Printing and Packaging Enterprises

Yueyue Xing, Peng Liu(✉), Pengchao Dou, and Hao Wan

School of Printing Packaging and Digital Media, Xi'an University of Technology, Xi'an, Shaanxi Province, China
liupeng@xaut.edu.cn

Abstract. Regenerative Thermal Oxidizer is the main equipment to treat VOCs waste gas in printing and packaging enterprises, and the regenerator in regenerator chamber is the key component that affects the efficiency of equipment. Therefore, it is of great significance to analyze the heat transfer performance of regenerator. In this paper, SolidWorks software is used to establish the three-dimensional model and fluid domain model of the regenerator. The internal flow and heat transfer characteristics of the regenerator are studied by the local thermal non-equilibrium (LTNE) model of porous media in the software. The change of flue gas and air outlet temperature with time in the start-up stage of the regenerator is mainly simulated and analyzed; the influence of the length, pore size and wall thickness on the heat transfer performance of the regenerator, such as the outlet temperature, temperature efficiency and heat storage capacity. The effects of various parameters on the heat transfer performance of the regenerator are obtained. The numerical simulation results can be used for the structural optimization design of Regenerative Thermal Oxidizer.

Keywords: CFD · Regenerator · Porous medium · Heat transfer performance

1 Introduction

One of the equipments that can best meet the environmental protection requirements of VOCs treatment in printing and packaging enterprises is Regenerative Thermal Oxidizer (RTO), which mainly cools the high-temperature purified gas and preheats the VOCs waste gas through the regenerator inside the regenerator chamber [1]. Therefore, the regenerator is the key component of RTO, which accumulates in the regenerative oxidation reactor and directly contacts with high-temperature purified gas or normal-temperature VOCs waste gas to periodically absorb heat and release heat [2]. There are many factors that affect the heat transfer performance of regenerator, such as the shape, size and material of holes in regenerator, etc. Because of experimental research, it is difficult to obtain the parameters of velocity, temperature, pressure and so on in each position [3]. With the help of Computational Fluid Dynamics, this paper focuses on the

M. Xu et al. (Eds.): CACPP 2022, LNEE 991, pp. 384–392, 2023.
https://doi.org/10.1007/978-981-19-9024-3_48

heat transfer performance of regenerator [4]. As the research content, numerical simulation is carried out with Fluent software [5]. So as to obtain the factors that affect the heat transfer and flow characteristics of the regenerator, which has guiding significance for the design of regenerator chamber of Regenerative Thermal Oxidizer in enterprises.

2 Model and Boundary Conditions

2.1 Physical Model

The type of regenerator selected in this paper is quadrangular honeycomb ceramic regenerator, which is a common type of regenerator in RTO. The specification of a single regenerator is 150 * 150 * 300, the number of holes is 43 * 43, and the size of holes is 2.9 mm * 2.9 mm. In this paper, a quarter of the section of a single regenerator is selected as the original model, Fig. 1 is the original model established by SolidWorks. Because the original model has a big difference between the pore size and the overall size, and it is difficult to mesh, it is equivalent to a porous media model, Fig. 2 is the equivalent model.

Fig. 1. Original model

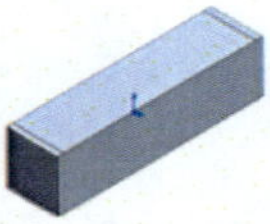

Fig. 2. Equivalent model

For the heat transfer model in porous media, the earliest scholars assumed porous media [6]. There is no heat exchange between the internal fluid phase and the solid phase, so the Local Thermal Equilibrium (LTE) model is constructed. The LTE model is also called the single equation model, which has the advantages of simple model and simple calculation [7]. Then some scholars suggested that if the temperature difference between the fluid phase and the solid phase in porous media is too large to be ignored, the LTE model is no longer applicable, and it is necessary to build energy equations for the fluid phase and the solid phase respectively, that is, the Local Thermal Non-Equilibrium (LTNE) model. LTNE model is also called two-equation model. Under this model, there is temperature difference and heat transfer between solid phase and mobile phase in porous media area [8]. Literature [9] points out that when the thermal conductivity of porous media solid is greater than that of fluid, LTNE model is more appropriate.

2.2 Mathematical Model

Combined with the porous media model and the coupling problem of fluid flow and heat transfer, the thermodynamic control equations inside the regenerator can be expressed as:

Mass conservation equation

$$\frac{\partial(\rho_f)}{\partial t} + \nabla \cdot (\varepsilon \rho_f \vec{u}) = 0$$

Momentum conservation equation

$$\nabla p = -\frac{\mu}{\alpha} \vec{u}$$

Energy conservation equation

$$\varepsilon(\rho c)_f \frac{\partial \vec{T}_f}{\partial t} + (\rho c)_f \vec{u} \cdot \nabla \vec{T}_f = \varepsilon k_f \nabla^2 \vec{T}_f + h(\vec{T}_s - \vec{T}_f)$$

$$(1 - \varepsilon)(\rho c)_s \frac{\partial \vec{T}_S}{\partial t} = (1 - \varepsilon) k_S \nabla^2 \vec{T}_S + h(\vec{T}_S + \vec{T}_f)$$

The $\vec{u}, p, \vec{T}_f, \vec{T}_s$ represents the velocity, pressure and temperature of fluid in porous media, and subscripts f and s represent fluid phase and solid phase respectively. $k, \varepsilon, \mu, \rho, c$ represents the thermal conductivity, porosity, fluid viscosity, density and specific heat capacity of porous media respectively.

2.3 Boundary Conditions

Heat release stage of regenerator: the gas inlet is a velocity inlet with a velocity of 1.5m/s and a temperature of 300 K, and the turbulence intensity and hydraulic diameter are set according to the fluid domain model. The outlet is set as pressure outlet, turbulence intensity and hydraulic diameter are set, and the wall is set as the thermal insulation wall.

Heat storage stage of regenerator: the gas inlet is a velocity inlet with a velocity of 1.5 m/s and a temperature of 1073.15 K The turbulence intensity and hydraulic diameter are set according to the fluid domain model. The outlet is set as pressure outlet, turbulence intensity and hydraulic diameter are set, and the wall is set as the thermal insulation wall.

3 Numerical Simulation and Result Analysis

3.1 Analysis of Start-Up Process of Regenerator

The temperature changes of flue gas and air outlet at the start-up stage are shown in Figs. 3 and 4.

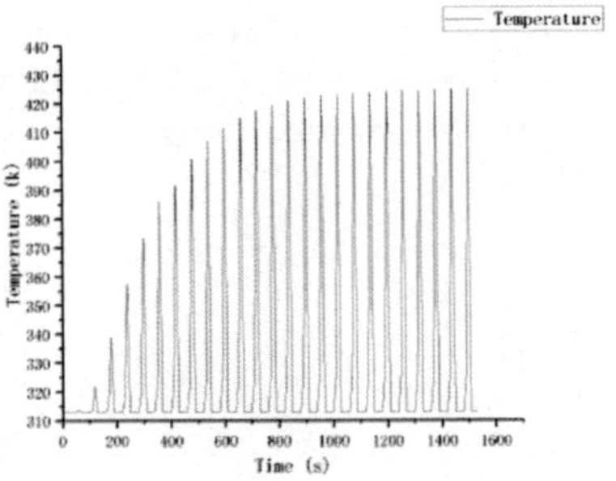

Fig. 3. Instantaneous temperature of flue gas outlet in start-up process

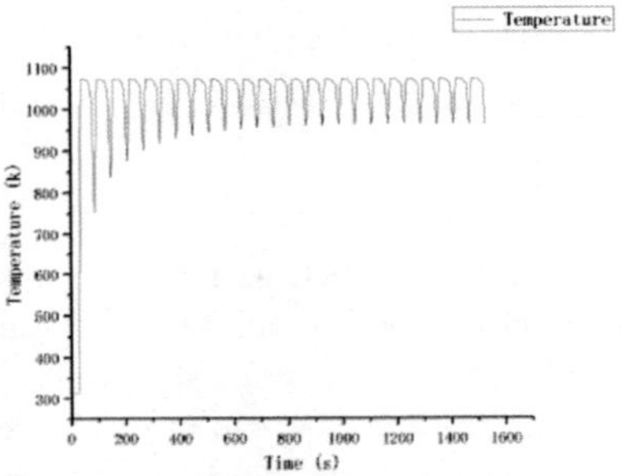

Fig. 4. Instantaneous temperature of air outlet in start-up process

As can be seen from the Figure, the flue gas outlet temperature gradually increases with the passage of time, and reaches a stable level after a period of time. At first, the temperature difference between the gas and the regenerator is large, the convective heat transfer is strong, and the flue gas outlet temperature is extremely low. With the increase of heat exchange times, the temperature difference gradually decreases, and the convective heat transfer gradually weakens. It can be seen from the Figure that the outlet temperatures of air and flue gas tend to be stable around 800s. The regenerator enters a stable operation state.

3.2 Analysis of Stable Operation Stage of Regenerator

Heat Storage Stage of Regenerator. As shown in Fig. 5, the temperature changes of various substances in the gas heat release stage are shown. Figure (a) shows the gas temperature changes in the regenerator channel at different times in a stable regenerator cycle. As the inlet temperature of different cycles is greatly affected by the boundary conditions, the curve range in the Figure is from the inlet to the outlet of the channel. As the heat absorption temperature of the regenerator rises during the gas heat release period, the heat released by the gas to the regenerator decreases with time, so the gas temperature in the channel tends to increase with time, and the solid temperature also tends to increase with time. At the same time, the gas outlet temperature increases with time, and the temperature changes by 100 K, because the regenerator absorbs the heat of the gas and heats up. As the regenerator absorbs heat and heats up, its heat storage capacity decreases, so the excess heat is discharged from the regenerator with flue gas, and the gas outlet temperature increases accordingly. Similarly, the solid temperature increases with time.

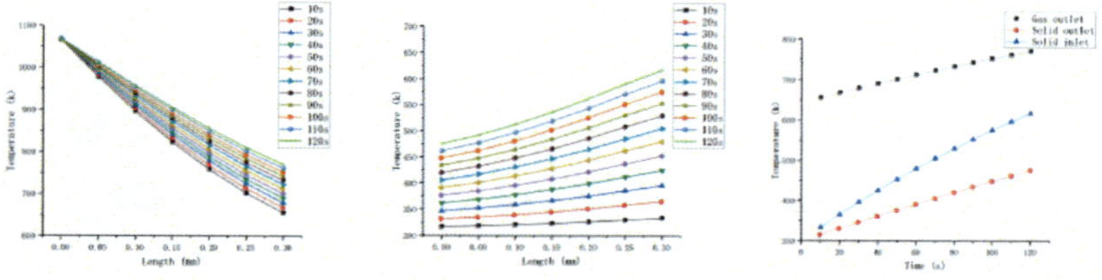

a.Curves of gas temperature b. Curves of solid temperature c.Curve of temperature

Fig. 5. Numericaysis results of regenerator in heat storage stage

Exothermic Stage of Regenerator. Figure 6 shows the temperature changes of various substances in the gas heating stage, and Fig. 6a shows the gas temperature changes in the regenerator channel at different times in one cycle of the stabilized regenerator. As above, since the inlet temperature of each cycle is greatly affected by the setting of boundary conditions, the curve range in the Figure is from the inlet to the outlet of the channel. In the heating stage, the regenerator releases the gas heat stored in the previous stage to the low-temperature VOCs waste gas, and the waste gas is heated continuously, so the heat absorption of the waste gas decreases, and the gas temperature in the channel decreases with time, and the solid temperature also decreases with time. At the same time, the gas outlet temperature decreases with time, and the temperature changes by 50 K, the reason is when the gas enters the regenerator, the gas will absorb the heat of the regenerator and heat up. As the gas absorbs the heat and heats up, the heat release capacity of the regenerator decreases. Thus the excess cold gas is directly discharged from the regenerator without absorbing heat, and the gas outlet temperature decreases accordingly. Similarly, the solid temperature increases with time. The inlet and outlet temperatures of solids also decrease with time.

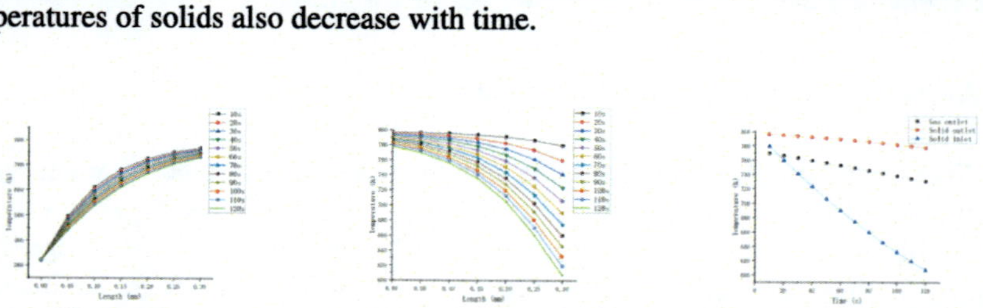

a.Curves of gas temperature b.Curves of solid temperature c.Curve of temperature

Fig. 6. Numerical analysis results of regenerator in heat release stage

4 Effect of Key Parameters of Heat Transfer Performance of Regenerator

In this chapter, taking 1.5 m/s gas velocity inlet, standard atmospheric pressure outlet, adiabatic wall and initial temperature of regenerator is 300 k as the basic settings, the influence of various parameters of regenerator on heat transfer and flow performance of regenerator during heat storage period is studied.

4.1 Regenerator Length

Draw the line chart of the relationship between regenerator length and outlet temperature, pressure drop, temperature efficiency and heat storage capacity as shown in Fig. 7.

a.Curves of outlet temperature and pressure drop

b.Curves of heat storage capacity and temperature efficiency

Fig. 7. Effect of regenerator length on performance

It can be seen from the Figures that with the increase of the length of the regenerator, the outlet temperature gradually decreases, the pressure drop increases, the temperature efficiency decreases, and the heat storage capacity decreases. The reason is with the increase of the length, the longer the gas stays in the regenerator, the larger the heat exchange area, the more sufficient the gas-solid heat exchange, and the lower the outlet temperature, that is, the higher the temperature efficiency. Because the heat storage capacity of the regenerator is certain, the increase of heat exchange area is greater than the increase of heat storage capacity, so after the length of the regenerator increases to a certain value, the outlet temperature changes slowly, and the temperature efficiency no longer increases. When the length of the regenerator is more than 600 mm, the outlet temperature is extremely low and close to the inlet temperature, and its temperature efficiency is as high as 99%. When the length of the regenerator is between 150 and 1500 mm, the pressure loss of the gas with the heat storage capacity between 4.7 and 83.7 kJ. The pressure drop in the process of gas flow is proportional to the length of its channel. When the gas inlet velocity is constant, the longer the length of regenerator, the greater the pressure drop change, and the smaller the pressure drop change. When the length of the regenerator is between 150 and 1500 mm, its pressure loss varies between 20.76 and 207.99 Pa.

4.2 Regenerator Pore Size

Draw the line chart of the relationship between the pore size of regenerator and outlet temperature, pressure drop, temperature efficiency and heat storage capacity as shown in Fig. 8.

a.Curves of outlet temperature and pressure drop

b.Curves of heat storage capacity and temperature efficiency

Fig. 8. Effect of regenerator pore size on performance

It can be seen from the Figures that with the increase of the side length of the small hole of the regenerator, the outlet temperature gradually increases, the pressure drop decreases, the temperature efficiency decreases, and the heat storage capacity increases. The reason is when the porosity and flow rate are fixed, with the increase of the pore size, the heat transfer area of the regenerator decreases, its heat transfer capacity weakens, and the heat transfer becomes insufficient, so the outlet temperature value of the regenerator increases and the temperature efficiency decreases. When the pore size is less than or equal to 3 mm, the outlet temperature and temperature efficiency change slowly, but when the pore size is longer than 3mm, the outlet temperature and temperature efficiency change sharply. When the pore size is between 2.5 and 4.5 mm, the heat storage capacity increases from 13.91kJ to 27.7kJ. When the diameter of the regenerator becomes larger, the number of holes will be smaller. At this time, when the gas flows through the regenerator, the change of the flow cross-section is smaller, and the friction and viscosity between the gas and the regenerator channel are smaller, so the pressure drop is smaller. When the pore size of the regenerator is between 2.5 and 4.5 mm, its pressure loss varies between 9.56 and 69.52 Pa.

4.3 Regenerator Wall Thickness

Draw the line chart of the relationship between the wall thickness of regenerator and outlet temperature, pressure drop, temperature efficiency and heat storage capacity as shown in Fig. 9.

It can be seen from the Figures that with the increase of the wall thickness of the small hole of the regenerator, the outlet temperature gradually increases, the pressure drop increases, the temperature efficiency decreases, and the heat storage capacity increases.

a. Curves of outlet temperature and pressure drop

b.Curves of heat storage capacity and temperature efficiency

Fig. 9. Effect of regenerator wall thickness on performance

The reason is with the increase of wall thickness, the heat that the regenerator can store increases, and the gas with the same speed can release more heat when flowing through the regenerator, but the heat exchange area decreases. Under the comprehensive influence, the outlet temperature increases and the temperature efficiency decreases. The wall thickness increases the heat storage capacity of the regenerator to a certain extent. When the wall thickness increases from 0.5 to 1.05 mm, the outlet temperature increases from 321.21 to 327.37 k, and the temperature efficiency increases from 98.1 to 98.9%. Heat storage capacity increased from 16.6 to 27.9 kJ. When the inlet gas flow rate is constant, the thicker the wall thickness of regenerator orifice, the greater the pressure loss, and the smaller the wall thickness of regenerator orifice, the smaller the pressure loss. When the hole wall thickness of the regenerator is between 0.5 and 1.05 mm, its pressure loss varies between 41.52 and 291.14 Pa.

5 Conclusion

In this paper, through modeling and simulation of the quadrilateral honeycomb ceramic regenerator, the temperature changes in the start-up stage and the stable operation stage of the regenerator are simulated. It is concluded that the time from the start-up of regenerator to stable operation is about 800 s. In the heat storage stage of the regenerator, the flue gas temperature is always higher than that of the regenerator, and in the heat release stage of the regenerator, the air temperature is always lower than that of the regenerator.

Acknowledgements. This research is supported by the Technological Innovation Guidance Plan of Shaanxi Province, China (No. 2020QFY03-02).

References

1. Liu Kai, Y., Hang, Y.Y.: Discussion on the transformation of two-chamber regenerative incinerator to meet the emission standards. Zhejiang Chem. Ind. **52**(12), 32–35 (2021)
2. Limiao, H., Feihui, H., Yongwei, W., Sixun, N., Guoyuan, C., Yanling, X.: Research status and application of honeycomb ceramics. Foshan Ceram. **31**(06), 32–39 (2021)
3. Guo, B., Wang H., et al.: Numerical simulation technology and application progress of volatile organic compounds treated by regenerative oxidizer. Mod. Chem. Ind. **38**(07), 44–47+49 (2018)
4. Yingjie, H., Zhiqiang, W., et al.: Chem. Ind. Eng. Prog. **37**(01), 319–329 (2018)
5. Heping, S., Jianxin, W., Zhang Qingyu, W., Qiming, W.J.: Numerical simulation analysis of regenerator and heat exchange time. Sci. Technol. Eng. **21**(04), 1382–1387 (2021)
6. Bo, W., Menghui, L., Qinglong, S., Chengxi, S.: Numerical simulation of resistance characteristics of honeycomb ceramic regenerator. Therm Power Eng. **35**(07), 152–158 (2020). https://doi.org/10.16146/j.cnki.rndlgc.2020.07
7. Zhang, Q.: Study on heat transfer characteristics of regenerator in regenerative burner. Inner Mongolia University of Science and Technology (2020).https://doi.org/10.27724/D.CNKI.GNMGK.20008.000000000005
8. Hou, H., Chen, X., Jassamyn, L., Xu, Z.: Study on the performance of regenerator of regenerative oxidizer. Digital Printing (02), 66–70+92 (2019). https://doi.org/10.19370/j.cnki.cn10-1304/TS.2019
9. You, X.: Especially prosperous study on the influence of boundary conditions and layered filling on flow and heat transfer in porous media channels under local non thermal equilibrium conditions. Huazhong University of Science and Technology (2016)

Research on Unwind and Splicing Based on Polynomial Interpolation

Zhijiang Yang[1(✉)], Fan Feng[1], and Hui Wang[2]

[1] Xi'an Aerospace-Huayang Mechanical and Electrical Equipment Co., Ltd, Xi'an, China
zhijiangyang@163.com

[2] China Academy of Aerospace Liquid Propulsion Technology, Xi'an, China

Abstract. Aiming at the problem that the initial roll diameter of the new web axis and the unwind speed after splicing cannot be determined due to the change of web diameter when the axis of the unwind unit of the flexographic press is switched. In this paper, firstly, the web diameter corresponding to the rotation angle of the cantilever is obtained by the simulation of CAD on the basis of keeping the angle of the splicing web and the cutter unchanged. Secondly, a polynomial interpolation algorithm for the rotation angle and web diameter of the cantilever is proposed. Newton interpolation polynomial algorithms are used respectively to analyze the functions obtained by different polynomial degrees, and the optimal approximate curve is obtained. Finally, the experimental verification is carried out on the production site. The simulation and experimental results show that the proposed algorithm can obtain the accurate value of the initial web diameter only according to the rotation angle of the cantilever when the new axis is pre-driven. The new method replaces the ultrasonic sensor detection unit, solves the problem of determining the unwind speed of the unwind cantilever, and avoids the tedious process of making the test mould. It reduces the waste of the material film, saves the debugging cost, optimizes the unwind process, realizes the functions of calculating the initial diameter and the pre-driven speed of the unwind.

Keywords: Flexographic press · Unwind and splicing · Web diameter detection · Newton interpolation algorithm

1 Introduction

When the pre-driven and non-stop splicing on unwind is adopted, in order to ensure new unwind axis precise speed, it is necessary to use ultrasonic sensor to detect on the new web diameter to obtain the same linear speed [1, 2], which will lead to the production cost higher. Meanwhile, ultrasonic signal is easy to be interfered, it makes the detected web diameter error.

New web diameters are frequent changing which is difficult during debugging site. Unwind new web diameter will directly influence the splicing success rate. If provided new web diameter has any deviation, it will influence the web tension and even result in failure of web splicing, which effects the whole machine running. Therefore, in order to

M. Xu et al. (Eds.): CACPP 2022, LNEE 991, pp. 393–401, 2023.
https://doi.org/10.1007/978-981-19-9024-3_49

maximize the splicing success rate and printing quality to largest extent, it is necessary to study the calculation of unwind splicing new web diameter.

This paper proposal an algorithm to calculate the new web diameter by detecting the rotation angle of the unwind turret, which will simplify the current web diameter detection method.

2 Unwind Turret Rotation Position Model and Geometric Relationship Derivation

Unwind turret rotary angle is associated with new axis web diameter. Referring to Fig. 1, new axis rotating from B to the position under the nip roller, the sensor detects web and stop on the B', then finish splicing. Unwinding turret rotary angle is detected by rotational potentiometer. The new axis rotation angle is associated with the new axis diameter. In order to find the mapping relationship between them, polynomial interpolation method can be used. The data measured from experiment are used as interpolation points, by calculating the polynomial of the interpolation point.

In summary, this paper uses polynomial interpolation method to calculates the web diameter according to the know rotation angle of unwind turret to carry out the correct splicing. Firstly, standard CAD model is drawn to find the corresponding new web diameter data of the rotation angle of unwind turret, then the geometric derivation method is adopted to obtain the corresponding relationship. Thirdly, the interpolation polynomial is used as the approximate curve the new web diameter. Finally, experimental verification is carried out at the production site.

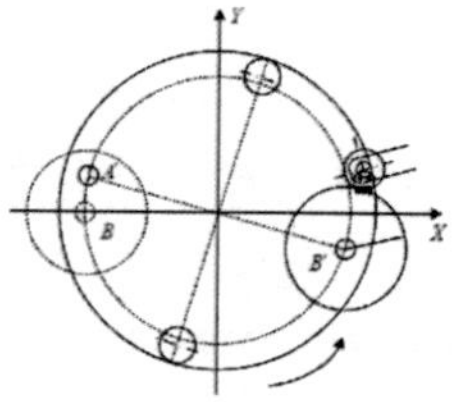

Fig. 1. Unwind turret mechanical structure

2.1 Unwind Turret Splicing Position Model

Use CAD model to build the model of unwind and splice, as show in Fig. 2, then obtain the data of actual unwind turret rotary angle and precise diameter of corresponding new web.

From Fig. 2, it displays the relationship between unwind diameter range 200–800 mm and new unwind turret rotary angle status. Based on the changes, 13 groups data is listed as Table 1.

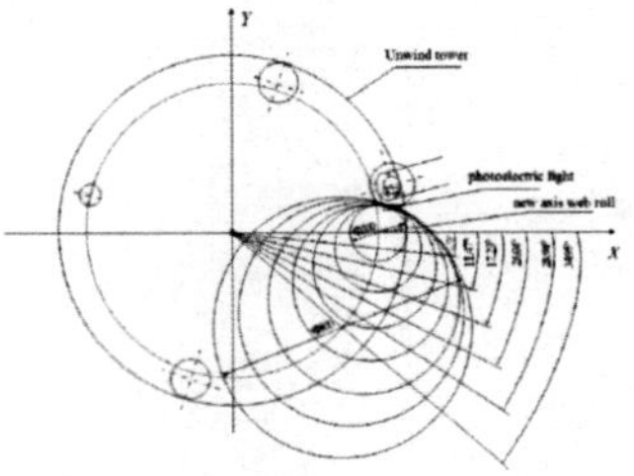

Fig. 2. Unwind splicing CAD Modeling

Table 1. Web radial and positioning angle CAD modeling measure

Cantilever angle angle/(°)	0	2.86	5.73	8.60	11.47	14.34	17.23
Unwind radius r/(mm)	200	250	300	350	400	450	500
Cantilever angle angle/(°)	20.12	23.04	25.97	28.94	31.92	34.95	\
Unwind radius r/(mm)	550	600	650	700	750	800	\

2.2 Unwind Turret Geometric Relation Derivation

The relationship between the splicing angle and the web diameter is difficult to determine even if the practical new web diameter data is accrue, which is measured by CAD model. And hence, the geometric relation between unwind turret new diameter and the splicing angle should be derived firstly. Then the model of unwind turret and new axis is established.

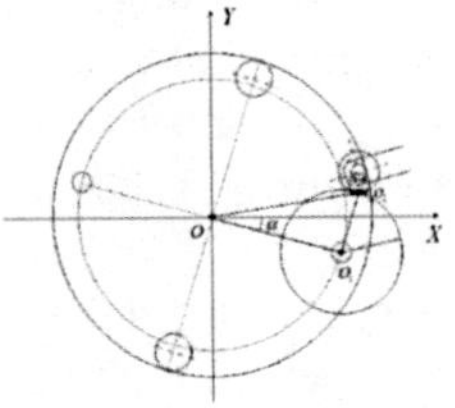

Fig. 3. Unwind turret geometric relation derivation process

Figure 3 unwind turret rotation axis is centered at the original point O of the coordinate system XOY, the radius R of unwind turret is known, during web change and splice, unwind turret rotating and new axis closed the nip roller, when photoelectric sensor under the nip roller detected new axis web, unwind turret stop rotating, at this time unwind turret rotating potentiometer gets the angle α between R and X.

Refer to Fig. 4 new web diameter calculation process to gain individual the polar coordinate equation of O_1 and O_2. Then put it into distance equation of O_1 and O_2. The straight-line distance from point O_1 to point O_2 is the radius r of the new web, the

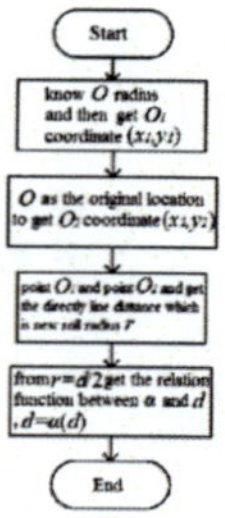

Fig. 4. Unwind turret geometric relation char

equation for the new diameter can be obtained:

$$d(\alpha) = 2 \times \sqrt{(518.4 - 500 \times \cos\alpha)^2 + (98.6 + 500 \times \sin\alpha)^2} \tag{1}$$

This equation displays the non-linear relationship between changing web splicing angle and new web diameter, and the equation $d(\alpha)$ is long and difficult to solve, which is not suitable for the actual PLC control. In order to quickly solving out unwind new web diameter and PLC, this paper introduces polynomial interpolation method to solve the approximate curve.

3 Interpolation Algorithm Based on Polynomial

During the engineering practice, the research function $y = f(x)$ is often complex. However, a set of observation data (sampling points) of the function can be obtained by means of experimental observation, and through the function table of the discrete date which is a simple function $\varphi(x)$ to replace the original function $f(x)$, which is called interpolation method [3, 4]. It commonly used Lagrange interpolation and Newton interpolation. But there is the difference at the amount of calculation. Newton interpolation method to overcome the Lagrange interpolation method adds a node, the whole calculation work to start the disadvantage of saving multiplication, division times, and use of difference, difference quotient concept, it is easy for programming calculation.

Therefore, Newton interpolation method is chosen to study the exact calculation of new web diameter.

In order to find the best interpolation polynomial, the data of order 3 ~ 7 and 13 are selected for analysis and comparison. And it will be obtained all orders (3–7, 13) interpolation polynomials as showed in Table 2.

Interpolation curve as show in Fig. 5.

Figure 5 displays the interpolation curves of each order in the same dimension for comprehensive comparison, to discuss the effected for more interpolation points and less interpolation points [5], locally magnify points 1st, 2nd, 12nd and 13rd to compare their parameters.

Figure 6 is the error between interpolation curve and geometric expression of each order. When the order is 13, the interpolated function is high degree polynomial (12 times) and produces oscillation, and the maximum error reaches 5.5mm and has the

Table 2. Orders (3–7, 13) interpolation polynomials

n	Interpolation polynomials
3	$\varphi(x) = -1.3776e^{-2} \cdot x^2 + 1.7649e \cdot x + 200$
4	$\varphi(x) = -4.1418e^{-4} \cdot x^3 + 7.7523e^{-3} \cdot x^2 + 1.7402e \cdot x + 200$
5	$\varphi(x) = -4.4341e^{-6} \cdot x^4 - 1.1976e^{-4} \cdot x^3 + 1.8756e^{-3} \cdot x^2 + 17.437e + 200$
6	$\varphi(x) = -2.0047e^{-7} \cdot x^5 + 2.2377e^{-5} \cdot x^4 - 1.3531e^{-3} \cdot x^3 + 2.5803e^{-2} \cdot x^2 + 1.7251e + 200$
7	$\varphi(x) = 1.0348e^{-8} \cdot x^6 - 6.1023e^{-7} \cdot x^5 + 2.3201e^{-7} \cdot x^4 + 1.6694e^{-4} \cdot x^3 - 4.1968e^{-3} \cdot x^2 + 1.7471e + 200$
13	$\varphi(x) = -1.1082e^{-12} \cdot x^{12} + 2.2894e^{-10} \cdot x^{11} - 2.0844e^{-8} \cdot x^{10} + 1.1007e^{-6} \cdot x^9 - 3.7327e^{-5} \cdot x^8 + 8.4999e^{-4} \cdot x^7 - 1.3196e^{-2} \cdot x^6 + 1.3883e^{-1} \cdot x^5 - 9.6279e^{-1} \cdot x^4 + 2.7207e \cdot x + 200$

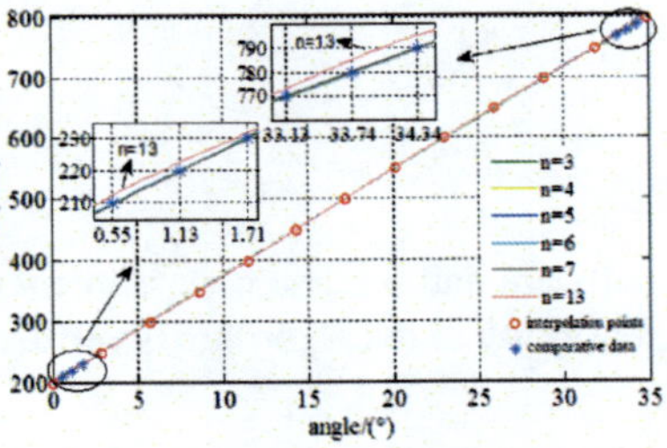

Fig. 5. Newton interpolation method difference order interpolation curve comparison

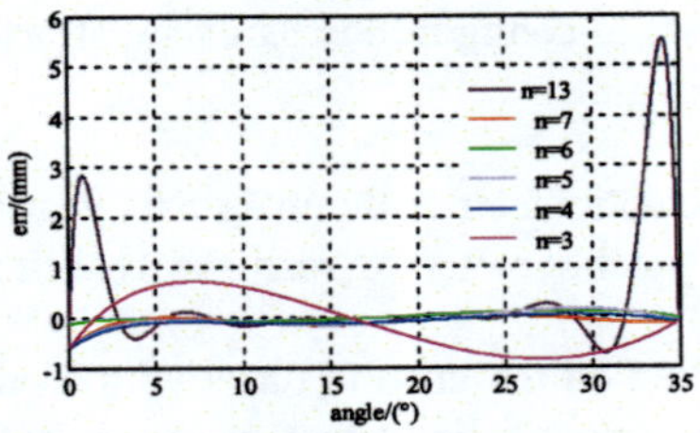

Fig. 6. Newton interpolation methods for difference order interpolation errors comparison

Runge phenomenon, that is, the interpolation effect of high degree polynomial is not sure better than low degree polynomial. In the process of interpolation, the error is diffused or amplified in the process of difference between the interpolation function

$\varphi(x)$ and original function $f(x)$, the high degree polynomials stable worse. In this case, it should be avoided selecting the high degree polynomials.

Above all, after comprehensive analysis and comparison, the third order polynomial function is selected:

$$\varphi(x) = -4.1418e^{-4} \cdot x^3 + 7.7523e^{-3} \cdot x^2 + 1.7402e \cdot x + 200 \quad x \in [0, 35] \qquad (2)$$

As unwind cantilever rotation position and web diameter interpolation function. X means angle, the change range 0 ~ 35°, $\varphi(x)$ is new web diameter. It is obtained interpolation function, the angle data needs to be substituted into the third order polynomial function to verify the accuracy of the algorithm. Edit the M file in MATLAB, collect and compare with CAD model simulation data, details check the following Table 3.

Table 3. Compare simulation with interpolation

X	0	2.86	5.73	8.6	11.47	14.34	17.23
Simulation data	200	250	300	350	400	450	500
Interpolation data	200	249.82	299.89	349.97	400	449.92	500.02
X	20.12	23.04	25.97	28.94	31.92	34.95	
Simulation data	550	600	650	700	750	800	
Interpolation data	549.89	599.99	649.90	700.07	749.90	799.99	

4 Application

This section establishes software and hardware experiment environment, and basis to production line site. The feasibility of the algorithm is verified according to the field test data.

4.1 Site Test Steps

Figure 7 Unwind turret splicing control topology.

Before the test, it needs to configure the hardware structure of the production line and write the software program:

1) Install the rotating potentiometer at the main axis gear of unwind turret, the production line adopts potentiometer to read the real-time unwind turret rotation position.
2) Detect the cam installed on the unwind turret with a proximity switch, as "zero" of the rotation angles of the unwind turret, as shown in O_3 position on Fig. 8, and $\alpha + \beta = 135° \quad \alpha \in [0°, 35°]$.
3) Three-phase synchronous motor is directly driven unwind turret.
4) Programming exact positioning program, trace curve and PID control on the main program. The flow chart is shown in the Fig. 9. Figure 10: unwind unit for production line.

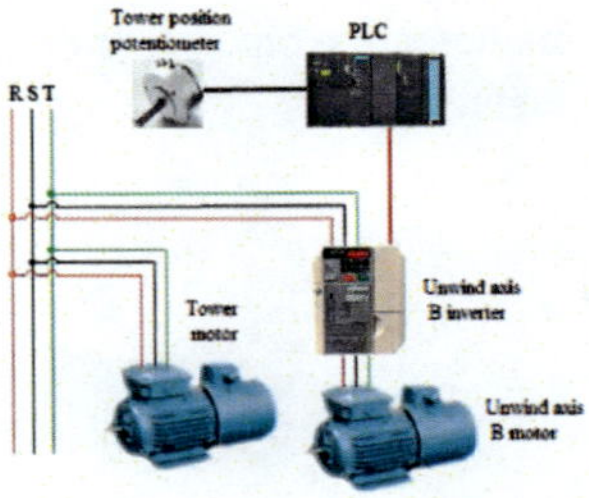

Fig. 7. Unwind turret splicing topology structure

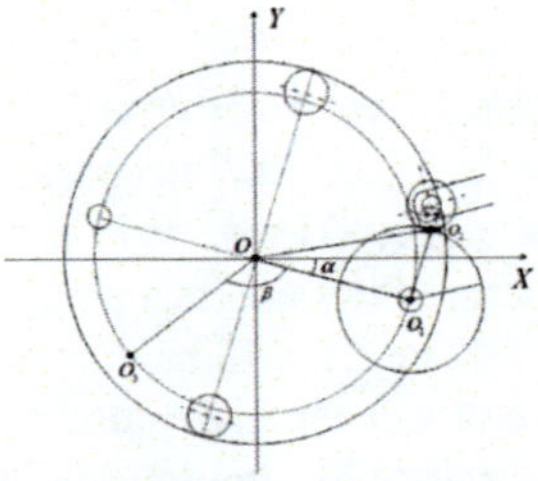

Fig. 8. Unwind turret zero detection

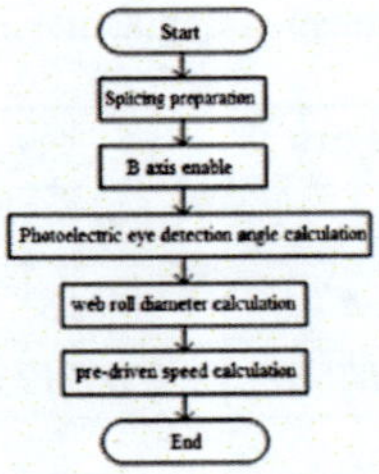

Fig. 9. Software process chart

Fig. 10. Production line unwind unit

4.2 Test Results and Analysis

Use the diameter 800 mm to test and verify unwind turret rotation angle and initial web diameter calculation. The production site trace curve is shown in Fig. 11, the red curve means calculated initial diameter, the blue curve means axis speed, the green curve means unwind turret rotating angle.

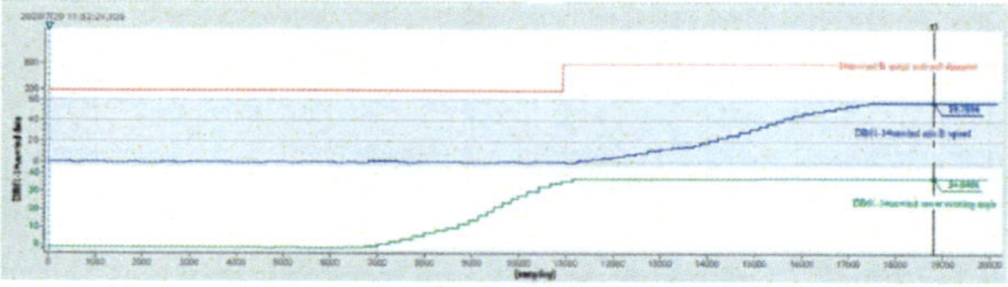

Fig. 11. Diameter 800 m trace curve

Refer to Fig. 11, when machine speed is 60 m/min, press the web splicing preparation button, unwind turret start to rotate about 7000 ms pass the zero, then start to detect the rotation angle. When it arrives at 11,000 ms, web is detected by the sensor and the unwind turret stop rotation. Operate the same test steps, individual more times test for 200 mm, 400 mm, 600 mm and 800 mm, it will be obtained data. Table 4 is interpolation method calculated the difference cantilever angle corresponding the diameter, the actual diameter are all integer values obtained by simulation of CAD model, then calculate the error between the actual diameter and calculated diameter.

Table 4. Groups turret angle corresponding actual diameter and calculated diameter

Turret rotating angle β/(°)	134.917	123.582	111.762	100.153
Actual diameter r/(mm)	200	400	600	800
Calculated diameter r/(mm)	201.57	399.13	603.41	798.29
Error ε	0.00785	0.002175	0.005683	0.002137

From Table 4, the error less than 1% between calculated diameter and actual diameter and it meets the technology requirement.

5 Conclusion

This paper focuses on the difficulties in measuring the diameter of new web diameter during unwind splicing of flexographic press. According to the data feature of unwind turret rotation position and corresponding unwind new web diameter, use CAD model for geometrical analysis and design the based on polynomial interpolation a calculation method of the new web diameter. By analyzing Newton interpolation algorithm, and control polynomial times to get the best approximate curve, which will be solved the calculation problem of unwind turret new web diameter, the further calculate unwind new web diameter rotation speed to ensure the unwind function with accuracy and reliability.

The experimental data shows that PLC can quickly calculate the new axis diameter by using the polynomial interpolation method proposed in this paper, according to new web diameter to calculate new web diameter speed and complete the unwind synchronization. At the actual operation, unwind realizes smooth change web and the dynamic performance of the system is improved, which verifies that the control strategy is correct and effective.

References

1. Qian, C., Junwei, Q., Yaocheng, L.: Analysis on the development status of flexo printing in china's flexible packaging. Shanghai Packag **5**, 3 (2020)
2. Essanhaji, A., Errachid, M.: Lagrange multivariate polynomial interpolation: a random algorithmic approach. J. Appl. Math. **2022** (2022)
3. Yajing, L., Chengcong, W., Mingyue, L.: Research on improved local polynomial interpolation weight function. Geodesy Geodyn. **40**(7), 5 (2020)
4. Jinming, Z., Yongqiu, C.: Research on robot trajectory interpolation method based on genetic algorithm. Combined Mach. Tool Autom. Mach. Technol. **5**, 4 (2020)
5. Li, C., Shan, L., Huang, T., et al.: Path planning method for industrial manipulators based on piecewise multi-order polynomial interpolation. CN111002308A (2020)

Information Engineering and Artificial Intelligent Technology

A Development Research of Self-guided Robot Based on Radio Frequency Identification

Yanbin Wei(✉) and Peng Liu

Faculty of Printing, Packaging Engineering and Digital Media Technology, Xi'an University of Technology, Xi'an, China
yb_wei@163.com

Abstract. Objective to solve the problem of navigation and obstacle avoidance of paper conveying robot in the environment of changeable dynamic but the stations relatively fixed for printing workshop, and to provide a practicable reference for the obstacle avoidance and self-guidance robot in dynamic changing environment. Methods using radio frequency identification technology and ultrasonic sensor technology, it has studied paper conveying robot with free path planning and autonomous obstacle avoidance in printing workshop. With obstacle avoidance system, control system, path planning system and driving system all designed and developed, the paper conveying robot experimental prototype was made and simulated in the laboratory environment. Results the robot system completed the work of obstacle avoidance and self-guidance, realized the automatic paper feeding process in the experimental environment. Conclusion the paper conveying robot can be guided and avoided by RFID and ultrasonic. It has advantages of path change flexibly, less vulnerable to environmental impact and cheap cost, and its application prospect is wide.

Keywords: Robot · Path planning · Radio frequency identification · Ultrasound obstacle avoidance

1 Introduction

Entering the 21st century, more attention has been paid to develop the robot technology at home and abroad. The United States, the European Union and Germany have all formulated corresponding research projects in robotics [1, 2]. China has also attached great importance to robot research and development plan, such as the National Natural Science Foundation of China, the major national science and technology projects and "made in China 2025" have also paid great attention to it [3–5]. With the sensor technology and artificial intelligence continue developing, robot technology also developing towards intelligence, high precision and speed, and networking [6]. Robot technology has become one of foundations for high-tech and emerging industries in the future.

In the printing and packaging industry, with increasing labor cost and the developing of mechanical equipment automation, using robot to replace manual to transport materials between stations is inevitable in the future.

M. Xu et al. (Eds.): CACPP 2022, LNEE 991, pp. 405–411, 2023.
https://doi.org/10.1007/978-981-19-9024-3_50

Over recent years, a lot of studies have emerged on indoor mobile and navigation of robots such as printing workshops [7–9], Zhang Shuliang used multi-sensor fusion method for indoor environment map modeling and navigation for an example, Lin Yizhong used machine vision for indoor environment modeling and navigation design, and Yang Kai used improved artificial potential field method to study robot guidance, etc. However, these studies all use visual sensors for map modeling, and navigation obstacle avoidance depends on image sensors, which has the problems of large amount of computing data and high manufacturing cost. Aiming to the situation of printing workshop that the equipment stations are relative fixed but the dynamic environment is changeable, using radio frequency identification and ultrasonic positioning technology, this paper studied the path planning, the path recognition, autonomous obstacle avoidance and guidance for four-wheel mobile robot, and the experimental prototype and robot platform is build for printing workshop.

2 Analyze Working Environment of the Paper Conveying Robot

The main equipment in printing workshop is mostly placed side by side, and the general structural layout is shown as Fig. 1.

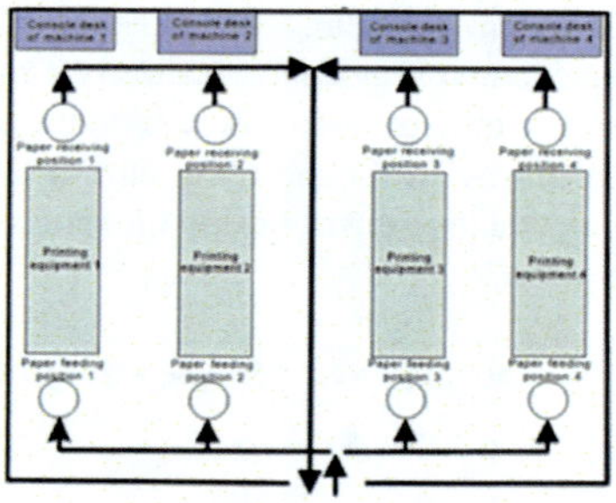

Fig. 1. Structural layout of printing workshop

In Fig. 1, two ends of printing machine are paper feeding end and paper receiving end respectively. The substrate to be printed should be placed near the paper feeding while the printed matter placed at the receiving, and the workshop transport corridors are located at the two ends of machines. During work, paper conveying robot needs to transport blank paper from paper warehouse to some position of printing workshop or transport printed matter from paper receiving position to next process position outside, as shown by the arrow in Fig. 1. Therefore, the robot route is relatively clear and the dwell station is fixed.

On the other hand, in printing workshop, the main flow obstacles affecting working process of paper conveying robot are workshop staffs. Therefore, when robot encounters workshop staff moving on the transportation channel, the robot should avoid flow obstacles actively.

3 Path Planning and Navigation Obstacle Avoidance

3.1 Path Planning of Paper Conveying Robot

Path planning refers to the robot perceives environment according to sensors and plans the best route independently. The commonly used path planning algorithms mainly include the grid method, the artificial potential field method, the genetic algorithm and the neural network and so on. Among these algorithms, the artificial potential field method is a relatively mature and efficient algorithm and is widely used because of the characteristics of small amount of calculation, short planning time and easy to realize the bottom control in the path planning of mobile robots at present [10, 11].

In this paper, the paper conveying robot's task is to transport paper in different fixed stations. The robot should plan several different paths and choose the best one according to the stations before moving. On one hand, considering the difference of paper specifications in stations, the primary selected path should not be easy changed during execution. On the other hand, when robot encounters a dynamic obstacle in process of moving, it needs to choose whether to wait for the obstacle to leave or plan a new path. Considering that in the actual working condition, the dynamic obstacle is mainly of the walking staffs in workshop, it is more operable to remind the staffs to get out of the way.

3.2 Paper Feeding Robot Navigation

The guidance methods of mobile robot are generally divided into the fixed path guidance and the free path guidance. With advantages of relatively simple control, easy implementation and low cost, the fixed path guidance generally becomes first choice [12, 13]. Early tape navigation, electromagnetic navigation and optical navigation and so on are all part of fixed route navigation. The free path guidance includes inertial guidance and GPS positioning guidance. The inertial guidance has the advantages of strong anti-interference ability, convenient combination, strong compatibility and wide application fields, but the technology is relatively complex and the cost is high. Using satellites to navigate and track, and with advantages of long guidance distance and flexible path change but disadvantages of high cost and low guidance accuracy, The GPS positioning guidance is mainly used for outdoor navigation but suitable for robots indoor.

Comparing various kinds of guidance, it can be seen that free path guidance has high cost and is mainly suitable for outdoor, while fixed path guidance is fit to indoor environment. Considering characteristics of printing workshop with relative fixed transportation channel, more mechanical devices and limited space, the subject selects RFID in fixed path guidance as the path guidance mode for paper conveying robot.

3.3 Paper Feeding Robot Obstacle Avoidance

In working, robot needs to rely on sensors to sense the external environment and act in response obstacle avoidance measure quickly; so, obstacle avoidance ability is one of the important indexes of intelligent robots.

The robot obstacle avoidance and ranging mainly rely on laser ranging sensor, infrared ranging sensor, ultrasonic ranging sensor and visual ranging sensor. The laser ranging has advantages of cheap cost, strong anti-interference ability but long-distance ranging error large and invalid for transparent materials. The infrared ranging has advantages of low price, good concealment and low power consumption but easy disturbed by light source and weak penetration. The ultrasonic distance measurement obtains the distance of the object to be measured by measuring the ultrasonic return time. It has advantages of large penetration ability, good directivity, high frequency, high resolution and low energy consumption. Using triangular ranging method, the vision sensor obtains the measured object's distance is accurate, but the real-time performance is poor because of the large amount of computation.

Compare the advantages and shortages of the four obstacle avoidance methods and combine the characteristic of the printing workshop, the ultrasonic sensor is selected as the obstacle avoidance scheme for the paper conveying robot.

4 Design of Robot Structure and Control System

4.1 Robot Structure Design

The paper conveying robot is designed with a double-layer frame structure. The upper layer is the transport bearing layer, the lower layer is the control system layer where the control unit, detection and communication system module, drive system and power supply unit are placed. The layout is shown as Fig. 2.

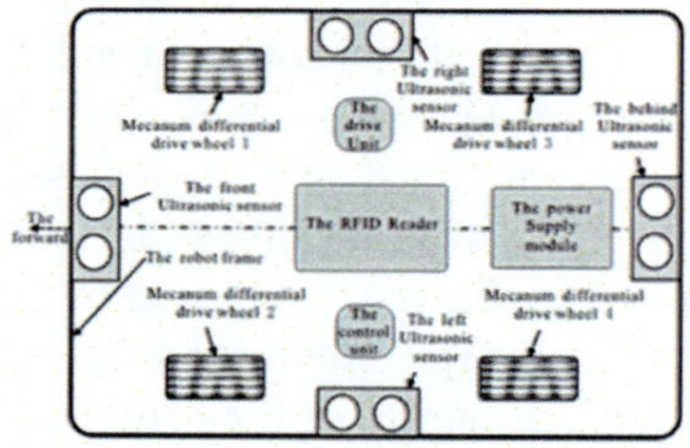

Fig. 2. Overall structure diagram of paper feeding robot

Four mecanum drive wheels are designed to realize omni-directional movement of transport vehicle, including front and back straight travel, left and right lateral movement, in-situ rotation, turning with specified radius, linear movement in any direction and so on, which can realize the precise movement of robot car.

4.2 Robot Control System Layout

The robot control system includes the navigation system, the obstacle avoidance system, the central control system, and the drive system and power supply unit.

The central control is the vital part of the paper feeding robot, which is in charge of the collection of various sensor signals and operation processing, and sending control instructions to relevant equipment in other side. In this study, the control core is selected as Arduino mega2560 with a crystal frequency of 16 MHz.

The navigation is realized by radio frequency identification technology. Each RFID tag is pasted on each point of the trolley travel route, and the RFID reader is installed in the center of the lower layer of the frame, as shown in Fig. 2. During the transportation process, the trolley shall identify labels in turn according to the planned route for navigation.

Four ultrasonic sensors are selected for obstacle avoidance, which are respectively installed at the front, rear, left and right middle positions of lower layer of the frame, as shown in Fig. 2. The robot detects obstacles in any direction during the moving process. When the obstacles are found to be less than the safe distance, it will quickly decelerate and judge what kind of obstacles the obstacles are. If they are dynamic obstacles, it will beep for warning and stop to wait for the obstacles to be cleared; otherwise, it will quickly detect the surrounding label information and correct the moving direction according to the map.

5 Software Design and Prototype Experiment

5.1 System Software Design

The development environment provided by the control core can directly complete path planning, ultrasonic obstacle avoidance and other functions, realize the movement and guidance of the paper feeding robot through algorithm control, and transmit it to the control core for execution through USB interface.

After the system is started, the paper feeding robot plans the path and starts driving according to the map information and task requirements pre stored in the RFID electronic tag. During driving, the transmitting module continuously sends ultrasonic waves. When encountering obstacles, the reflected signal returns. The receiving module transmits the feedback information to the system, and calculates the distance between the robot and the obstacles by using the time difference between transmission and feedback. Once the space is smaller than the safe range, the robot will decelerate quickly and give a warning. Considering that the main obstacle on the mobile channel of the robot in the workshop is the problem that the workers move freely from time to time, the robot slows down to stop and waits for 10 s. If the obstacle cannot disappear after 10 s, the robot will re plan the route. The system path guidance planning process is shown as Fig. 3.

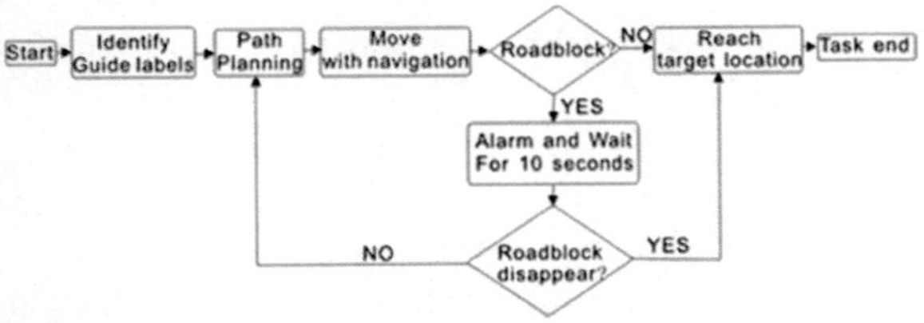

Fig. 3. System path guidance planning flow chart

5.2 Prototype Experiment

After completing the structure design, control system, hardware system and software system design of robot, the paper feeding robot prototype is built, and compile the control program, path planning algorithm and obstacle avoidance program according to needs. The prototype of the paper feeding robot is shown as Fig. 4.

Fig. 4. Model machine of paper feeding robot

At the time of experiment, RFID tags are pasted at sites as required, the experimental prototype can reach the station as required. At the same time, if obstacles are encountered during driving, the robot can stop in time to replan the route, so as to realize the design function.

6 Conclusions

USing radio frequency identification technology and ultrasonic sensor technology, this project realized path identification planning and obstacle avoidance control for paper feeding robot. The designed prototype simulates the transportation process of paper stacks in printing workshop, and realizes the automatic paper feeding function of robot. The results of the study are summarized as follows:

1) Using RFID technology and ultrasonic obstacle avoidance, the paper feeding robot is guided to the path and avoided obstacles autonomously, and the path planning and autonomous navigation functions of the robot are realized.
2) With advantages of low cost, path change flexibly, and not vulnerable to environmental impact, RFID technology has a bright future in navigation applications.
3) The paper feeding robot mainly uses indoor fixed route navigation, and the obstacle avoidance algorithm is relatively simple. Future study should consider it has more application value of navigation based on visual environment modeling.

References

1. Barack, H.O.: National Robotics Initiative (NRI). US National Science Foundation, Washington (2011)
2. Cui, Z.: Robot Future Strategy 2022. South Korean Ministry of Knowledge Economy, Seoul (2012)

3. Industry 4.0 Working Group: Securing the Future of German Manufacturing Industry Recommendations for Implementing the Strategic Initiative INDUSTRY 4.0. German Federal Ministry for Education, Berlin (2013)
4. National Manufacturing Power Construction Strategy Advisory Committee: Technology Roadmap in Key Areas of Made in China 2025. Green Paper on Technological Innovation in Key Areas of Made in China 2025, Beijing (2015)
5. Tianmiao, W.A.N.G., Yong, T.A.O.: Research status and industrialization development strategy of Chinese industrial robot. J. Mech. Eng. **50**(9), 1–13 (2014)
6. Feng, G.A.O., Weizong, G.U.O.: Thinking of the development strategy of robots in China. J. Mech. Eng. **52**(7), 1–4 (2016)
7. Zhang, S., Tan, X., Wu, W.-Q.: Indoor mobile robot localization based on multi-sensor fusion technology. Transducer Microsyst. Technol. **40**(08), 53–56 (2021)
8. Yizhong, L.I.N., Kai, M.A.: Design of autonomous navigation system for indoor mobile robot. Autom. Instrum. **36**(6), 38–42 (2021)
9. Kai, Y.A.N.G., Jia, L.O.N.G., Xueyan, M.A., et al.: Research on mobile robot path method based on improved artificial potential field. Modern Electron. Tech. **43**(7), 141–145 (2020)
10. Jin, S.H.I., Yao, D.O.N.G., Zhengdong, B.A.I., et al.: Research and implementation of mobile robot path planning method. J. Comput. Appl. **37**(11), 3119–3123 (2017)
11. Zhi, D.E.N.G., Haichao, L.I.: Survey of research on mobile robot autonomous navigation technology. Sci. Technol. Inf. **33**, 142–144 (2016)
12. Zhang, L., Xu, Y., Xia, C.: Design of AGV navigation system based on RFID positioning. Electron. World (05), 67–69 (2017)
13. Shili, P.A.N.G.: Design of fixed path guidance scheme for omnidirectional moving AGV. China Plant Eng. **09**, 153–154 (2019)

Digital Twin Fault Diagnosis Method for Complex Equipment Transmission Device

Jiahui Chen, Jinda Zhu(✉), Zhiying Qin, Yuejing Zhao, Fuxiang Zhang, and Fengshan Huang(✉)

College of Mechanical Engineering, Hebei University of Science and Technology, Shijiazhuang, Hebei, China
zhu_jinda@163.com, hfs_high@126.com

Abstract. Aiming at the problems of state monitoring delay and insufficient data sources in traditional PHM health management methods, a digital twin fault diagnosis method for complex equipment transmission devices is proposed. Firstly, a fault experiment platform of complex equipment transmission device is set to measure the running data of transmission device. Then, features of physical monitoring perception data are extracted based on the GA-BP neural network and fault diagnosis rules are quickly obtained. Combined with digital twin technology, visualization of physical monitoring perception data is realized. Finally, the fault diagnosis system of complex equipment transmission device is built with the combination of virtual reality technology and digital twin technology, verifying the feasibility and effectiveness of the proposed method.

Keywords: Digital twins · GA-BP · Complex equipment · Fault diagnosis

1 Introduction

Traditional PHM health management method is to collect all kinds of data of equipment, and to use various algorithms to monitor, analyze, diagnose and evaluate the health status of equipment [1, 2]. However, with the development and advancement of computer information technology, the complexity and intelligence of equipment are constantly improving. Therefore, the traditional health management method has exposed the problems of time lag of off-line monitoring and insufficient data sources [3], which is not the optimal health management scheme.

In recent years, the Digital Twin technology has been widely concerned by the manufacturing industry. Digital twinning technology makes full use of the physical model of equipment, sensor monitoring data and historical operation data of equipment, establishes high-fidelity mapping of physical equipment in virtual space, and constructs digital twinning of physical equipment in virtual space [4–6]. Synchronization of virtual and real interaction refers to monitor the running state of physical entities in virtual space, and interact with physical entities through control instructions [7, 8]. Based on digital twin technology, a fault diagnosis system for complex equipment transmission

M. Xu et al. (Eds.): CACPP 2022, LNEE 991, pp. 412–419, 2023.
https://doi.org/10.1007/978-981-19-9024-3_51

device is designed and developed, which can effectively monitor the operation process of complex equipment and solve the problem of monitoring lag existing in traditional PHM mode.

2 Complex Equipment Physical Monitoring Technology

During the operation of complex equipment, the health status of its transmission device is a key part to make the equipment run properly. Therefore, the study of health status monitoring of transmission device becomes particularly important when studying the health status monitoring of complex equipment. Take the transmission device of complex equipment as the research object to build an experimental platform. The experimental platform is shown in Fig. 1. The experiment platform includes driving motor, timing belt, gear box, acceleration sensor and toner brake.

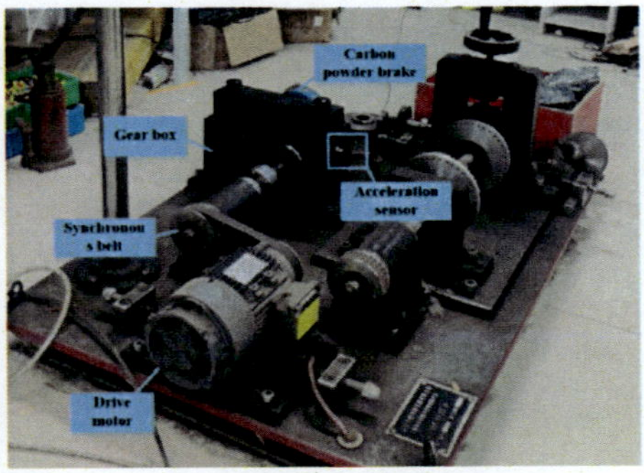

Fig. 1. Experimental platform for complex equipment transmission device

3 Fault Diagnosis Technology of Complex Equipment

The monitoring scheme of transmission device is designed, in which acceleration sensors are respectively deployed at both sides of transmission device box, at the input shaft bearing of transmission device and at the output shaft bearing of transmission device. Due to the limitation of physical equipment sensor installation, they cannot be installed above the input shaft bearing and the output shaft bearing. Thus, they are installed above the gearbox box. The photoelectric sensor is used to measure the rotational speed of the input shaft, and the reflective sheet is pasted on the surface of the input shaft for signal transmission. The sensor receives the signal to measure the rotational speed. The measuring equipment used is shown in Table 1.

The physical monitoring and fault diagnosis system of transmission device is built based on Labview platform, which receives the signals measured by acceleration sensors in real time, converts the sensor signals, and obtains the characteristic vectors of data after time domain analysis and frequency domain analysis, as shown in Fig. 2.

Table 1. Physical monitoring scheme

Number	Sensor type	Sensor measuring point position	Measuring signal
1	Acceleration sensor	Input shaft	Input shaft vibration signal
2	Acceleration sensor	Output shaft	Output shaft vibration signal
3	Acceleration sensor	Both sides of box body	Box vibration signal
4	Speed sensor	Input shaft	Input shaft speed
5	Voltage sensor	Above the motor	Working voltage
6	Temperature sensor	Above the motor	Operating temperature

Fig. 2. Transmission device physical detection and fault diagnosis system

3.1 GA-BP Neural Network Algorithm

BP neural network has strong nonlinear adaptive ability and self-learning ability. With its simple structure and fast running speed, BP neural network has been widely used in information classification and pattern recognition [9]. However, the disadvantage of traditional BP neural network is that it can't select the optimal weight and threshold during training calculation, and the selection of weight and threshold has a great influence on the learning effect of neural network. Because of the instability of BP neural network learning, genetic algorithm is used to optimize BP neural network, and find the optimal weights and thresholds for sample training.

The genetic algorithm finds the individual corresponding to the optimal fitness value through selection operation, crossover operation and mutation operation, and applies the weight threshold corresponding to the optimal individual to the BP neural network so as to improve the training accuracy and prediction effect of the neural network. The algorithm flow of genetic algorithm optimization BP neural network is shown in Fig. 3.

The key steps in the process are fitness function solution, selection operation, crossover operation and mutation operation. The fitness function is used to evaluate

Fig. 3. Neural network algorithm flow chart

individuals in the population, and the fitness value F is calculated as follows:

$$F = k\left(\sum_{i=1}^{n} abs(y_i - o_i)\right) \tag{1}$$

where n is the number of output nodes of the neural network; y_i is the expected output of the i node of the neural network; o_i Is the forecast output of the i node; k Is the coefficient.

Roulette is adopted as the selection method in this paper, which is a selection strategy based on fitness ratio. The better the fitness is, the greater the selection probability, and the selection probability of each individual is p_i.

$$p_i = \frac{F_i}{\sum_{i=1}^{N} F_i} \tag{2}$$

where F_i is the fitness value of the F_i-th individual; N is the number of individuals in the population.

Then, crossover operation is performed on the selected fitness individuals. Crossover operation is to cross and combine two chromosomes of two individuals through any two individuals in the sample population, and change the gene coding structure of two donors, so as to generate new excellent individuals. Then, mutation operation is carried out. In the process of replication, one point of an individual's chromosome is selected for gene mutation operation to produce a new individual. Genetic algorithm relies on continuous reproduction to produce excellent individuals, so as to optimize and improve the accuracy of prediction and classification. After the operation, the optimal weights and thresholds of BP neural network are obtained, used for neural network training.

3.2 GA-BP Neural Network Design

The construction of BP neural network generally includes the number of layers in the input layer, the output layer and the hidden layer, the number of ganglia in each layer and the transfer function between layers. Many faults will occur in the running process of complex equipment transmission device. Taking the transmission device gear fault as the research object, the feature vector dimension of the transmission device gear fault sample data collected in this paper is 6, and the number of nodes in the input layer should be equal to the feature vector dimension, so the number of nodes in the input layer is 6. The number of output nodes should be selected according to the output nodes, and binary coding is used to specify that the normal expected output of the gear should be (1,0,0). The expected output of gear broken teeth is (0,1,0); The expected output of gear crack is (0,0,1), the expected output of three fault types are all three-dimensional vectors, and the number of nodes in the selected output layer is 3; Select the number of hidden layer nodes based on empirical formula, which is shown in Formula (3);

$$k = \sqrt{m + n} + a \tag{3}$$

where

k is the number of hidden layer nodes;
M is the number of nodes in the input layer;
N is the number of output layer nodes;
A is an adjustment constant between 1 and 10.

According to Formula (3), where m = 6, n = 3, a is chosen as the constant 10, the calculated k is 13, and the number of nodes in the selected hidden layer is 13.

3.3 GA-BP Neural Network Results Analysis

In this paper, a total of 114 sets of fault data samples of complex equipment transmission device were measured. After repeated selection, 33 sets of samples were selected for fault sample training. In this paper, Sigmoid function was selected as the activation function of neurons, and GA-BP neural network was set to train 1000 times with the training accuracy of 0.001. The maximum iteration number of genetic algorithm was set to 50, the population size was 40, and the crossover probability and mutation probability were 0.7 and 0.01 respectively.

The training effect of GA-BP neural network is shown in Fig. 4. When the neural network is iteratively calculated for the 14th time, the fitting curve converges, and the neural network reaches the training goal with a fast convergence speed. Some test sample outputs are shown in Table 2. Table 2 shows the relative errors between the actual output and the expected output after the neural network training. According to the data in Table 2, the average relative errors between the actual results and the expected results are all below 10%, and the BP neural network has a good fault classification effect.

Fig. 4. GA-BP neural network training error curve

Table 2. Relative error analysis of GA-BP neural network

Number	Gear mode	Actual output	Average relative error (%)
1	Healthy	(0.976, 0.018, 0.022)	2.1
2	Healthy	(0.987, 0.036, 0.103)	5.1
3	Tooth broke	(0.107, 0.932, 0.043)	7.3
4	Tooth broke	(0.059, 0.921, 0.058)	6.5
5	Crack	(0.073, 0.115, 0.973)	7.2
6	Crack	(0.019, 0.051, 0.957)	3.8

4 System Verification

4.1 System Interface

The fault diagnosis system of complex equipment transmission device developed by Unity3D platform has its main interface as shown in Fig. 5. The main control interface is divided into three areas, namely, the functional panel area, the virtual scene display area and the equipment monitoring area. The module mainly includes operation manual, environmental parameters, fault analysis, monitoring of measuring points and other functions.

4.2 Real-Time Monitoring

As shown in Fig. 6, set up the monitoring and diagnosis analysis panel of measuring points, on which users can view the measuring points of three acceleration sensors. Users can click the numerical update button, in order to search for the latest data of each measuring point in the database, update the line chart of each measuring point data in real time. The system can evaluate the health status of the measured data based on GA-BP

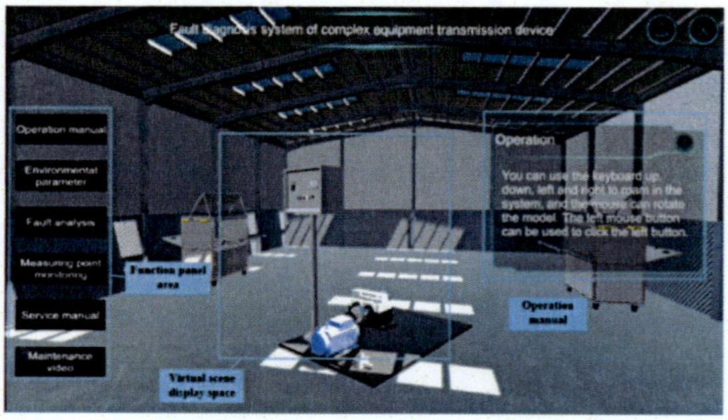

Fig. 5. System main control interface

Fig. 6. Measuring point monitoring interface

neural network, which is convenient for users to know the health status of equipment in time.

5 Concluding Remarks

Aiming at the problem that the traditional PHM health management method of complex equipment can't meet the demand of the real-time monitoring of equipment health status, this paper sets up the experimental platform of complex equipment transmission device, designs the sensor deployment scheme, measures the running data of transmission device, and collects vibration signals based on three types of gear failures. In the aspect of fault diagnosis, the measured data are analyzed on the basis of GA-BP neural network. Compared with the traditional BP neural network fault classification method, the addition of genetic algorithm not only solves the problems of the local optimal solution and slow convergence in the traditional BP neural network algorithm, but also improves the fault diagnosis accuracy of BP neural network.

Finally, a digital twin system is set up, so that maintenance personnel can remotely monitor the running status of equipment and visually present the monitoring data in the system. Compared with the traditional PHM health management method, it can effectively improve the real-time monitoring of equipment running status. The further step is to discuss the mapping relationship between physical equipment and virtual equipment on the basis of real-time monitoring, enrich the content of equipment status warning and maintenance guidance, and improve the digital twin system. I hope this paper can provide support and reference for the intelligent development of complex equipment maintenance.

Acknowledgements. This research is supported by key research and development projects in Hebei province (No. 21351801D and No. 22351801D).

References

1. Lei, Z., Yongxiang, Z., Danchen, Z.: Review on rolling bearing fault diagnosis and prognostic for complex equipment. China Measur. Test **46**(03), 17–25 (2020)
2. Guangning, W., Xiaonan, L., Yan, Y., et al.: Research progress of fault prediction and health management for on-board traction transformers. High Voltage Eng. **46**(03), 876–889 (2020)
3. Sen, Y.: Analysis of equipment PHM architecture based on condition maintenance. Comput. Sci. **44**(09),17–22+33 (2017)
4. Hao, L., Haoqi, W., Gen, L., et al.: Concept, system structure and operating mode of industrial digital twin system. Comput. Integr. Manuf. Syst. **27**(12), 3373–3390 (2021)
5. Fei, T., Weiran, L., Jianhua, L., et al.: Digital twin and its potential application exploration. Comput. Integr. Manuf. Syst. **24**(01), 1–18 (2018)
6. Datong, L., Kai, G., Benkuan, W., et al.: Summary and perspective survey on digital twin technology. Chin. J. Sci. Instrum. **39**(11), 1–10 (2018)
7. AlAli, A.R., Ragini, G., Tasneem, Z.B., et al.: Digital twin conceptual model within the context of internet of things. Future Internet **12**(10), 1–15 (2020)
8. Joanna, H.: Digital twin-driven approach towards manufacturing processes support. J. Phys: Conf. Ser. **2198**(1), 1–11 (2022)
9. Zhou, Z., He, D., Chen, Y., et al.: Optimization model and application of preventive maintenance of key components of train based on GA-BP neural network. J. Railway Sci. Eng. **18**(06), 1382–1391 (2021)

Modeling Method of Guide Roller Manufacturing Information Based on Ontology Modeling

Keqiang Shi, Shanhui Liu(✉), Zengqiang Zhang, Song Qian, and Han Zhang

Faculty of Printing, Packaging Engineering and Digital Media Technology, Xi'an University of Technology, Xi'an, China
shanhuiliu@xaut.edu.cn

Abstract. Cloud manufacturing is the direction for guide roller manufacturing, and building its information model is a prerequisite for cloud manufacturing. Therefore, this paper proposes an ontology-based modeling approach for cloud manufacturing information modeling of printing press guide roller. Firstly, the structure of the typical guide roller of printing press is analyzed, and the manufacturing process and characteristics of the guide roller are studied. Secondly, according to the basic principle of ontology modeling method, an ontological model of the cloud manufacturing information of the printing press guide rolls is established. Finally, the evaluation method of cloud manufacturing information ontology model of printing press guide roller is proposed. The cloud manufacturing information ontology model of guide roller established in this paper lays a foundation for cloud manufacturing of printing press guide roller.

Keywords: Guide roller · Manufacturing information · Ontology modeling

1 Introduction

Guide roller is a central component of printing press and is used in a wide variety of presses. At present, the production of guide roller still adopts the traditional manufacturing mode, with low resource utilisation rate and high cost, so it needs to be transformed and upgraded in the direction of intelligence. Cloud manufacturing has been developed on the basis of "manufacturing as a service" using the idea of cloud computing, which can improve resource utilisation and achieve manufacturing intelligence. The establishment of a unified and sharable information model is the key to cloud manufacturing, therefore, how to establish a guid roller manufacturing information model is to be solved.

For a long time, researchers at home and abroad have long studied the method of cloud manufacturing information ontology modeling [1]. Cao [2] applied ontology to model the design knowledge of the launch vehicle's guidance and attitude control system. Dong et al. [3] used the command information system ontology modeling technique to develop a scene-aware ontology modeling method of intelligent command information system. Aiming at some problems existing in the current equipment acceptance

M. Xu et al. (Eds.): CACPP 2022, LNEE 991, pp. 420–424, 2023.
https://doi.org/10.1007/978-981-19-9024-3_52

methods, Li et al. [4] proposed an intelligent equipment acceptance method based on ontology technology. In order to solve the complex workshop information problem, Tu et al. [5] uses protégé to construct the manufacturing process information ontology model. Zhuang and Yin [6] proposed an ontology modeling method of machine tool equipment resources based on metadata, and completed the establishment of the model. Wen et al. [7] constructed a multilingual ontology knowledge based on a lexical-level representation of descriptions to address the problem of inconsistent knowledge representation. Sridhar and Nori [8] proposed an ontology based framework for systematic modeling of different aspects of instructional design knowledge based on domain patterns. Asuquo and Usip [9] proposed an ontology modeling approach based on the Semantic Web, which is used to express the user's activities and social roles on mobile devices, using protégé to implement the model. El-Sappagh and Elmogy [10] combined ontologies with fuzzy logic reasoning to create a fuzzy case-base ontology for the medical domain. Trokanas et al. [11] proposed an ontology-based approach to modelling manufacturing resources and developed a corresponding resource ontology model. The cloud manufacturing information model for guide roller is in a missing state, limiting its cloud manufacturing implementation.

In this paper, an ontology-based information modeling method is proposed for guide roller. Section 2 analyzes the guide roller mechanism of printing press and its manufacturing process. Section 3 realizes the establishment of the guide roller information model. Section 4 proposes the evaluation method of the guide roller information model. Section 5 summaries the research content and significance of this paper.

2 Structure and Manufacturing Process of Guide Roller

As the core component of the printing press, the guide roller plays roles in supporting the substrate and ensuring its stable operation. There are many types of guide rollers. Through field research on printing press guide roller production enterprises, the main types of guide rollers are summarized: with steel shaft guide roller, without steel shaft guide roller, cooling guide roller, dance roller, etc., which has a steel shaft guide roller as an example to show the guide roller processing technology are shown in Fig. 1.

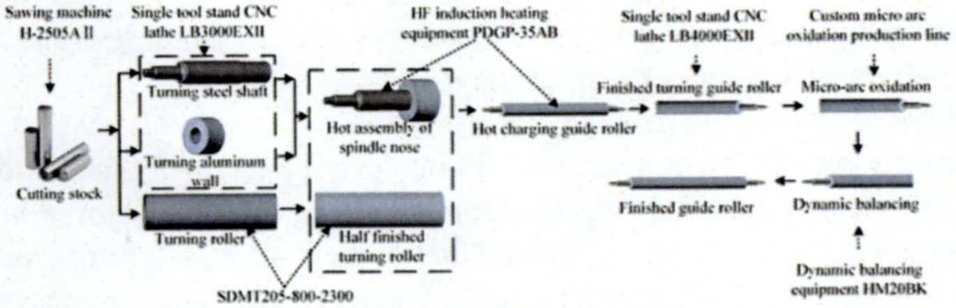

Fig. 1. Guide roller processing technology

3 Ontology Model of Guide Roller Manufacturing Information

Based on the analysis of the guide roller mechanism and manufacturing process, Combined with ontology modeling method, the roller guide cloud manufacturing information is modeled, including:

1) Clarify the purpose of the model and the information involved
The guide roller cloud manufacturing information model is the basis for cloud manufacturing, was created to organise knowledge in this area using ontological ideas and description languages. Solving the deficiency of the model can provide a basis for the subsequent work. The guide roller cloud manufacturing information model includes manufacturing resource information and manufacturing demand information of manufacturing process.
2) Determine classes and attributes
The guide roller cloud manufacturing information model includes process requirements category and equipment manufacturing resource service category. The parent attributes of process requirements include: process release, process processing, and process constraint. The parent attributes of the equipment manufacturing resource service include: equipment service, equipment processable scope, equipment service constraint. The child-attributes of process publishing include: requirement name, deadline for completion, minimum batch, required address range, maximum cost and payment method. The child-attributes of process processing include: processing method, processing weight, processing materials, processing drawings, processing size. The child-attributes of process constraint information include: machining accuracy, dynamic balance gramrange, hot charging temperature and material supply mode. The child-attributes of equipment service include: device name, model, type, address, status, service fee, time, payment method, and the starting batch to accept. The child-attributes of the machinable range of the equipment include: available machining method, maximum machinable weight, machinable material, and range of machinable size. The sub-attributes of equipment service constraint include: maximum dynamic balance grams, hot charging temperature, and maximum machining accuracy.
3) Establish the model framework
Building a model framework based on classes and attributes. Figures 2 and 3 show the guide roller process requirements framework and the equipment manufacturing resource services framework respectively.
4) Formal description of the model
To better aid knowledge sharing and communication, the model framework is to be described formally with the help of a computer language. The representation is in the form of a collection, with the initial letter of the class and its attributes as a code, and the information about the class and its parent attributes is represented in the form of capital letters, with the remaining sub-attributes in the form of initial capital letters and remaining lower case letters. The following is a formal description of the framework of the guide roll process requirements model as an example: PPD = {IR, PI, CI} (PPD: process demand, IR: process release, PI: process processing, CI: process).

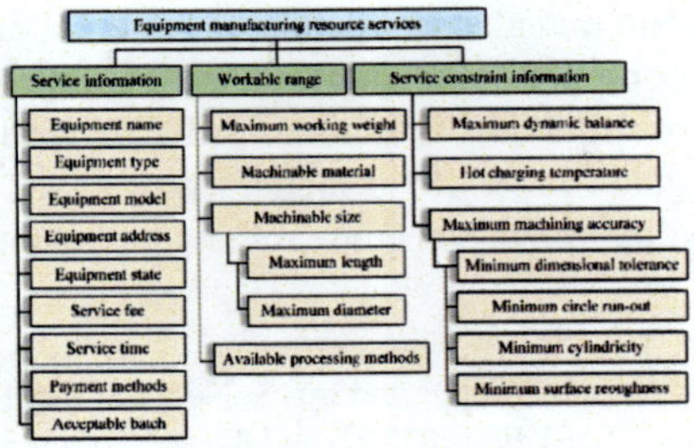

Fig. 2. Process requirement model framework

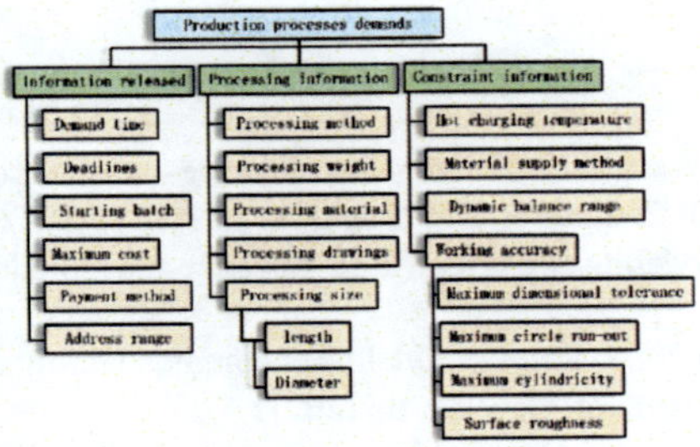

Fig. 3. Process requirement model framework

4 Evaluation of Manufacturing Information Ontology Model

The manufacturing information ontology model evaluation is shown in Fig. 4.

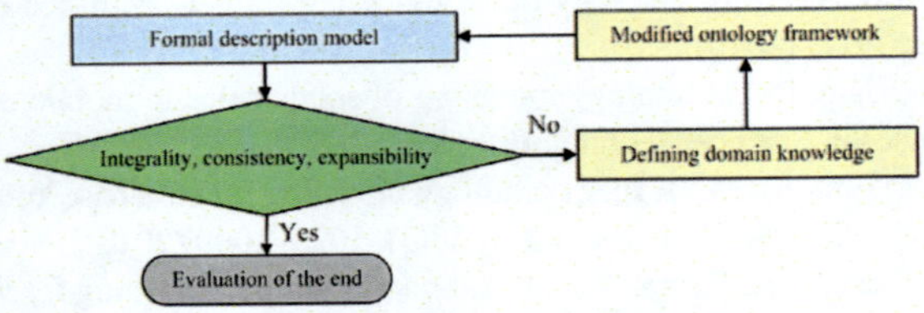

Fig. 4. Information model evaluation process

The evaluation process is a multi-level decision problem and the main evaluation indicators include: completeness, consistency and scalability. A corpus-based concept evaluation model for completeness, based on a concept semantic similarity algorithm to evaluate the consistency and scalability of relationships between concepts.

5 Conclusion

Aiming at the lack of information model for cloud manufacturing of guide roller, this paper analyses the types of guide roller and manufacturing process. The ontology idea is introduced into the cloud manufacturing field of guide roller, and an ontology-based cloud manufacturing information modeling method is proposed for guide roller, and the evaluation process of the model is summarized. The model effectively solves the bottleneck problem of guide roller cloud manufacturing, and lays the foundation for clustering, matching and optimizing of guide roller cloud manufacturing resources and services.

Acknowledgements. This project is supported by the National Key Research and Development Program of China (Grant No. 2019YFB1707200) and the technology innovation leading program of Shaanxi Province (Grant No. 2020QFY03-08).

References

1. Li, B., Zhang, L., Wang, S.: Cloud manufacturing—a new service-oriented networked manufacturing model. Comput. Integr. Manuf. Syst. **16**(01), 1–7 (2010)
2. Cao, X.: Ontology modeling of carrier rocket guidance and attitude control system, Southeast university (2018)
3. Dong, Q., Si, W., Zhu, W.: Ontology modeling of command information system for situational awareness. Comput. Technol. Develop. **06** (2021)
4. Li, Z., Meng, C., Wang, C.: Research on equipment intelligent Inspection and acceptance method based on ontology technology. Acta Armamentarii. **03** (2020)
5. Tu, F., Huang, H., Yao, L.: Research on ontology modeling for discrete workshop manufacturing information. Mech. Design Manuf. **02** (2018)
6. Zhuang, P., Yin, C.: Lightweight modeling method of machine tool equipment resources based on metadata. Modular Mach. Tools Autom. Mach. Technol. **12** (2019)
7. Wen, L., Li, J., Liu, Z., Jin, Y.: Knowledge representation and ontology modeling based on concept hierarchical network. J. Chin. Inform. Sci. **04** (2018)
8. Sridhar, C., Nori, K.V.: An ontology based modeling framework based on ontology. Smart Learn. Environ. **7**(1) (2020)
9. Asuquo, D.E., Usip, P.U.: Ontology modeling of social roles in mobile computing environments. Advances in Science, Technol. Eng. Syst. **3**(2), 319–328 (2018)
10. El-Sappagh, S., Elmogy, M.: A fuzzy ontology modeling for case base knowledge in diabetes mellitus domain. Eng. Sci. Technol. Int. J. **20**(3), 1025–1040 (2017)
11. Trokanas, N., Cecelja, F., Raafat, T.: Semantic input/output matching for waste processing in industrial symbiosis. Comput. Chem. Eng. **66**, 259–268 (2014)

Research on the Evaluation Method of Chinese Character Writing Quality Based on Machine Vision

Guirong Dong(✉), Zhixing Han, Yanqi Gu, Fuqiang Zhang, and Pihong Hou

Faculty of Printing, Packaging Engineering and Digital Media Technology, Xi'an University of Technology, Shaanxi, China
dongguirong2005@xaut.edu.cn

Abstract. In recent years, with the increasing emphasis on traditional culture, writing robots have gradually become a hot spot in the field of robotics research. In this paper, a method for evaluating the aesthetics of written Chinese character was proposed. By comparing the writing results with the strokes of the standard regular script, calculating the two aesthetic indicators of balance and inclination, and finally the grade and relative comments were given to the writing results. The experimental results show that the influencing factor for the balance indicator is the type of stroke. While the inclination indicator is more dependent on the type of font structure. This method of evaluating the aesthetics of Chinese characters helps to further improve the level of written Chinese character and also provides a theoretical basis for the study of Chinese character strokes.

Keywords: Calligraphy robot · Machine vision · Trajectory tracking · Chinese character evaluation · Aesthetic indicators

1 Introduction

Chinese characters are mainly made up of a series of basic strokes and can be formed by varied strokes. Distortion and deformation of Chinese characters are easily caused by changing strokes. The key point of improving Chinese character writing is to enhance the quality of strokes. Aesthetic indicators for the evaluation of Chinese characters should be established because judging the quality of writing by human vision may cause subjective results. The traditional evaluation methods of handwritten Chinese character include the inflection point Angle evaluation algorithm based on independent strokes of Chinese characters and the evaluation method based on "米" lattice form [1, 2]. However, the above methods are complicated to operate and the evaluation effect is limited. In this paper, the influence of strokes on the overall writing quality of Chinese characters has been estimated, as well as the interrelationship between strokes has been analyzed, a standard method for evaluating the quality of written Chinese characters has been eventually developed to guide the process of written Chinese characters.

M. Xu et al. (Eds.): CACPP 2022, LNEE 991, pp. 425–432, 2023.
https://doi.org/10.1007/978-981-19-9024-3_53

2 Introduction of the Robot Writing Platform

2.1 Robotic Writing Platform

The robot writing platform based on vision recognition used in this paper consists of a hardware system and a software system. The hardware system contains modules such as the Aubo-i5 collaborative robot, the 3D vision camera Kinect 2.0, and the end-effector. The software system is an intelligent robot operating system named COBOTSYS which is compiled in C++ to complete specific tasks with hardware. The robot writing platform and the end-effector were shown in Fig. 1.

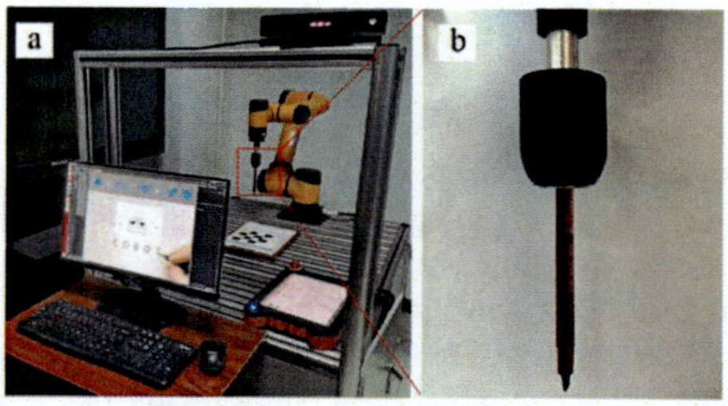

Fig. 1. (a) Robotic writing platform. (b) End-effector

2.2 Process of Chinese Characters Writing

The process of robotic Chinese character writing is as follows: firstly, a calibration board was used to write Chinese characters in the air, at the same time, the writing motion trajectory information was obtained by the Kinect vision camera in real-time [3]. Then, the data of stroke trajectories were converted into the joint values of the manipulator by the kinematic algorithm. Finally, 2D trajectory data was sent to the manipulator and the end-effector outputs the writing results [4].

3 Experimental Methods

Recently, most evaluations of written Chinese characters mainly rely on the subjective perceptions of estimators, a large scale of evaluations will be given to the same writing character. Therefore, a standard evaluation system was essential to be established. In this paper, two evaluation indicators of balance and inclination [5] were adopted as aesthetic parameters to estimate the effectiveness of written Chinese characters.

The quatrain "Mountain Xing" by the Tang Dynasty poet Du Mu was selected as the test sample [6]. The poem includes 28 characters which contain the basic strokes of Chinese characters. Considering the standardization and normalization, the regular script was set as the target font to estimate the writing results through the aesthetic indicators of balance and inclination.

3.1 Information Extraction of Regular Script Stroke

The extraction of regular script strokes information was divided into two parts, pre-processing and information extraction, where pre-processing includes image denoising, binarization, skeletonization and deburring [7]. Information extraction includes corner point detection. By comparing and analyzing the final Chinese character information, the correct regular script stroke information was obtained.

The flow of Chinese character image processing was shown in Fig. 2. In step 2, corner points were obtained to split the Chinese characters through the Harris corner point detection method, aiming to form independent stroke segments [8, 9]. The tilt angle of two adjacent stroke segments was calculated separately in terms of corner point coordinates. According to the tilt angle, if the difference value between them was less than a set threshold value, the two-stroke segments were combined and considered as one stroke, and vice versa. Therefore, the Information extraction of regular script strokes was realized by detaching independent strokes. In this paper, the character "于" was taken as an example to carry out stroke extraction, as shown in Fig. 3. The lengths and tilt angles of strokes are calculated and noted as l_i^* and θ_i^* respectively.

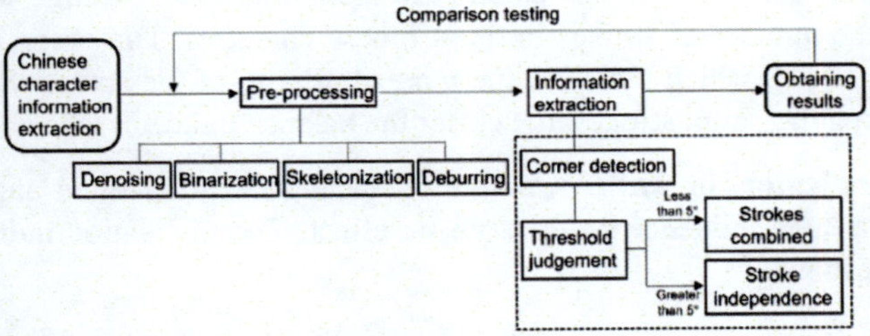

Fig. 2. Information extraction process

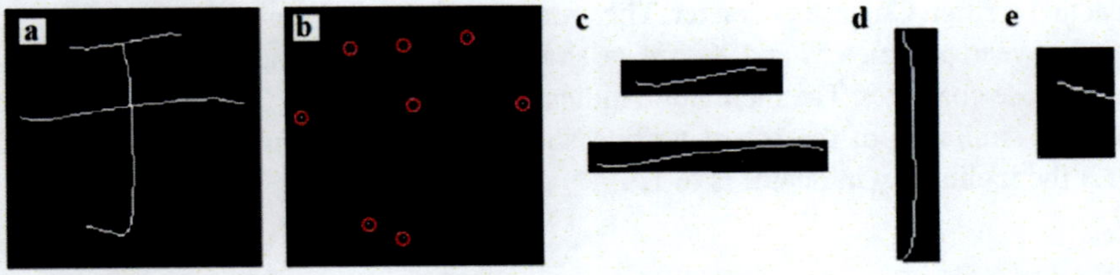

Fig. 3. Stroke extraction process of the Chinese character "于". (a) Chinese character skeleton. (b) Corner points. (c) Horizontal stroke. (d) Vertical stroke. (e) Hook stroke

3.2 Information Extraction of the Actual Written Chinese Characters

After the preliminary treatment of written Chinese characters trajectory data points collected by the Kinect camera, the corner points at the corresponding positions were extracted concerning the regular script, then the strokes were detached through the method introduced in Sect. 2.1. The lengths and inclination angles of stokes were noted as l_i and θ_i respectively.

3.3 Assessment of Similarity in Chinese Character

In this paper, two aesthetic indicators, balance and inclination, were chosen to evaluate written Chinese characters. Balance was used to evaluate the coordination between the lengths of adjacent strokes, and inclination was used to evaluate the degree of aesthetic of Chinese characters.

Balance indicator. Generally, by enlarging or reducing all the strokes in one Chinese character at the same time, the overall stylistic effect of the character will not be affected. But a large change in any one stroke, a significant effect would happen and the character can even change into another one. Therefore, an aesthetic evaluation indicator, balance, influenced both accuracy and aesthetic of Chinese characters. The balance indicator can be expressed as follows:

$$I_b = e^{-\frac{\sum_{i=2}^{N}\left|\frac{l_i}{l_{i-1}} - \frac{l_i^*}{l_{i-1}^*}\right|}{N-1}} \tag{1}$$

where l_i^* is the length of the ith stroke in the regular script. l_i is the length of the ith stroke in the actual written Chinese character. The parameter l can be obtained by measuring the distance between corners. The parametric model of stroke "Heng" was shown in Fig. 4. N is the number of strokes in one Chinese character. The range of balance is between 0 to 1 [10], which indicates the more similarity of the actual writing strokes length to the regular script strokes, the closer the balance indicator is to 1.

Inclination indicator. In written Chinese character, each tiny tilted angle of stroke would cause a huge influence on the stylistic effect. The inclination indicator can be expressed as follows.

$$I_c = e^{-\frac{\sum_{i=1}^{N}|S_i^* - S_i|}{N}} \tag{2}$$

where S_i^* is the tilt of the ith strokes in the regular script. S_i is the tilt of the ith strokes in the actual written Chinese character. The parameter S can be obtained by measuring the angle between parameter l and X-axis, as shown in Fig. 4. N is the number of strokes in one Chinese character. The inclination indicator also varies from 0 to 1, which indicates the more similarity of the actual writing strokes tilt to the regular script strokes, the closer the inclination indicator is to 1.

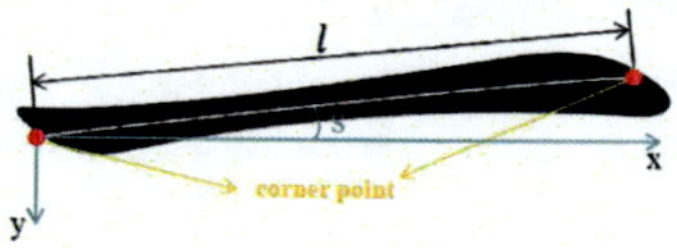

Fig. 4. Parametric model of stroke "Heng"

Evaluation of written Chinese characters. Weighting and evaluation were carried out according to the calculation results of the two aesthetic indicators including the balance indicator and inclination indicator. The weighted proportion of the two indicators is 0.5 each. Besides, the relevant grade and commentary were given in Tables 1 and 2.

Table 1. Definition and comments of balance indicator I_b

Evaluation indicators	Score	Grade	Commentary
Balance I_b	0.7–1.0	Excellent	The strokes are symmetrical
	0.4–0.7	Medium	The proportion of strokes need to be improved
	0.0–0.4	Worse	The coordination between strokes is poor

Table 2. Definition and comments of inclination indicator I_c

Evaluation indicators	Score	Grade	Commentary
Inclination I_c	0.7–1.0	Excellent	Chinese characters are standardized
	0.4–0.7	Medium	The tilt of the strokes need to be improved
	0.0–0.4	Worse	The standardization of the strokes is poor

4 Experimental Results and Analysis

There are 28 Chinese characters in the poem "A Journey to the Mountains". These characters classify the four structural types.

The upper and lower structure characters include: 寒, 家, 爱, 霜, 有, 花, 云;

The left and right structure characters include: 径, 斜, 停, 枫, 林, 晚, 叶, 红;

The semi-encircled structure characters include: 远, 处;

The regular script characters include: 上, 山, 石, 白, 生, 人, 车, 坐, 于, 二, 月.

The writing trajectory for the Chinese characters 云, 叶, 远, 于 were shown in Fig. 5. Specific scorning and commentary of selected Chinese characters were shown in Table 3.

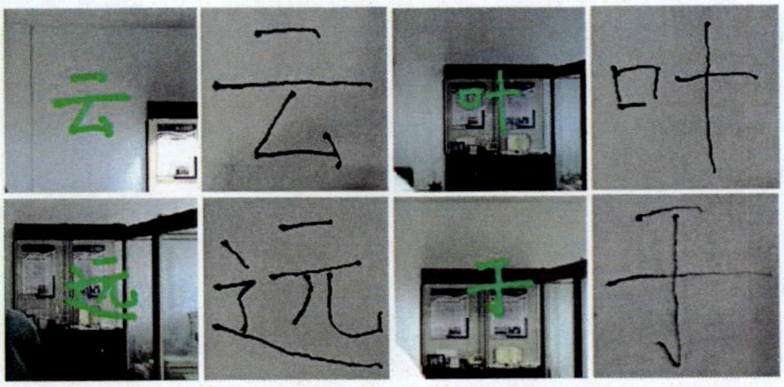

Fig. 5. Trajectory point acquisition and robot writing trajectory for the Chinese characters '云', '叶', '远' and '于'

Table 3. Evaluation of selected Chinese characters

Chinese characters	Balance	Inclination	Rating
云	0.7386	0.7469	Excellent
花	0.6947	0.6178	Excellent
叶	0.7133	0.7867	Excellent
林	0.6741	0.6184	Medium
远	0.5752	0.2520	Worse
处	0.5895	0.5110	Medium
于	0.8548	0.7577	Excellent
山	0.7845	0.7871	Excellent

4.1 Analysis of Balance Similarity Results

Balance values of all structural characters ranging from 0.65 to 0.85 were shown in Table 3, the results mean that the scores were related to the stroke types at all, rather than the structural type of the characters. In other words, when a character contains strokes such as left falling down, Left falling down, hook, vertical hook and horizontal hook, even if the subject had mastered the proportional relationship between the lengths of the target strokes before writing, it was still difficult to control them accurately at the desired level during the writing process, resulting in the poor overall coordination of the Chinese characters. Therefore, it can be considered that the balance indicator depends on not only the writing level of the subjects but also the stroke types. The smoother the stroke, that is, the higher the degree that the points in the stroke are in the same straight position, such as horizontal and vertical strokes, the better the balance of Chinese characters writing results. The more changes in the stroke, that is, the lower the degree that the points in the stroke are in the same straight position, such as left falling down, Left falling down and vertical hook strokes, the lower the balance of Chinese characters writing results. This shows that the balance values of Chinese characters depend more on the complex strokes in Chinese characters.

4.2 Analysis of Inclination Similarity Results

Based on the data in Table 3, the inclination value of "远" and "处" of the semi-enveloped structure are 0.2520 and 0.5110, respectively. It was clear that the inclination value of the lower-left semi-enveloped characters was generally lower than other structures.

Therefore, the strokes of Chinese characters with this structure were less normative. The main reason was that the lower-left semi-encircled characters generally had a long stroke and can be divided into several stroke segments. The fitted curve was comprised of many stroke segments that would directly decrease the value of the inclination indicator. Furthermore, for semi-enveloped structures, it was difficult to ensure that the strokes were written at the same angle of tilt as in a standard regular script, whether written by a human hand or robot. Therefore, unlike the balance indicator, the inclination indicator of Chinese characters is related to the type of font structure.

5 Conclusion

Focus on robotic Chinese character calligraphy, a method for evaluating the aesthetics of written Chinese character based on machine vision was proposed. In this method, writing trajectory points were collected and compared with the standard regular script to evaluate the writing results. There were two aesthetic indicators proposed in the evaluation including the balance indicator and inclination indicator, and then the grade and relative comments were given to the writing results.

The results showed that the balance indicator depends on stroke types, nevertheless, the inclination indicator of Chinese characters is related to the type of font structure. The two indicators can evaluate Chinese characters from different dimensions. The proposed Chinese character quality evaluation method is consistent with the subjective effect of human vision, indicating the effectiveness of the proposed Chinese character writing quality evaluation method. It is of great significance to further improve the writing level of Chinese characters and provide a theoretical basis for the study of Chinese strokes.

Although the proposed method can effectively evaluate written Chinese characters, the evaluation results are not comprehensive enough. More aesthetic evaluation indicators, such as writing strength, can be added to make the evaluation effect more in line with human aesthetics. Besides, the writing robot in this paper only wrote on a two-dimensional plane, it should be considered to increase the writing force on the Z-axis, so as to enrich the width of the strokes when writing Chinese characters and realize soft brush writing.

References

1. Sun, J.H.: Research and implementation of an intelligent evaluation method for calligraphy Chinese Characters, Hubei University of Technology (2020)
2. Yu, K., Wu, J.Q., Zhuang, Y.T.: Calligraphic characters retrieval based on skeleton similarity. J. Comput.-Aided Design Comput. Graphics **21**(6), 746–751 (2009)
3. Zhang, D.Z.: Research on real-time writing control of robots. Nanjing University of Posts and Telecommunications (2019)
4. Chao, F., Huang, Y.X., Zhang, X.: A robot calligraphy system: From simple to complex writing by Human Gestures. Eng. Appl. Artif. Intell. **59**, 1–14 (2017)
5. Xing, S.M., Liang, D.T., Liang, D., et al.: Research on robotic calligraphy copying based on style migration technology. Mech. Manuf. **56**(07), 42–47 (2018)
6. Huang, F.: Research on the evaluation method of hard-pen Chinese character writing quality based on template matching. Nanjing Normal University (2015)

7. Huang, Y.G.: A research on NAO-based robotic soft-pen calligraphy handwriting. Jiangxi University of Technology (2020)
8. Zhang, L.T., Huang, X.L., Lu, L.L.: Fast algorithm for Harris corner point detection based on grayscale differencing with templates. J. Instrum. **39**(02), 218–224 (2018)
9. Ma, Z., Su, J.B.: Aesthetics evaluation for robotic Chinese Calligraphy. IEEE Trans. Cogn. Dev. Syst. **9**(1), 80–90 (2017)
10. Liang, D.T., Liang, D., Xing, S.M., Li, P., Wu, X.C.: A robot calligraphy writing method based on style transferring algorithm and similarity evaluation. Intel. Serv. Robot. **13**(1), 137–146 (2019). https://doi.org/10.1007/s11370-019-00298-3

Robot Vision Recognition System Based on Improved YOLOv3 Algorithm

Yichen Gao, Zhenqing Gao(✉), Xinhao Chen, and Zhen Zhang

College of Mechanical and Electrical Engineering, Beijing Institute of Graphic Communication, Beijing, China
gaozhenqing@bigc.edu.cn

Abstract. With the rapid development of the logistics business, a visual recognition system is being developed to speed up the transportation, packaging, and categorization of goods. This system can enhance identification speed and accuracy. First, the experimental platform was set up, and the experiment was performed on a six-axis robot with a camera lens, light source, and acquisition card. The receptive field module was then used to construct an updated YOLOv3 model based on the original YOLOv3 model to enhance the in-depth features and increase the feature representation ability. Experiments were conducted on VOC data sets. The mAP values of twelve different types of objects, the visual recognition effects of five different groups of pictures, and the visual recognition values of the original and revised models were compared. The results showed that the improved YOLOv3 model outperformed the original model in terms of mPA, visual recognition image, and recognition speed.

Keywords: Visual recognition · Improved YOLOv3 · Intelligent robot · Experimental platform

1 Introduction

With the development and progress of the logistics business, it will encounter a wide range of items of all forms and sizes. Robots with visual recognition functions are frequently utilized in today's industry to increase the speed and accuracy of the logistics industry.

By varying the size of the model structure, the YOLOv3 algorithm, commonly employed in visual recognition robots, can balance speed and accuracy. However, the accuracy of YOLOv3 in small target recognition and detection is limited. This study will create a robot vision identification and grasping operating platform and experimental platform based on the improved YOLOv3 target detection algorithm, recognize the objects in diverse pictures, and validate its improved accuracy.

2 Research Status of Target Recognition

AlexNet, proposed by Krizhevsky et al., won the ILSVRC Challenge image classification task with a considerable advantage.

M. Xu et al. (Eds.): CACPP 2022, LNEE 991, pp. 433–439, 2023.
https://doi.org/10.1007/978-981-19-9024-3_54

Girshick et al. proposed R-CNN in CVPR 2014 and used a convolutional neural network to extract features [1]. The following year, they introduced the Fast RCNN algorithm. Ren et al. presented a Faster R-CNN network in 2016, considerably increasing detection frame generation speed [2].

In 2016, Joseph Redmon first published YOLOv1 in CVPR. Despite its high accuracy, the algorithm ran slowly [3]. Then, in 2017, Redmon introduced YOLO9000, which used a novel backbone network model to accelerate convergence [4]. The YOLOv3 model was obtained in 2018. In 2020, Bochkovskiy et al. proposed the YOLOv4 algorithm, which improved data processing, activation function, and backbone network [5].

In this paper, the receptive field module is added based on the classical yolov3 algorithm, and the input is convolved or pooled by the Inception module, and the jump connection finishes the multi-scale receptive field.

3 Construction of Experimental Platform

The upper computer module, relatively simple robot vision system module, six-axis robot, and robot drive controller module comprise the robot vision recognition and grasping platform developed in this study. The camera model used in the visual recognition system is the Logitech C270i, and it may be performed under enough lighting. Figure 1 depicts the experimental platform.

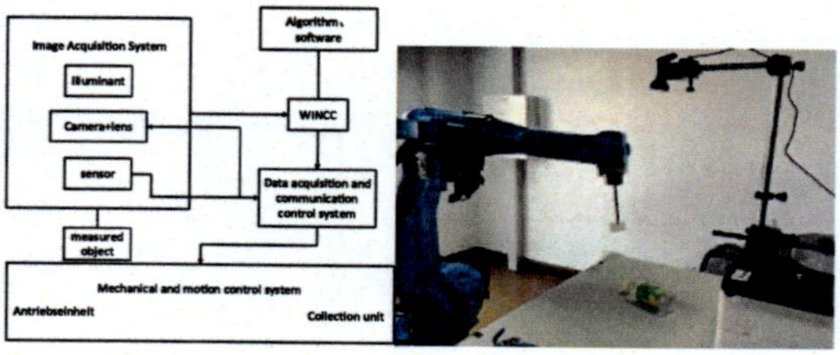

(a) (b)

Fig. 1. Composition diagram and physical diagram of the visual recognition system

The teaching device directs the experimental platform to capture an image of the experimental object and then transmits it to the higher computer, which recognizes the captured image using an algorithm.

4 Principle of Software and Algorithm

The YOLO series algorithm is a common one-stage target detection algorithm that integrates classification with the regression problem of target location by using anchor boxes, delivering excellent efficiency, flexibility, and good generalization performance.

4.1 Introduction of the YOLOv3 Algorithm

The model of YOLOv3 is substantially more complex than the prior model, and it features a superior primary categorization network [6]. YOLOv3 is comprised of three sections: the DarkNet-53 feature extraction network, the Feature Pyramid Networks (FPN), and the detection layer. The network structure of YOLOv3 is depicted in Fig. 2 [7].

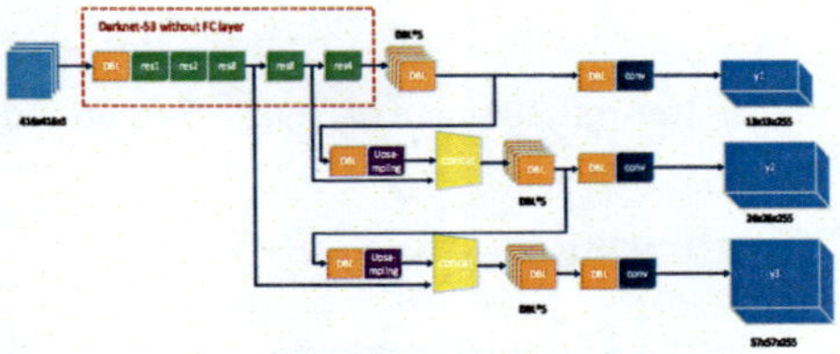

Fig. 2. Network structure of YOLOv3

Train YOLOv3 with the training set provided in Table 1. YOLOv3 has flaws, such as insufficient receptive field and low accuracy in recognizing and detecting small targets. Hence, this study enhances YOLOv3.

Table 1. Training configuration

Optimizer	Original learning rate	Weight decline rate	Varia-ble	Enter dimensions
Batch gradient de-scent (BGD)	0.001	0.005	0.68	230 × 230

4.2 Application of Target Detection Based on Improved YOLOv3

Based on YOLOv3, some decoder modules are provided to enhance the in-depth features and increase the feature representation ability, called YOLOv3-Improved. The optimized network structure is depicted in Fig. 3.

Because this network contains two decoder modules, the integration approach must be combined to decrease the network's computation. In the decoder, the receptive field module (RFB) is adopted. RFB is a module inspired by the human visual system's multi-scale group receptive field structure. It consists of two components: a multi-branch convolution layer and a hole convolution layer. It primarily simulates the structure of population Receptive Fields (pRFs) in the human visual system, increasing receptive fields without sacrificing resolution [8]. The input is convolved or pooled by the Inception module, or the convolution is expanded, and the jump connection completes the multi-scale receptive field.

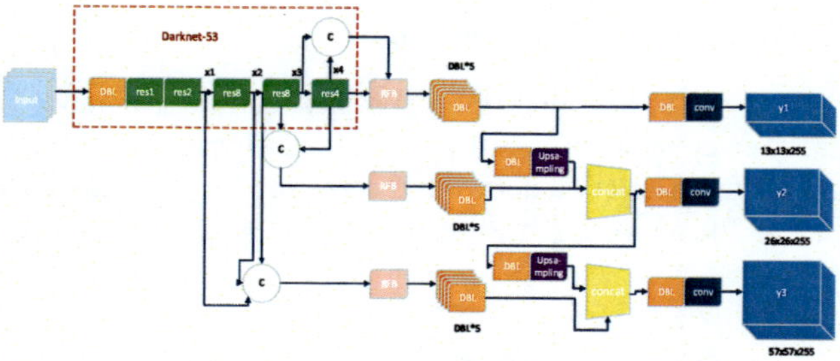

Fig. 3. Network structure of YOLOv3-improved

5 Experimental Confirmation

The enhanced YOLOv3-Improved model is compared to network models such as Fast R-CNN, Faster R-CNN, R-FCN, SSD, and YOLOv3.

5.1 MAP Comparison of Different Categories

The data set in this experiment is the VOC data set, and YOLOv3 and the enhanced YOLOv3-Improved are used to identify and analyze the 12 types of objects listed in the table below, with 80 identification experiments performed for each type of object. The experimental results are compared, and the comparison outcomes are displayed in Fig. 4.

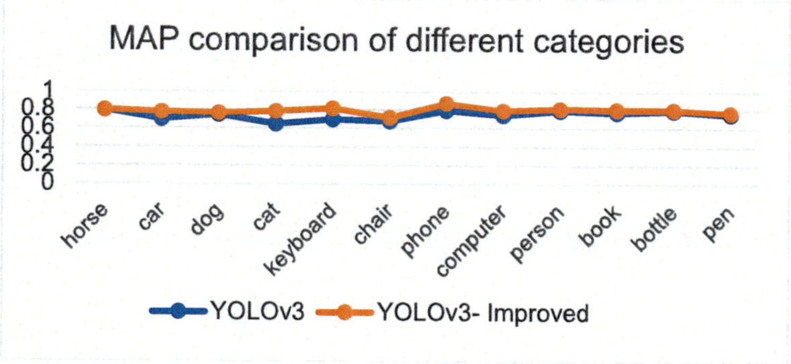

Fig. 4. MAP comparison between YOLOv3 and YOLOv3-improved

Figure 4 shows that blue represents YOLOv3, orange represents YOLOv3-Improved, and when $IOU = 0.5$, the mAP identified and detected by YOLOv3-Improved is better than or equivalent to the mAP of the YOLOv3 algorithm.

5.2 Comparison of Visual Recognition Effects

The recognition results of YOLOv3 and YOLOv3-Improved are compared for five groups of pictures, as shown in Fig. 5.

Fig. 5. Comparison chart of visual recognition effect

The recognition result using YOLOv3 is shown on the left, while the YOLOv3-Improved is shown on the right. The comparison of the first group of pictures shows that the YOLOv3-Improved model can recognize more objects in low-light situations. Similarly, when the following four groups of pictures are compared, it is evident that the visual recognition effect of the YOLOv3-Improved model is slightly better than the YOLOv3 model.

5.3 Comparison of Visual Recognition Values

Table 2 compares the detection speed indicators of the current mainstream algorithm and the improved algorithm. It is possible to conclude that the accuracy and speed of detection using the YOLOv3-Improved algorithm have been improved over YOLOv3.

Table 2. Numerical comparison

Detection algorithm	mAP/%	FPS
Fast R-CNN	70.3	0.7
Faster R-CNN	74.5	7.2
R-FCN	79.8	9.0
SSD	77.3	19.3
YOLOv3	80.6	32.2
YOLOv3-improved	82.5	32.4

As a result of the preceding three comparisons, it is possible to deduce that the YOLOv3-Improved model outperforms other algorithms in recognition and detection accuracy and speed, and outperforms YOLOv3 in target detection accuracy.

5.4 Comparison of Image Recognition in the Actual Situation

The following is an example of the recognition effect of the two algorithms on real situation objects and a confidence comparison of multiple objects as shown in Fig. 6 and Table 3.

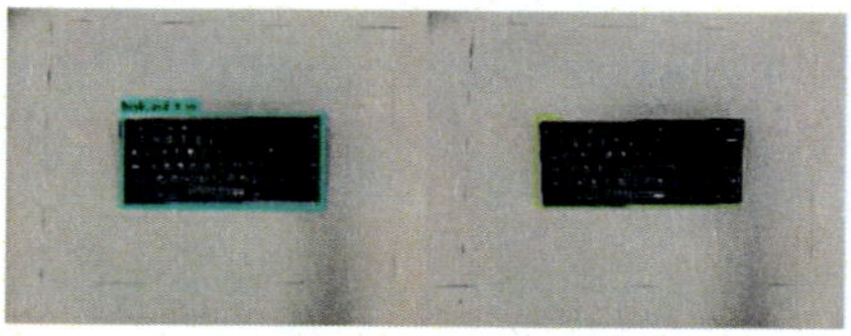

(a) (b)

Fig. 6. Example of actual recognition effect

Table 3. Confidence degree comparison

	Keyboard	Scissor	Bottle
YOLOv3	0.66	0.69	0.87
YOLOv3-Improved	0.91	0.91	0.69

6 Summary

To increase the operating efficiency of the logistics industry, this study proposes an improved target detection algorithm based on the YOLOv3 algorithm, which adds a receptive field module, convolutes or pools the input, and improves the speed and accuracy of object detection. The mPA values acquired by identifying different items and the experimental findings obtained by identifying five groups of pictures reveal that the revised algorithm is slightly faster and more accurate than the unmodified YOLOv3 algorithm. However, there are still several issues that have not been adequately addressed in this paper and will be investigated further.

References

1. Girshick, R., Donahue, J., Darrell, T., et al.: Rich feature hierarchies for accurate object detection and semantic segmentation. In: Proceedings of the Ieee Conference on Computer Vision and Pattern Recognition. IEEE, Columbus, USA, pp. 580–587 (2014)

2. Ren, S., He, K., Girshick, R., et al.: Faster R-CNN: towards real-time object detection with region proposal networks. IEEE Trans. Pattern Anal. Mach. Intell. **39**(6), 1137–1149 (2016)
3. Redmon, J., Divvala, S., Girshick, R., et al.: You only look once: Unified, real-time object detection. In: Proceedings of the IEEE Conference on Computer Vision and Pattern Recognition, pp. 779–788 (2016)
4. Redmon, J., Farhadi, A.: YOLO9000: better, faster, stronger. In: Proceedings of the IEEE Conference on Computer Vision and Pattern Recognition, pp. 7263–7271 (2017)
5. Bochkovskiy, A., Wang, C.Y., Liao, H.: YOLOv4: Optimal Speed and Accuracy of Object Detection, 04, 23 (2020)
6. Zhang, L.D., Deng, C.: Multi-scale fusion of YOLOv3 crowd mask wearing detection method. Comput. Eng. Appl. **57**, 283–290 (2021)
7. Lin, T.Y., Dollár, P., Girshick, R., et al.: Feature pyramid networks for object detection. In: Proceedings of the IEEE Conference on Computer Vision and Pattern Recognition, Honolulu, 21–26 July 2017, pp. 936–944 (2017)
8. Xiaopei, H., Dianhua, W., Zhijian, Q.: An improved YOLOv3 model for Asian food image recognition and detection. Open J. Appl. Sci. **11**(12), 1287–1306 (2021)

Intelligent Design of Agricultural Product Packaging Layout Based on Reinforcement Learning

Jianing Wang[1], Yuan Zhang[1], and Lei Zhu[2(✉)]

[1] School of Mechanical and Electrical Engineering, Beijing Institution of Graphic Communication, Beijing, China
[2] Postal Technology R&D Center, Beijing Institution of Graphic Communication, Beijing, China
zhulei@bigc.edu.cn

Abstract. Along with the rapid development of e-commerce logistics, the volume of business in online sales of agricultural products is also growing, but e-commerce for agricultural products also suffers from a lack of funding for packaging design and difficulties in matching the packaging image with the product itself. One of the solutions to these problems is to create an intelligent design platform through a machine learning paradigm that can quickly and inexpensively meet the personalised design needs of users. One of the key technologies is intelligent layout design. In this paper, a deep deterministic strategic gradient (DDPG) algorithm is used to output the layout of the packaging layout by taking the graphic design elements of the packaging as input data, while taking into account constraints such as aesthetic principles, to achieve automatic layout and automatic output of the layout of the packaging appearance. Finally, through a user survey, it is found that the packaging layout designed using the algorithms in this paper can basically meet the requirements.

Keywords: Packaging of agricultural products · Automatic layout · Reinforcement learning

1 Introduction

In recent years, along with the rapid development of e-commerce and logistics, the volume of agricultural products online sales business is also growing, and the transaction scale of China's fresh produce e-commerce market has reached 458.5 billion yuan in 2021 [1]. With the growth of packaging demand, the speed of traditional manual design has gradually failed to meet the needs of agricultural packaging design, and there is an urgent need for intelligent platforms to carry out intelligent layout design for customer needs.

Packaging design is a complex issue that encompasses layout, color, typeface, outline and style. In this paper, we will first consider the issue of layout generation. Many scholars have done a lot of research in the area of automatic layout of graphics, which consists of two main aspects: one is to make relevant layouts based on aesthetic principles [2, 3],

M. Xu et al. (Eds.): CACPP 2022, LNEE 991, pp. 440–445, 2023.
https://doi.org/10.1007/978-981-19-9024-3_55

and the other is to use machine learning algorithms to make layouts by learning from existing layouts [4, 5]. However, these methods are basically only for the layout of one page, and still lack applicability in the field of packaging design.

This paper adopts the three-stage method to realize the automatic layout of packaging: the first stage uses the LayoutTransformer model to complete the initial layout; the second stage adopts the DDPG algorithm to optimise the initial layout; and the third stage fills the elements in the wire frame representing the elements to realize the automatic generation of packaging layout. It is an effective exploration for the packaging field to carry out intelligent design of packaging.

2 Evaluation Standard of the Layout Quality

Considering the packaging layout types and layout elements comprehensively, and according to the relevant research in the [2, 6], we have defined the following three criteria for assessing the quality of the layout:

Alignment refers to the alignment of individual elements, as well as the alignment of individual elements or clusters of elements throughout the layout, including left-right alignment, top-bottom alignment, centre-alignment, etc.

$$e_{al}^{a} = \sum_{i=1}^{n}\sum_{j=1}^{n} A_{ij}^{a}$$
$$E_{al} = \max E(e_{al}^{a}) \tag{1}$$

A_{ij}^{a}ija is an a-alignment between element i and element j. $A_{ij}^{a} = 1$when both are aligned in a-alignment.

Overlap between design elements is very common in design, and a reasonable overlap will make the layout more aesthetically pleasing, while an unreasonable overlap will detract from the look and feel.

$$e_{gt} = \frac{\sum_{i=1}^{m} A_{gt-i}}{wh}; e_{gg} = \frac{\sum_{j=1}^{n} A_{gg-j}}{wh}$$
$$E_{gt} = E(e_{gt}); E_{gg} = E(e_{gg}); \tag{2}$$

$A_{gt\text{-}i}$ represents the region where pictures and text overlap, and $A_{gg\text{-}j}$ represents the region where pictures and pictures overlap.

Balance mainly considers the symmetry of the size of various elements on the page. Referring to the design, we design the balance as the upper and lower balance and the left and right balance:

$$e_{lr} = \frac{\min(A_l, A_r)}{\max(A_l, A_r)}; e_{tb} = \frac{\min(A_t, A_b)}{\max(A_t, A_b)}$$
$$E_{lr} = E(e_{lr}); E_{tb} = E(e_{tb}); \tag{3}$$

In the end, we decided to use the following formula as an assessment score for the quality of the layout. k_1, k_2, k_3, k_4 and k_5 are the weights of each item and are constant values.

$$E = k_1 * E_{al} + k_2 * E_{gt} + k_3 * E_{gg} + k_4 * E_{lr} + k_5 * E_{tb} \tag{4}$$

3 Layout Generation Mode

The use of a three-stage approach in this paper is an innovative attempt to improve layout efficiency and save layout time by performing an initial layout followed by an optimised layout, compared to using only one algorithm.

3.1 Initial Layout

LayoutTransformer is an autoregressive model for generating structured layouts that can be used to generate completely new layouts or to complete existing partial ones. As shown in Fig. 1, it uses the decoder structure of Transformer model, which mainly includes embedding layer, masked multi-head self-attention layer, feedforward layer, add and norm layer, anti-embedding layer, Softmax layer, etc.

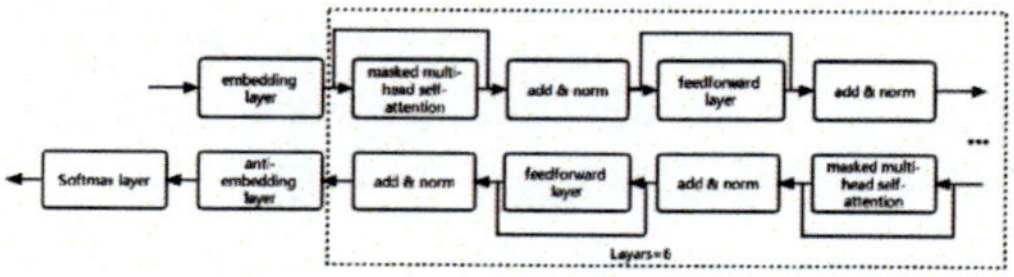

Fig. 1. LayoutTransformer model structure

The loss function in the LayoutTransformer model is determined as:

$$L = E_{\theta\ \text{Disc}}[D_{KL}(\text{SoftMax}(\theta^{L})\|p(\theta'))] + \lambda E_{\theta\ \text{Cont}}\left[\left\|\theta - \theta'\right\|\right] \quad (5)$$

The number of grid cells $n_{anchors}$ is set to 32 * 32, the model dimension d input to using the embedding layer conversion is set to 512, the number of self-attention layers n_{layers} to 6, the heads n_{heads} in each masked self-attention layer to 8, the cells d_{ff} to 2048, and to regularize the discard rate $P_{dropout}$ at 0.1 at the end of each feed-forward layer.

3.2 Layout Optimization

As shown in Fig. 2, the DDPG algorithm includes four networks: Actor online network, Actor target network, Critic online network, and Critic target network.

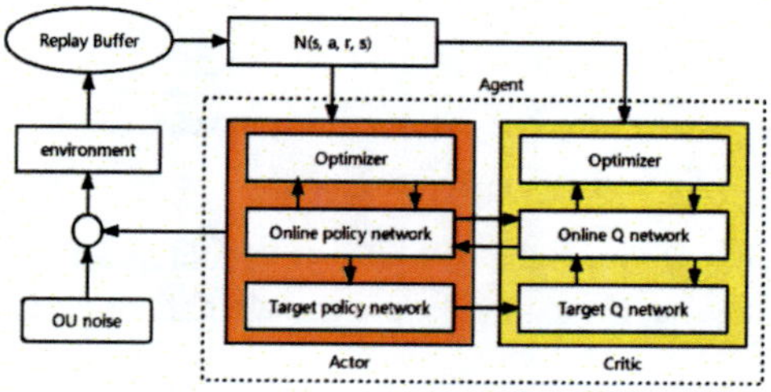

Fig. 2. Flow chart of the DDPG algorithm

The Value network objective function is defined as:

$$J_\beta(\mu) = E_\mu[r_1 + \gamma r_2 + \gamma^2 r_3 + \ldots + \gamma^{n-1} r_n] \tag{6}$$

where E is a function expectation, and γ is a discount factor.

Policy network parameters are updated as follows:

$$\nabla_{\theta^\mu} J = E_{s_1\ \rho^\beta}\left[\nabla_{\mathrm{a}} Q(s, a; \theta^Q)|_{s=s_t, a=\mu(s_t)}\right] \tag{7}$$

Reward is one of the key links of reinforcement learning, which plays an important role in determining the performance of the algorithm.In this work, the layout assessment metrics can be used intuitively.be defined as:

$$R = \begin{cases} E(s_{t+1}), E(s_t) + \alpha \le E(s_{t+1}) \\ E(s_t), E(s_t) + \alpha > E(s_{t+1}) \end{cases} \tag{8}$$

Finally, adjust the font size and picture size according to the wireframe size during element filling.

4 Experiments and Results

First of all, we collected a series of agricultural products packaging plane design drawings from the network, and used labelme (5.0.1) software to label each design element. The label types mainly include pictures, text, coding, elements in the standard library, etc. Later, we combined all the annotated images for training data set.The results trained using the LayoutTransformer model are shown in Fig. 3.

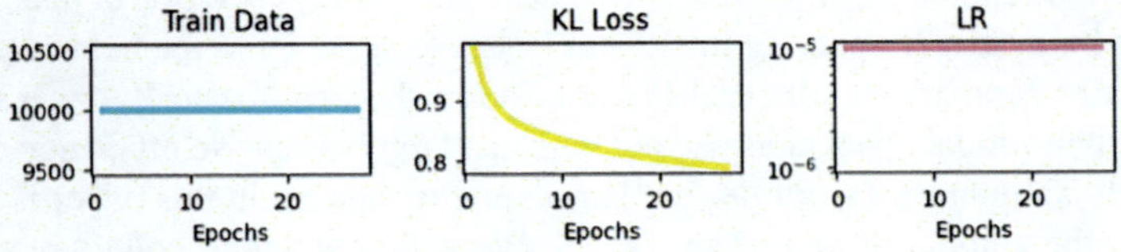

Fig. 3. Training results for the initial layout

Secondly, in the process of optimizing the layout, we conducted 100, 200, 500, and 1000 generations, respectively, as shown in Fig. 4, and 500 generations had the best results.

The final training results and after the filling elements are shown in Fig. 5. After the evaluation of professional designers, our design results can initially meet the layout needs of designers for the design elements.

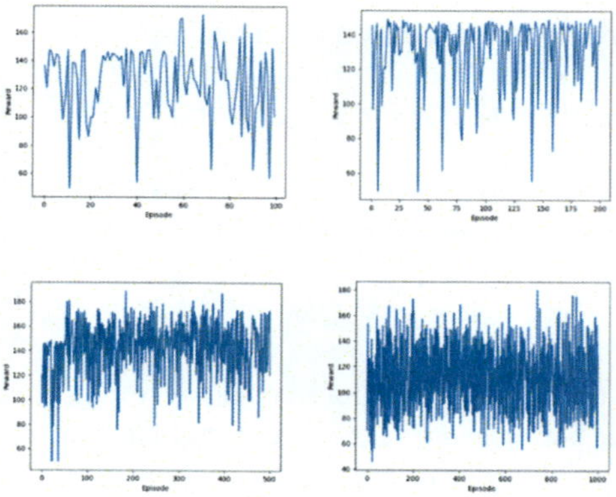

Fig. 4. Training results for the optimized layout

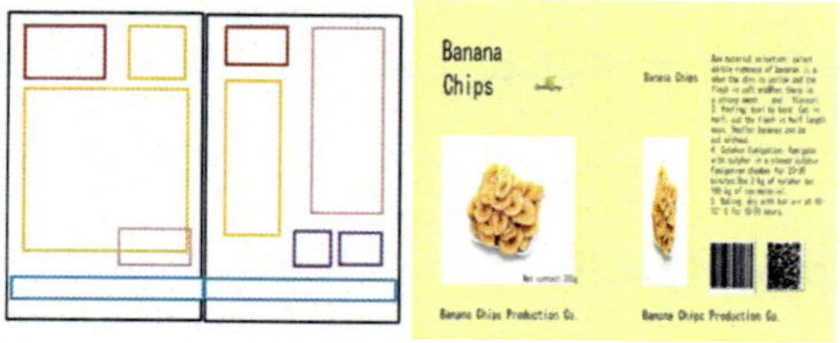

(a) (b)

Fig. 5. Optimization layout and results

5 Conclusion

This paper uses a three-stage approach to achieve automatic generation of pouch packaging, while innovatively applying intelligent layout technology to the field of packaging design. After comparison with real layouts, the method can basically meet the needs of agricultural product packaging layout, and can largely improve efficiency compared with traditional manual layout design. However, the data set in this paper is relatively single and the amount of data is also small. We will continue to collect more images of agricultural packaging designs, and will include data sets of different box types and specifications to build a larger data set, which can support better and faster generation of layouts.

References

1. Mo, D.: China's fresh food e-commerce trading scale to exceed 450 billion in 2021. Comput. Netw. **47**(18), 4 (2021)
2. O'Donovan, P.: Learning design: Aesthetic models for color, layout, and typography. Ph.D. Thesis, University of Toronto, Canada (2015)
3. Yang, X., Mei, et al.: Automatic generation of visual-textual presentation layout. ACM Trans. Multimedia Comput. Commun. Appl. **12**(2), 33.1 (2016)

4. Li, J., Yang, J., Zhang, J., et al.: Attribute-conditioned layout GAN for automatic graphic design. IEEE Trans. Visualization Comput. Graph. **99**, 1–1 (2020)
5. Gupta, K., Lazarow, J., Achille, A., et al.: Layout transformer: Layout generation and completion with self-attention. International Conference on Computer Vision (2021)
6. Hu, H., Zhang, C., Liang, Y.: A study on the automatic generation of banner layouts. Comput. Electr. Eng. **93**(2), 107269 (2021)

Design of Workshop Material Management System of Printing Enterprises Based on Modularization

Xiujie Chen, Wenjie Yang(✉), Zaining Lin, and Xuebin Zuo

Printing and Packaging Engineering School, Beijing Institute of Graphic Communication, Beijing, China
ywj@bigc.edu.cn

Abstract. Development and design based on modularization is currently an efficient production method for small and medium-sized enterprises. The paper is designed to optimize the efficiency of workshop material management, improve the completion and standardization of each workshop process, integrate and reorganize the different functional modules developed for workshop management, and further optimize the efficiency of the entire printing enterprise management operation. The workshop material management module is subdivided into basic module units using a modular development approach, and a customized workshop material management system is quickly and easily implemented based on enterprise characteristics and functional requirements. A combination of Java and SQL Server databases is used to achieve a customized functional design of the workshop material management system and the B/S architecture is used to facilitate the maintenance and development of the management system, and improve the efficiency of the entire workshop production management. With the help of this system, enterprises can not only reduce the workload of workshop managers but also improve real-time and accurate production management.

Keywords: Workshop materials · Management system · Modularization · Java

1 Introduction

Multi variety, variable batch, and individualized production are the primary production characteristics of printing enterprises, making it difficult to design enterprise management systems with a unified management model [1]. The system maximizes the timely sharing of material information among the warehouse, workshop, and machine, reduces material transportation loss and cost, and improves the workshop's production efficiency, as a result, the overall printing enterprise's production management efficiency [2].

The enterprise workshop material management is decomposed into minimum module units, the data interfaces defined between the module units, and a workshop material management system suitable for printing enterprises is designed according to the current production characteristics of printing enterprises, which is the rapid integration and reorganization of the workshop material management process, and customized design

M. Xu et al. (Eds.): CACPP 2022, LNEE 991, pp. 446–451, 2023.
https://doi.org/10.1007/978-981-19-9024-3_56

of the workshop material management system applicable to printing enterprises [3, 4]. Compared with the existing MES systems in enterprises, this paper focuses more on the functional optimization of each module and the advantage of designing the data interface between each module unit that can be quickly reconstructed.

2 System Analysis

2.1 System Functional Analysis

By decomposing the workshop material management of a printing enterprise into different module units, and defining the data interface between the module units, a workshop material management system suitable for a printing enterprise is quickly customized.

The system disassembles and establishes the data interface for the workshop material management functional modules, which are production receipt, replenishment, consumption, and inventory. The following functions are primarily implemented by the system: workshop material module, process material receiving module, process consumption module, and process inventory module. The system use case diagram is shown in Fig. 1.

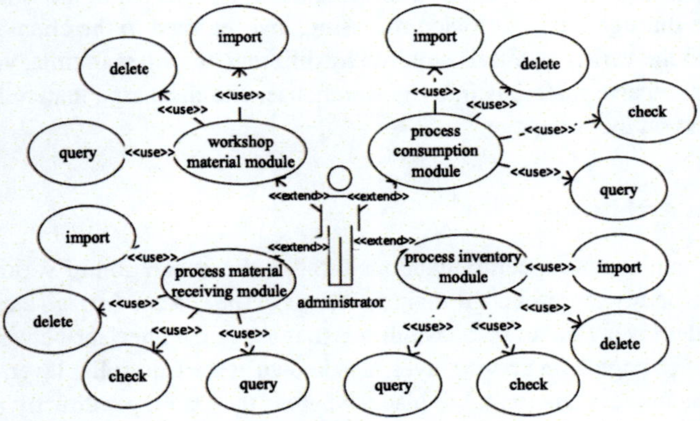

Fig. 1. Diagram of system's use case analysis.

2.2 Workshop Material Function Module Design

The system is organized into four functional components based on the demand analysis of printing workshop material management. For example, it's primarily designed to manage the workshop material inlet and outlet, process material collection and replenishment management, process consumption, process material inventory, workshop material inlet and outlet statistics, and other management modules according to the production daily. Each functional module is developed and designed independently, and different functional modules can be combined for different printing businesses.

2.3 Overall Business Model

In this paper, the system uses the Spring MVC design pattern, the entire system is divided into three functional levels Model, View, and Controller, where the model layer is the main part of the application, including the business logic layer and data access layer [5, 6]. The system is developed using Spring open-source framework, which makes the system easier to build. The overall framework model of the system based on business analysis is shown in Fig. 2.

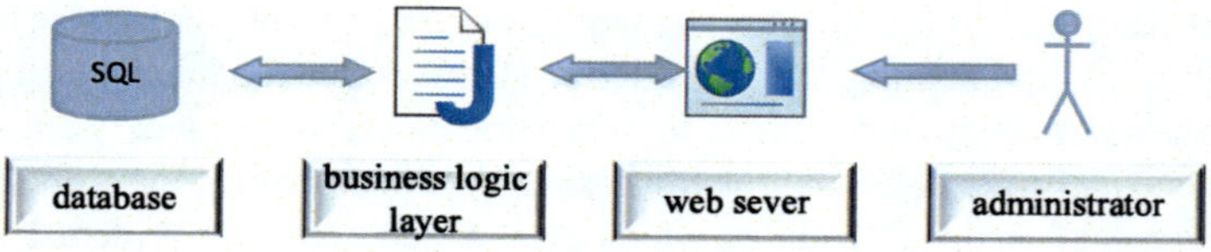

Fig. 2. Overall framework model

The workshop material management department's staff can log into the system for quick access to basic workshop material information, modification, and maintenance. And the workshop process's material information is transferred to the server, saved in the database through business logic processing, and returned to the client in real-time. According to the varied materials required by different processes in time, you can make up materials, receive materials into the warehouse, and discharge materials consumed in the process on the web page.

2.4 System Architecture

The system implements specific functions through the integration of various modules, and this part mainly represents the main functions of the system through class diagrams. The Controller layer's major responsibility is to manage the interface between the back-end and the web page. The Service layer, also known as the Controller layer, is where the core business implementation takes place, whereas the Dao layer directly manages the database and serves the Service layer. Figure 3 shows the workshop material management system module's class diagram design.

The system allows for the preservation and summary management of the workshop materials' outgoing, incoming, and inventory information, as well as the basic operations of adding, updating, querying, and deleting data. The Controller class handles requests for front-end material information processing for workshop material management, where the request operation of material information includes adding, importing, editing, deleting, and querying material, among other things. Through the methods in the Service class, the controller class realizes the processing of material information. The add(), update(), delete(), and inputItem() methods in the Service class are used to add, delete, change, check other operations of materials.

Figure 4 shows a sequence diagram of workshop material management. The manager will use the query method to get the basic material information from the Item table, then

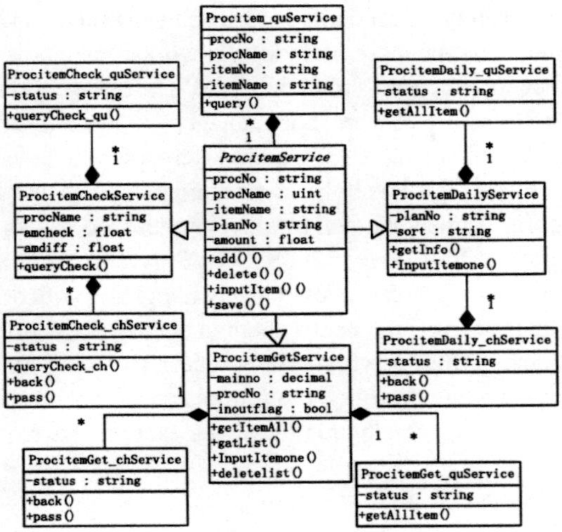

Fig. 3. Implementation class diagram

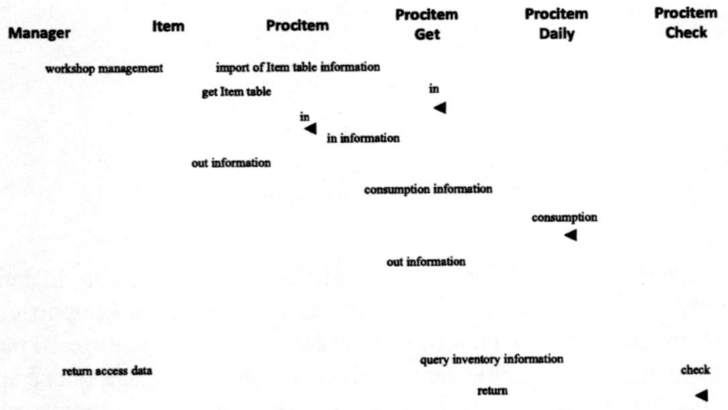

Fig. 4. Workshop material management sequence diagram

use the inputItem method in the ProcitemGet class to get the information list request, complete the receiving materials into the warehouse operation, and return the information from the incoming list to the Procitem class. By using the query check function, the ProcitemCheck class receives a list of incoming, outgoing, and consumed materials in the workshop and performs timely maintenance on workshop materials information [7].

3 Implementation of System Functions

The running environment of the system is J2SDK, Tomcat, and the application layer, it is mainly divided into three parts: server, workshop management client and database. The server includes the workshop server deployed in each workshop site and the WEB server deployed in the cloud. The database includes the workshop database and the database used for real-time display synchronized in the cloud. The three are closely coordinated to form the three-tier architecture of workshop management system software. The application layer server-client software is developed in Microsoft Windows environment using Java language, which is a high-level object-oriented programming language, based on Framework open source framework, and the development tool is Dorado7. The database software uses mainstream SQL Server 2008. SQL Server is scalable, has a simple and intuitive user interface, is easy to use, and is closely integrated with other server software.

The workshop material management system is mainly divided into four main functional modules: workshop inventory material, process receiving material, process consumption, and process inventory. Since the process and function of other modules are similar, we take the process receiving module as an example here.

Fig. 5. Interface for receiving process materials

Process material maintenance, process material audit, and process material inquiry are the three functional points of the process material module. It may import, delete, save, print, and submit basic material information in batches during the material maintenance process. The material information entered on the maintenance page can be approved or refused during the material audit procedure. The process material inquiry page allows you to search for information on process materials. Figure 5 shows the process material module's interface.

4 Summary

The system uses a modular development method, decomposing workshop material management into a single module unit, defining data interaction data information between different module units, and designing and implementing a workshop material management system for printing enterprises by integrating and reorganizing module units to meet the needs of printing enterprises for various functional modules, resulting in a technically feasible solution for the development of similar systems. It offers a practical technical solution for the creation of similar management systems for small and medium-sized printing businesses.

Acknowledgement. This research is supported by the National Key Research and Development Program of China (No. 2019YFB1707202).

References

1. Wenjie, Y.: ERP Technology for Printing Enterprises. Beijing Art and Science Electronic Publishing House, Beijing (2013)
2. Liang, W.J., Zhou, X.: Design and implementation of workshop management system. Logistics Eng. Manag. **40**(07), 147–148 (2018)
3. Chunming, L.: A workshop management system design method for discrete manufacturing enterprises. Electron. Mech. Eng. **37**(02), 45–49 (2021). https://doi.org/10.19659/j.issn.1008-5300.2021.02.011
4. Tong, Q., Ming, X., Zhang, X.: The realization for automated warehouse based on the integration of ERP and WMS. In: ICAL 2020: 2020 the 7th International Conference on Automation and Logistics (2020)
5. Yinling, Z.: Research and design of web-based material management system for small and medium-sized enterprises. J. Econ. Res. **18**, 11–12 (2017)
6. Nan, R.: Development and use of library management system based on B/S architecture. Digital Commun. World **11**, 45–47 (2021)
7. Xiaojiao, H.: Research on SAP-based warehouse management system. Finance Econ. **30**, 27–28 (2020)

Design of Intelligent Decision Support System for Production Collaboration in Flexible Packaging Printing Enterprises

Zaining Lin, Wenjie Yang(✉), Xiujie Chen, and Xuebin Zuo

Printing and Packaging Engineering School, Beijing Institute of Graphic Communication, Beijing, China
ywj@bigc.edu.cn

Abstract. Multi variety, small batch and short cycle are the production characteristics of printing enterprises. How to assign personalized orders to machines in multiple factories to achieve production collaboration is a difficult decision for printing enterprises. The intelligent decision support model of production collaboration is established, and the workflow from customer personalized order to production collaborative plan generation is described. Considering production time and transportation time between collaborative enterprises, a production planning calculation model for collaborative enterprises is established with the shortest production cycle as the objective function. The Genetic Algorithm is used to optimize and calculate the optimal production collaboration scheme, which realizes the production planning of multiple jobs assigned to multiple collaborative factories, thus solving the core problem of intelligent decision-making for production collaboration.

Keywords: Flexible packaging printing · Production collaboration · Genetic algorithm

1 Introduction

Multi variety, small batch, short cycle and cross regional demand are the main product characteristics of flexible packaging printing enterprise. Its production job includes multiple processes, such as plate making, printing, compounding, slitting, bag making, packaging, etc.. When the production capacity of a process in the factory is insufficient or the production cost is too high, it is necessary to select outsourced factories for production. The most difficult decision is how to assign the production job from personalized orders of customer to the machines of some factories for production cooperation.

A flexible job-shop scheduling model in the label printing production is given, and an improved Genetic Algorithm was used to get the minimizing production time [1]. The minimum processing time is used as the objective function to schedule the work order of a printing enterprise using Genetic Algorithm [2]. The shortest printing time is taken as the objective function, and the idle rate of the process is taken as the constraint to optimize the production scheduling [3]. In these articles, the production scheduling

M. Xu et al. (Eds.): CACPP 2022, LNEE 991, pp. 452–456, 2023.
https://doi.org/10.1007/978-981-19-9024-3_57

of printing enterprises does not consider the transportation time, and they are all single factory production mode. The production collaboration model of multi factories and multi orders on cloud platform is established, the cost is taken as the objective function and the adaptive simulated annealing Genetic Algorithm is used to realize the production collaboration planning [4]. The scheduling of distributed flexible job shop considering transportation time is realized by using improved Genetic Algorithm [5]. These two articles consider the factors of collaborative factory and transportation time, but not the production mode of printing enterprises. In many articles, the characteristics of Genetic Algorithm in optimization decision are shown, and good results are given.

Combined with the characteristics of flexible packaging printing production, the Genetic Algorithm is used to make decision for multiple jobs assigned to machines in different factories, so the production collaboration scheme is given.

2 Decision Support Model of Production Collaboration

When a factory receives a customized product order, the job is decomposed to get the task set that can be completed by the factory independently or by collaborative factories. The collaborative production task including production, delivery and other information is published through the collaborative production network system. Then, according to the basic information database, model database and knowledge database of the system, the collaborative factory production plan model is established, the Genetic Algorithm (GA) is used to generate the optimized production scheme, and the production job is assigned to the most suitable collaborative factory, as shown in Fig. 1.

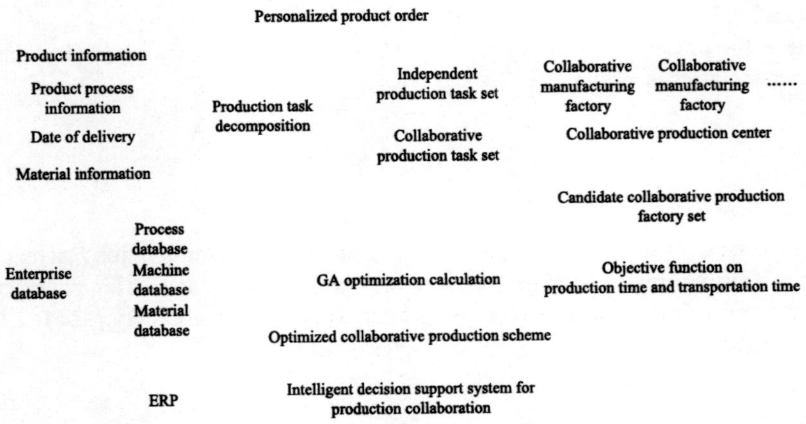

Fig. 1. Diagram of decision support model for production collaboration

The system is designed with Java series technologies and MySQL database, and browser/server (B/S) mode is used to achieve customer, order, plan, material, equipment and other management functions. Among them, plan management is mainly to use

Genetic Algorithms to assign jobs to the production machines of the collaborative factories. This is the core of intelligent decision support system for production collaboration in flexible packaging printing.

3 Objective Function in Decision Support Model

The key problem of realizing intelligent decision in production collaboration is how to reasonably assign multiple production jobs to the machines in factories for production. The objective function can be cost, time, machine utilization, or multi-objective. Here, the shortest production cycle of multiple orders as the objective function is considered. The production cycle includes the machine production time and the transportation time between collaborative enterprises.

It is assumed that:

1) The processing operations of a job are in set sequence.
2) Only one job can be processed on one machine at the same time, and the next job can only be processed after one job is finished.
3) The processing process of a job in one operation is not interrupted until the processing of the operation is completed before entering the next operation.
4) For a production process, different machines in different enterprises have the same production opportunities to get jobs.

There are N jobs, which are $J_1, J_2, \ldots, J_N$. Each job is produced according to M processes from process P_1 to process P_M, in which process P_i has Li machines, their numbers are $M_{i1}, M_{i2}, \ldots, M_{iLi}$ and $1 \le i \le M$. These machines belong to different factories $F_1, F_2, \ldots$ respectively.

Here, the objective function f is the production time of N jobs by M processes in different factories is minimized.

$$f = \min\left(\sum_{j=1}^{N}\sum_{i=1}^{M}(A_{jik}{*}V_{ik} + TP_{ik} + T_i)\right) \quad (1)$$

Here k is a random number among [1, Li], A_{jik} is processing quantity of job J_j at machine M_{ik}, V_{ik} is average production speed of machine M_{ik}, TP_{ik} is preparation hour of machine M_{ik}, and $A_{jik} * V_{ik}$ is machine processing hour. T_i is transportation hour for P_{i-1} to P_i processs, $T_0 = 0$, $1 \le j \le N$, $1 \le i \le M$, $1 \le k \le Li$.

4 An Example by Genetic Algorithm

First, the flexible packaging printing jobs and processes are coded, and each gene in the chromosome represents a job and its production process. Planning jobs and all production processes form a complete chromosome. Then the objective function is calculated after selection, crossover and mutation in the generated population, and the optimal result is output after multiple generations of inheritance.

The characteristic of packaging and printing production is that the machines between different processes cannot be used universally. In Genetic Algorithm, the calculation of the machine should be carried out in the same process, so that each chromosome generated can be guaranteed to be a feasible production scheme.

There are 9 jobs of food packaging bags, and each job includes printing P_1, compounding P_2, slitting P_3 and bag making P_4 processes. The production length of job is 20, 18, 18, 19, 20, 20, 12, 15 and 16 km respectively. The average production speed unit of the machine is km/h, and the preparation hour is h. The factories are F_1, F_2 and F_3 (Table 1).

Table 1. Machine production rate/preparation hour table in each process

P_1	P_2	P_3	P_4	
8/1	9/0.5	8/0.2	10/0.2	F_1
7/1	8/0.5	9/0.2	9/0.2	F_1
9/1	9/0.5	7/0.2	8/0.2	F_2
8/1	9/0.5	7/0.2	9/0.2	F_2
8/1	8/0.5	8/0.2	8/0.2	F_3

For simple calculation, the transportation time between processes in the same factory is 0, and the transportation time between different factories is 2 h. In Genetic Algorithm calculation, after some tests, the selection probability is 0.8, the crossover probability is 0.3, the mutation probability is 0.2, the gene population size is 100, and the evolutionary iterations is 100. The plan diagrams of each factory and job generated by GA calculation are shown in Fig. 2. The vertical coordinate in the left is the factory-machine, the vertical coordinate in the right is the job, and the horizontal coordinates of both are the time. The color block represents the job processing time in the process.

Fig. 2. Production plan by factory(left) and by job(right)

With 64 bit operating system Windows11 and Intel Core i7-1165G7 processor, the calculation time is only 48 s. The 9 jobs are assigned to machines of different processes in 3 factories, the whole production cycle is 23.11 h. From the right figure, Job5 is arranged in the same factory and the production process is closely connected, which is ideal. Job6 is arranged in different factories, the delay between processes is longer, and the end time of production is the latest.

5 Conclusion

The management mode of multi factory production collaboration is studied. Combining the characteristics of flexible packaging printing production, considering the transportation time between collaborative factories, the minimization of production cycle is defined as the objective function, and Genetic Algorithm is used to calculate the allocation of multiple jobs to machines in different factories. The optimal production collaboration plan is given, which solves the core problem of intelligent decision-making for production collaboration in printing factories.

Acknowledgement. This research is supported by the National Key Research and Development Program of China (No. 2019YFB1707202).

References

1. Xiaomei, M., Fei, H.: Label printing production scheduling technology based on improved genetic algorithm. J. Comput. Appl. **3**, 860–866 (2021)
2. Zexin, Z., Huirong, Y.: The optimal design for the production and scheduling of printing enterprises. Mod. Comput. **14**, 42–47 (2019)
3. Xin, L.: Application of cloud computing technology in the optimization of production process in printing enterprise workshop. China Comput. Commun. **20**, 173–175 (2021)
4. Wang, J., et al.: The model and solution for collaborative production planning with order splitting in cloud manufacturing platform. J. Shanghai Jiao Tong Univ. **12**, 1655–1662 (2018)
5. Hongliang, Z., et al.: Distributed flexible job shop green scheduling with transportation time. China Mech. Eng. **11**, 1–10 (2021)

Application Research of Self-powered Technology in Smart Packaging

Ben Yang[1], Yuan Zhang[2], and Lei Zhu[2](✉)

[1] School of Mechanical and Electrical Engineering, Beijing Institute of Graphic Communication, Beijing, China
[2] Postal Industry Technology R&D Center, School of Mechanical and Electrical Engineering, Beijing Institute of Graphic Communication, Beijing, China
2215152007@qq.com

Abstract. In view of the drawbacks of the power supply method for smart packaging, this paper proposes a scheme to apply electromagnetic vibration energy harvesting technology to smart packaging. To verify the feasibility of self-powered devices applied to smart packaging, the energy output of the experiments under 2 g excitation and 0.5 g excitation was tested experimentally. The results show that the self-powered device can reach peak-to-peak output voltage of more than 4 V in the sweep experiment under 2 g acceleration excitation. The output voltage exceeds 2 V at 0.5 g fixed frequency excitation, which can provide a maximum output power of 1 mW. The experiment verifies that the energy output of the electromagnetic vibration energy harvesting device can meet the work of low-power sensors. The analysis concludes that the electromagnetic vibration energy harvesting device has the potential to be applied to smart packaging.

Keywords: Smart packaging · Self-powered technology · Electromagnetic · Energy harvesting device

1 Introduction

Smart packaging refers to the use of emerging technologies on the basis of traditional packaging to enable the packaging to sense, monitor, and record information related to the environment in which the product is located by adding components with certain functional characteristics to the packaging [1].

The realization of various functions of smart packaging requires various types of sensors, electronic tags, and other electrical devices. Batteries, as the most common power supply method for smart packaging, can meet the electrical energy supply of smart packaging, but the frequent battery replacement work raises the labor cost of enterprises, reduces the efficiency and increases the pressure on the environment.

As a new energy technology, self-powered technology is used to drive low-power electronic devices by converting various energy in the surrounding environment into electrical energy. The power supply devices made by self-powered technology are environmentally friendly, lightweight, low-cost, and have a long service life. If self-powered

M. Xu et al. (Eds.): CACPP 2022, LNEE 991, pp. 457–463, 2023.
https://doi.org/10.1007/978-981-19-9024-3_58

technology can be applied to smart packaging will reduce the cost of enterprises, reduce environmental pollution and improve the endurance of smart packaging.

Vibrational energy harvesting technology is one of the self-powered technologies. The main vibration energy harvesting techniques are piezoelectric conversion [2, 3], electrostatic conversion [4], electromagnetic conversion [5, 6], magnetostrictive conversion, frictional nano-conversion [7], and various composite methods of energy harvesting. The electromagnetic type has a simple structure, low cost, low material requirements, and higher output current. The presence of vibration and the use of low-power sensors in smart packaging provides the basis for the application of vibration energy harvesting techniques.

In this paper, a self-powered device for smart packaging based on the principle of electromagnetic induction is designed, and the device is selected for vibration energy harvesting with fixed coil and motion magnet types of vibration structures. In this work, a prototype of the smart packaging self-powered device is fabricated and the feasibility of the device is experimentally demonstrated. Section 2 describes the structural components of the self-powered device. Section 3 establishes the dynamics model of the device. Section 4 describes the specific experimental procedure and analyzes the experimental results. Section 5 summarizes the possible challenges of applying the self-powered device to smart packaging and an outlook on future work.

2 Structural Composition of Self-powered Device for Smart Packaging

The electromagnetic type of vibration energy harvesting technology has low material requirements and low cost, and is the primary choice for vibration energy harvesting methods. As shown in Fig. 1, for the composition of smart packaging power supply system.

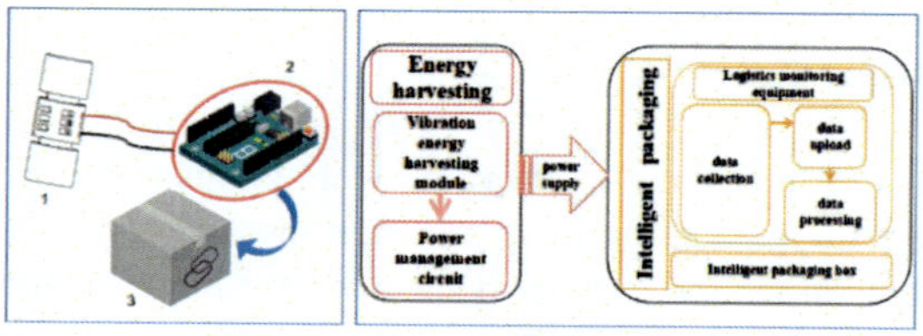

(a) hardware module composition (b) functional module composition

Fig. 1. Smart packaging power supply system.

Figure 1(a) is the vibration energy collection part. Figure 1(b) is the power circuit management part.

The energy conversion part of the smart packaging power supply device designed in this paper is of the type of motion magnet, and the internal structure is shown in Fig. 2.

A cylindrical magnet fixed between the spacers provides a stable magnetic field. The cylindrical moving magnets move up and down in the hollow sleeve along with

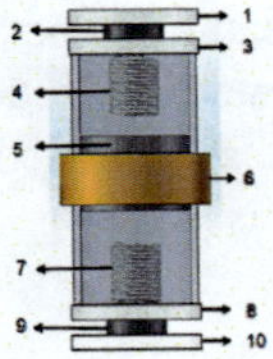

Fig. 2. Composition of the energy conversion part. 2, 5, 9 are permanent magnets; 1, 3, 8, 10 are spacers; 6 are fixed coils; 4, 7 are fixed springs.

vibrations in the environment. The up and down movement of the moving magnets produces displacement relative to the fixed coil, which changes the magnetic flux through the fixed coil, and creates an induced electric potential.

3 Dynamic Modeling

The force on the suspended magnet of the device is shown in Fig. 3.

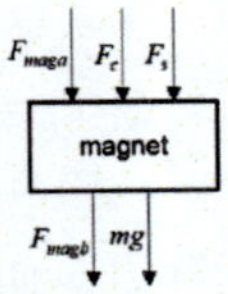

Fig. 3. Force on suspended magnet

The suspended magnet is subjected to gravity, repulsive force of the magnet, spring force of the spring and damping force during the motion.

The repulsive forces F_{maga} and F_{magb} of the fixed magnets are calculated using the magnetic dipole model, as shown in Fig. 4, where a and b are fixed magnets and c is a suspended magnet. The position of the device at rest is chosen as the initial position to establish the coordinate system.

The repulsive forces F_{maga} and F_{magb} of the magnets are calculated using the magnetic dipole model, as shown in Fig. 4. The coordinate system is the position of the device at rest.

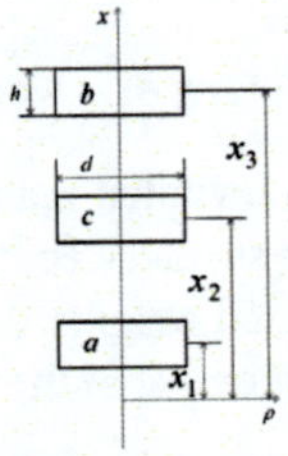

Fig. 4. Magnetic polariton drawing

During the operation of the device, the magnet repulsion acts mainly along the axial direction, so the location of the requested magnetic field strength can be selected as on the axis. The magnetic field along the axial direction of the moving coil magnet is the integral over the volume, the formula is [8]:

$$B_{sm} = \frac{p_{mc}\mu_0}{2V_{sm}}\left(\frac{h-2(x-x_2)}{\sqrt{d^2+[h-2(x-x_2)]^2}}+\frac{h+2(x-x_2)}{\sqrt{d^2+[h+2(x-x_2)]^2}}\right) \tag{1}$$

Similarly, the magnetic field of a fixed magnet along the axial direction can be expressed as:

$$B_a == \frac{p_{ma}\mu_0}{2V_a}\left(\frac{h-2(x-x_1)}{\sqrt{d^2+[h-2(x-x_1)]^2}}+\frac{h+2(x-x_1)}{\sqrt{d^2+[h+2(x-x_1)]^2}}\right) \tag{2}$$

$$B_b == \frac{p_{mb}\mu_0}{2V_b}\left(\frac{h-2(x-x_3)}{\sqrt{d^2+[h-2(x-x_3)]^2}}+\frac{h+2(x-x_3)}{\sqrt{d^2+[h+2(x-x_3)]^2}}\right) \tag{3}$$

The component of the total repulsive force on the suspended magnet in the axial direction is expressed as:

$$\begin{aligned} F_1(x_2) == p_{mc}d^2\mu_0\left[\frac{p_{ma}}{V_a}\left(\frac{1}{\left(d^2+[h+2(x_2-x_1)]^2\right)^{\frac{3}{2}}}-\frac{1}{\left(d^2+[h-2(x_2-x_1)]^2\right)^{\frac{3}{2}}}\right)\right] \\ +p_{mc}d^2\mu_0\left[\frac{p_{mb}}{V_b}\left(\frac{1}{\left(d^2+[h+2(x_2-x_3)]^2\right)^{\frac{3}{2}}}-\frac{1}{\left(d^2+[h-2(x_2-x_3)]^2\right)^{\frac{3}{2}}}\right)\right] \end{aligned} \tag{4}$$

where μ_0 is is the vacuum permeability, Pm is the dipole moment vector of the magnet, V is the volume of the magnet, h is the thickness of the magnet, d is the diameter of the magnet, and x_1, x_2, and x_3 are the displacements of the magnets a, b, and c, respectively.

Assuming that the lateral motion of the levitated magnet can be neglected, the equation of motion of the levitated magnet is:

$$\begin{cases} F_1+F_s+F_c+mg=-m_{magc}\ddot{x}_2 \\ F_s=k_0x_x \\ F_c=-c(\dot{x}_2-\dot{x}_1) \end{cases} \tag{5}$$

where c is the damping factor of the levitated magnet, Fs is the linear spring force, Fc is the damping force, k_0 is the spring stiffness coefficient, x'_2, and x''_2, are the first and second order derivatives of the levitated magnet displacement, respectively, and x'_1, is the first order derivative of the displacement at the fixed end of the device.

4 Experiment and Result Analysis of Self-powered Device

Smart packaging is transported by road in most cases, and after reviewing the data, it was found that in road transport, four transport vehicles, heavy trucks, vans, cars and two-wheeled electric vehicles, have similar power spectrum peak frequencies, with the most intense vibration in the vertical direction, and the vibration frequency is mainly concentrated between 1–40 Hz [9].

The experiments set the simulated vibration frequency at 0–80 Hz. During the whole experiment, the energy harvesting device was vibrated under the excitation of different frequencies, and its output is shown in Fig. 5.

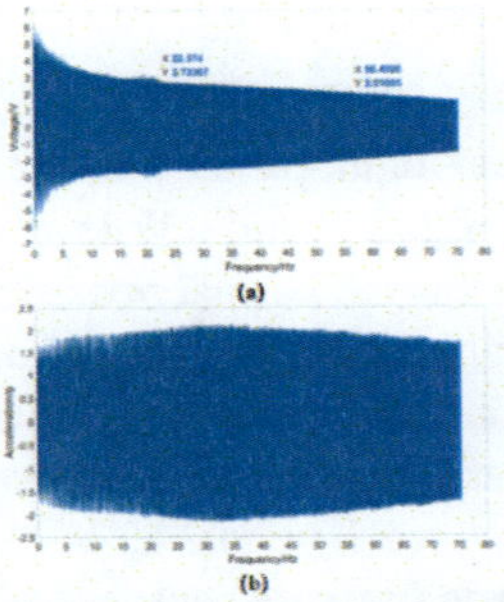

(a) Vibration acceleration (b) 0-80Hz output

Fig. 5. Energy output at different frequencies under 2 g excitation

It can be seen from Fig. 6 that under the excitation of 2 g acceleration, the peak-to-peak value between 0–55 Hz in the frequency sweep experiment can reach more than 4 V, which is similar to the main vibration frequency range of the four transport vehicles. The vibration intensity in the vertical direction is mainly simulated by the signal generator, power amplifier and vibration exciter. In the frequency range of 1–40 Hz, the vehicle exhibits the most intense vibration in the vertical direction, which is in line with the working environment of the energy harvesting device.

Under the excitation of 0.5 g, fixed-frequency experiments were carried out at frequencies of 8, 10, 12, and 14 Hz, as shown in Fig. 6.

The test results show that under the excitation of 0.5 g, in the low frequency range of 8–14 Hz, the peak-to-peak output of the energy harvesting device can reach more than 4 V. After measurement, the internal resistance of the coil is 1.13 kΩ. According to the maximum power transmission theorem, when the load resistance is equal to the coil internal resistance, the maximum power is allowed to be provided, and the maximum power transmission occurs, such as Formula (6):

$$P_{max} = U_0^2/4R_0 \tag{6}$$

where U_0 is the voltage across the coil and R_0 is the internal resistance of the coil.

At 10 Hz frequency, the output is sinusoidal alternating current. The peak value of the sinusoidal AC is $\sqrt{2}$ times the root mean square (RMS), and the peak value of the

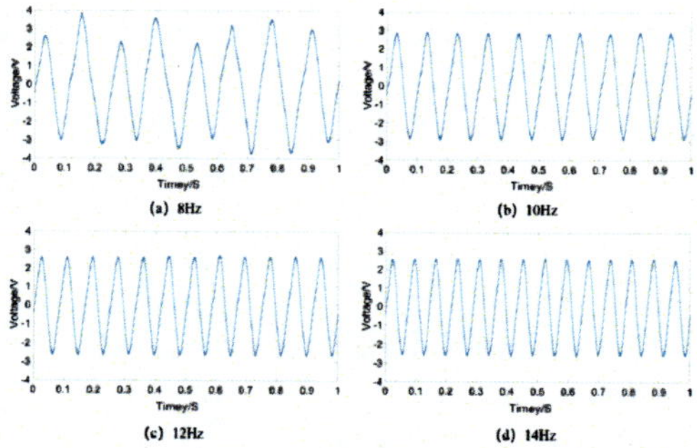

Fig. 6. Fixed frequency experiment under 0.5 g excitation

output voltage at 10 Hz is 3 V, with an RMS of about 2.12 V. Substituting the RMS of the current into Eq. (1) gives a maximum power transfer of about 1 mw. For low-power sensors with operating voltages of 2–3 V and μW levels, the energy harvesting device can meet the power requirements for the normal operation of low-power sensors.

The common operating voltage range of BMP390 atmospheric pressure sensor is 1.65–3.6 V, and the required current size is 3.2 μA. Under 0.5 g excitation, 8–14 Hz low frequency range, 1.13 kΩ coil internal resistance, and maximum output power, the output voltage of self-powered device can reach up to 2.82 V, and the current is 1.2 mA, which can meet the BMP390 atmospheric pressure sensor work requirements. Therefore, the introduction of electromagnetic vibration energy harvesting device in the field of smart packaging can theoretically meet the demand for electrical energy of smart packaging well.

5 Conclusions

The article analyzes the possibility of applying electromagnetic vibration energy harvesting devices to smart packaging from two aspects: the working environment of smart packaging and the application of low-power sensors in smart packaging. However, the choice of materials for smart packaging, the functional characteristics of the packaging and the installation location of the device will have an impact on the performance of the energy harvesting device.

1) The high magnetic permeability material used in smart packaging affects the magnetic field distribution within the energy harvesting device, thus affecting the energy conversion efficiency of the device.
2) Packaging is used to protect the product from damage. The vibration dampening effect of the packaging can lead to a conflict in the role of smart packaging and electromagnetic type vibration energy harvesting devices.
3) There is no vibration generation when going through a static logistics link such as warehousing. A long stay in a static storage link can cause the energy harvesting device to lose its original function.

How to realize the application of self-powered technology in smart packaging is the main part of the next work. The possibility of self-powered technology in smart packaging is further determined by experimental verification of powering low-power sensors with energy harvesting devices, followed by further improving the energy harvesting performance of energy harvesting devices by changing the structure or parameters of energy harvesting devices.

References

1. Changhai, H., Chao, L.: Talking about intelligent packaging and future development trends. J. Shanghai Packaging. **10**, 25–27 (2018)
2. Francesco, P., Konstantinos, G.: Piezoelectric energy harvesting from vortex shedding and galloping induced vibrations inside HVAC ducts. J. Energy Build. **158**, 371–383 (2018)
3. Wu, Y., Hu, Y., Huang, Z., et al.: Electret-material enhanced triboelectric energy harvesting from air flow for self-powered wireless temperature sensor networksens. J. Sensors Actuators A Phys. **271**, 364–372 (2018)
4. Sezer, N., Ko, M.: A comprehensive review on the state-of-the-art of piezoelectric energy harvesting. J. Nano Energy. **80**(105567), 1–25 (2020)
5. Aouali, K.: Efficient broadband vibration energy harvesting based on tuned non-linearity and energy localization. J. Smart Mater. Struct. **29**(10), 10–16 (2020)
6. Wang, B., Long, Z., Hong, Y., et al.: Woodpecker-mimic two-layer band energy harvester with a piezoelectric array for powering wrist-worn wearables. J. Nano Energy. **89**, 106385 (2021)
7. Li, R., Yu, Y., Zhou, B., et al.: Harvesting energy from pavement based on piezoelectric effects: fabrication and electric properties of piezoelectric vibrator. J. Renew. Sustain. Energy **10**(5) (2018)
8. Aldawood, G., Nguyen, H.T., Bardaweel, H.: High power density spring-assisted nonlinear electromagnetic vibration energy harvester for low base-accelerations. Appl. Energy **253**, 113546 (2019)
9. Hao, Z.: Research on random vibration analysis and simulation method of road transportation packaging. D. Jinan University (2017)

Printing Material and Related Technology

Factors Impacting Optical Properties of Mirror-Like Silver Ink Printed Paper

Xiaoyang Qu[1], Ling Yang[1,2], Haihu Tan[1,2], Xiaochun Xie[1], and Duo Ding[3](✉)

[1] Hunan Luck Printing Co., Ltd., Changsha, China
[2] School of Packaging and Materials Engineering, Hunan University of Technology, Zhuzhou, China
[3] China Tobacco Hunan Industrial Co., Ltd., Changsha 410000, China
16926998@qq.com

Abstract. To further improve the metallic flash effect of mirror-like silver ink printed paper (silver printing paper) and make its optical performances close to those of vacuum plating release paper (transfer paper), the influence of factors including substrate paper, priming coating process, ink viscosity, and dryness on the surface optical properties of silver printing paper was systematically studied. Simultaneously, the process and methods to improve the smoothness, uniformity, and coverage of aluminum pigments arranged on paper were summarized. The results show that the brightness of the silver printing paper is significantly improved with high whiteness, excellent smoothness, and strong light reflection of the substrates. Remarkably, the viscosity and volatility gradient of the mirror-like silver inks are controlled by mixed solvents, which is beneficial for the directional alignment of the aluminum pigments as well as leveling of the ink film. With excellent material performance and an optimized printing process, the silver printing paper presents an idealized specular gloss, which is expected to partially replace the transfer paper in the application of cigarette packaging.

Keywords: Mirror-like silver ink · Silver printing paper · Printing process · Optical properties

1 Introduction

Vacuum aluminized release paper is widely used in cigarette and other commodities packaging to improve product quality by imparting metallic luster to the packaging surface [1–3]. However, vacuum aluminized release paper with high production costs and poor environmental friendliness due to the complex production process, high energy consumption, and consumption of non-degradable transfer film (PET or BOPP) [4–6]. Mirror-like silver ink uses high-brightness thin aluminum as a pigment, which can form a dense aluminum layer with a mirror effect on the substrate, has been widely used in home appliances, automobiles, and other fields. The introduction of mirror-like silver ink into paper prints can achieve a metallic luster effect that exceeds traditional silver ink prints [7]. This requires the optimized surface state of the substrate and the regulated

M. Xu et al. (Eds.): CACPP 2022, LNEE 991, pp. 467–474, 2023.
https://doi.org/10.1007/978-981-19-9024-3_59

printability of the mirror-like silver ink to achieve the aluminum pigment a uniform and flat arrangement on the surface of the paper print like on the metal or film surface [8]. Due to its comparable optical properties with vacuum aluminized release paper, mirror-like silver ink paper prints are expected to be a substitution.

The uniform and orderly arrangement of the thin-layer aluminum sheets in the mirror-like silver ink on the paper surface is the basis for the good optical performance of the printed matter. The properties including viscosity, solid content, initial dryness, the flatness of the substrate surface, porosity, surface tension, etc. of the ink have a certain impact on the arrangement of the aluminum sheets during the transfer and drying process of the ink, resulting in different light reflection effect [9–11]. In this work, the effects of the paper, primer, initial drying of the ink, ink viscosity, and other conditions on the surface flatness, brightness, and spectral reflectance of mirror-like silver ink printed matter were systematically evaluated. The main factors of its optical properties were analyzed. These printing process parameters can guide the production of high gloss silver ink printing paper.

2 Materials and Equipment

Calendered paper (80 g/m^2), coated paper (200 g/m^2), white cardstock (220 g/m^2), photo paper (230 g/m^2), glass cardstock (200 g/m^2), and imitation glass cardstock (200 g/m^2) were purchased from Dongguan Kemei Paper Products Co., Ltd. Schlenk mirror silver paste 2004# and the special resins for Schlenk mirror-like silver ink were supplied by Guangzhou V-Chain Trading Co., Ltd. Waterborne polyurethane, printing varnish, waterborne varnish, and polyacrylic acid supplied by Hunan Hulida Coating Technology Co., Ltd. Were used as paper coating material. The n-propyl ester (AR), isopropanol (AR), ethyl acetate (AR), and butyl acetate (AR) were purchased from Hunan Huihong Reagent co., Ltd.

Gravure ink prototype (Dongguan Hengke Automation Equipment Co., LTD, HK-310B) was used for mirror-like silver ink printing. Roughness meter (IMI, 58-06-00-0001), spectrophotometer (X-rite, Ci64), whiteness meter (Hangzhou Zhibang, ZB-B), Zahn viscometer 4# cups, and scraper viscometer were used to characterize the ink and the printing paper. The 5 different viscosity silver inks were obtained by adjusting the mass ratio of silver paste, resin, n-propyl ester as 3:1.5:5.5, 3:2:5, 3:2.5:4.5, 3:2.7:4.3, 3:3:4. Then, these inks were printed on different papers by the gravure ink prototype and dried in the natural environment (25 °C, RH 65%).

3 Results and Discussion

3.1 Effect of Ink Properties on Optical Properties of Mirror-Like Silver Ink Printed Matter

The silver inks with different viscosity were prepared by changing the relative ratio of resin and solvent in the ink system. The viscosity of the five groups of inks was measured by Zahn 4# cup viscometer to be 10, 15, 20, 22, and 30 s, respectively. These five types of inks were used for gravure proofing with white cardstock as the substrate.

Four different printing areas with mesh depths of 40, 45, 50, and 55 μm were set on the gravure printing plate. The optical properties of the post-printing surfaces in the four areas were tested and the measured lightness values are shown in Table 1. Using the X-Rite Ci64 spectrophotometer to measure its brightness value. The reflection spectrum and the L * a * b value which include the specular reflection effect (SPIN) and the specular reflection effect exclusion (SPEX) were obtained during the test with the integrating sphere spectrophotometer. As a printed silver product, its brightness and specular reflection effect determine the metallic luster texture that will be presented in the subsequent application process, so the brightness value is selected as the main evaluation index of the optical performance. The SPIN brightness values (L_1) and SPEX brightness values (L_2) of the printed product area of the same ink with different mesh depths and of the printed product area of the same mesh depths of different inks show irregularity. However, from the perspective of human visual evaluation, the printed areas of the same ink with deeper mesh depths (55 and 50 μm *vs* 15 and 40 μm) have better gloss and brightness. This feature is consistent with the changing trend of the difference values ($L_2 - L_1$) between the SPIN brightness value and the SPEX brightness value. Previous research has shown that the difference values reflect the level of visual gloss and brightness [12]. The luster and brightness received by the human eye are the reflections of the overall intensity of the specular reflection light. From the test principle, the difference values correspond to the brightness changing value.

Taking into account the specular reflection effect of printed products in different mesh depths regions, regardless of the ink viscosity, higher mesh depths (50 or 55 μm) was easier to obtain a better specular reflection effect. It may be that deeper mesh depths could obtain a larger amount of silver ink transfer and achieve better coverage on the surface of the printed product, thereby obtaining a more specular reflection effect. There was little difference in the specular reflection effect of the printed area with mesh depths of 50 or 55 μm. Further increasing the mesh depths may easily hinder the complete transfer of ink and eventually lead to blockage of the printing plate.

Compared with the printing matter based on inks with a viscosity of 15, 22, and 30 s, the ink with a viscosity of 10 and 20 s have a slightly better specular reflection effect. Some research results showed that the low viscosity of silver ink is beneficial to the regular arrangement of aluminum flakes [13]. This may be because the content of aluminum powder in inks with different viscosities was fixed under the conditions of gravure proofing, and the ink viscosity did not play a decisive role in the hiding power of silver ink and the arrangement of aluminum pigments. The viscosity of silver ink should be fixed within this range to avoid blockage and turbulence (Fig. 1).

Four kinds of silver ink with different initial drying properties were prepared by using isopropanol, n-propyl ester, ethyl acetate, and butyl acetate with different volatility as the solvent of silver ink respectively. The initial dryness of the four inks was measured according to GB/T 13217.5-2008 liquid ink initial dryness test method as 23.3 mm, 20.3 mm, 19.2 mm, and 12.7 mm, respectively. The four types of silver inks were proofed on white cardboard. The tested results including L_1 value, L_2 value, and the difference value ($L_1 - L_2$) of the samples were similar, which demonstrateed that they also show a similar specular reflection effect. Although the experience of silver ink printing production showed that the slow initial drying of the ink is conducive to the

Table 1. Brightness values of printed matter with different viscosities of specular silver inks

The viscosity of inks (s)	Mesh depths (μm)	SPEX brightness (L_1)	SPIN brightness (L_2)	$L_2 - L_1$
10	40	69.23	89.15	19.92
	45	66.27	90.52	24.25
	50	63.57	90.99	27.42
	55	65.27	92.15	26.88
15	40	69.69	89.33	19.64
	45	67.31	90.58	23.27
	50	66.63	91.27	24.64
	55	67.71	91.85	24.14
20	40	70.85	88.75	17.90
	45	66.41	89.91	23.50
	50	64.11	90.37	26.26
	55	64.37	91.84	27.47
22	40	72.51	88.36	15.85
	45	67.29	89.52	22.23
	50	64.86	89.82	24.96
	55	64.86	91.24	26.38
30	40	69.72	88.76	19.04
	45	67.33	90.74	23.41
	50	65.24	90.28	25.04
	55	64.45	91.17	26.72

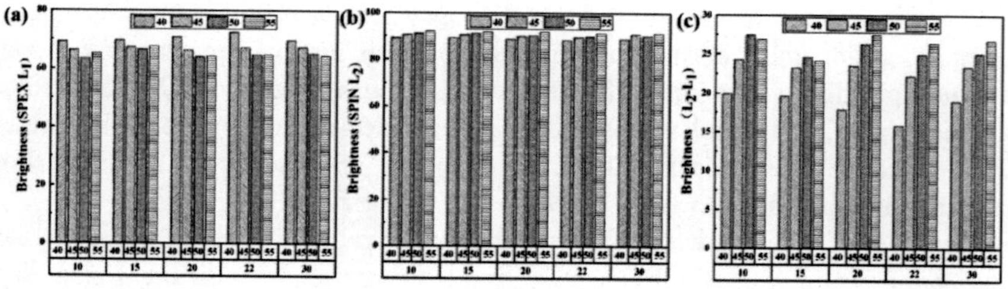

Fig. 1. SPEX brightness values (a), SPIN brightness values (b), and Difference values between SPIN brightness values and SPEX brightness values (c) of different viscosities mirror-like silver ink prints

regular arrangement of the aluminum powder during the drying process, the initial drying of the ink in the range of 12.7–23.3 mm will not affect the arrangement of the aluminum powder in the research process of this paper (Table 2).

Table 2. Brightness values of printed matter with different mirror-like silver inks

Initial drying (mm)	Mesh depths (μm)	SPEX brightness (L_1)	SPIN brightness (L_2)	$L_2 - L_1$
23.3	40	71.02	89.69	18.67
	45	66.97	90.18	23.21
	50	66.84	90.96	24.12
	55	68.38	92.28	23.90
20.3	40	69.23	89.15	19.92
	45	66.27	90.52	24.25
	50	63.57	90.99	27.42
	55	65.27	92.15	26.88
19.2	40	68.77	89.64	20.87
	45	67.25	91.13	23.88
	50	65.06	91.43	26.37
	55	68.38	92.46	24.08
12.7	40	68.62	89.46	20.84
	45	65.32	91.14	25.82
	50	64.59	91.33	26.74
	55	65.08	92.13	27.05

3.2 Influence of Substrate on the Optical Properties of Printed Products

The different surface flatness and solvent absorption between various papers affect the arrangement of aluminum pigments during the transfer and drying process of silver ink resulting in the different optical properties of mirror-like silver ink prints. Calendared paper, glass cardboard, photo paper, imitation glass cardboard, white cardboard, and coated paper were selected as substrates for gravure proofing with mirror-like silver ink. The surface roughness of the 6 kinds of paper is 1.08, 2.43, 0.90, 1.18, 0.64, and 0.51 Ra, respectively, which is closely related to the fiber thickness, papermaking pressure, and the postprocessing process. In addition, silver ink printing also has different influences on the roughness of various papers. The surface roughness of glass boards and imitation glass board after silver printing are reduced. While the surface roughness of the glossy paper, photo paper, white cardboard, and coated paper is improved after a silver printing. The surface of the last four kinds of the paper itself has good flatness. Because the aluminum pigment in the silver ink is larger, the surface will be unevenly arranged after silver printing and the surface roughness is improved. Although glass cardboard and imitation glass cardboard have high surface roughness, silver ink can fill the unevenness of the surface after transfer. In addition, the surface state of these two kinds of paper can make the aluminum pigment in the silver ink achieve a better smoothness arrangement.

From the tested results of the surface brightness of 6 kinds of silver-printed paper, white cardboard has the highest SPEX brightness, and the silver-printed samples of all papers have similar SPIN brightness. The whiteness for 6 kinds of paper is 1.15, 0.85, 2.10, 0.55, 1.22, and 0.71, respectively. The results of the spectral reflectance test (Fig. 2b) show that several kinds of silver printing paper have high reflectivity. According to naked eye observation, the white cardboard-based silver ink prints with dull metallic luster, and the imitation glass cardboard-based silver ink prints have an excellent mirror reflection effect with a bright metallic luster. The result exactly matches the difference value ($L_1 - L_2$). Combined with the analysis of the surface roughness changes before and after silver printing on different papers, it can be found that the flatness of the substrate surface does not determine the flatness of the silver-printed products. The surface tension and the absorption characteristics of the paper together affect the regular arrangement of the aluminum powder. Therefore, coating and modifying paper are two effective ways to improve the optical properties of mirror-like silver ink prints.

Fig. 2. Roughness values of different papers before and after silver ink printing (a), reflection spectrum curves (b), and brightness values (c) of different printing silver papers (1—calendered paper, 2—glass cardboard, 3—photo paper, 4—imitation glass cardboard, 5—white cardboard, 6—coated paper)

3.3 Influence of Primer Coating Process on the Optical Properties of Printed Products

Among the six types of paper, the white cardboard exhibits an unsatisfactory mirror-silver ink printing effect. However, white cardboard is the most commonly used substrate in cigarette packages. Therefore, the technique for improving the metallic flash effect of white cardboard baed silver-ink printing has a significant practical application value. Five functional materials commonly used in gravure printing, including polyurethane (PU) resin, special resin for mirror-like silver ink, grinding oil, water-based varnish, and polyacrylic (PAA) resin, were selected as coating materials to modify the surface of white cardboard. Then, the mirror-like silver ink was printed on the modified white cardboard. Figure 3 is the tested result of the difference value of $L_1 - L_2$ of the 55 μm printed silver area after five different primer treatments. It can be seen from the tested results, that the optical properties of the paper coated with the polyurethane resin and the special resin are not significantly improved compared with before the coating. White cardboard though varnish, water-based varnish, and polyacrylic resin coating can obtain a better specular reflection effect. The white cardboard coated paper with polyacrylic resin as the coating has the largest difference in brightness, and also has a good specular reflection effect in the visual perception evaluation. After coating, the surface roughness of the paper decreases to a similar extent and the surface coating forms a functional layer with a certain viscosity during the drying process of the paper surface. The coating modification improves the optical properties of mirror-like silver ink prints due to the changes in the absorbency of the substrate to silver ink.

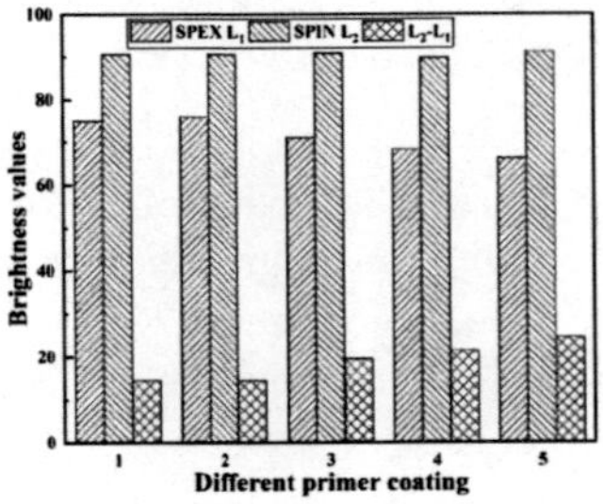

Fig. 3. Brightness values of mirror-like silver ink prints based on different primer coating modified paper (1—PU, 2—special resin, 3—grinding oil, 4—varnish, 5—PAA)

Furthermore, the effect of the viscosity of polyacrylic resin coating on the optical properties of printed products was explored. Polyacrylic coatings with viscosities of 8, 16, and 20 s (pure polyacrylic resin) were prepared by adjusting the ratio of polyacrylic acid to n-propyl ester in the coating system. Three polyacrylic resins with different viscosities were printed on white cardboard as gravure inks. After coating materials were completely dried, gravure proofing with mirror-like silver ink was performed. Measure the brightness values of SPEX and SPIN of the printing area with a cell depth of 55 μm cell, and used the brightness difference as an index to evaluate the specular reflection effect of the printed product. The results showed that when the viscosity of the primer coating is 8 s and 16 s, a better effect of improving the optical properties of mirror-like silver ink prints can be obtained. Polyacrylic paint with high viscosity forms an uneven surface after coating on the white cardboard due to the poor leveling performance of the paint. It can be verified from the surface roughness of the three silver printing papers, which are 0.70, 0.85, and 1.14, respectively (Fig. 4).

4 Conclusion

The high-gloss printing silver paper was prepared by using mirror-like silver ink and has a good prospect to replace the transfer paper used in cigarette packages. The specular emission effect of high-gloss prints can be evaluated by the difference between the SPIN and SPEX brightness values. The printing effect of mirror silver ink on paper needs to be controlled by systematically regulating the characteristics of the ink and the surface characteristics of the substrate. Mirror-silver ink with low viscosity, deep gravure printing cells are used, and the paper with smooth surface could obtained a better mirror

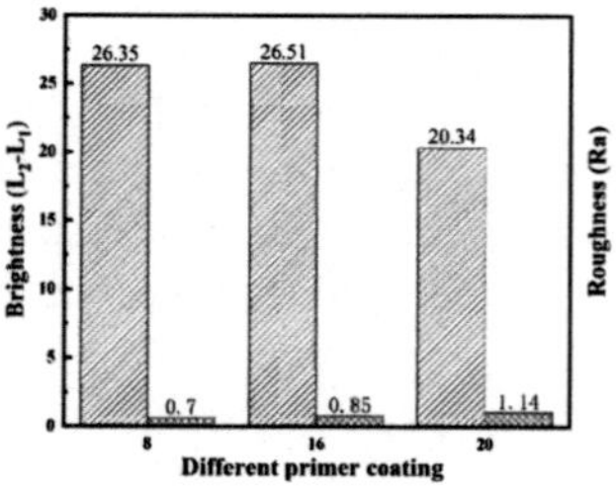

Fig. 4. Brightness and roughness values of mirror-like silver ink prints based on different viscosity PAA primer coating modified paper

reflection printing effect. The specular reflection effect of white cardboard that is not suitable for mirror silver ink printing can be greatly improved by a proper PAA primer coating process.

References

1. Zhao, W.: Application of environmental protection material in cigarette packaging printing. Ind. Design **04**, 104–105 (2017)
2. Zhu, J.: Analysis of vacuum aluminized paper and its suitability for smoke printing. Print. Qual. Stand. **10**, 44–49 (2013)
3. Yi, Y.: Production and quality control of vacuum aluminized paper by transfer method. Hunan Packag. **32**(01), 109–111 (2017)
4. Wang, C.: Analysis of the current situation of vacuum aluminized paper. Heilongjiang Pulp Paper **43**(01), 30–31 (2015)
5. Liu, H.: Using the properties of talc to improve the packaging performance of vacuum aluminized lining paper. Heilongjiang Pulp Paper **49**(02), 1–5 (2021)
6. Huang, L.: Application research of water-based laser coating on vacuum metalized paperboard. J. Yunnan Univ. **41**(S1), 76–82 (2019)
7. Chen, D.: Application of mirror ink in IMD technology. Screen Printing **11**, 11–13 (2005)
8. Qi, C.: Characteristics of mirror ink and screen printing technology. Screen Printing **9**, 22–23 (2005)
9. Zhang, J.: A study on the ink adhesion fastness on the surface of cast coated paper. China Pulp Paper Ind. **41**(12), 32–35 (2020)
10. Peng, C., Xing, J., Liu, Y., et al.: Influencing factors of particle size of water-based silver ink paste. Packag. Eng. **41**(11), 145–150 (2020)
11. He, B., Guo, L., Wang, Y., et al.: Composite process parameters of silver ink lined paper. Packag. Eng. **41**(09), 124–129 (2020)
12. Ke, Y., Yang, P., Gao, Na.: Measurement and evaluation of color properties of gold and silver cardboard. Inf. Rec. Mater. **15**(02), 21–24+28 (2014)
13. Wang, M.: How to improve the quality of silver ink printing. Shanghai Packag. **12**, 39–40 (2013)

Research on Reproduction of Scroll Painting Based on Xuan Paper Pre-lamination Technology

Xue Song[1], Jinglin Ma[1](✉), and Qi Zeng[2]

[1] Art Design Department, Shandong Communication and Media College, Jinan, China
371725079@qq.com

[2] Information Engineering Department, Shandong Communication and Media College, Jinan, China

Abstract. The purpose of this paper is to research the positive effect of xuan paper pre-lamination technology on improving the reproduction quality and efficiency of scroll painting by experiment. Xuan paper is an important material for copying scroll painting by inkjet printing, especially in the process of copying scroll painting using long range reel xuan paper, because of the poor flatness of paper surface, low paper stiffness, and local expansion of paper caused by ink, it will cause the paper to wrinkle or rub dirty, then leads to replication failure. This paper is based on experimental research, laminating xuan paper and backing paperusing hot melt adhesive film and starch aqueous solution, and detect the stiffness change of the laminated materials. Detect the safe passing length of single-layer xuan paper and laminated xuan paper respectively by printing color samples. Because the pigment particles attached to the surface of xuan paper have color loss in the process of heating laminating and water-soluble laminating, Detect the difference of color restoration degree between post-lamination and pre-lamination xuan paper with color densitometer. By testing the stiffness, safe passing length and color restoration of pre-lamination xuan paper, and comparing it with the conventional printing and reproduction methods of coated xuan paper, it is determined that the pre-lamination xuan paper technology has a positive promoting effect on improving the printing quality and reproduction efficiency of scroll painting.

Keywords: Pre-lamination · Pass the machine · Color restoration

M. Xu et al. (Eds.): CACPP 2022, LNEE 991, pp. 475–480, 2023.
https://doi.org/10.1007/978-981-19-9024-3_60

1 Introduction

Ink jet printing is the main mode of copying Chinese long scroll paintings and calligraphy, The material used in the process of copying and printing is reel xuan paper. In the process of production and processing, because of to the large length, which is generally 50–100 m, the surface flatness of the paper is low, and the tension of the reel xuan paper is uneven during the winding process. These factors will cause the failure of the reel xuan paper in the printing and copying process. In addition, due to the low thickness and loose surface of the xuan paper, it will swell and deform after absorbing a large amount of ink, causing the drum phenomenon, causing the printing nozzle to rub the xuan paper, resulting in printing failure. In this paper, according to the framed form of long scroll calligraphy and painting, xuan paper and backing paper are covered in advance, the stiffness change of composite materials is detected, and the difference in safe passing length between single-layer xuan paper and laminated xuan paper is compared. After printing, the difference of color restoration between pre lamination and post lamination was detected and compared to determine the positive promoting effect of xuan paper pre lamination technology on improving the printing quality and replication efficiency of long scroll of calligraphy and painting.

The use of pre-lamination technology for scroll painting has not been used in the industry, and only a few enterprises have adopted a similar mode when using offset press to print small format paintings, which has greatly improved the printing quality and efficiency [1].

2 Fault Manifestation of Xuan Paper Passing Machine

2.1 Longitudinal Wrinkling

The phenomenon of longitudinal wrinkling is caused by the low thickness and surface flatness of the paper, and the longitudinal wrinkling is formed under the extrusion of the paper pick-up roller of the printer. This longitudinal fold will not disappear by itself, and will gradually increase with the longitudinal transport of the paper, resulting in printing failure at last. The shape of longitudinal wrinkling is shown in Fig. 1.

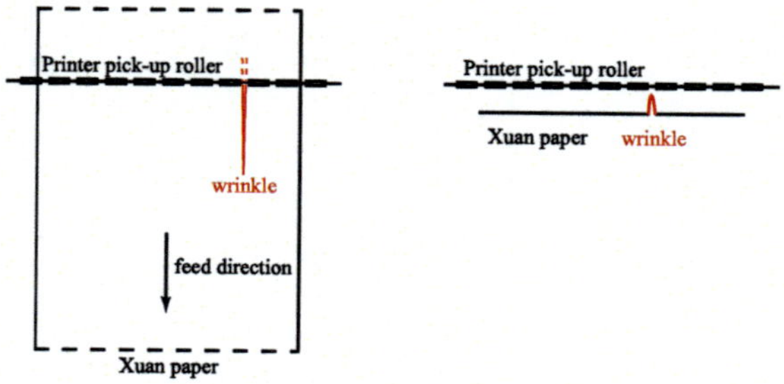

Fig. 1. Longitudinal wrinkling shape of xuan paper

2.2 Horizontal Smudging

Horizontal smudging is caused by the swelling of xuan paper after a large amount of ink is covered, which is caused by the horizontal movement of the printing nozzle. Xuan paper because of the low flatness caused by the extrusion wrinkle will also form the phenomenon of rubbing dirty. The shape of horizontal smudging is shown in Fig. 2.

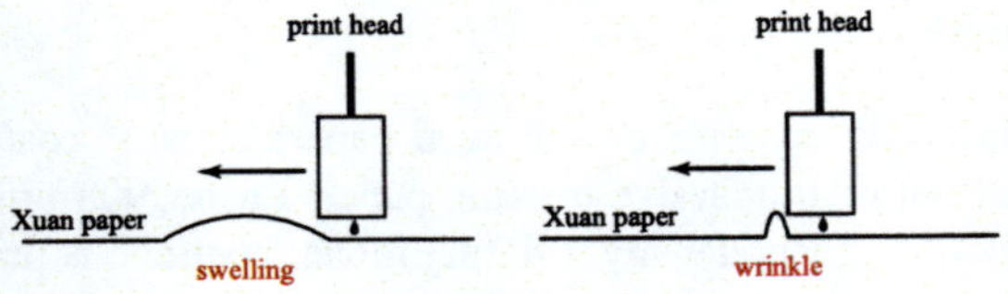

Fig. 2. Horizontal smudging shape of xuan paper

3 Passing Experiment of Xuan Paper Pre-laminating

3.1 Lamination Mode

Laminate the xuan paper with the back paper is a necessary process for the mounting of calligraphy and painting. In order to improve the stiffness and surface flatness of the xuan paper, putting the lamination stage before printing can improve the safety and success rate of the xuan paper. The backing paper can be divided into two types: precoated hot melt adhesive and non adhesive, the lamination mode of precoated hot melt adhesive backing paper is shown in Fig. 3.

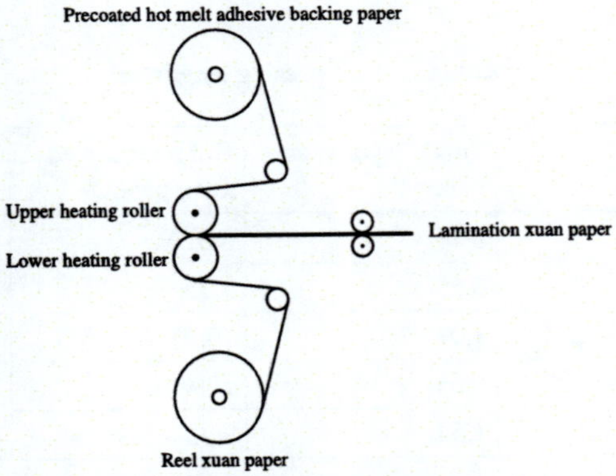

Fig. 3. Xuan paper lamination mode

When non adhesive backing paper is used for lamination, the backing paper and the solid hot-melt adhesive film need to be laminated in advance. The lamination mode is the same as that in Fig. 3. Then it is laminated with the roll xuan paper to form the laminated xuan paper. The use of starch aqueous solution is a traditional Chinese painting and calligraphy mounting mode. This mode does not need heating, and can be peeled off again for secondary mounting. It is suitable for mounting expensive and valuable paintings and calligraphy. The mounting mode using starch aqueous solution takes a long time and requires a large space for fixed drying, which is generally limited to less than 3 m. It is difficult to mount long format calligraphy and painting.

3.2 Passing Experiment

Experimental materials. 55 g/m^2 coated xuan paper; 55 g/m^2 coated xuan paper + 40 g/m^2 precoated hot melt adhesive backing paper; Epson9908 printer, K3 pigment ink, 720 dpi × 1440 dpi print density. Printing mode: continuous printing, print head spacing: standard, air suction level: − 1. The test sample is shown in Fig. 4.

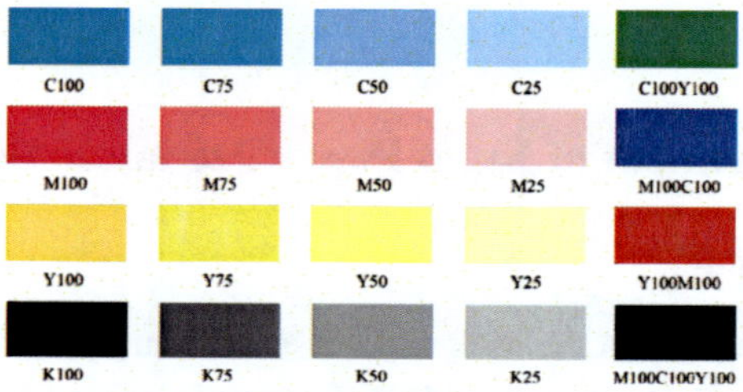

Fig. 4. Print test sample (55 cm × 30 cm)

Safe passing length test. The safe passing length refers to the length of xuan paper without wrinkling and rubbing during printing, which is used to measure the printability of xuan paper (Table 1).

Table 1. Safe passing length test

Material type		Single layer xuan paper	Hot melt coated xuan pape	Starch coated xuan paper
Safe passing length (m)	1	0.74	8.02	3(*)
	2	0.55	9.33	3(*)
	3	1.07	10(#)	3(*)
	4	0.65	10(#)	3(*)
	5	1.54	9.73	3(*)

#The sample of the hot-melt coated xuan paper shall be 10 m, and there shall be no wrinkle or dirt within 10 m

*Starch coated xuan paper can only reach 3 m, and there is no wrinkle or dirt within 3 m

According to the test, the single-layer xuan paper has low safe passing performance. Due to the different surface characteristics of xuan paper, there is a high probability of failure within 1 m. The surface flatness of the coated xuan paper have been greatly improved, and the safe passing length has been greatly improved.

4 Color Restoration Experiment

4.1 Experiment Object

Sample1#: materials laminated with hot melt adhesive film from printed xuan paper and back paper.

Sample2#: materials laminated with starch aqueous solution from printed xuan paper and back paper.

Sample3#: sample printed with lamination material using of hot melt adhesive film.

Sample4#: sample printed with lamination material using of starch aqueous solution.

4.2 Experiment Equipment

KeMei FD-5 BT Spectrodensitometer.

4.3 Experiment Result

The experiments show that the pigment on the surface of xuan paper will be lost and the saturation will be reduced in the process of hot-melt coating; The coating process of xuan paper in starch aqueous solution will also cause the loss of pigment on the surface, resulting in the decrease of saturation; The color samples printed with precoated xuan paper have the highest reduction (Table 2).

5 Conclusions

Pre-lamination technology can significantly improve the flatness and stiffness of xuan paper, and greatly improve the safe passing length. Pre-lamination technology can also significantly improve the color restoration performance and improve the replication quality. The xuan paper lamination can be operated by the existing double-sided lamination machine on the market. It is only necessary to customize the roll precoated hot melt adhesive backing paper with the specified aperture. So the equipment cost has not been significantly improved, and the material cost will be reduced becouse of instead of reduction of scrap rate. Due to the increase of the safe passing length, the Pre-lamination technology can significantly improve the replication efficiency and reduce the replication cost, which is suitable for the replication of long scroll paintings and calligraphy.

Table 2. Color restore experiment

	1# (Lab)	2# (Lab)	3# (Lab)	4# (Lab)
C100	58, – 35, – 42	59, – 36, – 44	58, – 40, – 51	59, – 37, – 48
C75	69, – 23, – 32	69,– 24, – 33	67, – 27, – 39	69, – 25, – 36
C50	80, – 14, – 20	80, – 15, – 21	79, – 17, – 25	80, – 16, – 23
C25	91, – 7, – 10	91, – 7, – 10	90, – 8, – 12	90, – 7, – 11
M100	52, 66, – 4	52, – 69, – 4	49, 78, – 4	51, 73, – 4
M75	64, 47, – 4	63, 48, – 4	61, 55, – 5	63, 51, – 5
M50	77, 28, – 4	76, 29, – 4	75, 33, – 5	76, 31, – 4
M25	89, 13, – 2	89, 14, – 2	88, 15, – 3	89, 14, – 2
Y100	88, – 6, 82	89, – 7, 85	92, – 8, 99	92, – 8, 91
Y75	91, – 6, 59	92, – 6, 61	94, – 7, 71	94, – 7, 65
Y50	94, – 5, 37	95, – 5, 38	96, – 6, 44	97, – 5, 41
Y25	97, – 3, 17	98, – 3, 18	98, – 3, 20	98, – 3, 19
K100	24, 2, 1	20, 3, 2	10, 5, 4	10, 5, 4
K75	50, 0, 0	48, 0, 0	43, 1, 0	43, 1, 0
K50	69, 0, 0	68, 0, 0	66, 0, 0	68, 0, 0
K25	86, 0, 0	86, 0, 0	85, 0, 0	86, 0, 0
C100Y100	55, – 61, 25	55, – 64, 26	52, – 78, 28	54, – 70, 27
M100C100	28, 22, – 47	26, 23, – 49	21, 27, – 56	25, 25, – 52
Y100M100	52, 60, 51	52, 62, 52	49, 72, 57	51, 66, 55
M100C100Y100	24, – 3, – 1	26, – 2, 0	19, – 7, – 4	21, – 5, – 3

Reference

1. Yan, L.: How to print calligraphy and painting paper with ordinary offset printing machine. Print. Technol. **17**(4), 51 (2010)

Preparation and Properties of Food Wrapping Paper Coated on a Complex Sizing Agent of Oxidized Nanofibrillated Cellulose/Cationic Guar Gum

Yong Lv[1,2](✉), Ci Song[1,2], Yunfei Bao[2], and Deng Ye[2,3]

[1] School of Light Industry and Engineering, South China University of Technology, Guangzhou, Guangdong, China
530301120@qq.com

[2] School of Engineering and Information, Yiwu Industrial and Commercial College, Yiwu, Zhejiang, China

[3] Zhejiang Lanyu Digital Technology Co., Ltd., Yiwu, Zhejiang, China

Abstract. With the promotion of the looming restrictions and the refuse classification, the green environmental protection and barrier performance of paper packaging materials has become a research hotspot of packaging materials. In this study, oxidized nanofibrillated cellulose (ONFC) was prepared by the nanofibrillated cellulose (NFC) oxidized by sodium periodate. The complex sizing agent of oxidized nanofibrillated cellulose (ONFC)/cationic guar gum (CGG) was prepared by combinding oxidized nanofibrillated cellulose (ONFC) with cationic guar gum. The obtained sizing agent was employed to coat on food packaging base paper. Water vapor permeability (WVP) and the digital printing suitability of the coated paper was evaluated. The results showed that when the content of ONFC and NFC increased from 0 wt% to 0.75 wt%, the WVP of the coated paper decreased from 3.12×10^{-11} g $cm^{-1}s^{-1}pa^{-1}$ to 1.63×10^{-11} g $cm^{-1}s^{-1}pa^{-1}$ and 2.08×10^{-11} g $cm^{-1}s^{-1}pa^{-1}$, respectively. It shows that ONFC and NFC can improve the film-forming performance of the sizing agent, which can improve the water vapor barrier performance of the coated paper. For the digital printing proofing of the coated paper, the ink-jet printing method has a larger color gamut. The tone reproduction has a better linear relationship. It presents better digital printing proofing performance. The coated paper coated by the complex sizing agent has broad application prospects in the field of personalized packaging materials and labels.

Keywords: Oxidized nanofibrillated cellulose (ONFC) · Cationic guar gum (CGG) · Sizing agent · Food wrapping paper

1 Introduction

Due to the promotion of the looming restrictions and the refuse classification, the recyclable package materials have become science's most notable subjects of waste classification and resource recycling [1]. In order to achieve completely recyclable materials,

M. Xu et al. (Eds.): CACPP 2022, LNEE 991, pp. 481–485, 2023.
https://doi.org/10.1007/978-981-19-9024-3_61

packaging materials used by the same or similar materials, have becomes particularly important in resource recycling [2, 3]. At present, paper packaging materials are the most common packaging materials in commodity packaging and logistics packaging. Howerver, paper packaging materials with a single paper fiber barrier, could not meet the packaging requirements. Therefore, the paper-based materials are usually modified on the surface (such as vacuum aluminized, polymer coating and other processes) to meet the barrier performance requirements. Although the barrier properties of the above materials is excellent, it is difficult to fully recycle them. It brings a burden to environmental pollution [4]. The paper-based materials coated by natural biomaterials is an important research direction for the preparation of completely recyclable paper packaging materials at present.

Cationic guar gum (CGG), prepared by cationic modification of guar gum, has good barrier properties [5]. However, it requires a larger surface coating amount to achieve barrier properties. It leads to high cost and limits its use in paper packaging materials. Nanofibrillated cellulose (NFC), with the biological characteristics of natural polymers and the dual characteristics of nano materials, could be prepared from the natural biomass in nature [6]. It also has excellent barrier properties. In order to further improve the cross-linking performance of NFC and the barrier performance of coated paper, NFC was oxidized to prepare oxidized nanofibrillated cellulose (ONFC). The paper coated by the complex sizing agent of ONFC/CGG, coluld improve the water vapor barrier performance of coated paper. At the same time, the digital printing proofing of the coated paper expands its application prospect of the package materials, such as special food packaging materials, food labels, personalized packaging and other fields.

2 Experimental

2.1 Materials and Methods

Nanofibrillated Cellulose (the NFC content is 2 wt%. The length of NFC is 100–1000 nm. The diameter of NFC is 20–50 nm) is purchased from Zhejiang jinjiahao green nanomaterials Co., Ltd. Base paper of food packaging is from Zhejiang Jinchang Special Paper Co., Ltd. Cationic guar gum is provided by Wuxi Jinxin Group Co., Ltd. Sodium periodate ($NaIO_4$), sodium hydroxide (NaOH), ethylene glycol, hydrochloric acid (HCl) and other reagents are purchased from Sinopharm Chemical Reagent Co., Ltd.

2.2 Preparation of the Complex Sizing Agent ONFC/CGG

Sodium periodate aqueous solution (with the molar ratio of $NaIO_4$ to the dehydrated glucose unit of NFC 1:1) was employed to the oxidative modification reaction of NFC.The degree of oxidation is 0.31 by titration of sodium hydroxide solution. The modified ONFC was obtained by vacuum drying at 4 °C. Adding 4.0 g CGG into 96 g deionized water, The 2 wt% CGG aqueous solution was obtained. Adding different quality ONFC solutions into CGG solution, the ONFC/CGG complex sizing agent could be obtained from homogenizing and stirring in 95 °C water bath for 1h. The obtained cmplex sizing agent was used to coat on the surface of food packaging base paper by rod coating method. The coating amount determined by the coating times.

2.3 Determination of the Performance of Water Vapor Barrier Performance and Digital Printing Performance

The weighing method is used to test water vapor barrier performance. Epson7910 inkjet printer and Xerox DC2100 digital printer are selected respectively to digitally proofing on ONFC/CGG coated paper. The spectrophotometer (X-Rite 528) was used to test the L * a * b value and density value on the color block of the digital printing sample.

3 Results and Discussions

3.1 Effect of the Content of ONFC and NFC on Water Vapor Barrier Performance

In food packaging, water vapor barrier performance has an important impact on the storage performance. Figure 1 shows the effect of different content of ONFC and NFC in the coplex sizing agents on the water vapor permeability of coated paper.

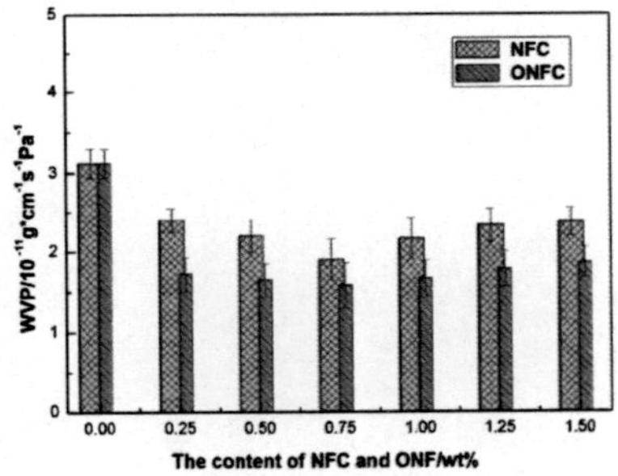

Fig. 1. The relationship between WVP and the content of ONFC and NFC

When the content of ONFC and NFC increased from 0 wt% to 0.75 wt%, the WVP of the coated paper decreased from 3.12×10^{-11} g $cm^{-1}s^{-1}pa^{-1}$ to 1.63×10^{-11} g $cm^{-1}s^{-1}pa^{-1}$ and 2.08×10^{-11} g $cm^{-1}s^{-1}pa^{-1}$, respectively. It shows that ONFC and NFC could improve barrier properties of the coated paper. ONFC contains a large amount of hydroxyl and aldehyde groups. It has a good film forming properties, which can reduce the water vapor permeability and Oil-resistance properties.

3.2 Digital Printing Performances of the Coated Paper Sized by the ONFC/CGG Complex Agent

Color gamut refers to the range of colors that the paper reproduces under the conditions of printing process. The larger the color gamut, the stronger the color reproduction ability of printing. As shown in Fig. 2, the color gamut of ink-jet method is larger than that of xerograph, which shows ink-jet printing can have a larger color reduction range.

The reproduction of color images of the coated requires not only accurate reproduction of colors, but also the accurate tone reproduction. As shown in Fig. 3, there is a linear relationship between the gradient density and the dot percentage, the tone reproduction could meet the basic requirements of digital printing proofing.

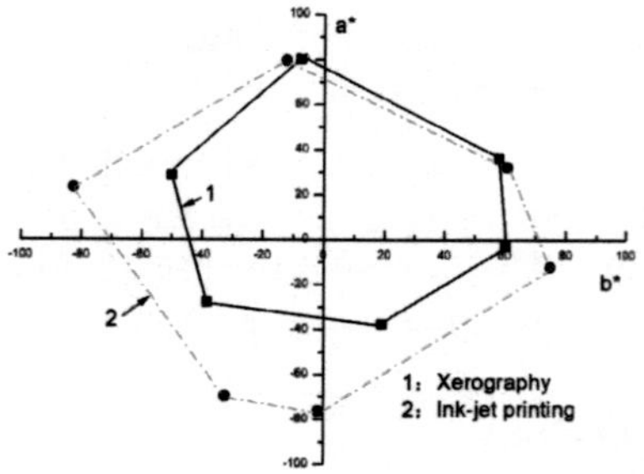

Fig. 2. Famut of the coated paper with digital printing

References

1. Ashish, K., Bhardwaj, N.K., Singh, S.P.: Cationic starch and polyacrylamides for alkenyl succinic anhydride (ASA) emulsification for sizing of cellulosic fibers. Cellulose **26**(18), 9901–9915 (2019)
2. Tyagi, P., Lucia, L.A., Hubbe, M.A., Pal, L.: Nanocellulose-based multilayer barrier coatings for gas, oil, and grease resistance. Carbohyd. Polym. **206**(13), 281–288 (2018)
3. Yong, L., Zheng-wei, X., Song Ci, X., Qiuqian.: Preparation and properties of the composite sizing agent of dialdehyde chitosan/cationic starch. Digital Print. **199**(1), 94–99 (2019)
4. Hirvikorpi, T., Vaehae-Nissi, M., Mustonen, T., Iiskola, E., Karppinen, M.: Atomic layer deposited aluminum oxide barrier coatings for packaging materials. Thin Solid Films **518**(10), 2654–2658 (2010)
5. Iqbal, D.N., Nazir, A., Iqbal, M., Yameen, M.: Green synthesis and characterization of carboxymethyl guar gum: application in textile printing technology. Green Process. Synthesis **9**(1), 212–218 (2020)
6. Koppolu, R., Lahti, J., Abitbol, T., Swerin, A., Kuusipalo, J., Toivakka, M.: Continuous processing of nanocellulose and polylactic acid into multilayer barrier coatings. ACS Appl. Mater. Interfaces **11**(12), 11920–11927 (2019)

Polyacrylate Latexes with Alkali-Soluble Resin as Surfactant: Effect of Functional Monomers, Detection and Control of Residual Monomers

Jie Liu, Fei Xia, Xiaoyu Li, and Haiqiao Wang(✉)

Beijing Engineering Research Center for the Synthesis and Applications of Waterborne Polymers, Beijing University of Chemical Technology, Beijing 100029, China
wanghaiqiao@buct.edu.cn

Abstract. TO solve the problems caused by the migration of small molecule emulsifier during film formation, in this paper, an alkali-soluble resin was used as polymer surfactant instead of small molecule emulsifier to synthesize polyacrylate latexes. The effects of three functional monomers, isobornyl acrylate (IBOA), acrylamide (AM) and hydroxyethyl acrylate (HEA), on the properties of latexes and the adhesion of the relative inks on biaxially oriented polypropylene (BOPP) films were investigated. Among these functional monomers, HEA endows the corresponding ink with the highest adhesion on BOPP thin films due to the fact that HEA can form hydrogen bonds with hydroxyl groups on the surface of BOPP films. The results of the residual monomer detection using gas chromatography (GC) showed that the latex had not only more unreacted residual monomers but also isooctanol due to the saponification of isooctyl acrylate. The residual monomers in the latexes can be nearly completely eliminated when the latexes were subjected to two post-elimination treatments.

Keywords: Alkali-soluble resin · Polyacrylate latexes · Inks · Properties · Residual monomer

1 Introduction

With increasing environmental protection requirements, it is highly desirable to develop water-based inks with non-toxicity and lower volatile organic compound (VOCs) content [1, 2]. Polyacrylate latex has become one of the most widely used binders for water-based inks due to its excellent adhesion on the surface of the substrate and low cost [3].

Traditional polyacrylate latexes are synthesized using small molecular emulsifiers. Non-reactive small molecular emulsifiers tend to migrate to the surface of the latex film during the film-forming process of the latex, which has a great impact on its performance (gloss and water resistance, etc.). Reactive small molecule emulsifiers can participate in the polymerization process, eliminating the disadvantage of poor performance caused by the migration of emulsifiers. However, due to their high cost, the application also has certain limitations [4]. As a polymer surfactant, alkali-soluble resin (ASR) generally has a molecular weight of several thousand or more and contains hydrophilic groups and

M. Xu et al. (Eds.): CACPP 2022, LNEE 991, pp. 486–493, 2023.
https://doi.org/10.1007/978-981-19-9024-3_62

lipophilic groups in its structure, which provides excellent emulsifying effects. The use of alkali-soluble resins as surfactants can not only eliminate the disadvantages caused by the migration of emulsifiers, but also have a lower cost. In addition, ASR can reinforce the mechanical properties of the dried film, so it can greatly broaden latex's application range [5].

Styrene (St) and methyl methacrylate (MMA) are the most commonly used hard monomers in the latex polymerization process, providing hardness and solvent resistance to the latex; n-butyl acrylate (BA) and isooctyl acrylate (EHA) are the most commonly used soft monomers, providing properties such as flexibility and gloss to the latex [6, 7]. However, polyacrylate latexes synthesized only with soft monomers and hard monomers generally have some disadvantages, such as low mechanical strength and poor adhesion on the substrate surface. Therefore, in the process of latex synthesis, a small number of functional monomers is often added to improve the performance of a certain aspect of the latex. In this paper, polyacrylate latexes were synthesize using isobornyl acrylate (IBOA), acrylamide (AM), and hydroxyethyl acrylate (HEA) as functional monomers respectively, and ASR as surfactant. The effects of functional monomers on the fineness, particle size, viscosity of the latexes and the adhesion of the inks on the BOPP film were studied.

In addition, the polymerization conversion rate of the latex is a very important consideration, as a low conversion rate means high monomer residue, which will lead to a large irritating odor of the latex, and will inevitable have a large negative impact on the health of operators and the environment [8]. In the polymerization of acrylic latexes with alkali-soluble resin as surfactant, the alkali-soluble resin forms a very thick barrier layer at the periphery of the latex particles, which has a very strong barrier effect on the entry of the water-soluble initiator into the interior of the latex particles, so the monomer conversion rate of this polymerization reaction is lower than that of the conventional emulsion polymerization with small-molecule emulsifiers. That is, these latexes have a more serious problem of irritating odor. However, to the best of our knowledge, there is no study on the residual monomers in the polymerization with alkali-soluble resin as surfactant. Herein, the content of residual monomers in the latex was detected by gas chromatography (GC), and the reasons for the high odor of the latex synthesized from alkali-soluble resin was investigated. Finally, a polyacrylate latex with almost no pungent odor and excellent comprehensive performance was obtained through optimization of the polymerization process and two post-elimination treatments.

2 Experiment

2.1 Materials

Alkali-Soluble Resin (ASR), J586, BASF (China) Co., Ltd.; Isooctyl Acrylate (2-EHA), Styrene (St), Isobornyl Acrylate (IBOA), Acrylamide (AM), Hydroxyethyl Acrylate (HEA) etc. are all industrial-grade products, which were used directly without treatment; yellow color paste, Xiamen Qinghe Chemical Co., Ltd.

2.2 Synthesis of Polyacrylate Latexes

The weighed ASR, deionized water and alkali neutralizer were added into a 2 L four-necked flask and dissolved for 60 min at 85 °C and under nitrogen atmosphere to get a transparent resin aqueous solution. A part of the mixed monomers and part of the initiator aqueous solution were then added, and the remaining mixed monomers and initiator aqueous solution were added dropwise when the reactants turn blue. The dripping time was controlled within 120 min, and the temperature was kept at 85 °C for 60 min after the dripping was completed. After that, tert-butanol hydrogen peroxide was added, and ascorbic acid was then added dropwise after 10 min, ensuring that the dropwise addition time was about 30 min. Finally, the temperature was lowered to 40 °C, ADH was added, and the latex was discharged after stirring for 30 min. The monomer formulation of emulsion polymerization is tabulated in Table 1.

Table 1. The recipe of the monomers

Ingredients	Amount (g)
ASR	100
EHA	200
St	200
DAAM	12
ADH	6
IBOA[a]/AM[b]/HEA[c]	12

[a] IBOA was added to the formulation only when the effect of IBOA on adhesion is studied
[b] AM was added to the formulation only when the effect of AM on adhesion is studied
[c] HEA was added to the formulation only when the effect of HEA on adhesion is studied

2.3 Preparation of Water-Based Ink

The yellow color paste and the latex was mixed in a mass ratio of 1:1, and it was stirred at a certain speed for 30 min. Then it was filtered with a 200-mesh filter to produce water-based ink.

2.4 Characterization

The fineness of the latex was measured using a scraper fineness meter; Z-average particle size was measured by dynamic light scattering (DLS, Zeta-sizer, UK); The viscosity of the latex was measured by a rotary viscometer (NDJ-5s digital display, Guangdong Sanfeng Precision Measuring Instrument Co., Ltd.); The latex adhesion was measured using 3M tape according to the test standard ISO 2409-2006. The residual monomer content in the latex was determined using a gas chromatograph (GC-14C, Shimadzu Shanghai Laboratory Equipment Co., Ltd.).

2.5 Determination of Residual Monomer Content by Gas Chromatography

2.5.1 Instrument and Test Conditions

The chromatographic column used here is SP™-2560 (Φ100 m × 0.25 mm × 0.2 μm). Temperature of gasification chamber is controlled at 260 °C; temperature of detector is 280 °C. Temperature of chromatographic column was programmed heated: initial temperature of 60 °C for 3 min, then rising to 260 °C at a heating rate of 20 °C/min and maintained for 27 min; split ratio: 40:1; injection volume: 0.2 μL.

2.5.2 Determination of Relative Correction Factors

The relative correction factor *Fi* of the residual monomers to the internal standard cyclohexanone is calculated according to the following formula:

$$F_i = \frac{m_i \times A_s}{m_s \times A_i}$$

m_i and m_s are the mass of residual monomer and internal standard, respectively; A_i and A_s are the peak areas of residual monomer and internal standard, respectively.

2.5.3 Calculation of Residual Monomer Content

The content W_i of the residual monomer to be measured can be calculated by the following formula, the letters represent the same meaning as the above formula:

$$W_i = \frac{F_i \times m_s \times A_i}{m_i \times A_s}$$

3 Results and Discussion

3.1 Influence of Functional Monomers on Latex Properties and Ink Adhesion on BOPP Films

The effects of functional monomers on the fineness, particle size, viscosity of polyacrylate latexes and ink adhesion on BOPP films were studied. The results are shown in Table 2. It can be seen from Table 2 that the latexes made from the three different functional monomers all have a fineness of 0 μm, as well as a similar particle size and viscosity. However, the adhesion of the ink formulated with the three latexes on BOPP films is quite different. The ink with AM as the functional monomer has poor adhesion, the one with IBOA has a slightly better adhesion, and the one with HEA has the best adhesion. The reason for the best adhesion of the ink containing HEA is that the hydroxyl groups of HEA can not only react with the amino groups of ADH to form covalently cross-linked networks, but also form intramolecular and/or intermolecular hydrogen bonds with polymerized carbonyl, ether or other electronegative acceptor functional groups.

Since the ink containing HEA as the functional monomer has the best adhesion on BOPP film, the effect of the amount of HEA on the performance of the relative latex

Table 2. Effects of functional monomer types on latex properties and ink adhesion on BOPP films

Functional monomer	Fineness (μm)	Particle size (nm)	Viscosity (mPa s)	Adhesion/BOPP (%)
IBOA	0	105.5	166	89
AM	0	108.6	134	82
HEA	0	105.6	129	100

was further investigated. It can be seen from Table 3 that the addition of HEA reduced the particle size of the latex, but as the amount of HEA increased from 1% to 3%, the particle size of the latex gradually increased, resulting in a gradual decrease in viscosity. The adhesion of the ink on the BOPP film increased with the increase of the amount of HEA, because the increase of HEA dosage introduces more hydroxyl groups into the system and forms a denser cross-linked network during the film formation process.

Table 3. Effect of HEA on basic properties of latex and adhesion of ink on BOPP films

Dosage of HEA (%)	Fineness (μm)	Particle size (nm)	Viscosity (mPa s)	Adhesion/BOPP (%)
0	0	113.2	215	95
1	0	85.20	177	98
2	0	92.59	138	99
3	0	105.6	129	100

3.2 Detection of Residual Monomer Content in Latexes

In the acrylic latexes, there are generally residual monomers that are not fully reacted, resulting in strong irritating odor. Due to the use of ASR as a polymeric surfactant, ASR forms a thick protective layer on the outer layer of the latex particles, which greatly prevents free radicals from entering the latex particles to initiate polymerization, resulting in more residual monomers content in the latex than in conventional latexes polymerized with small molecular emulsifiers. For this purpose, the residual monomer content of the latex was measured by gas chromatography. Figure 1 is a gas chromatogram of a polyacrylate latex with a neutralization degree of 120%. It is easy firstly to determine that the chemical substance corresponding to the peak at 2.268 min is the solvent acetone. Then, styrene, cyclohexanone and isooctyl acrylate were added to the sample in sequence, and the position where the peak area increased was the corresponding chemical substance. Therefore, it was determined that the peak at 9.385 min was styrene, the peak at 12.884 min was the internal standard cyclohexanone, and the peak at 26.796 min was isooctyl acrylate. The data was processed according to the peak area of each peak, and the content of unreacted styrene and isooctyl acrylate in the emulsion can be obtained, as shown in Table 4.

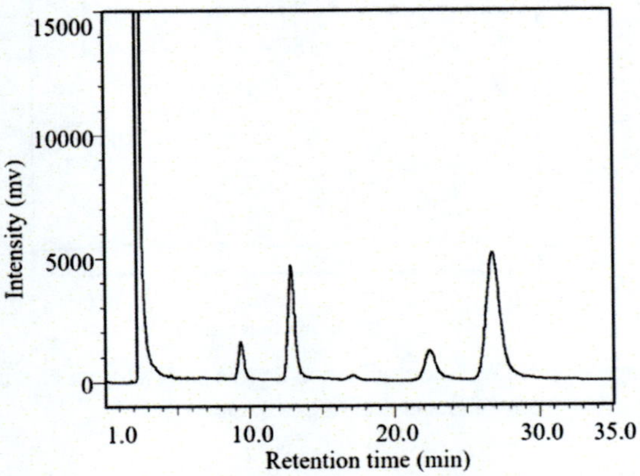

Fig. 1. Gas chromatogram of acrylate latex with 120% neutralization degree

Table 4. Content of unreacted styrene and isooctyl acrylate in latex

	Residual monomer content (%)
St	0.109
EHA	1.617

3.3 Confirmation of Isooctanol Produced During Latex Synthesis

In the gas chromatogram of the polyacrylate latex with a neutralization degree of 120%, in addition to the peaks of acetone, styrene, isooctyl acrylate and the internal standard cyclohexanone, an unidentifiable peak with a large area was found at 22.422 min. Considering that the emulsion polymerization here was carried out under more alkaline conditions, and oils and fats will undergo saponification reaction to generate alcohol and carboxylate under the catalysis of alkali, we guessed that isooctyl acrylate was saponified to form isooctanol during emulsion polymerization. Indeed, by adding isooctanol to the sample, the peak area at 22.422 min increased significantly, while the peak areas at other positions remained unchanged, which confirmed that the peak at 22.422 min is isooctanol (Fig. 2).

3.4 Effect of Post-elimination on Residual Monomer Content

Redox post-elimination reactions have been an effective technique for eliminating residual monomers from latex. In this paper, t-Bu was used as the post-elimination oxidant, and VC was used as the post-elimination reductant. The ratio of oxidant and reductant

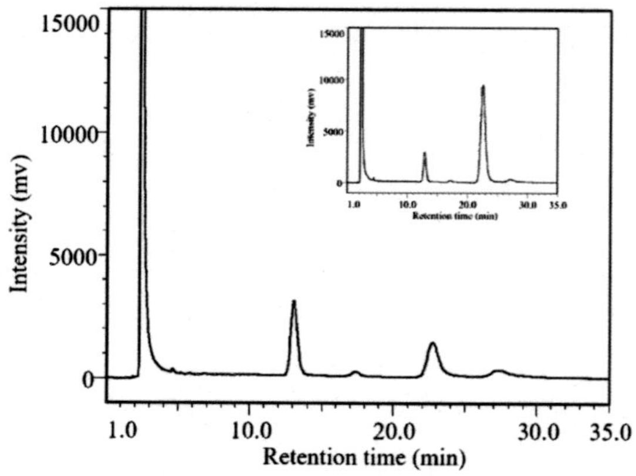

Fig. 2. GC of the sample with 120% neutralization degree and two post-eliminations (the inset is the GC after adding a few drops of isooctanol to this sample)

Table 5. Residual monomer content after one post-elimination and two post-eliminations

	St (%)	EHA (%)	Isooctanol (%)
One post-elimination	0	0.201	0.225
Two post-eliminations	0	0	0.228

was 2:1, and the effect of post-elimination on the residual monomer content was studied. The specific content of residual monomers was shown in Table 5. It can be seen that, after the latex underwent two post-elimination treatments, the amount of residual styrene and isooctyl acrylate was almost 0, but there was still a small amount of isooctanol. Theoretically, the hydrolysis of isooctyl acrylate can be reduced by reducing the degree of neutralization, thereby reducing the content of isooctanol. However, when the neutralization degree of the alkali-soluble resin is low, its emulsification effect on the polymeric monomer is poor, and the fineness of the latex prepared is large. The latex with larger fineness tends to cause some problems such as blocking during the printing process of water-based ink. Considering that the content of the isooctanol in the latex obtained at 120% neutralization degree of the ASR is within an acceptable range and that the latex has a very good fineness, it is appropriate to use polymerization conditions with alkali-soluble resins at a neutralization degree of 120%.

4 Conclusion

An alkali-soluble resin was used as polymer surfactant to synthesize polyacrylate latex binder for application in water-based ink. The effect of three functional monomers, that is AM, IBOA and HEA, on the properties of the latexes and corresponding water-borne inks were studied. It was found that when HEA was used as a functional monomer and its dosage was 3% of the total amount of monomers, the relative latex has the best comprehensive performance with a fineness of 0 μm, a particle size of 105.6 nm, a viscosity of 129 mPa s, and the adhesion of the corresponding ink on BOPP film reaches to 100%. The results of gas chromatography measurements showed that there are not only residual monomers but also isooctanol produced from the saponification reaction of isooctyl acrylate during the polymerization, which lead to a strong and irritating odor of the latex. The use of two post-elimination reactions can effectively eliminate the residual monomers in the latex, making latex virtually free of irritating odors.

References

1. Wada, T., Yasuda, M., Yako, H., et al.: Preparation and characterization of hybrid quaternized chitosan/acrylic resin latexes and their films. Macromol. Mater. Eng. **292**(2), 147–154 (2007)
2. Hu, W., Bai, Y., Zhang, C., et al.: Coating based on the modified chlorinated polypropylene latex for promoting printability of biaxially oriented polypropylene film. J. Adhes. Sci. Technol. **32**(1), 50–67 (2017)
3. Xu, Y.X., Yu, Q., Lin, T.T., et al.: Synthesis of styrene-acrylic latex modified with hydroxyl phosphate and its corrosion properties. Adv. Mater. Res. **233–235**, 1157–1161 (2011)
4. Zhang, J., Zhao, Y., Dubay, M.R., et al.: Surface enrichment by conventional and polymerizable sulfated nonylphenol ethoxylate emulsifiers in water-based pressure-sensitive adhesive. Ind. Eng. Chem. Res. **52**(25), 8616–8621 (2013)
5. Lopes Brito, E., Ballard, N.: Film formation of Alkali Soluble Resin (ASR) stabilized latexes. Prog. Org. Coat. **159**, 106444 (2021)
6. Akbulut, G., Bulbul Sonmez, H.: Synthesis of styrene and n-butyl acrylate latex polymers modified by functional monomers and their waterborne paint applications. J. Coat. Technol. Res. (2022)
7. Athawale, V.D., Kulkarni, M.A.: Preparation and properties of urethane/acrylate composite by latex polymerization technique. Prog. Org. Coat. **65**(3), 392–400 (2009)
8. Viljanen, E.K., Skrifvars, M., Vallittu, P.K.: Dendrimer/methyl methacrylate co-polymers: residual methyl methacrylate and degree of conversion. J. Biomater. Sci. Polym. Ed. **16**(10), 1219–1231 (2005)

Preparation of PANI/GO Electrode Material Modified by Non-ionic Surfactant TX-100

Qiqi Huang[1,2,3] and Fuqiang Chu[1,2,3,4](✉)

[1] Faculty of Light Industry, Qilu University of Technology (Shandong Academy of Sciences), Jinan, China
fqchu@126.com

[2] State Key Laboratory of Biobased Material and Green Papermaking, Qilu University of Technology (Shandong Academy of Sciences), Jinan, China

[3] Key Laboratory of Green Printing and Packaging Materials and Technology in Universities of Shandong, Faculty of Light Industry, Qilu University of Technology (Shandong Academy of Sciences), Jinan, China

[4] Kiev College, Qilu University of Technology (Shandong Academy of Sciences), Jinan 250353, China

Abstract. To explore Polyaniline graphene oxide composites (PANI/GO) composites, electrode materials suitable for flexible supercapacitors were prepared by modifying composites with the nonionic surfactant TX-100. Firstly, the improved Hummers method was used to prepare graphene oxide solution, PANI/GO composite material was prepared, and then the nonionic surfactant TX-100 was added to modify the solution to prepare a flexible electrode material. The experimental results show that when aniline (An) is 0.036 mol, it is compounded with ammonium persulfate (APS) of the same molar mass and graphene oxide (GO) of 120 mg, and the PANI/GO morphology is relatively excellent, and the conductivity is 62.5 S/m. Taking 0.1 g of the composite material supplemented by 1 ml of deionized water and 50 μl of TX-100, the optimal area specific capacitance of the conductive film prepared is 35.32 mF/cm^2, and the adhesion on the PET film has been significantly improved. Electrode materials with good porosity, large specific capacitance and excellent adhesion were prepared, which provided good method for the subsequent preparation of printed electronic components.

Keywords: TX-100 · PANI/GO · Hydrazine hydrate · Electrode material

1 Introduction

As a kind of conductive polymer, polyaniline (PANI) is widely used in the field of printed electronics. PANI has good environmental stability and conductivity under specific conditions. PANI is widely used in the preparation of sensors, supercapacitors and other fields. PANI has attracted much attention in the field of electrode materials because of its conductivity, easy synthesis and low cost [1]. Supercapacitor is a new type of green energy storage device [2–5] that can charge and unload quickly and has an ultra

M. Xu et al. (Eds.): CACPP 2022, LNEE 991, pp. 494–500, 2023.
https://doi.org/10.1007/978-981-19-9024-3_63

long lifespan. It is widely used in many aspects of life. The performance supercapacitors is directly related to the electrode materials. Currently, electrode materials used in the preparation of supercapacitors include primarily carbon materials, metal oxides and conductive polymer materials [6].

Because of its unique properties, PANI can undergo a unique doping response with acids. When it is used as the electrode super material of supercapacitor, it can provide a high pseudo capacitance [7]. However, researchers found that when PANI was used as an electrode material for supercapacitors, its energy storage effect did not reach the desired effect. In fact, when the PANI material is completely oxidized or completely reduced, it is not a conductor of electricity [8]. Consequently, when PANI is chosen as a supercapacitor electrode material, it is generally composed with other electrode materials [9] to prepare supercapacitors with high specific capacity and excellent performance.

Graphene is easy to agglomerate, insoluble in water or organic solvents, so it has certain limitations in practical application. Graphene oxide (GO) is a kind of chemically modified graphene [10]. The difference between it and graphene is that oxygen-containing polar groups are introduced between its layers, which makes it soluble in organic solvents and water [11]. GO has a large specific surface area and sheet layered structure, which is conducive to charge storage and ion passage, but its conductivity is poor. In this paper, different proportions of GO and PANI were compounded to prepare more stable electrode materials with good specific capacitance.

2 Experimental

2.1 Materials and Instruments

Main materials: graphite powder (325 mesh, 99.95%), Qingdao Huatai lubrication and sealing section; Aniline (AR, 98%) Tianjin Bodi Chemical Co., Ltd.; Tea Polyphenols (AR, 98%) Saen Chemical Technology Co., Ltd.; Ammonium persulfate (AR, 98%), hydrochloric acid (AR, 38%), concentrated sulfuric acid (AR, 98%), hydrogen peroxide (AR, 30%), potassium permanganate (AR, 99.5%), Sinopharm Chemical Reagent; Deionized water; Distilled water.

Main instrument: Plasma Flecto 30 plasma processing equipment, Plasma Technology GmbH; RTS-8 Four Probe Tester, Guangzhou Four Probe Technology; PGSTAT302N Electrochemical Workstation, Metrohm China Co., Ltd.; Polarized light microscope, Shanghai Liguang precision instrument; AUW120D Analytical Balance, Shimadzu, Japan; RET basic magnetic force, Aika (Guangzhou) instruments and equipment; DZF-6050 blower drying box, Shanghai quasi-quasi instruments and equipment.

2.2 Experimental Content

Preparation of Polyaniline/Graphene Oxide Composites. Firstly, GO was prepared. Added 1 g of toner and 23 ml of 1 mol/l H_2SO_4 into a four-port flask, and then mechanically stir it in an ice bath for 1 h. Then 3 g K_2MnO_4 was added to it for 30 min. After that, the temperature was increased to 35 °C, heated for 120 min, and 40 ml of deionized water was added to the four-mouth flask during the period, and the time was controlled at 30 min. Then the temperature was raised to 95 °C and kept for 10 min. After the solution was cooled to room temperature, H_2O_2 was added to neutralize the acid, and then it was allowed to stand for 12 h, and the supernatant was dumped. Under the action of deionized water, the neutral solution was finally obtained after five times of centrifugation and one time of ultrasound. Secondly, preparation of micro structured Polyaniline. Configure 1 mol/L of HCl solution. An (0.036 mol) was uniformly dispersed in HCl (1 mol/L) under the action of ultrasound. Under the action of ultrasonic wave, APS with different proportions were uniformly dispersed in four equal parts of HCl solution. Then, they were added into four equal parts of prepared An solution, each of which took 60 min. The obtained product was vacuum filtered until the filtrate was colorless, and the obtained product was vacuum dried at 60 °C for 24 h to obtain microstructure PANI. Thirdly, preparation of PANI/GO. Configure 1 mol/L of HCl solution. Added a certain amount of GO solution to 20 ml HCl solution, stir and ultrasonic disperse evenly. An solution was prepared by the same method. Taken quantitative APS according to the ratio of An:APS of 1:1, and prepared the APS solution according to the above method. Under the condition of the ultrasonic water bath, the configured APS solution was slowly added to the An/GO mixed solution, and the time was controlled at 60 min. The obtained product was vacuum filtered until the filtrate was colorless, and the obtained product was vacuum dried at 60 °C for 24 h to obtain PANI/GO.

Preparation of PANI/rGO Conductive Film. Switching to 2 × 2 cm PET slices were put in absolute ethanol ultrasonic treatment for 10 min, and then putted them in a blast drying oven for drying. They were taken out and cooled to room temperature, and then they were plasma treated to facilitate the preparation of conductive films. Added 0.1 g PANI/GO to the solvent containing 1 ml deionized water and 50 μl TX-100 to prepare the electrode material. Added 0.5 ml of prepared electrode material onto the PET plastic, and then placed it on the heating table to heat and dry at 60 °C. Putted the completely dried electrode material film in a vacuum drying oven and reduced it with hydrazine hydrate at 80 °C for 10 min to obtain PANI/rGO conductive film.

2.3 Testing and Characterization

Graphene Oxide Solid Content Test. The net weight of weighing bottles were recorded as M_1, M_2, M_3 and M_4. Dropped 1ml of GO solution respectively, and recorded the mass as m_1', m_2', m_3', and m_4' after drying. Calculated the solid content n of the prepared graphene oxide solution according to Eq. 1.

$$N = \frac{\frac{[m_1 - m_1' + m_2 - m_2' + m_3 - m_3' + m_4 - m_4']}{4}\text{mg}}{\text{ml}} \tag{1}$$

Conductivity Test. The conductivity of the microstructure PANI prepared with different ratios of An/APS was measured by a four probe tester. When An/APS is 1:1, the conductivity of PANI/GO composites prepared with different ratios of An/GO is tested by four probe tester.

Scanning Electron Microscope (SEM) Test. Scanning electron microscope (SEM) is an important means to characterize the micro morphology of samples by scanning the surface of samples with a focused electron beam to show the magnified sample image. PANI/GO prepared with different An/GO ratios was adhered to the stage under the action of conductive adhesive, and then sprayed with gold for SEM test.

Electrochemical Performance Test. The prepared conductive films were tested by cyclic Voltammetry (CV), Constant current charge discharge (GCD) using electrochemical workstation. The area specific capacitance of the conductive film is calculated by Eq. 2.

$$C = \frac{\int idV}{A \cdot \Delta V \cdot S} \tag{2}$$

C is the area specific capacitance, a is the electrode area of the supercapacitor, ΔV is the voltage window and S is the scanning rate.

3 Results and Discussion

The solid content of GO prepared by Hummers method is 6 mg/ml according to Eq. 1. When preparing PANI/GO, the quantity of An was 0.036 mol. First, the conductivity of PANI synthesized with different ratio of An/APS was investigated. It was found that when the ratio of An/APS was 1:1, the conductivity of PANI was the best, which could reach 294.1 S/m. When An/APS is 1:1 and 120 mg GO is added, the PANI/GO composite is relatively stable with a conductivity of 62.5 S/m. The reason why the conductivity of the composite is smaller than that of PANI is that the conductivity of GO is very poor. With the addition of GO, the conductivity of the prepared composite is also affected (Fig. 1).

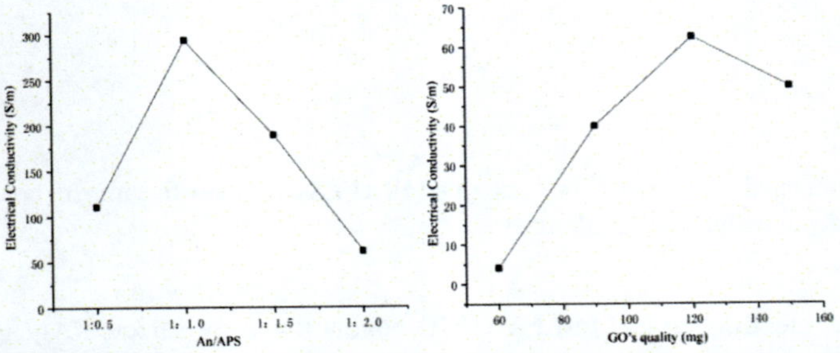

Fig. 1. Effect of different proportion of An/APS on conductivity of PANI and different content of GO on conductivity of PANI/GO composites

Because there are four existing states of PANI, which are converted to each other through oxidation-reduction reaction, PANI will lose its conductivity in excessive oxidation and over reduction states. The purpose of compounding PANI with GO is to improve its environmental stability and broaden its application fields. In the composite process, too much or too little GO cannot reach the target ideal state. When the amount of GO is too small, the cycle stability of PANI cannot be improved. When the amount of GO is too many, the cross-linking of PANI in the GO layer is weakened, and PANI is not formed on many surfaces. As shown in the SEM of Fig. 2, PANI without GO presents a porous irregular columnar structure on the microstructure; When 60 mg GO was compounded, a large amount of PANI could not be well cross-linked with the GO layer; when 90 mg GO was compounded, agglomeration occurred; when 120 mg GO was compounded, PANI wrapped the GO lamella, realizing good cross-linking growth; when 150 mg GO was compounded, some areas of GO lamella did not even have PANI growth.

Fig. 2. SEM of PANI/GO synthesized by GO of different quality a) PANI; b) 60 mg GO; c) 90 mg GO; d) 120 mg GO; e) 150 mg GO

According to the data analysis of the conductivity test and SEM test of the prepared composites, PANI/GO compounded with 120 mg GO with An/APS ratio of 1:1 was selected to prepare conductive films. As shown in Fig. 3, PANI/GO alone can't achieve good adhesion on PET film. When 50 μl TX-100 is added to PANI/GO, the composite material spreads well on PET film, and the adhesion is improved accordingly.

Fig. 3. a) Untreated conductive film; b) dropped 50 μl TX-10 conductive film; c) bent conductive film; d) conductive film bent for 50 times

Conduct electrochemical test PANI/GO conductive film, selected Ag/AgCl as reference electrode, Pt as counter electrode, and 1 mol/l H_2SO_4 as electrode solution. As shown in Fig. 4, according to Eq. 2, when scanning at 10 mV/s, it can be calculated that the area specific capacitance of the conductive film is 35.32 mF/cm^2; when scanning at 50 mV/s, the area specific capacitance is 30.79 mF/cm^2; when scanning at 100 mV/s, the area specific capacitance is 25.37 mF/cm^2; when scanning at 150 mV/s, the area specific capacitance is 19.42 mF/cm^2; when scanning at 200 mV/s, the area specific capacitance is 9.16 mF/cm^2. When the scanning rate is 50 mV/s, compared with 10 mV/s, the capacitance retention is 87.17%, and the change is not very large. Similarly, the corresponding capacitance retention rates of 100 mV/s, 150 mV/s and 200 mV/s are 71.83%, 54.98% and 25.93%. This is because the electrode material can be well activated when the scanning rate is low, and the ions in the electrolyte can migrate to the porous electrode more fully, so the capacitance is larger when the scanning rate is low. The constant GCD test of the conductive film shows that the curves obtained at different scanning rates are almost isosceles triangle, indicating that its charge discharge performance is excellent.

Fig. 4. Electrochemical measurement of conductive films

4 Conclusion

In this paper, the non-ionic surfactant TX-100 was added to PANI/GO composites to modify the composite solution, and the electrode materials with excellent adhesion were prepared, which laid a good foundation for the subsequent application in flexible supercapacitors. Take 0.036 mol An, APS with the same molar mass and 120 mg GO for compounding. Take 0.1 g of product, add 1ml of deionized water and 50 μl of TX-100, disperse uniformly by ultrasound, and prepare 2 × 2 cm conductive film. The optimum area specific capacitance is 35.32 mF/cm^2, and the adhesion of the modified composite on PET film is significantly improved, which provides unlimited possibilities for the subsequent preparation of printed electronic components.

Acknowledgement. This research is supported by Shandong Province Science and Technology Small and Medium-sized Enterprise Innovation Capability Improvement Project (No. 2021TSGC1168).

References

1. Beygisangchin, M., Abdul Rashid, S., Shafie, S., et al.: Preparations, properties, and applications of polyaniline and polyaniline thin films—a review. Polymers **13**(12) (2021)
2. Yu, G., Xie, X., Pan, L., et al.: Hybrid nanostructured materials for high-performance electrochemical capacitors. Nano Energy **2**(2), 213–234 (2013)
3. Zhu, Q., Zhao, D., Cheng, M., et al.: A new view of supercapacitors: integrated supercapacitors. Adv. Energy Mater. **9**(36) (2019)

4. Chen, G.Z.: Supercapacitor and supercapattery as emerging electrochemical energy stores. Int. Mater. Rev. **62**(4), 173–202 (2016)
5. Sun, K., Zhang, X., Li, B., et al.: The application and development of new carbon materials in supercapacitor energy storage technology under the background of energy internet. Funct. Mater. **49**(02), 2043–2053 (2018)
6. Lili, D., Chen, Y.: Research progress of electrode materials for supercapacitors. Shandong Chem. Ind. **50**(23), 65–67 (2021)
7. Lu, Zhou, X., Xu, L., et al.: Preparation and energy storage performance of flexible polyaniline/titanium nitride nanowire electrode materials. Acta Silicate Sinica **50**(07), 1909–1918 (2022)
8. Eftekhari, A., Li, L., Yang, Y.: Polyaniline supercapacitors. J. Sources **347**, 86–107 (2017)
9. Meng, Q., Cai, K., Chen, Y., Chen, L.: Research progress on conducting polymer based supercapacitor electrode materials. Nano Energy **36**, 268–285 (2017)
10. Singh, R., Ullah, S., Rao, N., et al.: Synthesis of three-dimensional reduced-graphene oxide from graphene oxide. J. Nanomater. **2022**, 1–18 (2022)
11. Dikin, D.A., Stankovich, S., Zimney, E.J., et al.: Preparation and characterization of graphene oxide paper. Nature **448**(7152), 457–460 (2007)

Printability Study of Electroluminescent Flexographic Inks

Yongjian Wu, Beiqing Huang(✉), Xianfu Wei, Hui Wang, Wan Zhang, and Linhong Huang

School of Printing and Packaging Engineering, Beijing Institute of Graphic Communication, Beijing, China
384573464@qq.com

Abstract. This paper uses Alq3 as the organic electroluminescent material in order to develop a high-performing electroluminescent flexo ink. The effects of single and mixed solvents, single and mixed resins on the viscosity, drying speed, surface tension, contact angle and other important printing parameters of the formulated EL inks were investigated. The results show that high printability can be achieved by using DMF and chlorobenzene as solvents and B-814 resin as link stuff.

Keywords: Formula experiments · Solvent · Resin · Electrofluorescence · Printability

1 Introduction

Organic materials are usually considered insulators. However, with the progress of science, scientists have not only invented organic materials that can conduct electricity, but also found that some organic semiconductor materials can be electroluminescent, which is electroluminescent materials [1]. After more than a decade of development, the performance indicators of electroluminescent materials continue to improve and enhance the market is usually used to produce OLED displays, but in addition, electroluminescent materials can be dispersed in organic solvents because of its characteristics, very suitable for the production of ink, so that the ink in the role of voltage to emit bright colour light, printed with electroluminescent ink, can be adjusted by adjusting the Luminous material and voltage to adjust the luminous colour and brightness, to the printed material to achieve a positive decorative effect. The flexographic printing technology, because it can be printed on large-format devices, printing materials, such as a wide range of advantages, you can print electroluminescent ink in conductive film, the production of electroluminescent printed materials. However, there is no available flexo luminescent ink on the market, therefore, the development of flexo luminescent ink is of critical importance.

Currently available electroluminescent materials can be divided into organic fluorescent materials and organic metal complex materials, with 8-hydroxyquinoline aluminum being a unique metal complex. Since it is able to do the light-emitting layer as well as

M. Xu et al. (Eds.): CACPP 2022, LNEE 991, pp. 501–508, 2023.
https://doi.org/10.1007/978-981-19-9024-3_64

electron transport layer, this material was chosen as the test material [2]. Printability directly affects the hue, saturation, gloss and reproduction effect of the printed material printed with the ink. In this paper, we aim to develop high quality flexo electroluminescent inks. The types and ratios of the key components of the inks - solvents and resins - are used in formulation experiments to carry out the corresponding research work in order to obtain good printability and to obtain electroluminescent inks suitable for flexo printing.

2 Experimental Section

2.1 Experimental Materials

Electroluminescent material: Tris-8-hydroxyquinoline Aluminum (Alq3).

Resin: B-814 resin, AZR resin, polyurethane resin.

Solvents: N-Methyl pyrrolidone (NMP), butyl acetate, o-xylene, N,N-dimethylformamide (DMF), dimethyl sulfoxide (DMSO), methanol, acetone, tetrahydrofuran, propylene glycol methyl ether (PMCAS), butyl acrylate (BA).

Auxiliaries: surfactant 410, surfactant 420, defoamer.

Printed material: ITO-PET film.

2.2 Experimental Equipment

Electronic analytical balance (FA2104N, Shanghai, Jing hai Instruments).

Flexographic proofing machine (FLEXIPROOF100, RK, UK).

Rheometer (AR2000ex, TA Instruments, USA).

Fully automatic surface tension meter (K100, KRUSS, Germany).

Multi-head magnetic heating stirrer (HJ-6, Jin tan, Jin nan Instruments).

Contact angle observer (DSA100, KRUSS, Germany).

2.3 Preparation of Inks

First, the resin needs to be pre-dispersed in a solvent. The resin is added to the organic solvent, placed on a magnetic stirrer and heated and stirred for 30 min to dissolve it completely. Various assistant agents are then added to obtain an electroluminescent ink.

2.4 Ink Performance Test Methods

Viscosity. Select the appropriate rotor for the rheometer according to the viscosity range of the ink to be measured. Take an appropriate amount of ink sample and add it dropwise to the centre of the rheometer measuring table. Adjust the height of the rotor with the software so that the rotor is just cling the ink sample. Then measure multiple viscosities of the ink sample to take an average value.

Surface tension. Mount the platinum plate on the surface tension meter test bench and place the measurement ink sample in the sample tray and then on the sample bench.

Adjust the height of the platinum plate above the sample liquid level and measure using the software's automatic measurement mode.

Angle of contact. Turn on the light source knob of the contact angle viewer and adjust the brightness of the light source according to the image displayed by the computer. Adjust the drip tip so that it appears in the middle of the image. Fill the glass syringe with the ink sample and mount it on the holder. Rotate the micrometer tip so that the liquid flows out and drips onto the glass on the table, measuring the contact angle between the ink sample and the glass. After waiting for 1 s, click on "Capture currently displayed image" to capture the image of the liquid on the solid surface and use the software to measure the contact angle of the liquid on the solid surface [3].

Drying rate. A small amount of ink will be placed in the scraper instrument slot of 100 μm, quickly scrape down, immediately open the stopwatch, after 30 s with a sheet, the lower end aligned with the zero scale, flat against the groove, with the palm of your hand quickly pressed, remove the paper, measure the length of the unsticky ink, expressed in millimetres, that is, the initial drying [4].

3 Results and Analysis

3.1 Influence of Solvents on the Printability of Inks and Their Determination

Prescreening of solvents. The production of electroluminescent inks, of which the electroluminescent material is one of the key components, requires additional considerations. The most basic premise of the solution method of film formation is that the solvent dissolves the solute as fully as possible [5]. The Alq3 electroluminescent material used here is a luminescent material that is insoluble in most organic solvents. Therefore, the solubility of Alq3 needs to be tested first to ensure that the solubility of Alq3 in the chosen solvent can be greater than 5 mg/ml.

Commonly used laboratory organic solvents with different polarities and boiling points were chosen to test the solubility of Alq3 and the final test results are shown in Table 1.

Table 1. Solubility of Alq3 in different solvents

Solvent	Methanol	Ethanol	DMSO	DMF	Chlorobenzene
Solubility	×	×	√	√	√
Solvent	Acetone	NMP	PMCAS	BA	Tetrahydrofuran
Solubility	×	√	×	×	×

As can be seen from Table 1, the solubility of Alq3 in DMF, DMSO, NMP and chlorobenzene meets the requirements and these four solvents can be selected for the preparation of the ink.

Effect of solvents on the printability of inks. Solvents play a role in solvent-based ink systems by dissolving solutes, resins and additives. The type of solvent directly affects the drying speed, surface tension and viscosity of the ink. And, in electroluminescent ink systems, the solvent evaporation process directly affects the thickness uniformity of the film and thus the luminescent properties of the printed electroluminescent ink [6]. According to the above screened solvents, the corresponding samples of electroluminescent inks were prepared. Their viscosity, surface tension, contact angle drying speed and other printability were measured, as shown in Table 2.

Table 2. Properties of electroluminescent ink samples prepared with different solvents

Solvent	Viscosity (Pa s)	Surface tension (mN/m)	Angle of contact	Drying rate (min)
DMF	0.79	22.41	15	3
DMSO	1.98	23.78	20	3
NMP	1.65	22.96	24	5
Chlorobenzene	0.36	20.43	23	1

As can be seen from Table 2, different solvents contribute differently to the various printing indicators of electroluminescent inks. DMF has the smallest contact angle, which facilitates the spreading of the ink on the substrate; chlorobenzene has the lowest boiling point, so the fastest drying speed and good film formation; the most suitable viscosity range for flexographic inks is 0.1–2 Pa s, and NMP has a moderate viscosity, which is conducive to ink transfer.

Determination of solvents. In order to obtain the best overall performance of the electroluminescent ink, the formulation test was used to design a mixed solution formulation. The number of solvents selected was m = 4. DMF was designed as x1, DMSO as x2, NMP as x3, chlorobenzene as x4. 15 solvent ratios were designed according to the simple gravity method. The designed experimental sequences and solvent ratios are shown in Table 3.

Since m = 4, the regression equation is as follows:

$$\begin{aligned} y = {} & b1x1 + b2x2 + b3x3 + b4x4 + b12x1x2 + b13x1x3 \\ & + b14x1x4 + b23x2x3 + b24x2x4 + b34x3x4 \\ & + b123x1x2x3 + b124x1x2x4 + b134x1x3x4 \\ & + b234x2x3x4 + b1234x1x2x3x4 \end{aligned} \tag{1}$$

According to Table 3, prepare the electroluminescent ink and test the printing properties as shown in Table 4. According to the comprehensive analysis of the needs of electroluminescent inks, the viscosity of the ink, surface tension, contact angle, and drying speed of four important parameters are set at 0.3, 0.2, 0.1, and 0.4. The best viscosity is not a specific value, but from the range in which it is determined. Therefore the viscosity indicators are first rated and then, together with the other indicators, the

Table 3. Ratio of solvents

Experimental serial number	X1	X2	X3	X4
1	1	0	0	0
2	0	1	0	0
3	0	0	1	0
4	0	0	0	1
5	0.5	0.5	0	0
6	0.5	0	0.5	0
7	0.5	0	0	0.5
8	0	0.5	0.5	0
9	0	0.5	0	0.5
10	0	0	0.5	0.5
11	0.33	0.33	0.33	0
12	0.33	0.33	0	0.33
13	0.33	0	0.33	0.33
14	0	0.33	0.33	0.33
15	0.25	0.25	0.25	0.25

data are processed. This is doneate the index affiliation of each performance indicator and the overall score for each formulation.

According to the weights given for calculating the composite score y obtained by the ink, and putting y, x1, x2, x3, x4 into the formula (1), to get 15 fourth-order regression equation. The regression equation can be solved by combining the regression equation of the overall performance of the electro-luminescent ink.

$$\begin{aligned} y = {} & 0.77x1 + 0.64x2 + 0.59x3 + 0.78x4 + 0.005x1x2 - 0.008x1x3 \\ & + 0.03x1x4 - 0.016x2x3 + 0.065x2x4 - 0.095x3x4 + 0.208x1x2x3 \\ & - 0.536x1x2x4 - 0.345x1x3x4 - 0.445x2x3x4 - 3.15x1x2x3x4 \end{aligned} \quad (2)$$

Through the programming solver can be obtained, when x1:x2:x3:x4 = 0.48:0:0:0.52, with DMF and chlorobenzene in the ratio of 0.48:0.52 resulting in the best performance of electroluminescent ink.

3.2 Effect of Resin on Ink Performance

The resin, as the linking material of the ink, plays a major role in connecting and allowing the ink to flow evenly during the transfer process, which can have a large impact on the viscosity and drying time of the ink. Three resins, B-814 resin, AZR resin and polyurethane resin, were selected to prepare samples of electroluminescent inks. The viscosity, surface tension and drying speed of the ink samples were tested and the results are shown in Table 5.

Table 4. Properties of electroluminescent ink samples prepared with different samples

Experimental serial number	Viscosity (Pa s)	Surface tension (mN/m)	Angle of contact	Drying rate (min)
1	0.79	22.41	15	3
2	1.98	23.78	20	3
3	1.65	22.96	24	5
4	0.36	20.43	23	1
5	1.41	22.47	15	3
6	1.26	22.45	18	4
7	0.52	20.89	18	2
8	1.86	23.14	22	4
9	1.03	21.19	21	2
10	0.85	21.07	24	4
11	1.27	23.09	21	4
12	0.84	23.05	20	3
13	0.73	21.88	22	4
14	1.13	22.86	23	4
15	1.09	22.39	22	4

Table 5. Properties of ink samples made of different resins

Experimental serial number	X1	X2	X3	Viscosity (Pa s)	Surface tension (mN/m)	Drying rate (min)
1	1	0	0	0.57	21.14	2
2	0	1	0	0.63	21.49	6
3	0	0	1	0.84	22.94	8

As can be seen from Table 5, there is little difference in surface tension and viscosity between the samples prepared from the three resins. The content of the resin in the ink system is much lower than that of the solvent and the surface tension and viscosity are more influenced by the solvent. The resin is not the decisive factor for surface tension and viscosity in the ink system. Of the three different types of resin, B-814 is the fastest drying, AZR is moderate and polyurethane is the slowest. This is because the B-814 resin is more soluble in the solvent and therefore the solvent evaporates more easily and dries faster.

It was concluded that the most effective performance of the electro-luminescent ink was obtained by using only B-814. Consequently, it is preferable to use only B-814 resin in the experiments.

3.3 Printability of the Developed Electroluminescent Flexographic Ink

Through the above experiments, the main properties of the inks were tested by making the corresponding ink samples according to the best solvent-based ink formulations obtained. The data are presented in Table 6.

Table 6. Performance of Ink sample after formulation optimization

Solvent	Resin	Viscosity (Pa s)	Surface tension (mN/m)	Angle of contact	Drying rate (min)
DMF:chlorobenzene = 0.48:0.52	B-814 resin	0.51	20.84	18	2

According to the data in Table 6, it can be seen that after the formulation test to study the suitability of electroluminescent ink printing is optimal.

4 Conclusion

In the solvent-based electroluminescent ink system, the solubility of different solvents for electroluminescent materials varies greatly, and the solvents have an effect on the viscosity, surface tension, drying speed and contact angle of the ink, and the prepared electroluminescent ink samples have optimal performance when the ratio of DMF and chlorobenzene is 0.48:0.52.

The resin in the solvent-based electroluminescent ink system mainly affects the drying speed of the ink, and has little effect on the surface tension and viscosity of the ink. The ink samples formulated with B-814 resin have better drying speed than those formulated with AZR resin and polyurethane resin, and the overall performance of the prepared ink samples is good.

The developed electroluminescent flexographic ink has good printability.

Acknowledgements. We gratefully acknowledge the support from BIGC Project (Nos. Ea202203, Ee202205) and Institute of Advanced Ink, Beijing Institute of Graphic Communication.

References

1. Xiao, L., et al.: Blue fluorescent small molecule electroluminescent materials. J. Opt. **30**(07), 1895–1903 (2010)
2. Wang, H.: Synthesis and properties of PPV-based electroluminescent polymers. Southeast University (2006)
3. Wang, K.: Study on thermal conductivity modification and preparation process of epoxy resin based on double impermeability effect. Shanghai Jiaotong University (2020). https://doi.org/10.27307/d.cnki.gsjtu.2020.001782

4. Peng, C.: Preparation and dispersion stability study of aqueous gravure silver ink. Nanjing Forestry University (2019). https://doi.org/10.27242/d.cnki.gnjlu.2019.000172
5. Gao, J.: Research on carrier balance and device performance enhancement in organic electroluminescent devices. Tianjin University of Technology (2017)
6. Liu, H.: Study on the preparation method of ink-jet printed electroluminescent films and devices. South China University of Technology (2016)

Study on the Influences of Surfactants in the Preparation of Thermally Expandable Microcapsules

Zhenzhen Li, Zhicheng Sun(✉), Zhitong Yang, Gongming Li, Chenyang Liu, and Yibin Liu

Beijing Engineering Research Center of Printed Electronics, School of Printing and Packaging Engineering, Beijing Institute of Graphic Communication, Beijing 102600, China
sunzhicheng@bigc.edu.cn

Abstract. The central core component of physical foaming inks is thermally expanded microcapsules (TEMs). In the preparation of TEMs by suspension polymerization, surfactants are usually added to the aqueous phase to promote emulsion compatibility between the core and wall monomers, allowing the oily core to be better encapsulated. In this study, sodium dodecyl sulfate and aliphatic alcohol ethoxylates, two different types of surfactants, were used to prepare TEMs, and the structure and properties of the prepared TEMs were characterized by the scanning electron microscopy (SEM), thermal expansion instrument (DIL) and thermogravimetric analyzer (TG). By studying the influencing factors of different surfactants on the microcapsule preparation process and further analyzing their swelling properties, the most suitable surfactants, as well as the microcapsule products with the best foaming effect, can be identified, thus laying a good foundation for the subsequent research on foaming inks.

Keywords: Thermally expanded microcapsules (TEMs) · Surfactants · Suspension polymerization

1 Introduction

The development of TEMs as an essential component of foaming [1] inks has been significant in recent years. TEMs prepared by suspension polymerization [2] are obtained by mixing the oil and aqueous phases after they have been prepared separately and reacted in a water bath for a while. In the formulation of the aqueous phase, the role of surfactants is of utmost relevance, as it is decisive for the amount of core material that the wall material can encapsulate. A number of different surfactants are currently available in the market, anionic surfactants, cationic surfactants, nonionic surfactants, amphoteric surfactants, compound surfactants and other surfactants, to name a few. In this paper, TEMs have been prepared using anionic surfactants in contrast to non-ionic surfactants. The better of the two has been explored to provide a reasonable basis for the subsequent development of multifunctional foaming inks [3] in construction, medicine, and clothing [3, 4].

M. Xu et al. (Eds.): CACPP 2022, LNEE 991, pp. 509–513, 2023.
https://doi.org/10.1007/978-981-19-9024-3_65

2 Experiment

2.1 Materials and Instruments

Sodium hydroxide, magnesium chloride hexahydrate, sodium chloride, n-hexane, sodium lauryl sulfate, 1,4-butanediol dimethacrylate, fatty alcohol polyoxyethylene ether, Shanghai Aladdin Biotechnology Co., Ltd. Acrylonitrile 99+%, Beijing Inno Chemical Technology Co., Ltd. Methyl methacrylate (stabilized with 6-tert-butyl-2,4-xylenol), methyl acrylate (stabilized with MEHQ), 2,2′-Azobis(isobutyronitrile), Tokyo Chemical Industry Co., Ltd.

The scanning electron microscope, Gemini 300, Germany; thermogravimetric analyzer, NETZSCH STA 449 F5/F3 Jupiter, Germany; thermal dilatometer, NETZSCH DIL 402 PC, Germany.

2.2 Preparation

The oil phase: Add acrylonitrile, methyl methacrylate, methyl acrylate, azobisisobutyronitrile, and n-hexane in a beaker with a mass ratio of 7:2:1, 2,2′-Azobis(isobutyronitrile), 1,4-butanediol dimethacrylate, stir evenly with a magnetic stirrer. The water phase: Sequentially add magnesium hydroxide, sodium lauryl sulfate or fatty alcohol polyoxyethylene ether, sodium hydroxide, sodium chloride, and a certain amount of deionized water into a three-necked flask and stir uniformly to obtain a reaction water phase. Make the oil phase thoroughly and evenly dispersed in the water phase. Heat to 65 °C, stir mechanically at 400–900 r/min, react for 10–20 h, cool, discharge, repeatedly wash and filter, and blast dry Bake in an oven at 45 °C for 24 h to obtain the white powder microcapsules.

3 Results and Discussion

3.1 Surface Topography

Figure 1 shows SEM pictures of the prepared microcapsules, (a) and (b) show the effect of the sodium dodecyl sulfate preparation, and (c) and (d) show the effect of the fatty alcohol polyoxyethylene ether preparation. From the figures, it can be seen that the TEMs prepared by the anionic surfactant have the initial morphology of unexpanded heat expanded microcapsules, while the morphology of the heat expanded microcapsules prepared by the non-ionic surfactant is incomplete, showing apparent broken hollow shell shape, indicating that the preparation effect of the non-ionic surfactant is poor. The effect is much lower than the preparation effect of the anionic surfactant.

3.2 Thermal Analysis

The graph above shows the characterization of the swelling properties of the microcapsules. Curve (a) shows the effect of sodium dodecyl sulfate preparation, and curve (b) shows the effect of fatty alcohol polyoxyethylene ether preparation. The trend of curve (a) decreases and then increases. The decrease represents the softening of the outer shell

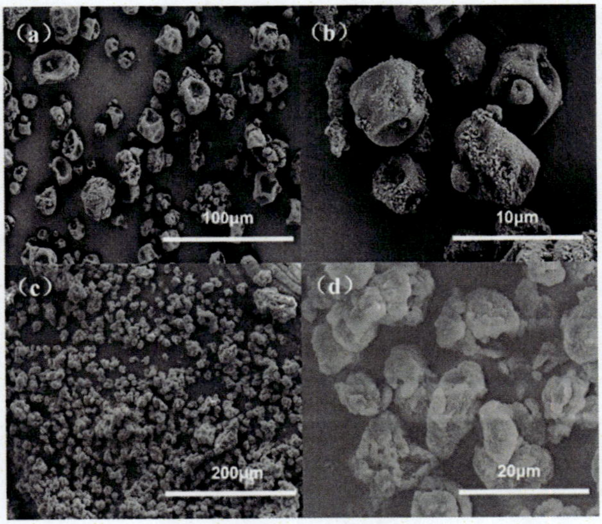

Fig. 1. SEM pictures of TEMs

of the TEMs as the temperature increases and the volume decreases; as the temperature increases further, the internal low boiling point core material begins to vaporize and expand, making the microcapsules larger, then the microcapsules rupture and decrease in volume rapidly and undergo further decomposition. The trend of the curve (b) is always decreasing and not increasing at the later stage, i.e., the microcapsules decomposed directly with the increase of temperature, indicating that the prepared TEMs did not encapsulate the core material and the preparation failed, which is consistent with the conclusion of the SEM pictures.

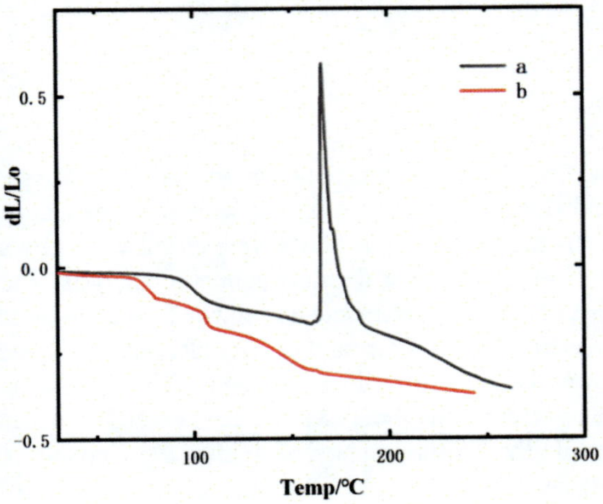

Figure 3 shows the TG analysis of TEMs. Curve (a) shows the effect of sodium dodecyl sulfate preparation, and curve (b) shows the effect of fatty alcohol polyoxyethylene ether preparation. This indicates that the microcapsules prepared with sodium dodecyl sulfate were successfully prepared, with the first decrease being due to the release of the vapor from the core material of the microcapsules as the temperature increased and the second stage being due to the decomposition of the shell. This means that the geometries did not cover the core material after the TEMs preparation of the TEMs, so the curve kept decreasing after the temperature increase, thus confirming the conclusions in Figs. 1 and 2.

Fig. 3. TG curves of TEMs

4 Conclusions

From the above data, we can learn that the two groups of the TEMs prepared by sodium dodecyl sulfate and aliphatic alcohol ethoxylate were compared, and the shell nucleus structure and specific properties of the TEMs were characterized by scanning electron microscopy, thermal expansion, and thermogravimetric analysis. The TEMs prepared by polyoxyethylene ether were incomplete and showed apparent broken hollow shells, indicating that the effect of the preparation of fatty alcohol polyoxyethylene ether was poor and much lower than that of sodium dodecyl sulfate. Observing the DIL and TG pictures, it can be concluded that the TEMs were successfully prepared using sodium dodecyl sulfate, while the TEMs prepared with aliphatic alcohols ethoxylated ethers did not encapsulate the core material, and the preparation failed. This indicates that the anionic surfactant sodium dodecyl sulfate is superior to the non-ionic surfactant aliphatic alcohol ethoxylates in preparing the TEMs.

Acknowledgments. This work was supported by the National Natural Science Foundation of China (Grant No. 22278037).

References

1. Hu, J., Zheng, Z., Wang, F., Tu, W., Lin, L.: Synthesis and characterisation of thermally expandable microcapsules by suspension polymerisation. Pigm. Resin Technol. **38**(5), 280–284 (2009)
2. Chen, S.Y., Sun, Z.C., Li, L.H.: Preparation and characterization of thermally expandable microspheres. Mater. Sci. Forum **852**, 596–600 (2016)
3. Zhang, Q., Sun, Z., Li, F., Wen, J., Huang, S.: Study on preparation and printing evaluation of composite aromatic expansion microsphere ink. In: Advances in Graphic Communication, Printing and Packaging Technology and Materials, pp. 646–650. Springer, Singapore (2021)
4. Zhang, J., et al.: Thermally expandable microspheres with excellent high-temperature expansion property. Express Polym. Lett. **16**(7) (2022)

Check for updates

Preparation of Flame Retardant Phase Change Microcapsules and Ink Application

Zhitong Yang, Zhicheng Sun(✉), Gongming Li, Zhenzhen Li, Yibin Liu, and Chenyang Liu

Beijing Engineering Research Center of Printed Electronics, School of Printing and Packaging Engineering, Beijing Institute of Graphic Communication, Beijing 102600, China
sunzhicheng@bigc.edu.cn

Abstract. In this study, a novel type of refractory phase change microcapsules was prepared by in situ polymerization using melamine resin as the wall material and paraffin and chlorinated paraffin as the mixed core material. The microcapsules, linking materials, color paste and additives were mixed in a certain ratio to formulate an ink with dual functions of flame retardant and phase change. The surface morphology, phase change properties and flame retardant properties of the ink were characterized by scanning electron microscopy (SEM), differential scanning calorimetry (DSC) and heat release rate (HRR). The results showed that the flame retardant phase change microcapsules were successfully prepared, showing a tiny core–shell structure with an average latent heat of phase change of 16.74 J/g, which indicated a good latent heat of phase change. Compared with ordinary inks, flame retardant phase change inks have higher flame retardant efficiency, which is of great significance for the research of functional inks.

Keywords: Flame retardant phase change microcapsules · Screen printing · Ink

1 Introduction

In recent years, with the rapid development of science and technology and society, the role of energy in human survival and development has become more and more significant, and the problem of energy shortage will seriously affect people's life. How to use energy efficiently has become a hot issue. Due to the depletion of non-renewable energy and the intensification of global warming, the development of renewable energy and the improvement of energy utilization have become the top priority. The exploration of new energy sources and the rational and effective use of existing energy sources are the only way to solve the increasingly serious energy problems. Known for their high thermal energy storage capacity through the latent heat of phase change, phase change materials have developed as an alternative solution to the energy utilization problem [1].

However, one of the main drawbacks of phase change materials is their flammability, which is highly susceptible to casualties and property damage in case of fire [2]. Therefore, proper fire protection and environmental safety-specific precautions must be taken when incorporating phase change materials into polymers. Hu et al. doped

M. Xu et al. (Eds.): CACPP 2022, LNEE 991, pp. 514–519, 2023.
https://doi.org/10.1007/978-981-19-9024-3_66

the phase change material with a flame retardant to achieve the flame retardant phase change performance of the material. This physical doping method will lead to the deterioration of the mechanical properties of the material [3]. Chlorinated paraffin plays an indispensable and important role in flame retardants and is the most used flame retardant in China. Compared with inorganic flame retardant materials, chlorinated paraffins can achieve the flame retardant efficiency of inorganic flame retardants with a small amount [4]. In this paper, we use microcapsule wrapping technology to combine phase change thermal storage technology with flame retardant technology, and prepare flame retardant phase change microcapsules by in situ polymerization with sodium dodecyl sulfate as the emulsifier, melamine resin as the shell, and phase change material paraffin and flame retardant material chlorinated paraffin as the core material, and apply the phase change microcapsules to the ink, which accelerates the progress of the printing field of phase change microcapsules and strongly broadens the microcapsule application area [5].

2 Experiments

2.1 Materials and Instruments

Main reagents: solid paraffin, Sinopharm Chemical Reagent Co., Ltd; chlorinated paraffin, Maclean's Reagent Co., Ltd; triethanolamine (AR), melamine, formaldehyde, sodium dodecyl sulfate, sodium hydroxide, anhydrous citric acid, Shanghai Aladdin Biochemical Technology Co., Ltd; petroleum ether, Beijing Chemical Reagent Factory; acrylic film-forming resin, Qingdao Linke Industry and Trade Co.

Main instruments: constant speed automatic stirrer, D2004, Shanghai Zhiwei Electric Co. Ltd.; Scanning Electron Microscope, SU8020, HITACHI, Japan; Electronic Balance, PX224ZH, OHAUS Instruments Co.

2.2 Preparation of Flame Retardant Phase Change Ink

2.2.1 Preparation of Flame Retardant Phase Change Microcapsules

Add 15 g of sodium dodecyl sulfate powder, 4.2 g of sodium hydroxide and 130.6 g of deionized water to a three-necked round-bottom flask, stir mechanically and keep constant temperature under the condition of water bath at 50 °C, and make sodium dodecyl sulfate emulsifier after 5 h, pour it into a beaker and set aside. Take another three-necked round bottom flask, add 10 g of sodium dodecyl sulfate emulsifier, paraffin and chlorinated paraffin, add 100 ml of deionized water, melt it in a water bath at 70 °C, after that emulsify it at 700 rpm/min for 2 h to get a stable O/W emulsion, adjust the pH of the emulsion to 4–5 with citric acid of 10% mass fraction. in the round bottom flask Add 6.6 g of formaldehyde and 3 g of melamine, add 25 ml of deionized water, and then adjust the pH value to 8–9 with triethanolamine at a mass fraction of 10%, stir the reaction mechanically at 500 rpm for 30 min, and place it in a constant temperature water bath at 70 °C while stirring to produce a melamine resin prepolymer solution. Under the condition of maintaining the mechanical stirring speed of 500 rpm, the melamine resin prepolymer solution was added dropwise to the aqueous phase at a rate of 0.5 ml/min. After the dropwise addition, the pH was adjusted to 9 by adding triethanolamine dropwise after three hours, and then, the product was washed three times each with petroleum ether and distilled water to obtain an extracted product, and dried under vacuum in a 50 °C oven to finally obtain Flame retardant phase change microcapsules were obtained.

2.2.2 Preparation of Flame Retardant Phase Change Ink

The mixed printing transparent paste (50%), PVA (30%), color paste (9%), flame retardant phase change microcapsules (10%), defoamer (0.2%), leveling agent (0.8%) are prepared in a certain mass ratio to obtain flame retardant phase change ink. For example, after the printed thermal expansion microcapsule ink is cured, the sample with flame retardant phase change effect can be obtained by heating at 120 °C for 1 min with a heating plate.

3 Results and Discussion

3.1 Morphology of Flame Retardant Phase Change Microcapsules

From Fig. 1, it can be seen that the flame retardant phase change microcapsules are regular spherical in shape, with intact encapsulation between the shell layer and the core material. The average diameter of the prepared microcapsules is 870 nm, and their particle size is small, which can be easily filled into the ink voids. The surface of the microcapsules is obviously observed to have folds, which are formed by the polymerization of the melamine resin. The existence of agglomeration of the microcapsules indicates that the microcapsules have the problem that nanoscale particles are prone to [6]. From the SEM image and the TEM image, it can be confirmed that the spherical flame retardant phase change microcapsules can be prepared by the in situ polymerization method.

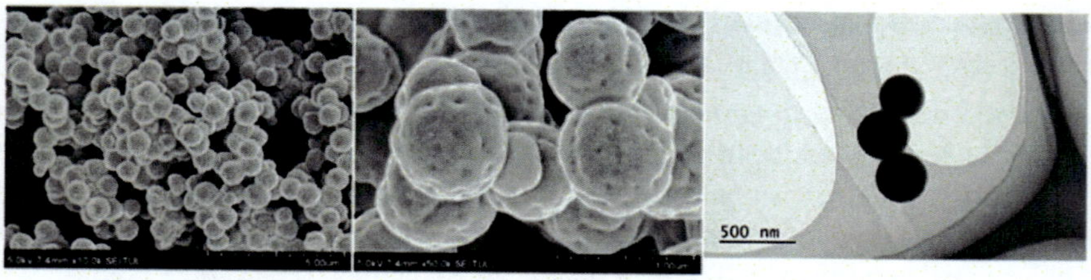

Fig. 1. Microscopic view of flame retardant phase change microcapsules

3.2 Thermal Properties of Flame Retardant Phase Change Microcapsules

Differential Scanning Calorimetry (DSC) is a thermal analysis method that measures the power difference input to a sample and a reference as a function of temperature under programmed temperature conditions [7]. From the DSC curves in Fig. 2, it can be seen that the flame retardant phase change microcapsule curves are two obvious absorption or exothermic peaks both at increasing and decreasing temperatures, which shows that the flame retardant phase change microcapsules successfully encapsulate paraffin in them. Among them, the enthalpy of melting (ΔH_m) of microcapsules is 18.11 J/g, and the enthalpy of crystallization (ΔH_c) of microcapsules is 15.35 J/g indicating that the microcapsules have excellent heat storage capacity of phase change.

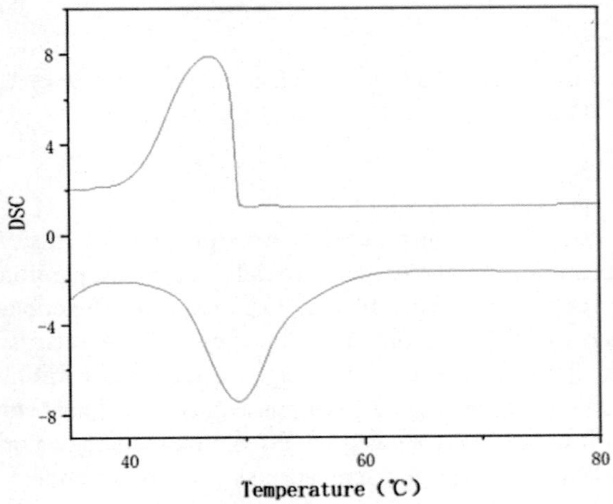

Fig. 2. Phase change latent heat curve of microcapsules

3.3 Flame Retardant Phase Change Ink with Flame Retardant Properties

HRR is the rate of heat release per unit area when the material is ignited under a given incident heat flow condition. HRR is a relatively important performance index, which reflects the intensity of fire. As shown in Fig. 3a, the HRR value of ordinary ink is 201 W/g, and the HRR value of flame retardant phase change ink is 148 W/g. Compared with ordinary ink, the HRR of flame retardant phase change ink is reduced by 26%, and the smaller the HRR value, the less heat is fed back to the material surface by combustion, which results in faster pyrolysis of the material and reduced generation of volatile combustibles, thus slowing down the propagation of flame. Figure 3b and c are the sliver printed by ordinary ink and flame retardant phase change ink, from which it can be seen that the sliver treated with ordinary ink start to burn violently upon ignition and are accompanied by the appearance of molten drops; the sliver treated with flame retardant ink have smaller flames and no molten drops after ignition, and the samples can be self-extinguishing after 60 s, indicating that the flame retardant phase change ink has good flame retardant effect.

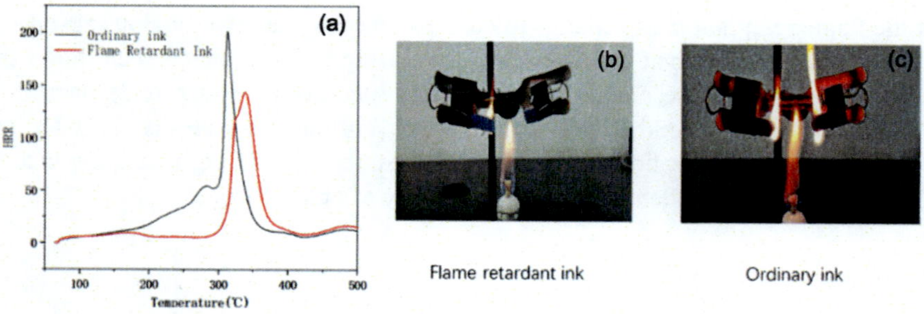

Fig. 3. Flame retardant phase change ink flame retardant effect diagram

4 Conclusions

Flame retardant phase change microcapsules were prepared by in situ polymerization synthesis. The melamine resin was used as the wall material and paraffin and chlorinated paraffin were used as the hybrid core material to prepare the bifunctional microcapsules with excellent thermal properties and high efficiency of flame retardancy. The surface morphology of the flame retardant phase change microcapsules was observed by scanning electron microscopy; the latent heat of phase change and heat storage capacity of the prepared microcapsules were studied by differential scanning calorimetry. The prepared flame-retardant phase change microcapsules were used as fillers, and the ink was made according to a certain ratio and printed on sliver by screen printing, and the flame-retardant phase change ink was tested with the help of micro calorimeter. The results show that the microcapsules have uniform morphology, good latent heat of phase change and high flame retardant efficiency, which can play an important role in building fire prevention and promoting energy and environmental sustainability, and have significant application potential.

Acknowledgements. This work was supported by the National Natural Science Foundation of China (22278037).

References

1. Han, K.T., Lhosupasirirat, S., Srikhirin, P., et al.: Development of flame retardant stearic acid doped graphite powder and magnesium hydroxide nanoparticles, material for thermal energy storage applications. J. Phys. Conf. Ser. (2022)
2. Normura, T., Sheng, N., Zhu, C., et al.: Microencapsulated phase change materials with high heat capacity and high cyclic durability for high-temperature thermal energy storage and transportation. Appl. Energy **188**, 9–18 (2017)
3. Hu, Z.T., Reinack, V.H., An, J., et al.: Ecofriendly microencapsulated phase-change materials with hybrid core materials for thermal energy storage and flame retardancy. Langmuir (2021)
4. Wang, Z., Hu, X., Wang, X., et al.: Synthesis of high compatibility modified chlorinated paraffin as flame retardant plasticizer and its application in polyvinyl chloride. Polym. Mater. Sci. Eng. (2019)

5. Liu, C., Chen, K., Ji, L., et al.: Preparation of epoxy resin coated ammonium polyphosphate microcapsules and their flame retardant effects on polypropylene. Acta Mater. Compos. Sinica **32**(3), 728–736 (2015)
6. Li, F., et al.: Preparation and performance of dual-functional magnetic phase-change microcapsules. Chem. Asian J. **16** (2021)
7. Chen, R., Ge, X., Li, X., et al.: Facile preparation method of phase change microcapsule with organic-inorganic silicone shell for battery thermal management. Compos. Sci. Technol. **228** (2022)

Research and Application Progress of Conductive Ink Based on Polyaniline

Shasha Li[1](✉), Xu Li[1], Lixin Mo[1], Zhiqing Xin[1], Luhai Li[1](✉), Meijuan Cao[1](✉), Xiuhua Cao[2], Jun Huang[2], and Yintang Yang[3]

[1] Beijing Engineering Research Center of Printed Electronics, Beijing Institute of Graphic Communication, Beijing, China
1947713985@qq.com, liluhai@bigc.edu.cn, meijuancao@126.com
[2] State Key Laboratory of Key Materials and Technology for Advanced Electronic Components, Zhaoqing 526040, China
[3] JiangXi LianSheng Electronic CO., LTD., Jingdezhen, China

Abstract. As a key material for printed electronics, conductive inks have received extensive attention. Conductive polymer polyaniline (PANI) as a conductive filler has the advantages of easy synthesis, low cost, high conductivity and good environmental stability, which provides a new way for the application and development of conductive inks. In order to fully grasp the progress of PANI conductive ink and its application, this paper comprehensively analyzes the research progress, application and preparation process of PANI conductive ink based on the literature and research work of conductive polymer ink in recent years. It is pointed out that the main development direction of PANI conductive ink is to compound PANI with other functional materials (such as metals and carbon materials) to form a stable, efficient, low-cost and environmentally friendly composite conductive ink, and the application prospect of PANI conductive ink is proposed.

Keywords: Conductive polymer · PANI · Conductive ink · Printed electronics

1 Introduction

Printed electronics (PE) is emerging with the development of electronic information technology. Its market volume is estimated to rise from US $6.8 billion in 2018 to US $13.6 billion in 2023 [1], showing great development potential. As one of the key materials of PE, conductive ink is often used to manufacture electrodes and circuit interconnects, meeting the requirements of interconnecting high conductive tracks in most electronic devices [2]. Its comprehensive performance is determined by conductive

M. Xu et al. (Eds.): CACPP 2022, LNEE 991, pp. 520–528, 2023.
https://doi.org/10.1007/978-981-19-9024-3_67

fillers. At present, conductive inks in the market mostly use metal or carbon materials as conductive fillers [3, 4]. But the metal is expensive and the curing or sintering conditions are harsh. Although some carbon conductive inks are much cheaper, their conductivity is not enough to meet various applications. Conductive polymers (CPs) provide a new way for the application and development of conductive ink. As conductive fillers, CPs have good mechanical properties and multi-functional processing [5].

Polyaniline (PANI) is considered as one of the most promising conductive polymers due to its simple synthesis, high conductivity, excellent environmental stability and unique doping/dedoping mechanism [6–8]. PANI conductive ink, as an organic ink, has great advantages in flexibility and processability [9]. It can print electrodes and conductive films by coating, screen printing and jet printing, etc., and has made some progress in the application of PE. In order to further study PANI conductive ink and its application progress, this paper summarizes the research progress, application and preparation process of PANI conductive ink based on the literature reports of conductive polymer ink in recent years and the research work of our research group. On this basis, the development direction of PANI conductive ink was discussed, and its application prospect was put forward.

2 Conductive Polymer Ink

Conductive inks are mainly composed of conductive fillers, binders, solvents and additives [10, 11]. According to conductive fillers, they can be divided into inorganic, organic and composite conductive inks [12]. In this paper, the conductive ink with conductive polymer as conductive filler is regarded as conductive polymer ink. CPs are organic polymers with π conjugation system, including polyacetylene (PA), polypyrrole (PPy), PANI, polythiophene (PTH) and its derivatives [13].

The author's research group developed an environment-friendly water-based conductive ink with PEDT based conductive polymer aqueous dispersion as conductive filler. In actual printing, the thickness of the conductive film is less than 3 μm, and the resistance value is lower than 100 k$\Omega/\square$. It can be stored for a long time at room temperature and has stable conductivity. Kraft et al. [14] prepared poly (3,4-ethylene dioxythiophene) (PEDOT: PSS) doped with polystyrene sulfonic acid into a conductive ink for printing stretchable connectors and electrodes. The electrical conductivity of the connector can reach 700 s/cm, can withstand more than 100% strain, and has good stability. In order to prepare high-performance inks and meet the actual printing needs, CPs are often mixed or reacted with other fillers (such as metal, carbon materials, etc.). Tsugita team [15] combined PPy and indium tin oxide (ITO) to obtain PPy-ITO conductive ink. ITO particles are used as dispersants improve the overall stability of ink. The ink does not need an annealing process, is responsive to pH value, and is still excellent in conductivity at low temperature. Arena et al. [16] mixed PPy and graphene nanosheets in dodecyl benzene sulfonic acid solution, combined the charge transport performance and high surface area of nanocarbon structure with the electrochemical activity of PPy. The area capacitance of the electrode prepared by this ink increased about 3 times compared with that of PPy electrode.

3 PANI Conductive Ink

3.1 PANI

PANI is a structural conductive polymer with complex and diversified structure. It is generally believed that PANI has both oxidation unit (benzene structure) and reduction unit (quinone structure). The conductivity of PANI depends on the doping rate of proton acid and the degree of oxidation. After being doped with different acids/dyes and other dopants, the conductivity of PANI can be improved by nearly ten orders of magnitude, changing from insulator to conductor, which is called doped PANI [17]. It can return to the insulator state again after contacting with the alkaline substance, that is, the doping/de doping state of polyaniline is reversible. PANI can be synthesized by chemical oxidation polymerization, electrochemical polymerization, polycondensation polymerization and other methods [18]. In actual production, PANI with different conductivity can be prepared according to different needs.

3.2 PANI Conductive Ink

PANI conductive ink is dispersion of PANI in solvents, including water-based, solvent-based, composite conductive inks, etc. Water-based and solvent-based PANI conductive ink is usually prepared into conductive ink after doping induction and structural modification of polyaniline powder or insoluble film to obtain water-soluble or soluble PANI. Composite conductive ink is generally composed by two or more materials to make up for the shortcomings of a single material in performance, so that it has higher detection sensitivity and selectivity [19] and improves the comprehensive performance of the ink.

PANI is often combined with metals and carbon materials, such as graphene and MWCNTs. Metal/metal oxides often use PANI as a carrier, which provide high conductivity and excellent electrocatalytic activity for ink. Polyaniline, as a support material, improves the durability of metal materials [20]. Graphene has unique physicochemical properties, mechanical properties and conductivity [7]. Under the synergistic effect, graphene can improve the conductivity and stability of ink; Polyaniline, as a spacer, improves the dispersion of graphene [21]. MWCNTs have good conductivity and excellent biocompatibility. The p-p interaction between the composite structures increases electrons, which can promote electron transfer in REDOX reactions and effectively increase electron transfer kinetics.

According to Table 1, with PANI as a single conductive filler, the conductive ink is simple to prepare, low cost and environmentally friendly, but it has poor performance in conductivity and printability. With composite materials as conductive fillers, the conductive ink performance is more better, which can meet the needs of more printing processes for ink, and has a wider range of applications. Considering the actual production benefits, the development direction of PANI conductive ink should be to composite with other materials to form composite conductive ink with better comprehensive performance.

Table 1. The summary of PANI conductive ink

Category	Conductive filler	Mechanism	Electrical conductivity	Advantage	References
Solvent-based (DMSO)	PANI-CSA/PANI-AMSA	Doping	The correlation between absolute conductivity and structural characteristics is poor	Low pollution, improved PANI solubility, simplified handling	[6]
Water-based	PANI-CSA		Electrical conductivity: 6.3 S/m; R_{min}: 67.6 Ω	Use at low temperature; Reduce recovery cost, rapid response, suitable for industrial production	[9]
	PANI-CSA		Electrical conductivity: 10^{-2} S/cm; at 0.5 A/g (dehydration state), specific capacitance: 386.9 F/g	Low cost, easy to manufacture, stable in air	[22]
Composite	C-MWCNTs/PANI nanosheets	Facilitate electron transfer	Area capacitance of a single electrode: 268.1 mF/cm^2	Air stability, easy to manufacture, fast electron transfer	[23]
	Ag NPs/PANI	Synergy	Resistance of the film: 10^{-6} Ω cm	Storage months without precipitation, suitable for industrial production	[24]
	GO/PANI/PEDOT: PSS		Electrode conductivity: ~ 64 S/cm, area of capacitance: 153.6 mF/cm^2	High concentration, high viscosity, good stability, no post treatment	[25]
	PANI/Ag/TPU		Sintering (80 °C), volume resistivity: 1.26×10^5 Ω m	Good printing suitability, good stability, high conductivity	[26]

4 Application of PANI Conductive Ink and Preparation

There are many applications of conductive inks, the most direct and simple of which is to replace the traditional ink in writing tools for direct writing. For example, Polysketch pen [27], which can draw simple circuits. In addition, conductive inks are often used to prepare conductive films and electrodes in production. Conductive thin films can be used to prepare a variety of devices in the field of printed electronics. The author's research group used DBD polymerization method to directly prepare 2,2′-dithiophene into polythiophene thin films on paper substrates to form conductive paper. In addition, conductive films can also be used for electrochromic layer [28], anode buffer layer (ABL) [29], etc. Electrodes have a wider range of applications, including glucose sensors [20], urea-pH sensors [30], heavy metal ion sensors [31] and supercapacitors [32] and other devices.

Coating, flexo printing, screen printing, inkjet printing, 3D printing and so on are commonly used in the preparation of conductive ink devices. Table 2 summarizes the current application of PANI conductive ink and preparation. According to Table 2, ink-jet printing and screen printing are still the mainstream printing methods in the field of printing electronics, and there are few types of PANI conductive inks suitable for flexography. With the pursuit of simple, low-cost, less waste, large-scale production and environmentally friendly manufacturing methods [33], the development of a PANI conductive ink suitable for large-scale printing process is still one of the focus of future research.

5 Conclusions and Prospects

PANI conductive ink is mainly used to prepare electrodes and conductive films. They are commonly prepared by screen printing and inkjet printing. It has a wide range of applications and broad prospects for development. PANI is mostly combined with other materials (such as metal and carbon materials) to form composite conductive ink with better comprehensive performance. In order to improve the comprehensive performance of PANI conductive ink and achieve the goal from experimental demonstration to practical application, the future research of PANI conductive ink will focus on how to manufacture it with low cost, low energy consumption and simple mass. Firstly, the preparation method of PANI should achieve green and improve its solubility. Secondly, PANI conductive ink and other materials composite and preparation of stable, efficient, low cost, environmentally friendly composite conductive ink for the development direction. Third, its application and preparation process will be extended to a broader printing platform, such as 3D printing and coating composite, to improve the paste performance and printing suitability, which is the only way to promote the industrialization and commercialization of PANI conductive ink.

Table 2. Application of PANI conductive ink and preparation

Conductive ink	Application	Preparation	Process characteristics	References
Nano-scale CNC-PANI	Conductive films	Flexography	High printing speed, low price, good compatibility; ink quality requirements are high	[34]
(2D) PANI sheet PANI-PAMPSA		IJP	High precision, high resolution; components with complex spatial configurations can be printed	[28, 29]
Ink A: aniline, phytic Ink B: Triton X-100, EG, APS		MRIJP	Simplify the preparation process, save consumables and reduce costs, expand the substrate type from porous to non-porous, good at micro-scale material synthesis	[35]
Cu^{+2}/PANI/rGO nanocomposite	Electrode	Drops of paint	Simple operation, flexible	20
MWCNT/PANI EDTA@PANI/MWCNT nanocomposite		Screen printing	Cost-effective, mass production, large surface area printing, modified electrodes	[29, 30]
PANI:PSS/PPy:PSS		IJP	High precision, high resolution	[36]
GO/PANI/PEDOT: PSS		Extrusion printing	High speed, high efficiency, high flexibility, scalable and large-scale operation	[25]

(continued)

Table 2. (*continued*)

Conductive ink	Application	Preparation	Process characteristics	References
PANI/GO		DIW	Making complex objects quickly; suitable for materials that can form a concentrated viscoelastic fluid with a specific solvent (ink)	[32]

Acknowledgements. Beijing Natural Science Foundation (jointly funded by the Municipal Education Commission) Research on the composition of the ink-absorbing layer and the sintering mechanism of conductive ink at room temperature (K Z202110 015019); Research project of Beijing Institute of Graphic Communication (Ef202002), open project of Fenghua High-tech Guozhong Laboratory (roll printing paste organic system development). China-Czech Scientific and Technological Cooperation Committee 43rd Regular Session Personnel Exchange Program (43-7); The National Natural Science Foundation of China (No. 22278037).

References

1. Abdolmaleki, H., Kidmose, P., Agarwala, S.: Droplet-based techniques for printing of functional inks for flexible physical sensors. Adv Mater **33**(20), e2006792 (2021)
2. He, P., Cao, J., Ding, H., et al.: Screen-printing of a highly conductive graphene ink for flexible printed electronics. ACS Appl. Mater. Interfaces **11**(35), 32225–32234 (2019)
3. Yu, Z., Zhang, Y.: Research progress of conductive ink. Colloids Polym. **02**, 80–84 (2021)
4. Hu, G., Kang, J., Ng, L.W., et al.: Functional inks and printing of two-dimensional materials. Chem. Soc. Rev. **47**(9), 3265–3300 (2018)
5. Guo, X.: Conducting polymers forward. Nat. Mater. **19**, 921 (2020)
6. Bocchini, S., Castellino, M., Della Pina, C., et al.: Inkjet printed doped polyaniline: navigating through physics and chemistry for the next generation devices. Appl. Surf. Sci. **456**, 246–258 (2018)
7. Saidina, D.S., Eawwiboonthanakit, N., Mariatti, M., et al.: Recent development of graphene-based ink and other conductive material-based inks for flexible electronics. J. Electron. Mater. **48**(6), 3428–3450 (2019)
8. Wang, H., Wen, H., Hu, B., et al.: Facile approach to fabricate waterborne polyaniline nanocomposites with environmental benignity and high physical properties. Sci. Rep. **7**(1), 1–12 (2017)
9. Tao, Y., Chang, X., Yao, S., et al.: Study on the properties of polyaniline doped by camphor sulfonic acid and its conductive inks. Chem. Eng. Equip. **08**, 11–16 (2019)
10. Li, J., Lu, J., Wang, Y., et al.: Research development of conductive inks and nanoparticles applied in conductive inks. Electron. Compon. Mater. **05**, 12–16+60 (2014)
11. Li, L., Mo, L., Ran, J., et al.: Conductive ink and its application technology progress. Imaging Sci. Photochem. **32**(4), 393 (2014)

12. Shi, T., Deng, Q., Wang, H., et al.: Research progress of conductive ink and its printing technology. Packag. Eng. **09**, 11–21 (2022)
13. Beygisangchin, M., Abdul Rashid, S., Shafie, S., et al.: Preparations, properties, and applications of polyaniline and polyaniline thin films—a review. Polymers **13**(12), 2003 (2021)
14. Kraft, U., Molina-Lopez, F., Son, D., et al.: Ink development and printing of conducting polymers for intrinsically stretchable interconnects and circuits. Adv. Electron. Mater. **6**(1), 1900681 (2020)
15. Tsugita, Y., Maeda, S.: Colloidal stability of polypyrrole-ITO conducting inks. Jpn. J. Appl. Phys. **61**(SE), SE1003 (2022)
16. Arena, A., Branca, C., Ciofi, C., et al.: Polypyrrole and graphene nanoplatelets inks as electrodes for flexible solid-state supercapacitor. Nanomaterials **11**(10), 2589 (2021)
17. Jangid, N.K., Jadoun, S., Kaur, N.: A review on high-throughput synthesis, deposition of thin films and properties of polyaniline. Eur. Polym. J. **125**, 109485 (2020)
18. Rahman, M.M., Mahtab, T., Mukhlish, M.Z.B., Faruk, M.O., Rahman, M.M.: Enhancement of electrical properties of metal doped polyaniline synthesized by different doping techniques. Polym. Bull. **78**(9), 5379–5397 (2020). https://doi.org/10.1007/s00289-020-03389-9
19. Tanguy, N.R., Thompson, M., Yan, N.: A review on advances in application of polyaniline for ammonia detection. Sens. Actuators B Chem. **257**, 1044–1064 (2018)
20. Anand, V.K., Bukke, A., Bhatt, K., et al.: Highly sensitive and reusable Cu^{+2}/polyaniline/reduced graphene oxide nanocomposite ink-based non-enzymatic glucose sensor. Appl. Phys. A **126**(7), 1–11 (2020)
21. Xu, Y., Schwab, M.G., Strudwick, A.J., et al.: Screen-printable thin film supercapacitor device utilizing graphene/polyaniline inks. Adv. Energy Mater. **3**(8), 1035–1040 (2013)
22. Chu, X., Chen, G., Xiao, X., et al.: Air-stable conductive polymer ink for printed wearable micro-supercapacitors. Small **17**(25), 2100956 (2021)
23. Chu, X., Zhu, Z., Huang, H., et al.: Conducting polymer ink for flexible and printable micro-supercapacitors with greatly-enhanced rate capability. J. Power Sources **513**, 230555 (2021)
24. Kariper, İA.: Conductive ink next generation materials: silver nanoparticle/polyvinyl alcohol/polyaniline. J. Inorg. Organomet. Polym Mater. **32**(4), 1277–1286 (2022)
25. Liu, Y., Zhang, B., Xu, Q., et al.: Development of graphene oxide/polyaniline inks for high performance flexible microsupercapacitors via extrusion printing. Adv. Funct. Mater. **28**(21), 1706592 (2018)
26. Hu, X., et al.: Silver flake/polyaniline composite ink for electrohydrodynamic printing of flexible heaters. J. Mater. Sci.: Mater. Electron. **32**(23), 27373–27383 (2021). https://doi.org/10.1007/s10854-021-07113-9
27. Prestowitz, L.C., Emery, J.D., Huang, J.: Polysketch pen: drawing from materials chemistry to create interactive art and sensors using a polyaniline ink. J. Chem. Educ. **98**(6), 2055–2061 (2021)
28. Huang, X., Chen, J., Xie, H., et al.: Inkjet printing of 2D polyaniline for fabricating flexible and patterned electrochromic devices. Sci. China Mater. 1–10 (2022)
29. Gribkova, O.L., Saf'yanova, L.V., Tameev, A.R., et al.: A water-soluble polyaniline complex for ink-jet printing of optoelectronic devices. Tech. Phys. Lett. **44**(3), 239–242 (2018)
30. Bao, Q., et al.: Printed flexible bifunctional electrochemical urea-pH sensor based on multiwalled carbon nanotube/polyaniline electronic ink. J. Mater. Sci.: Mater. Electron. **30**(2), 1751–1759 (2018). https://doi.org/10.1007/s10854-018-0447-5
31. Zhao, Y., et al.: All-printed flexible electrochemical sensor based on polyaniline electronic ink for copper (II), lead (II) and mercury (II) ion determination. J. Electron. Mater. **49**(11), 6695–6705 (2020). https://doi.org/10.1007/s11664-020-08418-x

32. Wang, Z., Zhang, Q.E., Long, S., et al.: Three-dimensional printing of polyaniline/reduced graphene oxide composite for high-performance planar supercapacitor. ACS Appl. Mater. Interfaces. **10**(12), 10437–10444 (2018)
33. Camargo, J.R., Orzari, L.O., Araujo, D.A.G., et al.: Development of conductive inks for electrochemical sensors and biosensors. Microchem. J. **164**, 105998 (2021)
34. Latonen, R.M., Määttänen, A., Ihalainen, P., et al.: Conducting ink based on cellulose nanocrystals and polyaniline for flexographical printing. J. Mater. Chem. C **5**(46), 12172–12181 (2017)
35. Teo, M.Y., Stuart, L., Devaraj, H., et al.: The in situ synthesis of conductive polyaniline patterns using micro-reactive inkjet printing. J. Mater. Chem. C **7**(8), 2219–2224 (2019)
36. Zea, M., Texidó, R., Villa, R., et al.: Specially designed polyaniline/polypyrrole ink for a fully printed highly sensitive pH microsensor. ACS Appl. Mater. Interfaces. **13**(28), 33524–33535 (2021)

Research Progress of Electron Beam Curing Ink

Xingyu Zhao, Beiqing Huang(✉), and Xianfu Wei

College of Printing and Packaging Engineering, Beijing Institute of Graphic Communication, Beijing, China
384573464@qq.com

Abstract. Electron beam curing is a form of radiation curing. Nowadays, countries in the world are pursuing the healthy development of green environmental protection and energy saving, and electron beam curing technology, a green radiation curing technology, is further applied. In the field of printing, electron beam curable inks are developing rapidly. This paper summarizes the research situation of electron beam curing ink, focusing on the curing mechanism of electron beam curing technology, the development status of electron beam curing ink, and the future prospect of electron beam curing ink.

Keywords: Electron beam curing · Curing mechanism · Ink

1 Development of Electron Beam Curing

Electron beam curing technology is a technology that uses high-energy electron beams to obtain energy through an electron accelerator, and then irradiates organic materials to cure them. Electron beam curing is characterized by environmental protection, fast curing speed, and less heat generation during curing. Compared with other radiation curing methods, electron beam curing is expensive, the consumption of inert gas is large, and the maintenance cost is high; but the performance of EB ink is excellent, the curing degree is more thorough, and it does not contain voc.

History of Electron Beam Curing [1, 2]: In the 1950s, electron accelerators for industrial applications began to develop. In the 1970s, with the emergence of electron beam curing equipment—high-power low-energy electron accelerators, electron beam curing technology began to develop. In the 1980s, the electron curtain accelerator appeared, which is the most ideal accelerator for electron beam curing, and electron beam curing began to develop rapidly.

The working process of the electron beam curing equipment [3]: the electron gun emits electrons, and the electrons pass through the electron accelerator, so that the electrons obtain energy. Multiple electrons with energy move in the same direction to form an electron beam, and the process of irradiating a liquid material with the electron beam to make it solid is electron beam curing. Generally common electron accelerators are divided into three categories [4]: low-energy electron accelerators, medium-energy electron accelerators, and high-energy electron accelerators. Electron accelerators that are used in electron beam curing are low-energy electron accelerators [5], and there are three major types [6]: scanning type, curtain type and multi-cathode type.

M. Xu et al. (Eds.): CACPP 2022, LNEE 991, pp. 529–534, 2023.
https://doi.org/10.1007/978-981-19-9024-3_68

One of the advantages of electron beam curing is low energy consumption, and its corresponding electron accelerator should also be a low-energy electron accelerator. With the green and healthy development of the world, the radiation curing technology of electron beam curing, which can cure liquid materials with very little energy [7], has outstanding advantages, and promotes the development of processing and manufacturing of coatings, inks, adhesives and composite materials [8–11].

The ink in which the electron beam is the radiation curing energy source is called electron beam curing ink. In recent years, some achievements have been made in the ink raw materials, properties and applications of electron beam curing inks. Electron beam curing technology will be more widely used.

2 Electron-Beam Curing Mechanism

The electron beam curing process is briefly described from a chemical point of view. The free radicals or ions formed by the interaction between high-energy electrons and organic matter cause a series of free radicals or ions to participate in chemical reactions, such as polymerization, cross-linking, backbone and side chain scission and grafting, etc. [12–14]. There are two types of electron beam curing mechanisms [15, 16], one is free radical curing, and the other is cationic curing. When a material is irradiated with electron beams, the monomer or oligomer gains energy, becomes an excited molecule, and then generates free radicals, which are capable of initiating polymerization in monomers or oligomers. The specific reaction process: free radical generation, chain initiation, chain growth, chain termination.

Free radical generation is the decomposition of a compound molecule into two primary free radicals under electron beam irradiation. Chain initiation is the initiation of free radicals by initiators, which combine with monomers to form monomer free radicals. Chain propagation is the combination of monomer free radicals and multiple monomers into chain free radicals. Chain termination is the continuous combination of multiple chain free radicals, and then inactivation, reaching a stable state to form a polymer.

In cationic curing, after the monomer or prepolymer is irradiated by electron beams, cations and anions are generated, and the ions initiate ionic polymerization. The specific reaction process: chain initiation, chain growth, chain termination. Chain initiation is the generation of ions by compound molecules under electron beam irradiation. Chain extension is the polymerization of ions with monomers. Chain termination is divided into two cases - coupling termination and single-radical termination, both of which balance the growing chain with electrons and protons, making it electrically neutral, reaching a stable state, and forming a polymer.

3 Preparation of Electron Beam Ink

3.1 Electron Beam Curing Equipment

There are three types of electron beam curing equipment [17]: low-energy electron beam curing equipment, medium-energy electron beam curing equipment and high-energy electron beam curing equipment. Generally, low-energy electron beam curing equipment is used for ink [18]. Electron energies range from 70 keV to 300 keV. The working process of electron beam curing equipment [19]: the tungsten wire generates electrons under the action of thermal current, the electrons are accelerated by an electric field into high-speed electrons, the electrons hit the surface of the material, the electron energy is transferred to the molecules, and further intramolecular chemical reactions occur, such as Polymerization, cross-linking, skeleton and side chain scission and grafting, etc. [20].

3.2 Electron Beam Curing Ink Raw Materials

The components of electron beam curing ink are oligomers, reactive diluents, pigments, fillers and other additives. In order to obtain inks with different functions, it is necessary to study the properties of the raw materials of the inks and prepare different electron beam curable inks [21]. Haitao Xu and his team devised an invention - Electron beam cured tin printing ink [22]. Its innovative features make it the ideal ink for low-cost, high-efficiency and environmentally friendly printing, and its electron beam curing iron printing ink has no solvent discharge, meets environmental protection requirements, reduces the environmental costs of the enterprise, and demonstrates the benefits of electron beam ink. Jingquan Wang invented a high-adhesion electron beam curing display cover ink, which can meet various printing methods, and has successfully applied the electron beam ink to improve the adhesion of the ink in the display revision [23].

After the electron beam ink is irradiated, the liquid becomes solid. After the irradiation, the properties and structure of the electron beam ink will change, and further research is required [24]. Pengfei Liu et al. synthesized four different urethane acrylates [25], and studied the structure and properties of these products after different doses of electron beam irradiation. Increasing the radiation dose under certain circumstances allows the system to cure faster.

3.3 Electron Beam Curing Irradiation Process

The electron beam irradiation process mainly studies the effect of the electron beam irradiation intensity and the temperature change during the irradiation process on the curing speed and effect of the electron beam curing ink [26]. Pengfei Liu et al. designed and prepared urethane acrylates with different functional groups, and further studied the effects of different electron beam irradiation doses and energy supply methods on urethane acrylates. The progress of the reaction was primarily observed through infrared spectroscopy, which confirmed the reaction on the carbon-carbon double bond of the acrylate. Gel permeation chromatography was measured to assess the degree of curing and crosslinking of the reactants [27]. It can be seen from the measurement results from various aspects that under a certain dose, the higher the electron beam irradiation dose, the better the curing effect. The effect of irradiation intensity on electron beam-cured ink raw materials is shown.

4 Electron Beam Curable Ink Applications

Compared with other radiation curing technologies, electron beam curing technology needs further research. The application field of electron beam curing has been developed in coatings, adhesives and composite materials, especially in electron beam curing inks. The use of electron beam ink has been developed in many fields [28–31]. Electron beam curing has significant applicability in food packaging. David A Biro and Jim Bishop summed up the electron beam curing technology for broadband flexible packaging printing [32], showing the green and pollution-free characteristics of electron beam curing ink. Electron beam curing has great advantages in color printing, which is conducive to color printing. Jingquan Wang developed a black electron beam ink [33]. The interaction between the components allows the ink to cure in a very short time.

5 Conclusion

The advantages of high efficiency, energy saving, environmental protection, wide adaptability and economy of electron beam curing ink have reflected the advanced nature of this radiation curing technology. This paper summarizes the development history of electron beam curing technology, curing mechanism, the development status of electron beam curing ink, and the future prospect of electron beam curing ink. It is hoped that electron beam curing technology will have further breakthroughs.

Acknowledgements. We gratefully acknowledge the support from BIGC Project (Nos. Ea202203, Ee202205) and Institute of Advanced Ink, Beijing Institute of Graphic Communication.

References

1. Grdanovska, S., Cooper, C.: Electron Beam Driven Industrial Chemistries. Fermi National Accelerator Laboratory: Illinois Accelerator Research Center. https://web.fnal.gov/organization/iarc/Shared%20Documents/Electron-Beam-Driven-Industrial-Chemistries.pdf. Accessed Nov 2020
2. Tong, J., Sun, Y., Yue, S., Liu, R.: Development of equipment for radiation curing coatings. China Coat. **36**(09) (2021)
3. Liu, P., Cheng, L., Liu, X., Liu, R.: Research progress of electron beam curing coating technology. Paint Coat. Ind. **51**(10), 73–79 (2021)
4. Scharf, W., Wieszczycka, W.: Electron accelerators for industrial processing—a review. AIP Conf. Proc. **475**(1) (1999)
5. Mondelaers, W.: Low-energy electron accelerators in industry and applied research. Nucl. Inst. Methods Phys. Res. B **139**(1–4), 43–50 (1998)
6. Jin, Y.: Performance and Application Manual of UV Curable Materials. Chemical Industry Press, pp. 519–524 (2010)
7. Du, Z., Janke, C.J., Li, J., Wood, D.L., III.: High-speed electron beam curing of thick electrode for high energy density Li-ion batteries. Green Energy Environ. **4**(4), 375–381 (2019)
8. Sánchez-Cadena, L.E., Tersac, G., Coqueret, X., Gamiño-Arroyo, Z.: Solvolysis of acrylate-urethane coatings cured by electron-beam and UV radiation. Prog. Org. Coat. **136** (2019)

9. Linzer, V., Laksin, M., Adhikari, P., Best, T., Modi, J., Chatterjee, S.: New electron beam cured liquid inks for the flexible packaging market. In: Technical Conference Proceedings—UV & EB Technology Expo & Conference, Charlotte, NC, United States (2004)
10. Singh, A.K., Niyogi, U.K., Sabharwal, S., Kowalczyk, A., Czech, Z., Mehra, D.S.: Shrinkage studies in electron beam curable polyurethane pressure-sensitive adhesive. J. Adhes. Sci. Technol. **27**(14), 1511–1524 (2013)
11. Coqueret, X., Krzeminski, M., Ponsaud, P., Defoort, B.: Recent advances in electron-beam curing of carbon fiber-reinforced composites. Radiat. Phys. Chem. **78**, 557–561 (2009)
12. Bandzierz, K.S., Reuvekamp, L.A.E.M., Przybytniak, G., Dierkes, W.K., Blume, A., Bieliński, D.M.: Effect of electron beam irradiation on structure and properties of styrene-butadiene rubber. Radiat. Phys. Chem. **149**, 14–25 (2018)
13. Naikwadi, A.T., Sharma, B.K., Bhatt, K.D., Mahanwar, P.A.: Gamma radiation processed polymeric materials for high performance applications: a review. Front. Chem. **10**, 837111 (2022)
14. Jiang, Q., Wang, S., Xu, S.: Preparation and characterization of water-dispersible carbon black grafted with polyacrylic acid by high-energy electron beam irradiation. J. Mater. Sci. **53**(8), 6106–6115 (2018). https://doi.org/10.1007/s10853-017-1966-9
15. Thiher, N.L.K., Schissel, S.M., Jessop, J.L.P.: Quantifying UV/EB dual cure for successful mitigation of oxygen inhibition and light attenuation. Prog. Org. Coat. **138** (2020)
16. Crivello, J.V.: UV and electron beam-induced cationic polymerization. Nucl. Inst. Methods Phys. Res. B **151**, 8–21 (1999)
17. Elmaaty, T.A., Okubayashi, S., Elsisi, H., Abouelenin, S.: Electron beam irradiation treatment of textiles materials: a review. J. Polym. Res. **29**, 117 (2022)
18. Berejka, A.J.: Prospects and challenges for the industrial use of electron beam accelerators. In: International Topical Meeting on Nuclear Research Applications and Utilization of Accelerators (ACCAPP'09), Vienna, Austria (2009)
19. Schwab, U.: A new accelerator for electron beam curing. Vacuum **62**, 217–224 (2001)
20. Sabharwal, S.: Electron beam irradiation applications. In: Proceedings of the 25th North American Particle Accelerator Conference, pp. 745–748 (2013)
21. Baikerikar, K.K., Tulchinsky, M.L., Argyropoulos, J.: UV curable, liquid diacrylate monomers based on (cis,trans)-1,3/1,4-cyclohexanedimethanol. J. Coat. Technol. Res. **7**(2), 175–188 (2010)
22. Xu, H., Liang, H., Xiong, L.: Electron beam cured tin printing ink. CN.201110262236.9, 18 Jan 2012
23. Wang, J.: High-adhesion electron beam cured display cover plate printing ink and preparation method thereof. CN201710429541.X, 24 Oct 2017
24. Vautard, F., Fioux, P., Vidal, L., Schultz, J., Nardin, M., Defoort, B.: Influence of the carbon fiber surface properties on interfacial adhesion in carbon fiber–acrylate composites cured by electron beam. Composites: Part A **42**, 859–867 (2011)
25. Liu, P., Mao, F., Webster, D.C., Liu, X., Liu, R.: Curing and performance stability of urethane acrylates with different main chains under electron-beam irradiation. Prog. Org. Coat. **152**, 106119 (2021)
26. Rizzolo, R., Walczyk, D., Kuppers, J., Montoney, D., Galloway, R.: Rapid consolidation and curing of advanced composites using electron beam irradiation. Proc. Inst. Mech. Eng. Part B: J. Eng. Manuf. **233**(4), 1168–1181 (2019)
27. Liu, P., Zhu, J., Cheng, L., Liu, X., Liu, R.: Curing and properties of urethane acrylates with different functionalities under electron-beam and ultraviolet irradiation. Prog. Org. Coat. **156**, 106252 (2021)
28. Yang, X.: Electron Beam Curable Color Ink for Automotive Tires. Wuhan Institute of Technology (2015)

29. Nakajima, A., Akase, T., Kaneshiro, S., Shiozaki, D.: Electron beam curable aqueous inkjet ink composition. JP. 201980067321.4, 04 June 2021
30. Zhai, Z., Wang, F., Mei, X., Zhang, Y., Cui, Y.: Preparation of graphene directly on liquid EB curing ink film by femtosecond laser. Optik **223**, 165485 (2020)
31. Galavis, P.E., Sanfilippo, N.J., Das, I.J.: Glamour, expression, and consequences of tattoos in radiation treatment. PLoS ONE **14**, e0220030 (2019)
32. Biro, D.A., Bishop, J.: Advances in electron beam curing in wide web flexible package printing. In: Proceedings of the Technical Association of the Graphic Arts, pp. 143–147 (2017)
33. Wang, J.: Electron-beam curing PET coating black ink and preparation method thereof. CN. 201710429406.5, 29 Sept 2017

Film and Related Material Technology

Experimental Study of Heat Storage Performance of 3D Printed Metal Foam and Phase Change Materials Composite in Packaging Applications

Chuan Zhang, Shengwei Yang, and Enyin Fang(✉)

Department of Printing and Packaging Engineering, Shanghai Publishing and Printing College, Shanghai, China
fensarying@163.com

Abstract. In food industry, phase change materials (PCMs) can be used as heat storage materials. However, their low thermal conductivity limits their application. In this study, an aluminum foam sample is prepared by metal 3D printing technology, and a heat storage module of aluminum foam and paraffin composite is obtained. An experimental comparison with the pure paraffin heat storage module is achieved. The experimental results show that the heat storage time of the aluminum foam-paraffin heat storage module is only 40.7% of that of the pure paraffin heat storage module. The addition of aluminum foam greatly shortens the heat storage time and improves the heat storage efficiency. The temperature distribution of aluminum foam-paraffin heat storage module is more uniform than that of the pure paraffin heat storage module. The temperature of aluminum foam-paraffin heat storage module reaches the phase transition temperature of paraffin during the heat storage process, which is beneficial to the application of paraffin in heat storage applications.

Keywords: Metal foam · Phase change material · Heat storage · 3D printing

1 Introduction

Phase Change Materials (PCMs) has been widely used in many applications such as food industry [1], building [2], battery [3], solar energy [4]. In food industry, many researchers investigated the impact of PCMs in food storage systems, especially in cooling and cold chain systems [5]. The quality of refrigerated food, such as meat [6] and ice cream [7], has been greatly improved. Furthermore, Floros et al. [8] found that the use of PCMs could reduce the temperature of hot water from 85 °C to 60 °C within 60 s and maintain the temperature of water in a desired temperature for a long period. During the age of fast food consumption and the COVID-19 pandemic, the services transporting hot food to the consumers' doorsteps are increasing. PCMs with phase transition temperature higher than ambient temperature and equal to the serving temperature have been used in many food or beverage containers, such as pizza package [9], cups [10]. Many recent studies

M. Xu et al. (Eds.): CACPP 2022, LNEE 991, pp. 537–544, 2023.
https://doi.org/10.1007/978-981-19-9024-3_69

have found that the addition of metal foam can improve the thermal conductivities of metal foam-PCMs composites [11], Xiao et al. [12] found that the thermal conductivity of the paraffin/nickel foam composite was nearly three times larger than that of pure paraffin. In this study, the PCM (paraffin) is filled in the pores of 3D printed aluminum foam forming a heat storage module, and its heat storage performance is experimentally studied by comparing with pure paraffin.

2 Experimental Method

The metal 3D printing technology was adopted to make an aluminum foam structure with a porosity of 95%, the size of the sample is 20mm × 100mm × 100mm, and the pore size is 5mm.

Figure 1 shows the aluminum foam-paraffin heat storage module, the paraffin was used as phase change material, which was completely filled in the pores of aluminum foam sample and vacuum sealed inside the plastic film to form a heat storage module. The paraffin has a phase transition temperature of 51–53 °C.

Fig. 1. Aluminum foam filled with paraffin

Figure 2 shows the sample of pure paraffin heat storage module. The same type of pure paraffin was used and the size of pure paraffin heat storage module is also 20 mm × 100 mm × 100 mm.

The heat storage module was placed at the bottom of an insulation box, and a water bag was placed above the heat storage module as shown in Fig. 3. The initial temperature of the water bag is 75 °C, the volume is 500 ml and the size is 130 mm × 250 mm.

During the test, the Omega K-type (TC#1, TC#2, TC#3, and TC#4) thermocouples ware applied to measure the temperature variation of the bottom surface of the heat storage module. The Omega K-type (TC#5, TC#5, TC#7, and TC#8) thermocouples were adopted to measure the surface temperature between the heat storage module and

Fig. 2. Pure paraffin module

Fig. 3. Module placement

water bag. The Omega K-type (TC#9, TC#10, TC#11, and TC#12) thermocouples ware applied to measure the top surface temperature of the water bag. The detailed positions of thermocouples are shown in Fig. 4. The initial temperature of the thermal storage module is the ambient temperature, 27.5 °C.

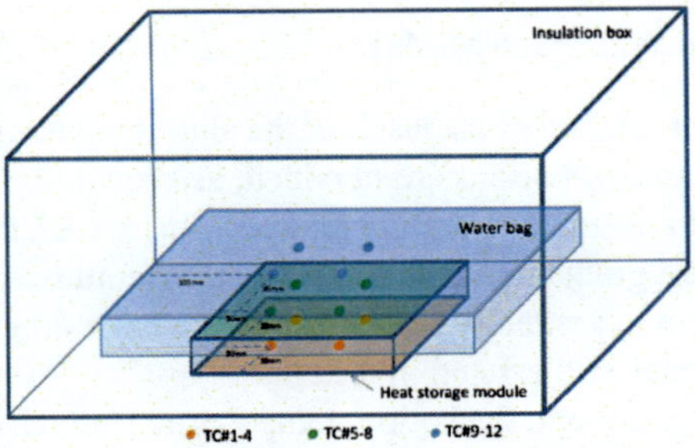

Fig. 4. Positions of thermocouples

3 Results and Discussions

3.1 Aluminum Foam-Paraffin Heat Storage Module

By averaging the temperature data of 4 thermocouples per surface (the bottom surface: the bottom surface of the heat storage module, the middle surface: the surface between the heat storage module and the water bag, the top surface: the surface above the water bag), three temperature variations are obtained. Figure 5 illustrates that the temperature at the bottom of the heat storage module increased rapidly after the start of the experiment. The temperature at the top of the water bag and the temperature between the water bag and the heat storage module gradually decreased. The final temperatures of three surfaces gradually approached. In the final stage of the experiment, the temperature at the bottom of the heat storage module was 52.3 °C, which proved that the paraffin inside the heat storage module completely melted.

Fig. 5. Temperature variations of aluminum foam-paraffin module

3.2 Pure Paraffin Heat Storage Module

Using the same analysis method as the result of the aluminum foam-paraffin heat storage module, three temperature variations are obtained, as shown in Fig. 6. It can be known from the figure that the bottom temperature gradually increased, the middle temperature and the top temperature gradually decreased. The temperature difference between the bottom and the middle was gradually reduced from the beginning of the experiment but the temperature difference still existed in the end.

At the end of the experiment, the bottom temperature is much lower than the top and middle temperature. In the final stage of the experiment, the bottom temperature was 44.5 °C, which did not reach the paraffin melting temperature.

Fig. 6. Temperature variations of pure paraffin module

3.3 Comparison of Heat Storage Time Between Two Modules

As shown in Fig. 7, it can be known that the aluminum foam-paraffin heat storage module took about 2400 s to complete the heat storage. However, the pure paraffin heat storage module completed the heat storage process in 5900 s. The heat storage time of the aluminum foam-paraffin heat storage module is 40.7% of the heat storage time of pure paraffin module.

Fig. 7. Heat storage time

3.4 Temperature Comparison of Two Modules

As shown in Fig. 8, when the aluminum foam-paraffin module completes the heat storage process, its temperature is 51.7 °C, and when the pure paraffin heat storage module completes the heat storage process, the temperature is 46 °C. The temperature of the paraffin in the aluminum foam reaches the melting temperature, so that the latent heat of phase change is used for heat storage in aluminum foam-paraffin module. The pure paraffin module does not reach the temperature of the phase change and fails to use the latent heat of phase change for heat storage.

Fig. 8. Temperature comparison of two modules

3.5 Internal Temperature Distribution of Two Modules

Two temperature variations are obtained by taking the difference of the temperatures of the middle and bottom surfaces, as shown in Fig. 9. It can be seen that the temperature difference of the aluminum foam-paraffin module decreases rapidly after the start of the experiment, and its decreasing speed is faster than that of the pure paraffin heat storage module. The temperature distribution inside the aluminum foam-paraffin heat storage module is more uniform than that of the pure paraffin module during the experiment. Aluminum has a higher heat transfer coefficient than paraffin, so that the heat can be easily transferred from the hot surface to the cold surface through aluminum foam, the heat storage time is much shorter and the heat storage efficiency is improved.

3.6 Temperature Comparison Between the Two Modules

Figure 10 shows that middle temperature of the aluminum foam-paraffin module is lower than that of the pure paraffin module. During the heat storage process of the pure paraffin module, the heat accumulated in the surface between the water bag and the pure paraffin. The heat cannot be quickly transferred by the natural convection of paraffin. Although the addition of aluminum foam limits the convection of paraffin, its high heat transfer coefficient reduces the accumulation of heat on the hot surface and promotes the heat storage process.

Fig. 9. Temperature difference between middle and bottom surfaces

Fig. 10. Temperature variations of middle surfaces

4 Conclusions

By simulating real usage scenarios, this paper studies the heat storage process of foam aluminum-paraffin heat storage modules and pure paraffin heat storage modules, and draws the following conclusions:

(1) The heat storage time of the foamed aluminum-paraffin heat storage module is 40.7% of that of the pure paraffin heat storage module, which greatly shortens the heat storage time and improves the heat storage speed.
(2) The temperature distribution of the foamed aluminum-paraffin heat storage module is more uniform than that of the pure paraffin heat storage module, which means that the paraffin far from the heat source melts more easily with the help of metal foam.

(3) The aluminum foam-paraffin heat storage module reached the phase transition temperature during the heat storage process. Hence, more energy is stored through phase change heat storage than pure paraffin.

Acknowledgements. This research is supported by Key Lab of Intelligent and Green Flexographic Printing (No. KLIGFP-02).

References

1. Gin, B., Farid, M.M.: The use of PCM panels to improve storage condition of frozen food. J. Food Eng. **100**(2), 372–376 (2010)
2. Memon, S.A.: Phase change materials integrated in building walls: a state of the art review. Renew. Sustain. Energy Rev. **31**, 870–906 (2014)
3. Ajour, M.N., et al.: The investigation of battery thermal management via effects of using phase change materials in the oval packages around the lithium-ion battery cells with an airflow. J. Energy Storage **53**, 105105 (2022)
4. Sikiru, S., et al.: Recent advances and impact of phase change materials on solar energy: a comprehensive review. J. Energy Storage **53**, 105200 (2022)
5. Zhao, Y., et al.: Research progress of phase change cold storage materials used in cold chain transportation and their different cold storage packaging structures. J. Mol. Liq. **319**, 114360 (2020)
6. Arjenaki, N.O., Soltanizadeh, N., Hamdami, N.: Designing an active phase change material package for thermal and qualitative protection of meat. Food Packag. Shelf Life **21**, 100362 (2019)
7. Leducq, D., Ndoye, F., Alvarez, G.: Phase change material for the thermal protection of ice cream during storage and transportation. Int. J. Refrig. **52**, 133–139 (2015)
8. Floros, M.C., Narine, S.S.: Latent heat storage using renewable saturated diesters as phase change materials. Energy **115**, 924–930 (2016)
9. Arjona, F., et al.: Application of the N-Alkane molecular alloys to thermally protected containers for catering. Boletin de la Sociedad Española de Ceramica y Vidrio **39**(4), 548–551 (2000)
10. Al-Jethelah, M., et al.: Latent heat storage for hot beverages. J. Mech. Eng. Sci. **13**(3), 5653–5664 (2019)
11. Huang, X., et al.: Thermal properties and thermal conductivity enhancement of composite phase change materials using myristyl alcohol/metal foam for solar thermal storage. Sol. Energy Mater. Sol. Cells **170**, 68–76 (2017)
12. Xiao, X., Zhang, P., Li, M.: Preparation and thermal characterization of paraffin/metal foam composite phase change material. Appl. Energy **112**, 1357–1366 (2013)

Development and Application of Chiral Nematic Cellulose Nanocrystalline Iridescent Films

Yunpeng Xie[1], Qi Zhu[1,2], and Guangxue Chen[1](✉)

[1] State Key Laboratory of Pulp and Paper Engineering, South China University of Technology, Guangzhou, China
chengx@scut.edu.cn
[2] Huagong-Liyan (Guangdong) New Material Technology Co.Ltd, Guangzhou, China

Abstract. Cellulose nanocrystals (CNCs), as a biorenewable resource, can form CNC chiral nematic iridescent nano-films by evaporation-induced self-assembly (EISA). CNC chiral nematic liquid crystal materials have become a hot spot in the current research on chiral photonic materials due to their rich sources, simple synthesis process, unique optical properties and good stability, and they are widely used in anti-counterfeiting, sensing, optical switches and other fields. CNC photonic materials have the problems of single color and non-uniformity. The CNC films are hard and brittle, which are not conducive to large-scale application. And they only reflect left-handed circularly polarized light, resulting in a dull color of the films. Currently, there is a large amount of prior researches that proposes corresponding solutions to different problems. This review mainly introduces the liquid crystal phase of CNCs, CNC iridescent films and the modification of CNC structural color materials. Finally, the application of CNC chiral nematic nanomaterials is summarized and prospected, which will be helpful for the follow-up research.

Keywords: Cellulose nanocrystals · Chiral nematic · Structural color · Iridescent films

1 Introduction

There are many natural structural color materials in nature, such as peacock feathers, cellulose in the fruit Polya Condensata, scales of lepidopteran adults, etc. With the continuous development of nanotechnology, researchers have succeeded in obtaining nanomaterials directly from biological resources. Cellulose nanocrystals (CNCs) are nanoscale 1D chiral nanorods obtained from cellulose. CNCs have excellent optical and electrochemical properties and can be used in many fields (Fig. 1) [1].

This review mainly introduces the preparation of CNC chiral nematic liquid crystal films and functional chiral nematic CNC film materials. Finally, the future development trend of chiral nematic CNC films is prospected.

M. Xu et al. (Eds.): CACPP 2022, LNEE 991, pp. 545–552, 2023.
https://doi.org/10.1007/978-981-19-9024-3_70

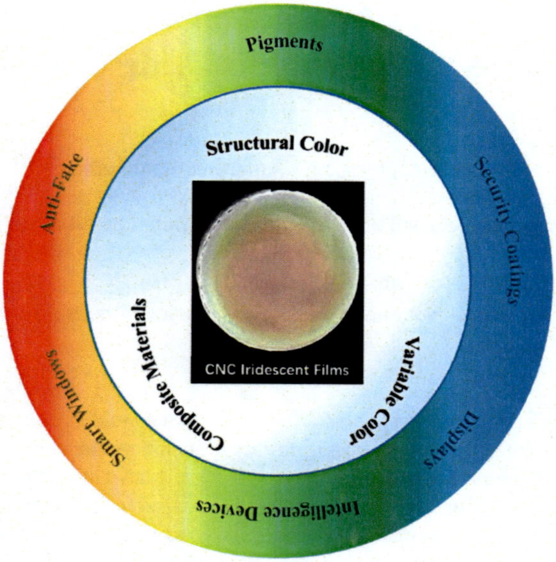

Fig. 1. Properties and applications of CNC films

2 CNC Iridescent Films

As early as 1959, Marchessault et al. first observed birefringence phenomenon after evaporation of CNC suspensions and found that CNCs were lyotropic liquid crystals [2]. In 1992, Revol et al. observed that the periodic structure in CNC suspensions and films was a helical structure, and identified the lyotropic liquid crystal behavior of CNCs as chiral nematic [3]. Afterwards, they retained the chiral nematic structure of CNC suspensions (≈3.5 wt%) in solid films by EISA [4]. The CNC films preparation was limited to the laboratory. In 2021, Droguet et al. scaled up the production of iridescent films via an industrial route by casting CNC suspensions on a commercial R2R coating unit [5]. The nanostructured CNC films produced by this method had good optical response. The prepared CNC films could be ground into tiny particles after further heat treatment. The particles also maintained optical response after one year and did not discolor and disperse in water. The preparation process of CNC films is shown in Fig. 2. Since the iridescent and birefringent properties can be observed in CNC films, they have great application potential and have received extensive attention.

As a nanostructured material and a chiral nematic liquid crystal material with photonic properties, CNCs have a broad application prospect and have received great attention. Researchers have been trying for decades to tune the color of CNC chiral nematic liquid crystal materials, improve the flexibility of CNC films, and tune the polarization of CNC films.

Fig. 2. Preparation process of CNC films

2.1 Adjust the Color of CNC Films

CNC suspensions can form the iridescence films through the EISA process, many factors will change the color of the films. Researchers have found that temperature, sonication or magnetic fields can affect the films' color during the evaporation process of CNC suspensions. Beck et al. found that increasing the temperature of drying CNC suspensions produced red-shifted films [6]. And when the CNC suspensions at the specific site were dried, a patterned membrane appeared. Guo et al. poured CNC suspensions on metal, metal alloy, polystyrene, stainless steel and other substrates to evaporate to form films [7]. They proposed the influence mechanism of three substrate effects and four shear effects on the structural color of the CNC films. Frka-Petesic et al. demonstrated that the magnetic field generated by ordinary NdFeB magnets could be used to control the self-assembly of CNC suspensions to form colored films [8]. They placed the CNC suspensions on a magnet for evaporation. When the applied magnetic field was perpendicular to the surface of the CNC films, a highly ordered chiral nematic structure was formed (Fig. 3). Tran and colleagues found that longer evaporation times of CNC suspensions produced blue-shifted films [9]. Meanwhile, patterned CNC films with gradient colors can be formed by differential evaporation. In addition, the use of a cellulose acetate (cellulose acetate itself does not significantly affect the pitch of the CNC films) mask confines evaporation to a defined area, resulting in a pattern with higher resolution.

The above works are all in changing the external environmental conditions during the evaporation of CNC suspensions. In fact, the photonic properties of CNC films can be improved and further tuned by using additives. Beck et al. prepared blue-shifted membranes by adding NaCl [10]. They showed that at higher salt concentrations, the

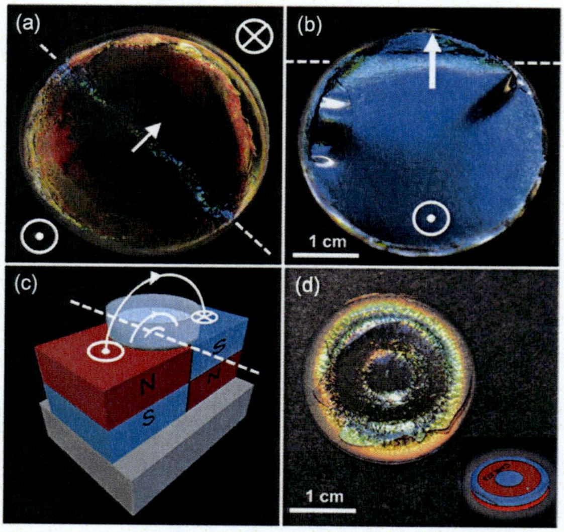

Fig. 3. Patterning CNC films by applying a magnetic field. A, B) Examples of films that were cast spanning two magnets, as indicated by the schematic in (C). D) CNC suspension was cast on a polydomain magnet. Interesting patterns are created upon film formation [8]. Copyright 2017, John Wiley & Sons, Inc.

reflection wavelength decreasesd from 550 to 390 nm. Mu et al. added glucose (GLU) to the CNC suspensions, which increased the pitch, resulting in a red-shift of the CNC films [11].

Through the above control methods, CNC iridescent films have new applications. Zhang et al. treated CNC films by ultrasonic, and the prepared CNC chiral photonic microparticles retained the optical properties of CNC chiral photonic films and could be mixed with polymer matrices to fabricate flexible coatings for anti-fake and identification [12].

In the introduction of the works in this section, although the researchers adjusted the color of the CNC films by changing the external environment during the evaporation of the CNC suspensions or adding additives such as salt and glucose, the prepared CNC films were hard, brittle and fragile. It made CNC chiral nematic liquid crystal materials unsuitable for some flexible materials.

2.2 Tuning CNC Films Flexibility

In addition to tunable optical color, the use of additives imparts new functionalities to CNC nanomaterials, including flexibility and crack-freeness. Guidetti et al. added the 3-(N,N-dimethylmyristylammonio)propanesulfonate (DMAPS) to the CNC suspensions, resulting in nanoscale CNC-DMAPS complexes [13]. The chiral nematic structure is still preserved through the CNCs EISA process, and the CNC films show tunable photonic properties and mechanical properties. Liu et al. added 1-allyl-3-methylimidazolium chloride (AmimCl) to the CNC films by vacuum filtration, and obtained the soft red-shifted

CNC films. The prepared AmimCl plasticized CNC films can be further toughened by hot pressing [14]. Nan et al. immersed the CNC films in sodium hydroxide. With the longer treatment time, the CNC films blue-shifted and the tensile strength and flexibility of the films increased [15]. Yao et al. added polyethylene glycol (PEG) to CNC suspensions and successfully fabricated flexible chiral nematic CNC/PEG composite membranes with uniformly tunable structural colors from blue to red [16]. Huang et al. introduced polyethylene glycol diacrylate (PEGDA) into the CNC suspensions, and the mixed solution of CNCs and PEGDA self-assembled in the dark. PEGDA was made into a three-dimensional network structure by UV light curing [17]. The PEGDA chain can form hydrogen bonds with the hydroxyl groups on the CNCs to improve the mechanical properties of the films. At the same time, the three-dimensional network structure constructed by PEGDA encapsulates the CNCs and improves the water resistance of the films. Kim et al. added 3,4,5-trihydroxyphenethylamine hydrochloride (TOPA) and polyethylene glycol (PEG) to the CNC suspensions under dark conditions, and formed the CNC films by capillary action after filling through the capillary [18]. Such films have uniform birefringence and can be rapidly dried to form films under restricted conditions. TOPA enhances the intermolecular hydrogen bonding ability. As a result, the generation and propagation of cracks are suppressed under capillary confinement, resulting in smooth, uniform, and crack-free birefringent films.

2.3 Tuning the Polarization of CNC Films

As mentioned earlier, the CNC films only reflect left-handed circularly polarized light and transmit right-handed circularly polarized light. Therefore, the maximum reflectivity of CNC chiral nematic films is only 50%, resulting in dull color of CNC films. The previously presented works did not alter this property of CNC films. So Fernandes et al. inserted the nematic liquid crystal 4-cyano-4′-pentylbiphenyl (5CB) as a λ/2 retardation plate in the middle of two left-handed cholesteric domains [19]. The CNC films can reflect the left-handed circularly polarized light and the right-handed circularly polarized light at the same time. The intensity of the reflected right circularly polarized light depends on the birefringence of the liquid crystal and can be reversibly adjusted by changing the temperature or applying an electric field. Anusuyadevi et al. added GLU to CNC suspensions to obtain CNC/GLU composite membranes [20]. Then 4-n-butyl-4’-methoxyazobenzene (BMAB) and 5CB were mixed to prepare a liquid crystal mixture and dissolved in chloroform. Then it was dropped on the CNC/GLU composite films, and the liquid crystal molecules remained on the surface of the composite films. The results show that specific wavelengths of reflected light can be generated in the visible spectral range by varying the GLU content. The reflected light color of the nanophotonic structure can be controlled by the polarization state of the illumination (LCP light or RCP light) (Fig. 4). Li et al. added ferroferric oxide (Fe_3O_4) and luminophores (e.g. Quinine sulfate, rhodamine b, etc.) to CNC suspensions, which were then placed in a magnetic field to self-assemble into films, resulting in a left-handed chiral nematic structure with perpendicular concentric helical orientations in a planar texture [21]. It is demonstrated that these unique left-handed chiral nematic structures are capable of selectively reflecting LCP light and converting self-luminescence to RCP emission on the film plane and sides.

(a) (b)

(c) (e) (f)

(d)

Fig. 4. A) Simplified schematic of two LC droplets (BMAB/5CB mixture) on top of a CNC/GLU film. B) Optical polarized images in reflectance mode of two LC droplets (black and white arrows) residing on top of a CNC/GLU = 39/61 film (red arrow), when illuminated by incident LCP (top) or RCP (bottom) light. Scale bars: 250 μm. C) Reflectance results for the CNC/GLU film, see red arrow in (B). D) Reflectance results for a region with an LC droplet, measured close to the center of the droplet (the black arrow in (B) points to the droplet) prior to UV exposure (nematic phase). E–F) Reflectance results after UV exposure as a function of time (t = 32, 60, 120, and 300 s) for incident LCP (E) or RCP (F) light. Measurements were performed on the same droplet as in (D). At t = 32 s, the LC is in the isotropic state and with time a gradual reversal back to the nematic phase is observed (steady state at 300 s) [20]. Copyright 2021, John Wiley & Sons, Inc.

3 Summary and Outlook

In this review, it is detailed that CNC suspensions can form chiral nematic liquid crystal film nanomaterials through EISA. CNC iridescent films have excellent optical properties. And the optical properties and mechanical properties of CNC iridescent films can be changed by a series of means. This liquid crystal materials prepared from renewable, sustainable, non-toxic, and low-cost biomass materials will have great application potential and commercial value in the future. It can be of great value in bionics research, anti-counterfeiting labels, security, sensing, optical applications, cosmetics, etc. However, the development time of cellulose-based liquid crystal materials is relatively short, so there are still many problems that need to be solved urgently. For example, the EISA process takes a lot of time, and it is necessary to find a convenient way to shorten the preparation time. Secondly, the mechanism research of these materials is still lacking. Finally, it is necessary to find a commercialized route to open up new directions, and then let these materials go out of the laboratory and into the public.

Acknowledgement. This work is supported by National Natural Science Foundation (No. 61973127), Guangdong Provincial Science and Technology Program (No. 2017B090901064), and Huagong-Liyan (Guangdong) New Material Technology Co. Ltd. (High-tech industrialization entrepreneurship team project of Foshan High-tech Zone, FSBG2021021).

References

1. Duan, C.L., Cheng, Z., Wang, B., et al.: Chiral photonic liquid crystal films derived from cellulose nanocrystals. Small **17**(30), 2007306 (2021)
2. Marchessault, R., Morehead, F., Walter, N.: Liquid crystal systems from fibrillar polysaccharides. Nature **184**(4686), 632–633 (1959)
3. Revol, J.F., Bradford, H., Giasson, J., et al.: Helicoidal self-ordering of cellulose microfibrils in aqueous suspension. Int. J. Biol. Macromol. **14**(3), 170–172 (1992)
4. Revol, J.F., Godbout, L., Gray, D.G.: Solid self-assembled films of cellulose with chiral nematic order and optically variable properties. J. Pulp Pap. Sci. **24**(5), 146–149 (1998)
5. Droguet, B.E., Liang, H.L., Frka-Petesic, B., et al.: Large-scale fabrication of structurally coloured cellulose nanocrystal films and effect pigments. Nat. Mater. **21**(3), 352–358 (2022)
6. Beck, S., Bouchard, J., Chauve, G., et al.: Controlled production of patterns in iridescent solid films of cellulose nanocrystals. Cellulose **20**(3), 1401–1411 (2013)
7. Guo, M.N., Li, Y., Yan, X.Y., et al.: Sustainable iridescence of cast and shear coatings of cellulose nanocrystals. Carbohyd. Polym. **273**, 118628 (2021)
8. Frka-Petesic, B., Guidetti, G., Kamita, G., et al.: Controlling the photonic properties of cholesteric cellulose nanocrystal films with magnets. Adv. Mater. **29**(32), 1701469 (2017)
9. Tran, A., Hamad, W.Y., Maclachlan, M.J.: Fabrication of cellulose nanocrystal films through differential evaporation for patterned coatings. ACS Appl. Nano Mater. **1**(7), 3098–3104 (2018)
10. Beck, S., Bouchard, J., Berry, R.: Controlling the reflection wavelength of iridescent solid films of nanocrystalline cellulose. Biomacromol **12**(1), 167–172 (2011)
11. Mu, X.Y., Gray, D.G.: Formation of chiral nematic films from cellulose nanocrystal suspensions is a two-stage process. Langmuir **30**(31), 9256–9260 (2014)

12. Chang, T., Wang, B.C., Yuan, D., et al.: Cellulose nanocrystal chiral photonic micro-flakes for multilevel anti-counterfeiting and identification. Chem. Eng. J. **446**(1), 136630 (2022)
13. Guidetti, G., Atifi, S., Vignolini, S., et al.: Flexible photonic cellulose nanocrystal films. Adv. Mater. **28**(45), 10042–10047 (2016)
14. Liu, P., Guo, X., Nan, F.C., et al.: Modifying mechanical, optical properties and thermal processability of iridescent cellulose nanocrystal films using ionic liquid. ACS Appl. Mater. Interfaces. **9**(3), 3085–3092 (2017)
15. Nan, F.C., Nagarajan, S., Chen, Y.W., et al.: Enhanced toughness and thermal stability of cellulose nanocrystal iridescent films by alkali treatment. ACS Sustain. Chem. Eng. **5**(10), 8951–8958 (2017)
16. Yao, K., Meng, Q., Bulone, V., et al.: Flexible and responsive chiral nematic cellulose nanocrystal/poly(ethylene glycol) composite films with uniform and tunable structural color. Adv. Mater. **29**(28), 1701323 (2017)
17. Huang, Y.Y., Chen, G.W., Liang, Q.M., et al.: Multifunctional cellulose nanocrystal structural colored film with good flexibility and water-resistance. Int. J. Biol. Macromol. **149**, 819–825 (2020)
18. Kim, M., Pierce, K., Krecker, M., et al.: Monolithic chiral nematic organization of cellulose nanocrystals under capillary confinement. ACS Nano **15**(12), 19418–19429 (2021)
19. Fernandes, S.N., Almeida, P.L., Monge, N., et al.: Mind the microgap in iridescent cellulose nanocrystal films. Adv. Mater. **29**(2), 1603560 (2017)
20. Anusuyadevi, P.R., Shanker, R., Cui, Y.X., et al.: Photoresponsive and polarization-sensitive structural colors from cellulose/liquid crystal nanophotonic structures. Adv. Mater. **33**(36), 2101519 (2021)
21. Li, P., Li, L., Jeong, K.J., et al.: Homeotropic concentric helix orientations in chiral nematic cellulose nanocrystal films by local magnetic fields. Adv. Opt. Mater. **10**(7), 2102616 (2022)

Preparation and Characterization of Nano-gold Titanium Oxide Composite Film for Photocatalytic Detection

Yi Fang[1,2](✉), Yuhan Zhong[2], Li An[2], and Yuguang Feng[2]

[1] Beijing Engineering Research Center of Printed Electronics, School of Printing and Packaging Engineering, Beijing Institute of Graphic Communication, Beijing, China
fangyi@bigc.edu.cn

[2] School of Printing and Packaging Engineering, Beijing Institute of Graphic Communication, Beijing, China

Abstract. As a key photocatalyst, Titanium oxide is commonly used, which has the advantages of non-toxic and good stability. However, due to its high photogenerated electron hole recombination rate, too wide band gap and other factors, its effective application in photocatalytic degradation is limited. Therefore, this work proposes a method of fabricating gold nanoparticles on the surface of titanium oxide film by using pulsed laser. Using the plasmon effect of gold nanoparticles, it is expected to obtain an efficient composite photocatalytic film. The results showed that under the effect of pulsed laser with a power of 5×10^7 W cm^{-1}, the gold nanoparticles were fabricated and evenly distributed. The composite film formed an obvious plasmon absorption peak near 550 nm, due to the effect of gold nanoparticles.

Keywords: Nano-gold · Titanium oxide · Composite film · Photocatalyst

1 Background

As a key photocatalyst, TiO_2 has been extensively studied during the past several years owing to its strong catalytic activity, high chemical stability, nontoxicity, and low cost [1–5]. However, only a small amount of the solar energy can be utilized and the conversion efficiency of photon to electron-hole pair is low, due to its wide bandgap [6]. Therefore, a new method, involving the surface plasmon resonance (SPR) of Ag or Au nanoparticles, has been investigated to improve photocatalytic efficiency for the widespread application of TiO_2. Surface plasmons are oscillations confined to the surfaces of metallic nanoparticles and interact strongly with light. Metal nanoparticles exhibit the surface plasmon resonance effect on specific absorption and scattering from visible to near infrared region, due to the electromagnetic fields excited by the surface free electrons [7]. Irradiating metal nanoparticles with light at their plasmon frequency generates intense electric fields at the surface of the nanoparticles. The frequency of this resonance can be tuned by varying the size and shape of metal nanoparticles. Recently,

M. Xu et al. (Eds.): CACPP 2022, LNEE 991, pp. 553–558, 2023.
https://doi.org/10.1007/978-981-19-9024-3_71

a vast majority of the publications in the literature discussing plasmon-enhanced photocatalysis have focused on reactions involving photocatalytic decomposition of organic compounds.

In order to obtain catalytic films by nano-gold titanium oxide composite with good performance and easy fabrication, a method of forming gold nanoparticles on the surface of titanium oxide film by pulsed laser is proposed in this work.

2 Experimental

After cleaning and drying with nitrogen, the quartz substrate is placed into the magnetron sputtering chamber. The chamber pressure is 3×10^{-5} Pa, and the ratio of argon and oxygen is 8:72 (flow ratio). Working power of the system is 210 W, voltage is 95 V, pre-sputtering time is 900 s, sputtering time is 3600 s, sweeping speed is 6 mm/s, distance is 300 mm.

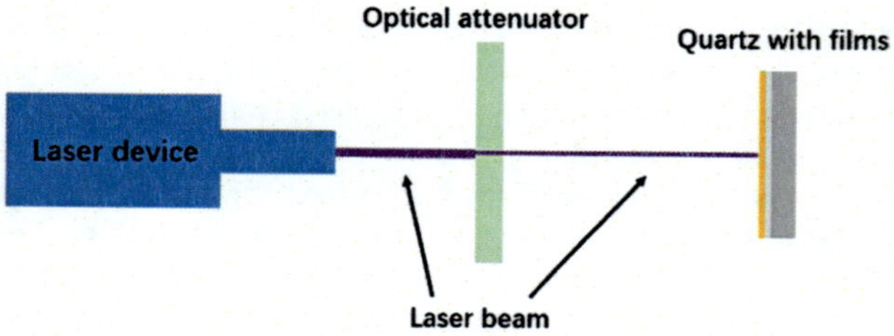

Fig. 1. Schematic diagram of gold nano-particles produced by laser beam.

After obtaining the TiO_2 layer, an Au layer of 4 nm was evaporated on this layer, and then the Au nanoparticles were prepared by irradiating with a 355 nm laser with an intensity of 5×10^7 W cm^{-1}. The schematic diagram of nano particles produced by pulsed laser is shown in Fig. 1. The laser light emitted by the laser device irradiated on the film after passing through the attenuator. Through the attenuator, the laser intensity can be adjusted to 5×10^7 W cm^{-1}. At this strength, the gold thin film can be prepared into nanoparticles. The information of materials and equipment used in the experiment is shown in Tables 1 and 2.

Table 1. Parameters of the materials

Category of materials	Purity (%)	Manufacturer
Gold wire	99.99	Beijing Chemical Reagent Factory
Ti target	99.999	Beijing Yannuoxincheng Technology Co., Ltd
Quartz substrate	99.999	Shanghai Meiruier Chemical Technology Co., Ltd

Table 2. Parameters of the devices

Category of devices	Model	Manufacturer
Magnetron sputtering chamber	GW-400	JP
X-ray diffractometer	D/max-2000 PC	Japanese science company
Scanned electronic microscope	S4800	Hitachi
Atomic force microscope	III a	Multimode nanoscope
UV-Vis spectrometer	UV-3101 PC	SHIMADZU
Pulsed laser	Nd: YAG	NEC Corporation

3 Results and Discussion

After titanium oxide films were prepared by magnetron sputtering, the samples were analyzed by X-ray diffractometer. It can be found from the picture (Fig. 2) that the fabricated titanium oxide was mixed phase.

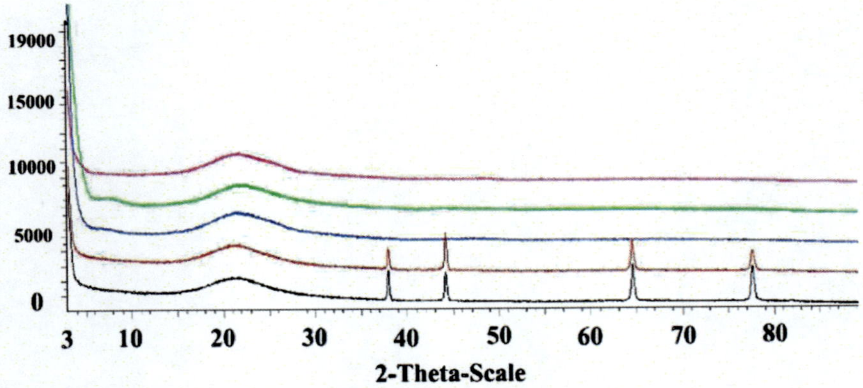

Fig.2. XRD pattern of prepared titanium oxide film.

Then Scanned Electronic Microscope (SEM) and Atomic Force Microscope (AFM) is used to characterize the titanium oxide film. It can be seen in Fig. 3 that there are a few raised bright spots in the film. Meanwhile, there is a "ghosting" effect on the AFM diagram for the raised bright spots.

After depositing a gold film of 4 nm on the titanium oxide film by vacuum evaporation, the gold film was characterized by SEM and AFM. In Fig. 4, it can see that the titanium oxide film is covered by a very fine film of gold particles. Similarly, large particles can be found. The large particles are further formed on the basis of the bright spot of titanium oxide film. After pulsed laser irradiation, we can clearly see in the SEM image (Fig. 5 a, c and e) that the gold film has become particles with a particle size

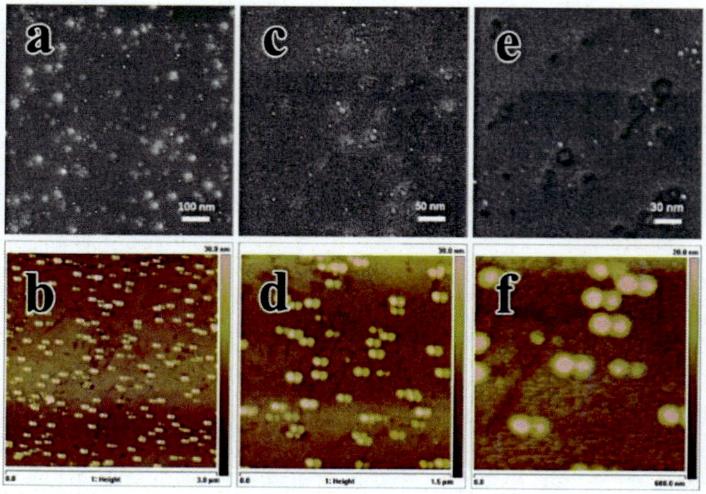

Fig.3. Morphology of titanium oxide film (a, c and e are the SEM pictures under different magnification, b, d and f are the AFM pictures under different magnification).

Fig.4. Morphology of gold film on titanium oxide film (a, c and e are the SEM pictures under different magnification, b, d and f are the AFM pictures under different magnification).

of tens of nanometers. The overall distribution of particles is relatively uniform. At the same time, AFM pictures (Fig. 5 b, d and f) also consistently show the distribution of gold nanoparticles.

Finally, we used UV-Vis spectrophotometer to characterize titanium oxide film, titanium oxide film coated with gold film and titanium oxide film with gold nanoparticles (Au NPs). Titanium oxide film is almost transparent in the visible light range, and there is no absorption peak. The titanium oxide film coated with gold film has slight

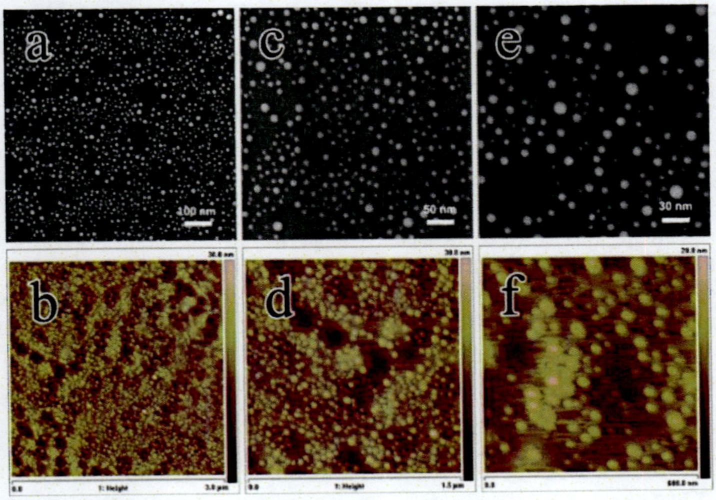

Fig.5. Morphology of gold nano-particles on titanium oxide film (a, c and e are the SEM pictures under different magnification, b, d and f are the AFM pictures under different magnification).

absorption in the red region. The titanium oxide film with gold nanoparticles shows an apparent plasmon resonance absorption peak, which is located near 550 nm. The resonance absorption of this plasmon will improve the absorption of the composite film and promote the catalytic reaction in photodegradation and photocatalysis (Fig. 6).

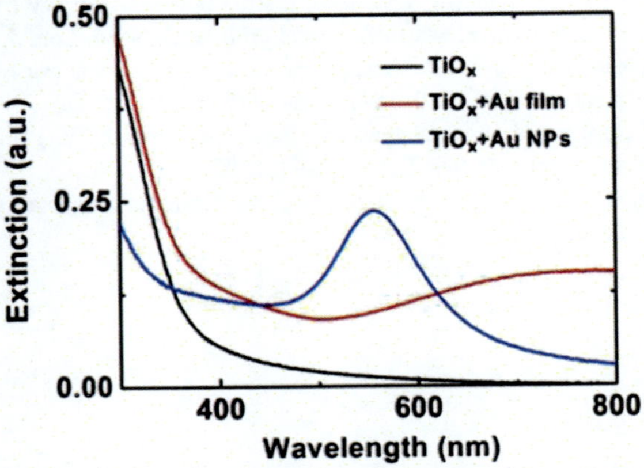

Fig.6. The UV-visible absorption spectrum of TiO_X (black line), TiO_X with Au film (red line) and TiO_X with Au NPs (blue line).

4 Conclusions

In this work, Titanium oxide films with uniformly distributed gold nanoparticles were obtained. Through the characterization of SEM and AFM, it can be found that the surface morphology of titanium oxide, gold film and nano gold particles are different. After pulsed laser irradiation, the gold film was instantly transformed into gold nanoparticles with a diameter of several tens of nanometers. Meanwhile, the composite film shows an absorption peak near 550 nm in the UV-Vis spectrum range, reflecting the plasmon effect of gold nanoparticles. This research is expected to provide an efficient method for preparing photocatalytic composite films.

Acknowledgement. This work is financially supported by the Beijing Municipal Education Commission's Science and Technology Program General Project (KM201910015010, KM201410015004), Printed Electronics Technology & Engineering Discipline Construction (III) (grant number 21090116001), the National Natural Science Foundation of China (21905028), Beijing Education Committee Key Item (KZ202110015019), Beijing Institute of Graphic Communication R & D Plan: (Ef202002), (Eb202102), (Ec202005).

References

1. Linsebigler, A.L., Lu, G., Yates, J.T.: Photocatalysis on TiO_2 surfaces: principles, mechanisms, and selected results. Chem. Rev. **95**, 735 (1995)
2. Mehr, M.E., Maleki-Ghaleh, H., Yarahmadi, M., Kavanlouei, M., Siadati, MH.: Synthesis and characterization of photocatalytic zinc oxide/titanium oxide (core/shell) nanocomposites. J. Alloys Comp. **882** (2021)
3. Wang, D., Zou, Y., Wen, S., Fan, D.: A passivated codoping approach to tailor the band edges of TiO_2 for efficient photocatalytic degradation of organic pollutants. Appl. Phys. Lett. **95**, 012106 (2009)
4. Jakob, M., Levanon, H., Kamat, P.V.: Charge distribution between UV-irradiated TiO_2 and gold nanoparticles: determination of shift in the Fermi level. Nano. Lett. **3**, 353 (2003)
5. Chu, J.Y., et al.: Mixed titanium oxide strategy for enhanced photocatalytic hydrogen evolution. ACS Appl. Mater. Interfaces. **11**(20), 18475–18482 (2019)
6. Jinxia, X., et al.: Enhanced photocatalysis by coupling of anatase TiO_2 film to triangular Ag nanoparticle island. Nanoscale Res. Lett. **7**, 239 (2012)
7. Pillai, S., Green, M.A.: Plasmonics for photovoltaic applications. Sol. Energy Mater. Sol. Cells. **94**, 1481–1486 (2010)

Novel Functional Material Technology

Research Progress and Prospect of Printed Batteries

Zihan Jiang and Guangxue Chen(✉)

State Key Laboratory of Pulp and Paper Engineering, South China University of Technology, Guangzhou, China
chengx@scut.edu.cn

Abstract. With the continuous progress of flexible functional materials, the intelligent development of portable devices, and the gradual popularization of the Internet of Things, society urgently need a new type of battery that is light, thin, environmentally friendly, and has high power and high energy density. The printed battery is prepared by printing technology, which has the characteristics of lightness, flexibility, environmental protection and low cost., which are in line with the development trend and have shown great application potential in flexible wearable devices, flexible displays, flexible RFID, sensors and other fields. This paper reviews the latest progress of printed batteries based on screen printing, inkjet printing and 3D printing technologies, and introduces the latest technologies of printed zinc-manganese batteries, printed lithium batteries and other types of printed batteries. Then, the opportunities and challenges faced by printed batteries are put forward, and the development prospects of the printed battery industry are prospected.

Keywords: Printed batteries · Lithium-ion batteries · Zinc-manganese battery · Screen printing · 3D printing

1 Introduction

Nowadays, mobile phones, computers, RFID devices, medical portable devices and electric vehicles have gradually become an indispensable part of daily life, and the matching batteries are one of the most important supporting components [1–3]. The global market value of lithium-ion batteries will reach $26 billion by 2023 [4].

In addition, batteries are also an important component of the hardware part of the Internet of Things technology. Various information sensors, scanners and other devices in the Internet of Things need an independent power supply that is portable, flexible and easy to integrate when communicating or processing information [5, 6]. Printed batteries are simple in process, low in cost, easy to customize for special application requirements [7, 8], and have become a brand-new power source for portable electronic devices, electric vehicles, etc.

If one or more components in a battery produced by printing technology, such as current collectors, electrodes, separators, electrolytes, then this battery can be defined as printed battery [8], the currently more mature type of printed battery are printed lithium-ion batteries and printed zinc-manganese batteries [8]. The main printing technologies for

M. Xu et al. (Eds.): CACPP 2022, LNEE 991, pp. 561–569, 2023.
https://doi.org/10.1007/978-981-19-9024-3_72

preparing printed batteries include screen printing and inkjet printing [9]. In view of the rapid development of 3D printing technology, we believed that 3D printing technology will be used more and more in printing batteries in the future [10].

2 Structure of Printed Batteries

Printed batteries are similar to ordinary batteries, consisting of positive and negative electrodes, an electrolyte solution separator, and an electrolyte solution. The structure is shown in Fig. 1, working principle is shown in Fig. 2.

For rechargeable printed batteries, the potential difference applied during charging causes ions in the electrolyte solution to move from the cathode to the anode through the diaphragm, while electrons move from the anode to the cathode through an external circuit, generating an electric current in the process to charge the battery. During the discharge process of the battery, the direction of movement of electrons and ions is opposite to that during charging [11]. At present, the more mature types of printed batteries are lithium-ion printed batteries and zinc-manganese printed batteries. The working principles and structures of these two batteries are similar, and the difference is mainly in the materials of the electrodes [12]. The most commonly used anode active material for Li-ion printed batteries is graphite [13], and the cathode material generally uses lithium alloy metal oxides, such as $LiCoO_2$, $LiFePO_4$ [14], $LiMn_2O_4$ [15, 16]. For zinc-manganese batteries, the anode material is zinc [17], and the cathode is composed of materials such as manganese dioxide.

Fig. 1. Schematic diagram of the printed zinc-manganese battery

3 Recent Advances in Printed Batteries

With the progress and development of the times, new electronic devices require high-performance power sources with different geometries [19], and printed batteries have become the most suitable choice, which is characterized by low cost, good scalability, and customization. Printed batteries can be divided into the following categories by differences in electrodes, separators, and electrolytes: lithium-ion batteries, zinc-manganese batteries, and other types of batteries. Research on printed batteries has focused on three

Fig. 2. Schematic diagram of the working principle of printed lithium battery

directions: the applicability of printing technology [20], the study of rheological and electrochemical properties of conductive inks [21], and the design of the shape and internal configuration of printed batteries [9].

Take screen printing technology as an example. Screen printing is to collapse the silk fabric mesh on the screen frame, and use the photochemical plate making method to make the printing plate. During printing, the paste is transferred to the surface of the substrate through the mesh on the printing plate through a certain pressure. Screen printing has many advantages such as simple equipment, convenient operation, uniform printing thickness and stable quality. The principle of screen printing coating is shown in Fig. 3.

3.1 Printed Li-Ion Batteries

Printed lithium-ion batteries are mainly prepared by two-dimensional printing techniques, such as screen printing or flexographic printing, which are prepared by a layer-by-layer printing process. In recent years, the use of additive manufacturing technology such as 3D printing technology to prepare printed lithium batteries has also gradually increased [35].

In 2019, L Yu et al. [22] prepared a nitrogen-doped MXene nanosheet by melamine formaldehyde template method. Nitrogen doping enhanced the electrical conductivity and redox activity of MXene, and then two MXene-N inks were prepared, which were suitable for screen printing and 3D printing, respectively, and provided technical support for the preparation of printed lithium-ion batteries.

Fig. 3. Screen printing coating principle

The assembly steps of Li-ion battery components also determine the electrochemical performance, safety and cyclability of Li-ion batteries to a certain extent. The use of printing technology can control the structure of the battery components from the micro and macro levels very accurately [23], including The thickness, outer dimensions, porosity of the components and designing special structures to obtain specific energy densities [24].

In 2016, RE Sousa et al. [25] developed a C-$LiFePO_4$ based ink and used screen printing technology to prepare the cathode part of the printed lithium-ion battery, the ink has elasticity on the order of 500Pa, shearing The apparent viscosity after yielding is on the order of 3Pa, the total resistance of the cathode is 450 Ω, and the diffusion coefficient is $2.5 \times 10 - 16\ cm^2 - 1$. The developed cathode has an initial discharge capacity of 48.2 $mAhg^{-1}$ at 5 C, and an initial discharge capacity of 39.8 $mAhg^{-1}$ after 50 cycles, reflecting the good performance of printed lithium batteries.

In 2019, SH Kim et al. [26] developed a bipolar all-solid-state lithium battery (asslsb). Bipolar asslsb is prepared by solvent-free drying, UV curing, and stencil printing process at room temperature, without the use of high temperature and high pressure steps such as sintering in the whole process. Using two thermodynamically immiscible and nonflammable gel electrolytes, methyl sulfone ethyl ester (EMS) and ethylene glycol dimethyl ether (TEGDME) to fabricate printed batteries, solves a long-standing problem with traditional inorganic solid electrolytes on grain boundary resistance, and the polysulfide shuttle effect in Li-S batteries.

In 2018, J Li et al. [27] proposed a new idea for fabricating battery electrodes, using additive manufacturing techniques such as 3D printing to simultaneously control the electrode structure from both macro and micro levels. Electrochemical tests show that these printed electrodes exhibit excellent performance in terms of specific capacity, areal capacity, and life cycle. The foldable printed lithium battery made by 3D printing technology has a discharge capacity of 1100 $mAh.g^{-1}$ at a rate of 0.1 C, which far exceeds the level that can be achieved by traditional lithium battery manufacturing technology, proving that additive manufacturing technology is effective in printing. The battery field has huge development potential.

3.2 Printed Zinc-Manganese Batteries

The technology of zinc-manganese battery is relatively mature, with the characteristics of safety and environmental protection, so it is widely used in printed batteries. In 2011, AM Gaikwad et al. [28] used a polyacrylic acid-based polymer gel electrolyte to fabricate a printed Zn-MnO_2 battery with a discharge capacity of 5.6 mAh.cm^{-2} at 0.5 mA. Two printed batteries in series are able to power the LEDs. Two years later, AM Gaikwad et al. [29] demonstrated a technique to strengthen battery electrodes by using high-strength, low-cost fibrous membranes that can be used as separators. This technique was used to fabricate high-energy-density, non-toxic Zn-MnO_2 batteries with printed current collectors. With a nominal potential of 1.5 V and an effective capacity of approximately 3 mA h cm^{-2}, the battery powers a light-emitting diode display integrating a strain sensor and microcontroller.

In 2019, X Wang et al. [30] demonstrated a Zn-MnO_2 battery fabricated based on screen printing technology, which exhibited excellent electrochemical performance, flexibility, flexibility, and modularity. The battery used Zn ink to print the anode, MnO_2 ink to print the cathode, and graphene ink with high electrical conductivity to print the current collector. The printed battery was tested to provide 19.3 mAh/cm^3 (393 mAh/g), and the volumetric energy density of 17.3 mWh/cm^3, which is better than that of lithium thin-film batteries ($\leq$10 mWh/cm^3). In addition, the Zn//MnO_2 printed battery still has a high capacity retention rate of 83.9% after 1300 cycles at 5 C, which is better than previously reported printed Zn//MnO_2 batteries.

3.3 Other Types of Printed Batteries

In addition to the two most common types of printed batteries above, other types of batteries can also be combined with printing technology. S Berchmans et al. [31] reported a screen-printed Ag-Zn battery, which is flexible and elastic, suitable for powering wearable devices. Characterization shows that its energy density is in the range of 1.3–2.1 mA h cm^{-2}. The battery exhibits a stable open circuit voltage of 1.5 V for 5 days with minimal performance degradation after repeated stretching and bending, and multiple cells can also be combined in series or parallel into a battery pack to obtain the desired capacitance or voltage value.

In 2017, R Kumar et al. [32] used polystyrene-block-polyisoprene -block-polystyrene (SIS) as a superelastic binder for printing inks and printed a high-strength Stretchable silver zinc oxide (Zn-Ag2O) battery. This fully printed, stretchable Zn-Ag2O rechargeable battery has a reversible capacity density of about 2.5 mAh cm^{-2} due to the use of SIS binder. This battery provides the highest reversible capacity and discharge current density for intrinsically stretchable batteries reported to date. This provides a new idea for the further development of wearable and stretchable electronic products.

The electrochemical performance of flexible zinc-air batteries has also been improved by printing a thin layer of aluminum oxide on the surface of zinc particles, making such batteries also applicable to electronic devices [33]. In 2018, Z Wang et al.

[34] fabricated a flexible nickel-zinc battery whose composite electrode was fabricated using a multi-walled carbon nanotube (CNT)-based ink combined with printing. The battery still shows good performance under bending conditions, and can be extended and applied by printing technology.

4 Application and Development Prospects of Printed Batteries

At present, the main application fields of printed batteries are sensor devices, flexible wearable devices, RFID devices, and portable medical devices. In these applications, battery is basically required to be thin, flexible, and controllable in shape, and most of batteries are disposable [8], which also puts forward higher requirements for the cost and environmental protection of printed batteries.

Printed batteries can be combined with pressure sensors, temperature sensors and other sensor devices and used in wearable devices. The flexible pressure sensor prepared with the new nano pressure-sensitive ink is thin and flexible, and can form a flexible pressure sensor that fits perfectly on the curved surface; the flexible circuit board is prepared by inkjet printing or screen printing, and combined with the printed battery, it can form a flexible pressure sensor from the energy supply to a complete system for data collection and uploading. Such systems are widely used in sports, health and monitoring. For example, smart insole, before the advent of flexible sensors and flexible energy, the realization of smart insole is quite difficult. Today, by encapsulating a flexible sensor into the insole, and embedding a flexible power source and an ultra-thin chip at the arch of the foot, a set of intelligent insoles for walking detection and collection are integrated. After being bent and deformed under pressure, the sensor can analyze the person's weight, walking gait and pressure distribution, and can monitor the person's motion state and amount of exercise, which greatly improves people's perception of their own walking posture analysis and exercise amount.

The temperature sensor labels in the sensor equipment are mainly used for the transportation and storage management of food and medicine in the cold chain of the Internet of Things. These sensor labels can monitor the temperature of the food or medicine from production to sales, and can be traced to the source and provided when necessary. Alarm function to ensure that the abnormal temperature of these foods and medicines can be detected in time in the process of inventory, transportation and sales, and possible quality problems can be dealt with in time to prevent the occurrence of unsafe food and medicine. At the same time, the detailed information of food and medicine, such as manufacturer, production date, expiration date, etc., can also be stored in the label for quality and logistics management. The realization of these functions depends on the help of printed batteries [18]. In June 2013, China's State Food and Drug Administration issued the "Quality Specification for Drug Operation", which put forward specific requirements for automatic temperature monitoring during drug storage and transportation and drug cold chain logistics management, which means that China will use such temperature sensors. The demand for labels has grown tremendously.

The printed battery is the product of the combination of functional ink and printing technology. The characteristics of the ink, the printing process parameters and the selection of printing equipment directly determine the performance of the printed battery.

Functional inks need to be matched to the printing process used, the rheological properties of the inks are critical, and in order to improve the scalability of the printed cells, the adhesion between the printed layer and the printed cell substrate is also a factor that needs to be evaluated in detail.

In order to improve the ionic conductivity and mechanical stability of printed batteries, printed battery separators and electrolytes are also an important research direction. The combination of laser technology, UV curing technology and printed batteries will be one of the directions worth studying in the future.

In the near future, 4D printing technology may turn a new page for printed batteries. The so-called 4D printing means that the structure printed by 3D technology can change its shape or structure under external stimulation, and directly connect materials and structures. The morphing design is built into the material. This will endow the printed battery with completely new functions such as self-healing, self-transmitting, etc.

In conclusion, the research on printed batteries has achieved gratifying results, but there are also a series of problems and challenges to be solved in the future. Printed batteries will develop synchronously with the printed electronics industry. Under the trend, printed batteries will also become one of the main growth points of the global economy in the next few decades.

Acknowledgement. This work has been financially supported by the National Natural Science Foundation (61973127), Science and Technology Program of Guangdong Province (2017B090901064), Foshan National High-tech Industrial Development Zone(FSBG2021021).

References

1. Ginley, D.S., Cahen, D.: Fundamentals of materials for energy and environmental sustainability, Cambridge University Press (2011)
2. Campanari, S., Manzolini, G., Garcia de la Iglesia, F.: Energy analysis of electric vehicles using batteries or fuel cells through well-to-wheel driving cycle simulations. J. Power Sour. **186**, 464–477 (2009)
3. Hall, P.J., Bain, E.J.: Energy-storage technologies and electricity generation. Energy Policy **36**, 4352–4355 (2008)
4. Lawes, S., Riese, A., Sun, Q., Cheng, N., Sun, X.: Printing nanostructured carbon for energy storage and conversion applications. Carbon **92**, 150–176 (2015)
5. Atzori, L., Iera, A., Morabito, G.: The internet of things: a survey. Comput. Netw. **54**, 2787–2805 (2010)
6. Gubbi, J., Buyya, R., Marusic, S., Palaniswami, M.: Internet of Things (IoT): a vision, architectural elements, and future directions. Futur. Gener. Comput. Syst. **29**, 1645–1660 (2013)
7. Sousa, R.E., Costa, C.M., Lanceros-Méndez, S.: Advances and future challenges in printed batteries. Chemsuschem **8**, 3539–3555 (2015)
8. Lanceros-Méndez, S., Costa, C.M.: Printed batteries: materials, technologies and applications, John Wiley & Sons (2018)
9. Choi, K.-H., Ahn, D.B., Lee, S.-Y.: Current status and challenges in printed batteries: toward form factor-free, monolithic integrated power sources. ACS Energy Lett. **3**, 220–236 (2018)

10. Dommati, H., Ray, S.S., Wang, J.-C., Chen, S.-S.: A comprehensive review of recent developments in 3D printing technique for ceramic membrane fabrication for water purification. RSC Adv. **9**, 16869–16883 (2019)
11. Crompton, T.P.J.: Battery Reference Book, Elsevier Science (2000)
12. Dell, R., Rand, D.A.J., Bailey, R., Connor, P.: Chemistry RSo. Royal Society of Chemistry, Understanding Batteries (2001)
13. Cassagneau, T., Fendler, J.H.: High density rechargeable lithium-ion batteries self-assembled from graphite oxide nanoplatelets and polyelectrolytes. Adv. Mater. **10**, 877–881 (1998)
14. Gören, A., Costa, C.M., Silva, M.M., Lanceros-Méndez, S.: State of the art and open questions on cathode preparation based on carbon coated lithium iron phosphate. Compos. B Eng. **83**, 333–345 (2015)
15. Mishra, A., Mehta, A., Basu, S., Malode, S.J., Shetti, N.P., Shukla, S.S., et al.: Electrode materials for lithium-ion batteries. Mater. Sci. Energy Technol. **1**, 182–187 (2018)
16. Takahashi, M., Tobishima, S., Takei, K., Sakurai, Y.: Characterization of LiFePO4 as the cathode material for rechargeable lithium batteries. J. Power Sour. **97–98**, 508–511 (2001)
17. Lao-atiman, W., Julaphatachote, T., Boonmongkolras, P., Kheawhom, S.: Printed transparent thin film Zn-MnO2 battery. J. Electrochem. Soc. **164**, A859–A863 (2017)
18. Zhang, X.C.: Paper batteries and printed electronics. China Mater. Prog. **36**(3), 186–188 (2019)
19. Seifert, T., Sowade, E., Roscher, F., Wiemer, M., Gessner, T., Baumann, R.R.: Additive manufacturing technologies compared: morphology of deposits of silver ink using inkjet and aerosol jet printing. Ind. Eng. Chem. Res. **54**, 769–779 (2015)
20. Suárez, L., Domínguez, M.: Sustainability and environmental impact of fused deposition modelling (FDM) technologies. Int. J. Adv. Manuf. Technol. **106**(3–4), 1267–1279 (2019). https://doi.org/10.1007/s00170-019-04676-0
21. Sun, C., Liu, S., Shi, X., Lai, C., Liang, J., Chen, Y.: 3D printing nanocomposite gel-based thick electrode enabling both high areal capacity and rate performance for lithium-ion battery. Chem. Eng. J. **381**, 122641 (2020)
22. Yu, L., Fan, Z., Shao, Y., Tian, Z., Sun, J., Liu, Z.: Versatile N-doped mxene ink for printed electrochemical energy storage application. Adv. Energy Mater. **9**, 1901839 (2019)
23. Gaikwad, A.M., Arias, A.C., Steingart, D.A.: Recent progress on printed flexible batteries: mechanical challenges, printing technologies, and future prospects. Energ. Technol. **3**, 305–328 (2015)
24. Chang, P., Mei, H., Zhou, S., Dassios, K.G., Cheng, L.: 3D printed electrochemical energy storage devices. J. Mater. Chem. A. **7**, 4230–4258 (2019)
25. Sousa, R.E., Oliveira, J., Gören, A., Miranda, D., Silva, M.M., Hilliou, L., et al.: High performance screen printable lithium-ion battery cathode ink based on C-LiFePO4. Electrochim. Acta. **196**, 92–100 (2016)
26. Kim, S.-H., Kim, J.-H., Cho, S.-J., Lee, S.-Y.: All-solid-state printed bipolar Li–S batteries. Adv. Energy Mater. **9**, 1901841 (2019)
27. Li, J., Liang, X., Liou, F., Park, J.: Macro-/micro-controlled 3D lithium-ion batteries via additive manufacturing and electric field processing. Sci. Rep. **8**, 1846 (2018)
28. Gaikwad, A.M., Whiting, G.L., Steingart, D.A., Arias, A.C.: Highly flexible, printed alkaline batteries based on mesh-embedded electrodes. Adv. Mater. **23**, 3251–3255 (2011)
29. Gaikwad, A.M., Chu, H.N., Qeraj, R., Zamarayeva, A.M., Steingart, D.A.: Reinforced electrode architecture for a flexible battery with paperlike characteristics. Energ. Technol. **1**, 177–185 (2013)
30. Wang, X., Zheng, S., Zhou, F., Qin, J., Shi, X., Wang, S., et al.: Scalable fabrication of printed Zn//MnO_2 planar micro-batteries with high volumetric energy density and exceptional safety. Natl. Sci. Rev. (2019)

31. Berchmans, S., Bandodkar, A.J., Jia, W., Ramírez, J., Meng, Y.S., Wang, J.: An epidermal alkaline rechargeable Ag–Zn printable tattoo battery for wearable electronics. J. Mater. Chem. A. **2**, 15788–15795 (2014)
32. Kumar, R., Shin, J., Yin, L., You, J.-M., Meng, Y.S., Wang, J.: All-printed, stretchable Zn-Ag_2O rechargeable battery via hyperelastic binder for self-powering wearable electronics. Adv. Energy Mater. **7**, 1602096 (2017)
33. Wongrujipairoj, K., Poolnapol, L., Arpornwichanop, A., Suren, S., Kheawhom, S.: Suppression of zinc anode corrosion for printed flexible zinc-air battery. physica status solidi (b). **254**, 1600442 (2017)
34. Wang, Z., Meng, X., Chen, K., Mitra, S.: Synthesis of carbon nanotube incorporated metal oxides for the fabrication of printable. Flexible Nickel-Zinc Batteries. Adv. Mater. Interf. **5**, 1701036 (2018)
35. Costa, G., Lopes, P.A., Sanati, A.L., Silva, A.F., Freitas, M.C., Almeida, A.T.D., Tavakoli, M.: Adv. Funct. Mater. **32**, 2113232 (2022)

Effect of Surfactants on Preparation of Perovskite Films

Lei Wang, Beiqing Huang, Xianfu Wei, Hui Wang, and Weimin Zhang(✉)

School of Printing and Packaging Engineering, Beijing Institute of Graphic Communication, Beijing, China
zhangweimin@bigc.edu.cn

Abstract. With the continuous development of perovskite solar cells, the preparation methods of thin film become diverse. Battery performance is further improved, and the efficiency is rapidly improved. The preparation of thin film also tends to be high-efficiency and large-area, including the spin coating, vacuum evaporation, inkjet printing and so on, which provides great help for the industrial application of perovskite batteries. For the preparation of perovskite film, there are still some problems that need to be solved urgently, such as the matching problem of ink and the hydrodynamic problem of ink. In this paper, the rheological properties (surface tension, viscosity, contact angle) of the ink were tested by adding different types and concentrations of surfactants into the configured perovskite ink, and the thin films were prepared by one-step spin coating method. The changes in the morphology of the thin films were explored, and the appropriate surfactants were selected, and the molar ratio of the addition was determined, so that the film has good morphology.

Keywords: Surfactant · Thin films · Perovskite · Spin coating

1 Introduction

With the continuous increase in people's demand for energy, energy shortage has become an important factor restricting economic development, and energy issues have been paid more and more attention. The application of cleaning energy to replace non-renewable energy is of great significance. Solar energy is a renewable, pollution-free energy, through solar cells can be converted into the energy people need. At present, researchers attach great importance to the research of solar cells, and are committed to developing solar cells with low cost and high efficiency. Among them, perovskite solar cells have become a research hotspot in the field of photovoltaic due to their low cost and high photoelectric conversion efficiency, and their preparation methods are also different. For rotary coating, the wetting and drying of ink will affect the performance of the battery. In order to improve the performance of devices, researchers have carried out a lot of research work in solvent engineering, interface engineering and additive engineering.

Additives are commonly used in the preparation of efficient and stable perovskite solar cells, and they are one of the effective means to optimize perovskite solar cells,

M. Xu et al. (Eds.): CACPP 2022, LNEE 991, pp. 570–576, 2023.
https://doi.org/10.1007/978-981-19-9024-3_73

which can be salt and polymer. The Mitzi team added lead thiocyanate in perovskite precursor ink can significantly increase the grain size of the film. There by reducing hysteresis of the battery and improving the photoelectric conversion efficiency. Kim et al. added MACl into $FAPbI_3$ perovskite without annealing, considerably increased particle size and crystal quality. Grätzel team reported the synergistic effect of deionized water additives and DMF vapor and prepared high-quality perovskite films by two-step spin coating method. The introduction of deionized water contributed to the formation of $MAPbI_3$ thick film, and the formation of larger grains by slowing the crystallization rate of perovskite. It can also monitor the dissolution of perovskite grains by the synergistic effect with DMF. Zhou Huanping team reported that the basicity of additives was an important parameter affecting the crystallization kinetics and photoelectric properties of perovskite film. A small amount of strong alkaline potassium hydroxide KOH) is sufficient to eliminate iodine I_2) in the precursor. The team also reported a kind of europium ion (Eu^{3+} $-Eu^{2+}$) that can serve as the "redox reaction" shuttle back and forth, and simultaneously oxidize Pbo and reduce I0 in periodic transformation. The addition of europium ions effectively inhibited the non-radiative recombination. Yang et al. reported three amino acid molecules with similar structure but different functional groups. Among them, 2-aminoethylphosphonic acid PA) molecules have high interaction energy with positively charged defects, resulting in passivation effect, reducing the defect density of perovskite film and improving the particle size of film. Snaith Team added ionic liquid 1-butyl-3-methylimidazolium tetrafluoroborate $BMIMBF_4$) into perovskite film, which greatly improved the photoelectric conversion efficiency and long-term stability of the device. After packaging at 70–75 °C for 1800 h under continuous simulated full-spectrum light, the photoelectric conversion efficiency decreases by only 5%, and it is estimated that the time required to reduce the device performance to 80% is about 5200 h, and the grain size of the film increases by 6 times.

Based on this, this study used surfactants as additives in perovskite ink, screened and tested the effects of several different surfactants on the physical and chemical properties of the ink, and rotated and coated the perovskite film. Combined with the experimental data, appropriate surfactants were selected to make the coated perovskite film have better surface morphology.

2 Screening of Surfactant and Preparation of Spin Coating

In this study, surfactant was added to two-dimensional perovskite ink to explore the change of ink performance, and the film was prepared by one-step spin coating method to observe the change of film morphology.

2.1 Experimental Materials and Instruments

Experimental materials: N, N-dimethylformamide (DMF, $\geq$ 99.5%), dimethyl sulfoxide (DMSO, Analytically pure), methyl ammonium iodide (MAI, $\geq$ 99.5%), butyl ammonium iodide (BAI, $\geq$ 99.5%), lead iodide (PbI_2, > 99.99%),polyethylene glycol (PEG), polyvinylpyrrolidone (PVP), sodium dodecyl benzene sulfonate (SDBS), sodium dodecyl sulfate (SDS), sodium dodecyl sulfate, and ITO conductive glass.

Experimental instruments: KW-4A/5 homogenizer, ultrasonic cleaner, drying oven, rheometer TA-AR2000ex, contact angle measuring instrument KRUSS-DSA100, Hande REVIEW series portable digital measuring instrument.

2.2 Sample Preparation

Firstly, the basic ink was prepared by fully dissolving BAI, MAI and PbI2 in DMF and DMSO binary mixed solvents at 60 °C in water bath. Then the surfactant was fully mixed with the ink according to the equigradient concentration of the molar percentage of the surfactant relative to the perovskite mass (Table 1), and the properties of the ink (surface tension, contact angle, viscosity) were investigated. Finally, the ink was made by one-step spin coating using a homogenizer to observe the change of the film. Finally, the type and the molar ratio of surfactant were determined to make the obtained film have better morphology.

Table. 1 Content of surfactant

Content (molar ratio)	Surfactant	Molecular weight	Addition				
100/1	PEG	200	0.0015	0.003	0.0045	0.006	0.0075
		4000	0.03	0.06	0.09	0.12	0.15
	SDBS	348.48	00025	0.004	0.006	0.008	0.010
	Sodium dodecyl sulfate (SDS)	272.38	0.002	0.004	0.006	0.008	0.010
	Sodium dodecyl sulfate	288.38	0.002	0.004	0.006	0.008	0.010
1000/1	PEG	20000	0.015	0.03	0.045	0.06	0.075
	PVP	10000	0.0075	0.015	0.0225	0.030	0.0375
10000/1	PVP	40000	0.003	0.006	0.009	0.012	0.015
		58000	0.004	0.009	0.013	0.017	0.022
		60000	0.0045	0.009	0.0135	0.018	0.0225

The ink with surfactant was spin-coated in one step by using a homogenizer, and the rotation speed and time were set respectively. Then, the ink was annealed at 100 °C for 15 min on the hot stage and observed by a microscope.

3 Result and Discussion

The change of ink properties with the type and the molar ratio of surfactants is different, which ultimately affects the quality of the film.

3.1 Effect of Surfactant Types on Ink Properties

The three properties (surface tension, contact angle and viscosity) of ink were measured and compared by contact angle measuring instrument and rheometer.

Fig. 1. Surface characteristic curves of ink (surface tension, contact angle, viscosity)

Different surfactants have different effects on rheological properties (Fig. 1). For polyethylene glycol, firstly, with the increase of concentration gradient, the change of tension showed a reliable linear change. With the increase of molecular weight, influence on surface tension gradually weakened. When the molecular weight of polyethylene glycol is 200 and the molar ratio is 2%, the rheological properties are the best (surface tension is 33.93 mN/m, viscosity is 6.02 m.pa/s, contact angle is 2.745°), which is more suitable for spin coating preparation. For polyvinylpyrrolidone, the viscosity of ink decreased greatly after adding, and the changes of surface tension and contact angle were not obvious. After adding dodecyl surfactants, with the increase in the molar ratio, the contact angle and viscosity decreased gradually, and the surface tension changed little.

3.2 Effect of Surfactants on Film Formation

Polyethylene Glycol. With the addition of polyethylene glycol, the rheological properties of ink were changed, and the uniformity of film compactness was better. When the ink contains PEG200 (Fig. 2A), the surface tension is low, and the contact angle decreases significantly, which is conducive to the wetting and spreading of the ink. At this time the crystal growth has a certain direction, the crystal is slender and dense. With the increase in concentration, the accumulation of film edge is aggravated. When PEG4000 was added (Fig. 2B), the change of surface tension was not obvious, and the contact angle increased. When the crystal grows, there are some noticeable gaps around the crystal nucleus. After annealing, the number of nuclei increased and the surface of the film was uneven. When PEG20000 was added (Fig. 2C), the surface tension changed significantly, and the contact angle decreased first and then increased. Crystals become dense, and there is a evident shrinkage after annealing.

Fig. 2. A. PEG200; B. PEG4000; C. PEG20000

Polyvinyl Pyrrolidone. After the addition of polyvinylpyrrolidone, the contact angle changed significantly, and the droplets spread well on the substrate surface, which could be broken down. After adding PVP10000 (Fig. 3A). Ink spreading can be adjusted. The higher the concentration is, the more obvious the grain boundary appears between crystals. With the increase in the molar ratio, the crystal becomes smaller and denser. With the increase of molecular weight, crystals accumulate and the grain boundaries gradually disappear. When the molecular weight is 40000 (Fig. 3B), the crystal grows from chaos to order, film color is shallow, the crystal radiates from the middle to the periphery, the thickness is altered, and the transparency increases. When the molecular weight is 58000 (Fig. 3C), the smoothness of the surface increases, but the disorder of the crystal increases, and the crystal shrinks after annealing. When PVP60000 was added (Fig. 3D), there were no obvious grain boundaries, but the morphology of the film was poor, and there was obvious boundary on the substrate. It speculates that coffee rang effect after annealing.

Polyvinyl Pyrrolidone After the addition of sodium dodecyl benzene sulfonate (Fig. 4A), the contact angle decreases with the increase of concentration, the spreading area on the substrate increases, the crystal length increases, and the order begins to increase. With the addition of sodium dodecyl sulfonate (Fig. 4B), the contact angle showed a linear change, first increased and then decreased. The spreading on the substrate surface was a very good thing. With the increase in the molar ratio, the crystal became short and dense, and the color of the film also deepened. Sodium dodecyl sulfate has little effect on the surface tension, the contact angle has a great change, and has a profound influence on the morphology of the film. With the increase in concentration, the accumulation of the film is serious, forming more islands, and the crystals begin to become dense, even overlap (Fig. 4C).

Fig. 3. A PVP10000; B PVP40000; C PVP58000; D PVP60000;

Fig. 4. A.SDBS; B. sodium dodecyl sulfonate (SDS); C. sodium dodecyl sulfate;

4 Conclusion

From the above experiments, polyvinylpyrrolidone had little effect on the surface tension, and the contact angle increased. The coating coverage rate of the film on the substrate surface was minimal, which was a serious waste of materials. Sodium dodecyl sulfate can reduce the contact angle to some extent, but the surface morphology of the film is weak. The crystal of sodium dodecyl sulfate is fine and dense, and the coverage rate is high. Polyethylene glycol can well decrease the surface tension and contact angle. When the molecular weight increases, the crystal length increases and the disorder increases. When the molecular weight of polyethylene glycol is 200 and the molar ratio is 2%, the rheological properties of ink are greatly improved, which is more suitable for spin coating preparation. The morphology of the film is better, and it can be uniformly spread on the substrate.

Acknowledgement. We gratefully acknowledge support from BIGC Project and Institute of Advanced Ink, Beijing Institute of Graphic Communication.

References

1. Ke, W., Xiao, C., Wang, C., et al.: Employing lead thiocyanate additive to reduce the hysteresis and boost the fill factor of planar perovskite solar cells. Adv. Mater. **28**(26), 5214–5221 (2016)
2. Kim, M., Kim, G.H., Lee, T.K., et al.: Methylammonium chloride induces intermediate phase stabilization for efficient perovskite solar cells. Joule **3**(9), 2179–2192 (2019)
3. Ye, F., Ma, J., Chen, C., et al.: Roles of MACl in sequentially deposited bromine-free perovskite absorbers for efficient solar cells. Adv. Mater. **33**(3), 2007126 (2020)
4. Chen, Y., Li, N., Wang, L., et al.: Impacts of alkaline on the defect's property and crystallization kinetics in perovskite solar cells. Nat. Commun. **10**(1), 1112 (2019)
5. Wang, L., Zhou, H., Hu, J., et al.: A Eu3+-Eu2+ ion redox shuttle imparts operational durability to Pb-I perovskite solar cells. Science **363**(6424), 265–270 (2019)
6. Zhao, Y., Zhu, P., Huang, S., et al.: Molecular Interaction regulates the performance and longevity of defect passivation for metal halide perovskite solar cells. J. Am. Chem. Soc. **142**(47), 20071–20079 (2020)
7. Bai, S., Da, P., Li, C., et al.: Planar perovskite solar cells with long-term stability using ionic liquid additives. Nature **571**(7764), 245–250 (2019)

Fully Printed Thin Film Transistors: Key Materials and Applications

Yun Weng, Zhaohui Yu, Lijuan Liang, Lianfang Li, Ti Wu, Shengzhen Liu(✉), and Sunhao Guo

School of Printing and Packaging Engineering, Beijing Institute of Graphic Communication, Beijing, China
lsz63371373@163.com

Abstract. Thin film transistors (TFTs) have been deeply studied as the core component of many information electronic systems and have made great progress in high-performance material research and multifunctional applications of devices. Due to the excellent flexibility and solution-processability, the preparation of TFTs can be highly integrated with a variety of printing technologies, which is convenient for manufacturing large-scale, low-cost, and large-scale TFTs. Many research groups are working hard to realize the application of flexible thin film transistors with high mechanical and electrical stability in emerging technology fields such as electronic skin and wearable devices. Fully printed TFTs have received extensive attention with the rapid development of soluble materials. Starting from the fully printed TFTs, this review discusses the printing compatible ink materials for different components of TFTs, including metal nanoparticle materials, carbon-based materials, and polymer materials. The applications of thin film transistors in the fields of flexible displays and sensors are also summarized. Finally, the challenges and future research direction for the full printed TFTs are discussed.

Keywords: Fully printing fabrication · Thin-film transistors · Organic semiconductors · Sensors

1 Introduction

TFTs are kinds of typical three-terminal semiconductor devices that can achieve signal conversion/amplification, logic switching, voltage regulation, rectification, and drive current. Thus, they have been widely used in areas such as planar displays [1] and sensors [2]. TFTs play a vital role in the modern semiconductor electronics industry, represented by integrated circuits and logic computing devices. However, the traditional manufacture of silicon-based TFTs requires complex lithography processes and vacuum systems, which primarily limit the development of TFTs towards large-area, high-resolution integrated electronic devices. As the demand for wearable or implantable devices increases, finding the suitable materials and fabrication methodology is highly desirable.

In recent years, the emerging printing electronic technology has become one of the most promising candidate technologies with the advantages of large-scale, low cost, and

M. Xu et al. (Eds.): CACPP 2022, LNEE 991, pp. 577–586, 2023.
https://doi.org/10.1007/978-981-19-9024-3_74

non-vacuum processing. Printing electronic technology, as an additive manufacturing process, can be roughly classified into screen printing, scraping printing, concave printing, inkjet printing, etc., which allows the fabrication of TFTs with stacked structures composed of a conductive layer, an insulating layer, and a semiconductor layer [3].

In the printing process of thin-film transistors, the full printing preparation of TFTs is accomplished by achieving inking of the material. Therefore, the components of TFTs, including source-drain electrodes, gate electrodes, dielectric layers, and active layers, should have solution processability. Therefore, in order to achieve competitive flexible TFTs, various functional materials are being widely explored. Metal materials with high electrical conductivity, such as gold nanoparticle ink, silver nanoparticle ink, and copper nanoparticle ink, have been used to print electrodes in TFTs. Many types of organic semiconductor materials, such as pentacene, PQT-12, F8T2, etc., have been reported to be applied to the printing process of TFTs.

In this review, we have conducted detailed and in-depth discussions on ink materials for printed electronics and the application of printed thin-film transistors in functional devices. Specifically, the various ink materials of thin-film transistors are summarized and analyzed, including metal nanomaterials, carbon-based materials, and organic semiconductor materials, and the selection of materials for different parts of transistors and their compatibility with printing methods are discussed. Next, we will make a brief study of the application progress of thin-film transistors in flexible displays, sensors, and other functional devices. Finally, we have a brief summary and outlook on this new technology.

2 Printable Functional Materials

Flexible thin-film transistors (TFTs) are key driver/switching components for wearable electronics. We need to select materials with solution-processability for the fabrication of each OTFT layer in the printing process. Materials dissolved in solvents are deposited on the substrate with high resolution to achieve full-print preparation of the device.

2.1 Metal Nanoparticle Materials

Conventional metal nanoparticles contain a metal core and ligand molecules with a size of about 1 to 100 nm, and the metal core is coated with ligand molecules to have better dispersion in organic solvents. Since the ligand molecules themselves do not conduct electricity, after the metal nanoparticles are deposited and cured, additional annealing or thermal sintering is required to remove the ligands so that they regain the metal conductivity.

Copper (Cu) nanoparticle inks have recently attracted attention with the advantage of low-cost and high conductivity. Norita et al. scattered the Cu nanoparticles with an average size of 70 nm in organic solvents to get the CU nano-granular ink for printed electrodes [4]. They successfully prepared TFTs using the inkjet-printed Cu layer as the electrodes and pentacene as the active layer with a good carrier mobility of 0.13 $cm^2V^{-1}s^{-1}$ and an on/off current ratio of 10^6.

Water, hydrocarbons, alcohols, and other oxygenated solvents are typical solvents for the preparation of conductive inks based on metal nanoparticles. Zhang et al. used a mixed solution with a mass ratio of 7:2:1 between ethylene glycol, ethanol, and water as solvents and prepared a silver nanoparticle conductive ink [5]. This ink contains high boiling point and low surface tension ethylene glycol, which can prevent ink from flowing to the edge of the ink droplet, prevent the formation of a "coffee ring" structure, and greatly improve the printing adaptability of the conductive ink. This silver nanoparticle conductive ink would have enormous potential for being optimized for TFT device applications.

Generally, the processing temperatures used by metal nanoparticle conductive inks during processing often cause significant distortion to the flexible substrate, thus affecting the accurate alignment between the different layers of the flexible device. So, Minari et al. reported a method for preparing fully printed OTFTs at room temperature [6]. They used metal-free phthalocyanine as a conductive ligand for gold nanoparticles (NPs). Orbital hybridization between the π orbitals of aromatic ligands facilitates charge transport between π junction NPs. In the π-junction NPs, orbital hybridization between the π orbitals of the aromatic ligand and the orbitals of the metal core contributes to charge transport among the NPs. A conductive film can be obtained by annealing the ink at room temperature without removing ligands. This sinter-free Au NPs ink exhibits a low resistivity of -9×10^{-6} Ω cm after printing and drying at room temperature. They then produced a fully printed OTFT on paper with an average mobility of about 2.5 $cm^2 V^{-1} s^{-1}$, showing potential applications in other thin-film devices such as organic photovoltaic cells.

2.2 Carbon Based Materials

In recent years, due to the low cost of carbon nanotubes (CNT) and graphene, and their unique electrical, mechanical, and optical properties, carbon conductive ink has a broad application prospect.

2.2.1 Carbon Nanotubes

Carbon nanotubes were first reported in the early 1990s by Lijima. Carbon nanotubes can be divided into single-wall carbon nanotubes (SWCNTs) and multi-walled carbon nanotubes (MWCNTs) according to the number of wall layers. Among them, single-walled semiconductor carbon nanotubes (s-SWCNTs) show ideal semiconductor properties due to their extremely high mobility and solution processing. In 2006, Chen et al. successfully constructed complementary metal-oxide semiconductor (CMOS)-type ring oscillators on 18 mm long SWCNTs, demonstrating the possibility of using carbon nanotubes in electronic circuits [7]. At present, sorting high-purity, large-diameter semiconductor single crystal carbon nanotubes for ink preparation is one of the important steps in the large-scale preparation of high-performance printed single crystal carbon nanotube technology. Lu and Franklin et al. chose purities of up to 99.9% CNT ink, achieve a printing speed of 8 mm·s−1 on the paper substrate by an aerosol jet printer at room temperature (23 °C) [8]. By successively printing 1D CNT semiconductor channel inks,

1D Ag NPs electrode inks, and 2D h-BN dielectric layer inks, a fully printed 1D-2D thin-film transistor is constructed. Figure 1 shows the process of preparing the 1D-2D TFTs on the polyimide lining, and each component can be deposited by aerosol jet printing. This printing process can be carried out on flexible substrates such as paper substrates, opening up a new direction for the fully print-in-place approach of flexible electronic devices.

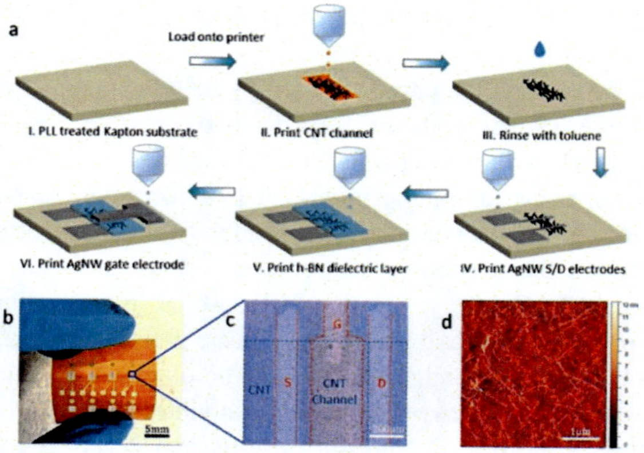

Fig. 1. (a) Preparation of 1D-2D TFTs process flow chart (b) Photos of TFTs (c) optical images of TFTs. (d) CNT channel atomic force microscopy image [7]

2.2.2 Graphene

As a representative material of two-dimensional nanomaterials, graphene has high stability and conductivity. Conductive inks prepared from graphene are compatible with inkjet printing methods and are widely used in the field of flexible printed electronics. Graphene ink can be roughly divided into two kinds: one is the prepared graphene mixed with other solvent additives, directly formulated graphene ink. The other is to use graphene precursors, mostly graphene oxide to make ink, and then use various means to restore the conductivity of graphene after printing, so that the printed product of graphene ink is finally obtained. Cho and co-workers prepared reduced graphene oxide (RGO)/polyvinyl alcohol (PVA) composites as electrode materials for thin-film transistors [8]. The presence of PVA can inhibit the aggregation of graphite sheets, improve the dispersion of RGO in solution, and effectively improve the stability and reproducibility of inkjet printing. After five inkjet prints, the RGO/PVA composite film has an 83% transmittance. Marks et al. applied graphene as a printing source/drain electrode material for IGZO TFTs with mobilities exceeding 6 $cm^2V^{-1}S^{-1}$. The resulting device has good stability to aging in ambient and excellent elasticity against thermal stress [9].

2.3 Organic Semiconductor Materials

As early as the 1970s, Organic semiconductor materials have received widespread attention from scientists, when polyacetylene challenged the notion that polymer materials can only be insulators by doping electricity [10]. Up to now, organic semiconductor materials have been widely used in many fields due to their diverse chemical structures and excellent properties such as light, electricity, and heat. The mesh or linear molecular structure of organic semiconductor materials gives them high tensile and stability. In addition, organic semiconductor materials also have a wide range of sources, good flexibility, and other characteristics, which are key advantages of flexible electronics such as TFTs. With the deepening of research work, a variety of p-type and n-type organic semiconductors have been reported in both small-molecule and polymer.

The small molecule organic semiconductor materials mainly include pentacene, 2,7-Dioctyl[1]benzothieno[3,2-b][1]benzothiophene (C8-BTBT), phosphine, and erythrofluorene. One of the most commonly used materials is pentacene and C8-BTBT. Anthony et al. reported 6,13-bis(triisopropylsilylethynyl)pentacene (TIPS-pen-tacene) substituted with silicon alkynyl at the C6 and C13 positions, greatly increasing its solubility in aromatic solvents [11].

Wang et al. proved that inkjet printing can control the crystal growth of TIPS-pentacene, where the surface-modified Au has a higher dispersion surface energy than the insulating material, and can achieve Au-induced TIPS-pentacene crystal orientation to obtain the best uniformity or higher electrical performance, which is conducive to obtaining high-performance, uniform surface morphology OTFT [12].

Minari et al. proposed a high-resolution printing technique based on π-junction Au nanoparticle inks, which can be used to print narrow Au lines with an interval of only 1 μm at room temperature [13]. In addition, they also printed C8-BTBT with polyxylene as the semiconductor layer material of the OTFT and prepared an OTFT device mobility of up to 3.5 $cm^2V^{-1}s^{-1}$. This room-temperature printing strategy will be promising in the fields of lithography-free, large-area, and high-resolution devices.

Polymer semiconductors are also considered promising candidates for printing OTFT due to their good solubility in inkjet printing solvents as well as their low-temperature processing properties. P3HT is compatible with the printing process and also has self-assembly properties to form polycrystalline structures, making it one of the most popular polymer semiconductors. Lai et al. prepared the semiconductor layer of the OTFT by inkjet printing or spin-coating P3HT, respectively, and comprehensively compared the morphologies of the two films obtained [14]. As shown in Fig. 2(a) and (d), the surface roughness of the spin-coated P3HT film (Ra = 0.9 nm) was smaller than that of the inkjet-printed P3HT film (Ra = 2.16 nm). The results show that the inkjet printed P3HT film is coarse, but the performance is comparable to that of the OTFT device prepared by spin coating, and the mobility is 0. 8 × 10^{-3} $cm^2V^{-1}s^{-1}$ and 1 × 10^{-3} $cm^2V^{-1}s^{-1}$.

Donor–acceptor (D–A) polymer semiconductors as an emerging class of semiconductive polymers have also begun to be applied to the channel active layer of printed electrons. Unlike polythiophene systems, the diversity of donor and acceptor units gives D–A conjugated polymers higher mobility and adjustable HOMO/LUMO.

Yang et al. developed a series of D–A conjugated polymer poly(benzimidazobenzophenanthroline) (BBL) as an n-type organic semiconductor material, after doping polyethyleneimide (PEI), an n-type conductive ink with ethanol as the solvent was prepared [15]. Pure BBL films have a conductivity as low as 10^{-5} S cm^{-1}, while n-type polymer films have a conductivity as high as 8 S cm^{-1}

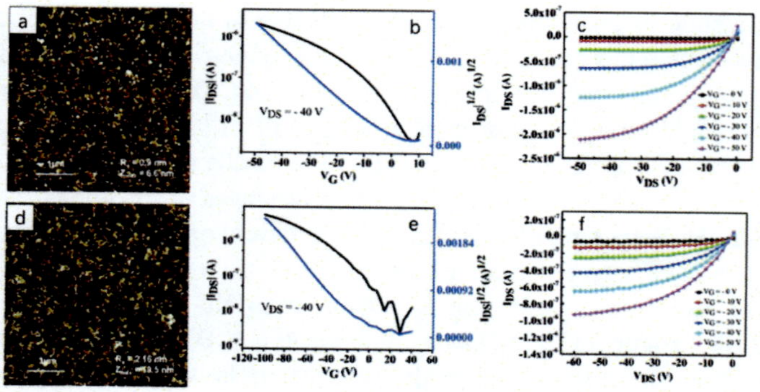

Fig. 2. (a) AFM images of spin-coated P3HT films (b, c) transfer and output characteristics of spin-coated P3HT-based OTFTs (d) AFM images of Inkjet printing P3HT films (e, f) transfer and output characteristics of the inkjet-printed P3HT-based OTFTs [14]

when PEI content reaches 50%. This n-type polymer ink prepares OTFT as an active layer by a simple spraying method with excellent thermal and air stability, providing hope for the realization of large-area, high-throughput electronic devices.

At present, inkjet printing TIPS-pentacene, C8-BTBT, P3HT, and other organic semiconductor materials has achieved excellent results, showing the superiority of the above materials and the possibility of practical application of OTFT in the future. Continuing to improve the performance of organic semiconductor materials and developing new excellent materials remain important challenges for future research.

3 Application of TFTs in Functional Devices

3.1 Flexible Display

With the advent of the intelligent era, people's requirements for displays are getting higher and higher, and at this stage of development, the breakthrough of flexible display technology has become the key to industry innovation. Flexible displays have the advantages of being lightweight, foldable, and easy to carry. The flexible TFT has become a research hotspot as a key device for flexible displays.

In 2017, China's first 2.2-inch high-pixel flexible AMOLED display was jointly developed by the research team of South China University of Technology and New Horizon Optoelectronics Technology Co., Ltd. They further optimized the active layer structure, and successfully produced TFTs with high carrier migration and excellent stability as the drivers of the AMOLED display. The AMOLED display features actively illuminating organic materials, flexible packaging technology, and a flexible substrate with an overall thickness of only 20 μm and bending radii as low as 3 mm [16]. In 2020, Xue et al. used inkjet printing technology to produce a 31-inch 4 K flexible active-matrix light-emitting diode display on a polyimide substrate, which has a pixel density of 144 ppi. In particular, they have prepared TG-TFT with both flexible and electrical stability. Two series-connected TFTs and dual low-voltage power structures form GOA circuits, which improve mechanical stability and successfully reduce the display capacitance load. This AMOLED can operate reliably for 500 h under accelerated test conditions of 60 °C and 90% humidity [17].

3.2 Sensors

In addition to being used in flexible displays, the unique advantages of printed TFTs also make them have great application prospects in sensors. With the rapid development of artificial intelligence and wearable devices, flexible sensors (magnetic sensors, gas sensors, pressure sensors, etc.) have attracted much attention in recent years.

Magnetic sensors are devices that convert changes in the magnetic properties of sensitive components caused by external factors such as magnetic fields, temperature, currents, stress strains, etc. into electrical signals to detect physical quantities such as displacement, angle, speed, and spacing. Kondo et al. developed a magnetic sensor matrix driven by a TFT-based P-type organic bootstrap shift register [18]. Compared with the power dissipation of other existing systems, the system has a much lower dynamic power consumption (−0.23 μW). Eventually, the dynamic power consumption is capable of operating at low supply voltages below 4 V and at high frequencies of about 100 Hz.

Gas sensors detect specific gases to protect people's health and help prevent the occurrence of hazards. OTFT-based gas sensors have powerful controllable chemical sensitivity and signal conversion amplification for high-performance, high-response multi-parameter detection. Wang et al. used the silver electrode prepared by screen printing as the source/drain electrode and prepared a printed SWCNT thin-film transistor on the PET substrate [19]. Figure 3(a) and (b) show the fabrication process and structure of flexible printed SWCNT TFTs. On this basis, a thin-film transistor-type NO_2 gas sensor was prepared (Fig. 3c). The sensor exhibits excellent electrical properties and mechanical flexibility at low leakage currents (~10^{-10} A).

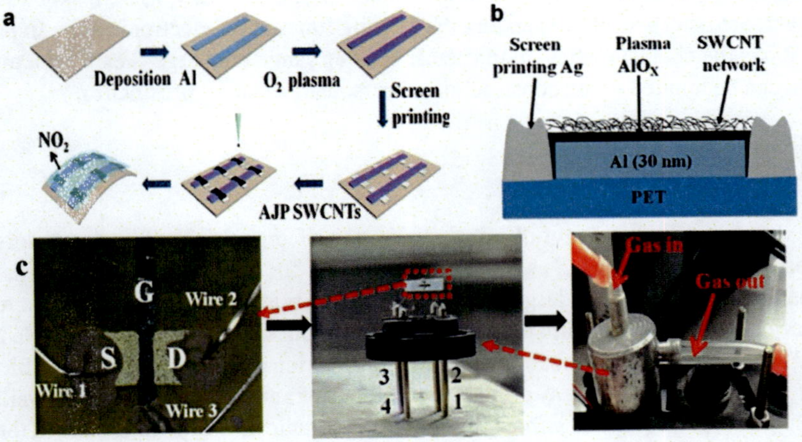

Fig. 3. (a) Diagram of the process of printing SWCNT thin-film transistors on a flexible substrate (b) Schematic of a bottom gate/bottom contact transistor (c) Optical image of the NO_2 sensor [19]

Lee et al. used inkjet printing technology to manufacture thin-film transistor (TFT) arrays with high yield and uniformity and integrated them with high-sensitivity piezoresistive sheets to prepare high-sensitivity, low-power pressure sensors [20]. The device can amplify the pulse signal in real-time, successfully generating a two-dimensional pulsed pressure wave map on the wrist.

3.3 Other Functional Devices

In the era of the Internet of Things (IoT), the development of wearable and implantable electronic products has always received attention from the academic community. Stretchable organic electronic devices have excellent mechanical properties, stable electrical properties, and low cost, easy integration, etc., which will provide more new applications in the field of personalized medical monitoring.

Park et al. have developed a soft, smart contact lens based on a glucose sensor [21]. The main mechanism of the sensor is the selective and sensitive detection of glucose through the redox reaction of glucose oxidase (GOD). The protons (H^+) produced during this process cause the positive charge transfer effect of graphene channels (p-type). Therefore, the carrier density is proportional to the glucose concentration and can be used to detect the glucose concentration. The sensor can detect an average glucose level of more than 0.9 mmol/L in the tears of diabetics. It has a sensitivity of approximately 22.72%/(mmol/L), a signal-to-noise ratio of 23.87, and a minimum detection concentration as low as 12.57 μmol/L (for the case of a signal-to-noise ratio of 3). It is worth noting that the device has negligible changes in response to 30% tensile strain.

The development of multifunctional sensors has a certain reference value for the realization of wearable and multifunctional electronic device applications. Wang et al. recently presented a highly stretchable and conformable matrix network (SCMN) [22]. They implemented three-dimensional (3D) integration of specific expandable sensor units on a structured polyimide network. The device can simultaneously perform multi-stimulus sensing including temperature, in-plane strain, humidity, light, magnetic field, and pressure, successfully expanding the sensing function of electronic skin. In particular, it has the characteristics of adjustable sensing range and large-area expandability, which can be applied to the construction of personalized smart prosthetics.

4 Conclusion

Over the past two decades, full-print preparation of organic thin-film transistors has gained widespread attention as a promising process for manufacturing low-cost, low-temperature processing and large-area electronics. This article examines the most recent advances in printable thin-film transistors, ranging from the properties of printable functional ink materials to applications of TFTs in functional devices such as flexible displays, sensors, and so on. Many printable functional inks are now commercially available, which greatly improves the printing efficiency of organic thin-film transistors in the full printing process and demonstrates the feasibility of industrializing thin-film transistor printing.

However, the realization of the full printing process of organic thin-film transistors still faces many significant challenges. 1) Different functional materials, such as metals, conducting polymers, carbon-based materials, etc., are usually transferred to the substrate by thermal evaporation or chemical gas deposition, which has shortcomings such as a cumbersome process and low material utilization rate. Therefore, more research is required to develop solution-processed materials by introducing functional groups, increasing side chains, and additives for achieving stable and high-resolution TFT printing. 2) Many printing strategies have been developed to fabricate high-performance TFTs, such as screen printing, gravure printing, inkjet printing, etc. However, these processing techniques still face significant challenges in patterning source/drain electrodes in TFTs, preparing high-performance insulating films to reduce gate voltage, and maintaining a low cost. We believe that if these issues can be addressed, printing technology will provide a promising path to low-cost, lightweight, and easily customizable wearable thin-film electronics.

Acknowledgments. This work was supported by Initial funding for the Doctoral Program of BIGC(No.27170121001/002), the general project of fundamental research of BIGC(No.Eb202001), Beijing Nateral Science Foundation(No.2202018), the general project of science and technology of Beijing Municipal Education Commission (No.KM202110015008 and KM202010015004), Talent introduction plan of BIGC(27170122008).

References

1. Mizukami, M., Oku, S., Cho, S. I.: A solution-processed organic thin-film transistor backplane for flexible multiphoton emission organic light-emitting diode displays. IEEE Electr. Dev. Lett. **36**(8), 841–843 (2015)
2. Sekine, T., Sugano, R., Tashiro, T.: Fully printed and flexible ferroelectric capacitors based on a ferroelectric polymer for pressure detection. Jpn. J. Appl. Phys. **55**(10S) (2016)
3. Chung, S., Cho, K., Lee, T.: Recent progress in inkjet-printed thin-film transistors. Adv. Sci. (Weinh) **6**(6), 1801445 (2019)
4. Norita, S., Kumaki, D., Yu, K.: Inkjet-printed copper electrodes using photonic sintering and their application to organic thin-film transistors. Organic Electr. 131–134 (2015)
5. Zhang, Z., Zhang, X., Xin, Z.: Synthesis of monodisperse silver nanoparticles for ink-jet printed flexible electronics. Nanotechnology **22**(42), 425601 (2011)
6. Minari, T., Kanehara, Y., Liu, C.: Room-temperature printing of organic thin-film transistors with π-junction gold nanoparticles. Adv. Func. Mater. **24**(31), 4886–4892 (2014)
7. CHEN, Z H., APPENZELLER, J., LIN, Y M.: An integrated logic circuit assembled on a single carbon nanotube. Science **311**(5768), 1735 (2006)
8. Lu, S., Cardenas, J. A., Worsley, R.: Flexible, print-in-place 1D–2D thin-film transistors using aerosol jet printing. ACS Nano. **13**(10), 11263–11272 (2019)
9. Lim, S., Kang, B., Kwak, D.: Inkjet-printed reduced graphene oxide/poly(vinyl alcohol) composite electrodes for flexible transparent organic field-effect transistors. J. Phys. Chem. C **116**(13), 7520–7525 (2012)
10. Secor, E. B., Smith, J., Marks, T. J.: High-performance inkjet-printed indium-gallium-zinc-oxide transistors enabled by embedded, chemically stable graphene electrodes. ACS Appl. Mater. Interf. 17428–17434 (2016)

11. Shirakawa, H., Louis, E. J., Macdiarmid, A. G.: Synthesis of electrically conducting organic polymers: halogen derivatives of polyacetylene, (CH) X. J. Chem. Soc. Chem. Commun. **1977**(16) (1977)
12. Park, S. K., Jackson, T. N., Anthony, J. E.: High mobility solution processed 6,13-bis(triisopropyl-silylethynyl) pentacene organic thin film transistors. Appl. Phys. Lett. **91**(6), 3514 (2007)
13. Wang, X., Qin, M., Yuan, M.: Au-Induced directional growth of inkjet-printed 6,13-Bis(triisopropylsilylethynyl) pentacene. J. Disp. Technol. **11**(5), 450–455 (2015)
14. Liu, X., Kanehara, M., Liu, C.: Ultra-high-resolution printing of flexible organic thin-film transistors. J. Inf. Disp. **18**(2), 93–99 (2017)
15. Lin, Y., Liu, C.-F., Song, Y.-J.: Improved performances of inkjet-printed poly (3-hexylthiophene) organic thin-film transistors by inserting an ionic self-assembled monolayer. RSC Adv. **6**(47), 40970–40974 (2016)
16. Yang, C. Y., Stoeckel, M. A., Ruoko, T. P.: A high-conductivity n-type polymeric ink for printed electronics. Nat. Commun. **12**(1), 2354 (2021)
17. Xu, Z.-ping.: Study on flexible polyimide substrate for active matrix organic
18. Light emitting diodes display. Guangzhou: South China University of Technology (2017)
19. Xue, Y., Wang, L., Zhang, Y.: 31-Inch 4K flexible display employing gate driver with metal oxide thin-film transistors. IEEE Electr. Device Lett. **42**(2), 188–191 (2021)
20. Kondo, M., Melzer, M., Karnaushenko, D.: Imperceptible magnetic sensor matrix system integrated with organic driver and amplifier circuits. Sci. Adv. **6**(4), eaay6094 (2020)
21. Wang, X., Wei, M., Li, X.: Large-area flexible printed thin-film transistors with semiconducting single-walled carbon nanotubes for NO_2 sensors. ACS Appl. Mater. Interf. **12**(46), 51797–51807 (2020)
22. Baek, S., Lee, Y., Baek, J.: Spatiotemporal measurement of arterial pulse waves enabled by wearable active-matrix pressure sensor arrays. ACS Nano **16**(1), 368–377 (2022)
23. Park, J., Kim, J., Kim, S-Y.: Soft, smart contact lenses with integrations of wireless circuits, glucose sensors, and displays. Sci. Adv. **4**(1), eaap9841 (2018)
24. Hua, Q., Sun, J., Liu, H.: Skin-inspired highly stretchable and conformable matrix networks for multifunctional sensing. Nat. Commun. **9**(1), 244 (2018)

Preparation and Properties of Self-healing Microcapsules Coatings

Chenyang Liu, Zhicheng Sun(✉), Yibin Liu, Zhenzhen Li, Gongming Li, and Zhitong Yang

Beijing Engineering Research Center of Printed Electronics, School of Printing and Packaging Engineering, Beijing Institute of Graphic Communication, Beijing 102600, China
sunzhicheng@bigc.edu.cn

Abstract. By incorporating the repairing agent using microencapsulation technology and preparing intelligent, responsive composite coatingsafter mixing, dispersing, and plastic coating with the layers, an externally aided self-healing microcapsule material is created. The microcapsules embedded in the coatings were ruptured when exposed to external forces such as chemical corrosion, outer mechanical strength, other external factors, structural damage, and microcracks caused by stress. The repairing agent was released to undergo a cross-linking reaction with the coating's curing agent for automatic repairing, prolonging thecoating's service life. This study used in-situ polymerization to create self-healing microcapsules using organic repairing chemicals as the core and different polymer resin materials as the wall. Scanning electronmicroscopy (SEM), Fourier transforms infrared spectroscopy (FTIR), and other techniques were used to study self-healing microcapsules' structure, shape, and characteristics. In addition, the repairability of the microcapsules was characterized by testing the change in tensile strength of the coating before and after repair.

Keywords: Melamine-Formaldehyde · Microcapsule · Self-healing · Smart Response coating

1 Introduction

The self-healing phenomenon is a kind of life phenomenon that exists widely in nature [1]. When external factors damage an organism in nature, its internal self-healing system will respond quickly and automatically repair the damaged part. As early as the end of the last century, this phenomenon has attracted many scholars' attention and proposed introducing self-healing systems into material science to obtain an intelligent material that can automatically repair damage in response to the external environment [2]. After nearly 40 years of continuous development, many achievements have been made in self-healing materials. At present, self-healing materials are mainly divided into intrinsic types and external types Intrinsic self-healing materials use reversible chemical reactions in the material matrix [3], including Diels – Alder reactions, dynamic covalent chemistry, desulphated bond reactions, and supramolecular structures containing hydrogen bonds,

M. Xu et al. (Eds.): CACPP 2022, LNEE 991, pp. 587–591, 2023.
https://doi.org/10.1007/978-981-19-9024-3_75

ionic polymerization, etc. The advantage of this system was that the healing response could be cycled many times [4], and the material properties change a little before and after healing. However, the structural strength of such materials was generally low [5]. The repair process requires a specific temperature or other external conditions. The repair time, the complex preparation process, and the high cost were extended. The external aid type introduces a structural material loaded with repairing agents in the material, such as microcapsules, hollow fibers, etc. When an external force cracks the matrix, the repair agent carrier packed in the material will break simultaneously. The repair agent is released from the capsule and fills the crack by a combination of its mobility and capillary action at the site of the damage. The reaction with the curing agent achieves the effect of repairing the crack.

In this study, we present an innovative introduction of a silicone-based microencapsulated self-healing system into the field of waterproof coatings. We embedded hydroxyl-capped silicone oil as a repairing agent in melamine-formaldehyde resin to obtain microcapsules. The microcapsule was characterized by its structure to prove that it has a stable shell-core system. The coating can be repaired by reacting with the restorer under the initiation of water when the layer is damaged.

2 Experiment

2.1 Materials

Formaldehyde Solution (37%) (stabilized with Methanol) was purchased from TCI. Melamine (99%), Resorcinol (99%), and Citric acid (99%) were purchased from Innochem Co., Ltd (Beijing, China). Sodium dodecyl sulfate, Triethanolamine (TEA), and methyltris(methylethylketoximino)silane (MOS) were purchased from Aladdin Chemical Reagent Company (Shanghai, China). Hydroxy-terminated Hydroxy terminated silicone fluid was obtained from Shenzhen Jipeng Silicon Fluorine Materials Co., Ltd. YH200 was purchased from Oriental Yuhong Civil Building Materials Co., Ltd.

2.2 Preparation of Self-healing Microcapsules

Mix 0.5 g of melamine and 1 g of formaldehyde solution, add a certain amount of deionized water to dilute, adjust the pH of the solution to 8.5 with triethanolamine, and then take out and cool after magnetic stirring in a water bath for 30 min to obtain a transparent melamine-formaldehyde resin prepolymer solution. Take 1.74 g of sodium dodecyl sulfate and 10.44 g of deionized water in a beaker, stir and dissolve fully, take 1.74 g of Hydroxy terminated silicone fluid dropwise into the emulsifier solution, and emulsify for 1 h at 500 rpm with a water bath magnetic stirrer. The prepolymer solution was dropped into the emulsion. After stirring for 10 min, add 0.1 g resorcinol, adjust the pH to 4.5, raise the temperature to 60 °C and react for 2 h. After the reaction is completed, filter, rinse with water/petroleum ether several times, and dry at 40 °C for 24 h to obtain microcapsules.

2.3 Preparation of Coatings Containing Self-healing Microcapsules

Add 10 wt.% of self-healing microcapsules, and 5 wt.% of methyltris(methyl ethylketoxime)silane (MOS) to the waterproof coating YH200 and mix well. Remove air bubbles by ultrasonic vibration for 10 min and then apply evenly to the surface of the glass sheet. Cure in an oven at 40° for 24 h. The repair is completed by scratching artificial scars on the coated surface with a knife and baking at 40 °C for 24 h.

2.4 Characterization and Measurements

The chemical composition of the microcapsules was analyzed using an FTIR (Nicolet IS10). The range of scanning wavelength is from 4000 - 400 nm^{-1}. The surface morphology of self-healing microcapsules and the surface of coatings were recorded by SEM (Hitachi SU8020). The accelerating voltage of SEM was 15.0 kV. The particle size distribution of the microcapsules was analyzed using a laser particle sizer (Mastersizer 2000, Malvern, UK).

3 Results and Discussion

3.1 Morphology and Structure of Self-healing Microcapsules

Figure 1 Shows the repair agent's SEM image of the melamine-formaldehyde resin shell self-healing microcapsules encapsulated with Hydroxy terminated silicone fluid. It can be seen that the microcapsules have a stable spherical structure. The melamine-formaldehyde resin forms a dense shell material on the surface of the core material droplet, which can effectively prevent the core material's leakage and ensure the microcapsule structure's integrity during coating processing. The surface was rough due to the presence of more melamine-formaldehyde resin particles. This rough structure increases the contact area between the microcapsules and the coating and improves the contact force between the microcapsules and the substrate, resulting in better compatibility of the microcapsules with the substrate. The resulting microcapsule particle size is mainly distributed in the range of 5 – 60 μm., with an average particle size of about 32.7 μm.

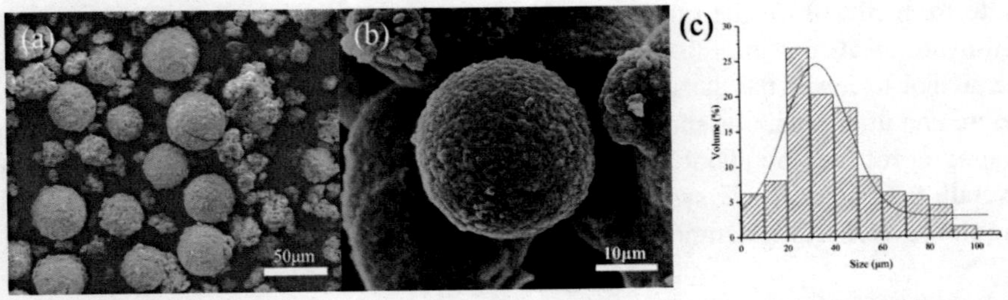

Fig. 1. (a, b) SEM images of self-healing microcapsules, (c) particle size distribution of self-healing microcapsules

Figure 2(a) shows the FTIR spectrum of the self-healing microcapsules; the curve shows a broad absorption peak at 3424 cm^{-1}, which is mainly attributed to the stretching vibrations of the O–H in the silanol band of the silicone oil at the hydroxyl capping end and the N–H group present in the melamine-formaldehyde resin. The distinct absorption peaks at 2924 and 2850 cm^{-1} are probably due to the stretching vibrations of the sp^3 heterodimer -CH_3, while 1343 cm^{-1} is due to the bending vibrations of CH_3. The regions at 1556 and 1488 cm^{-1} are due to coupled vibrations between $N - H_2$ shearing and $C = N$ stretching, demonstrating the presence of triazine rings in the structure. The absorption peak at 1194 cm^{-1} can be attributed to the antisymmetric and symmetric stretching vibration of $Si - O - Si$. The sharp absorption peak at 811 cm^{-1} may be due to the plane rocking oscillation of the base $- CH_3$ in $Si(CH_3)_2$. These peaks in the FTIR spectrum confirmed that the melamine-formaldehyde resin shell in the microcapsules successfully embedded the core of Hydroxy terminated silicone fluid.

Fig. 2. (a) FTIR images of self-healing microcapsules (b) load-displacement curves of self-healing coatings

3.2 Self-healing Mechanism of Coatings

Due to the self-healing microcapsules embedded in the coating, as well as the initiator MOS, when the coating is damaged and cracks appear, MOS reacts with moisture in the air to form silanol. At the same time, the microcapsules rupture and the repair agent, end-hydroxy silicone oil, fills the cracks with its fluidity and then cross-condenses with the silanol to repair the damage. By measuring the stress-strain curve of the coating before and after repair, as shown in Fig. 2 (b). After repair, the tensile strength of the coating is restored by about 70%, but the elongation length will be further reduced. Overall, the microcapsule system has good compatibility with the waterproof coating and shows excellent restoration results.

4 Conclusion

In this study, we embedded hydroxyl-capped silicone oil as a restorative agent into melamine-formaldehyde resin to obtain microcapsules with a regular spherical structure and an average particle size of about 32.7 μm. It has a very stable shell-core system and the rough microcapsule surface improves the compatibility of the microcapsules with the coating. Methyltris(methylethylketoximino)silane (MOS) can be stably present in the coating as a curing agent. The results show that the system exhibits a significant self-healing effect in epoxy-based coatings.

Acknowledgments. This work was supported by the National Natural Science Foundation of China (22278037).

References

1. White, S.R.: Autonomic healing of polymer composites. Nature **409**(6822), 794–797 (2001)
2. Kritika, C.: Self-healing nanofibers for engineering applications. Ind. Eng. Chem. Res. **61**(11), 3789–3816 (2022)
3. Wu, H.: Supramolecular engineering of nacre-inspired bio-based nanocomposite coatings with exceptional ductility and high-efficient self-repair ability. Chem. Eng. J. **437**(2), 135405 (2022)
4. Zhao, W.: High-strength, fast self-healing, aging-insensitive elastomers with shape memory effect. ACS Appl. Mater. Interfaces. **12**(31), 35445–35452 (2020)
5. Behera, P.K.: Self-healing elastomers based on conjugated diolefins: a review. Polym. Chem. **12**(11), 1598–1621 (2021)

Preparation and Application of Reversible Thermochromic Microcapsules

Gongming Li, Zhicheng Sun(✉), Chenyang Liu, Yibin Liu, Zhitong Yang, and Zhenzhen Li

Beijing Engineering Research Center of Printed Electronics, School of Printing and Packaging Engineering, Beijing Institute of Graphic Communication, Beijing 102600, China
sunzhicheng@bigc.edu.cn

Abstract. In this study, reversible thermochromic microcapsules with ternary complex as core material and polymer resin as wall material were synthesized by in-situ polymerization. The structure, morphology and properties of the reversible thermochromic microcapsules were characterized by differential scanning calorimeter (DSC), and scanning electron microscope (SEM). After mixing it with linseed oil and alkyd resin according to a certain proportion, a thermochromic ink with the dual response of temperature change and phase change can be obtained. When the external temperature changes, the core material structure of microcapsules will change, through solid-liquid transformation, so that the organic discoloration material in the core material will change color by electronic transition, to play a warning, decoration effect. X-rite spectrophotometer was used to detect the color change of the sample when the temperature changed, and quantitatively evaluate the discoloration effect of the microcapsule thermochromic system. Through the above analysis, it can be concluded that thermochromic ink has potential high-tech applications in smart fibers and textiles, wearable electronic products, energy-saving buildings, temperature-sensitive medical systems, safety clothing, aerospace engineering, and other fields in the future.

Keywords: Reversible organic discoloration · Thermochromism · Microcapsules · Phase change · Ink

1 Introduction

Microencapsulation of phase change materials (PCMs) is regarded as a promising technology to effectively solve the leakage problem of PCMs during phase change. This technology not only provides sufficient protection for encapsulated PCMs from surrounding materials and the environment but also provides them with a larger heat transfer area for improved thermal response and thermal energy storage efficiency [1]. Thermochromic materials are a class of temperature-sensitive materials containing color-changing compounds and other auxiliary components. Since the absorption spectrum of the color-changing substances changes during heating and cooling, its color will change with temperature changes within a certain range. For example, the phase change microcapsule system along with the reversible thermochromic indicator can display the phase

M. Xu et al. (Eds.): CACPP 2022, LNEE 991, pp. 592–596, 2023.
https://doi.org/10.1007/978-981-19-9024-3_76

change state of the PCM core material in the microcapsule through a color change, so this combination not only effectively improves the latent heat storage efficiency and thermal management, but also improve the thermal energy storage efficiency. The introduction of thermochromic materials can greatly enrich the functions and intelligence of materials. Reversible thermochromic material phase change microcapsules (RT-MPCMs) are a kind of smart materials with latent heat storage and thermal management properties obtained by microcapsule encapsulation technology based on a thermochromic system with thermochromic material as core material [2].

In this paper, using crystal violet lactone (CVL), bisphenol A (BPA), 1-hexadecanol as ternary compound core material, and urea-formaldehyde resin as wall material, the reversible thermochromic phase transition was prepared by the in-situ polymerization method. Energy storage material microcapsules. After characterizing and analyzing the properties of the microcapsules, the microcapsule powder was configured into a thermochromic ink, which was printed by a screen-printing process, and the printed samples were detected and analyzed. Thanks to its reliable and real-time thermochromic indication, this not only enables thermal energy storage and management for temperature display but also acts as a reminder and warning. It has potential high-tech applications in smart fibers and textiles, wearable electronics, energy-efficient buildings, temperature-sensitive medical systems, safety clothing, aerospace engineering, and more.

2 Experiments and Method

2.1 Experimental Materials

CVL (Sigma-Aldrich), BPA (> 99%; Sigma-Aldrich), 1-hexadecanol, gelatin, gum Arabic, AR; Acetic Acid, AR; Formaldehyde, AR; Urea, AR; n-hexane, sodium hydroxide, sodium chloride, hydrochloric acid, AR; The mineral oil; Alkyd.

2.2 Preparation of Thermochromic Microcapsules and Ink

Weigh a certain proportion of the core material and the emulsifier to prepare an emulsifier solution with a mass fraction of 5%, mix and stir evenly, and then emulsify it with a high-speed shear emulsifier for 10 min under the condition of 65 °C water bath. The prepolymer solution is slowly dripped into the core dispersion with a constant pressure drop funnel. After the prepolymer solution is added dropwise, then slowly add acetic acid to make the solution PH = 2.5. Stir at a uniform speed for 1 h after increasing the temperature of the water bath pot to 65 °C reactions for 1 h after taking it out, then filtering and washing twice with distilled water. The resulting sample is dried for 24 h to obtain a microcapsule product.

Inks were prepared concerning the production ratios in Table 1. The preparation of ink begins with the preparation of the binder, heating the linseed oil to about 125 °C, and slowly adding the alkyd resin. The microcapsule powder containing the thermochromic material is added at room temperature. After stirring evenly, the thermochromic ink can be obtained (see in Fig. 1e).

Table 1. Ink formulation

Ingredients	Quantity (%)
Linseed oil	40
Alkyd resin	25
Thermochromic microcapsules	30
Auxiliary	5

2.3 Thermochromic Ink Analysis and Testing

Use the X-rite color difference meter to test the CIELab value of the ink before and after discoloration, and use the following formula to calculate the color difference value ΔE of the ink.

$$\Delta E = \sqrt{(\Delta L^*)^2 + (\Delta a^*)^2 + (\Delta b^*)^2} \quad (1)$$

3 Results and Discussion

3.1 Thermal Properties of Thermochromic Microcapsules

Table 2. Thermal properties of microcapsules

Category	ΔH_M(J/g)	ΔH_C(J/g)	W_a%	E%	C%
1-hexadecanol	247.26	242.10	-	-	-
Ternary compound	218.45	213.68	88.3	88.2	99.8
Thermochromic microcapsules	135.68	132.96	62.2	63	99.7

ΔH_M: Enthalpy on DSC heating curve; ΔH_C: Enthalpy on DSC cooling curve; Wa: Actual core content; E: Energy storage Efficiency; C: Thermal storage capability.

Melting enthalpy (ΔH_M) and crystallization enthalpy (ΔH_C) are important parameters to evaluate the heat storage capacity of phase change materials during phase transition. As Table 2 shows, the hexadecanol was determined to have a rather high phase transition enthalpy. It is shown that the 1-hexadecanol and core compound can store high latent heat. The core material content of the microcapsules is 62.2%, and the energy storage efficiency value of the microcapsules is similar to the actual core material content value, indicating that the microcapsules can release almost all the potential heat energy during the phase transition process, and the heat storage capacity of all samples is high at 99%, this means that most of the core material has been well encapsulated, and most of the encapsulated compound core material can efficiently store and release latent heat through a phase transition, a property that contributes to reversible thermally induced discoloration Microcapsules have broad application prospects in energy storage and temperature regulation.

3.2 Microcapsule Testing and Characterization

The size and distribution of thermochromic microcapsules can be seen by scanning electron microscopy. It can be seen from Fig. 1a-d that the appearance structure of the microcapsules is regular spherical, the particle size is almost the same, the size is 50 microns, and the microcapsules are uniform and dispersive, and there is no obvious adhesion and damage. The surface of the microcapsule is slightly rough but there is no crack, and the density is good. The core material is completely covered by the wall material, and there is no core material attached to the surface of the microcapsule shell.

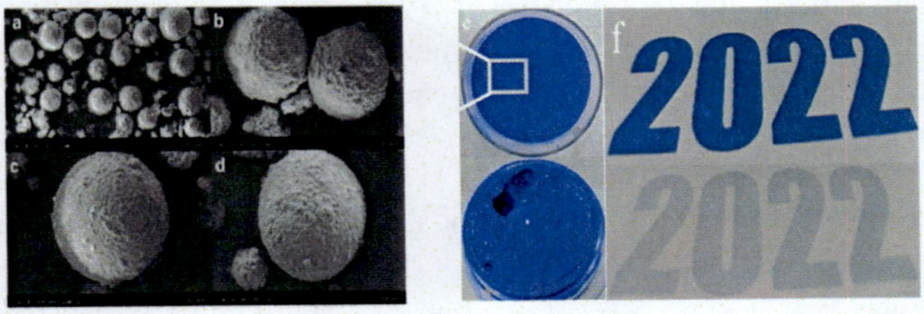

Fig. 1. (a-d) SEM image of thermochromic microcapsules; e) thermochromic microcapsules and Inks; f) color change of thermochromic ink before and after temperature change.

3.3 Thermochromic Ink Analysis and Testing

Table 3. Color variation of prints

Category	Before (15 °C)	After (55 °C)	Difference value Δ	ΔE
L	38	87	49	92
A	43	1	42	
B	−86	3	89	

The thermochromic ink is printed on the surface of the paper, the pictures before and after the sample change color are shown in Fig. 1f. The Table 3 shows the CIE Lab value of the obtained printed matter. According to the Lab value, the color difference ΔE between normal temperature and heating state is calculated to indicate the color change ability of the ink. After the thermochromic microcapsules are configured into ink, the color-changing ink prints change color quickly, accurately, and simply in the test, the color difference before and after the color change is large, and the ΔE value reaches 92, which has a high degree of recognition. The ink can be used in the following aspects potential high-tech applications in smart fibers and textiles, wearable electronics, energy-efficient buildings, temperature-sensitive medical systems, safety clothing, and aerospace engineering without any external auxiliary equipment.

4 Conclusion

In this paper, we successfully synthesized reversible thermochromic microcapsules with phase change characteristics by the in-situ polymerization process. SEM morphology analysis revealed that the capsules were monodisperse and homogeneous and their size was about 50 μm. DSC showed that reversible thermochromic microcapsules have broad application prospects in energy storage and temperature regulation. Thermochromic inks were produced from the obtained thermochromic microcapsules and linseed oil, alkyd resin as a reference, and successfully printed by screen printing. The color difference value ΔE value before and after the color change of the ink was obtained in the test of the printed samples and reached 92, which has a very good value. Reversible thermochromic microcapsules by this study exhibit a great potential for hi-tech applications in smart fibers and textiles, wearable electric devices, energy-saving buildings, and temperature-sensitive medical.

Acknowledgments. This work was supported by the National Natural Science Foundation of China (22278037).

References

1. Pielichowska, K.: Phase change materials for thermal energy storage. Prog. Mater. Sci. **65**, 67–123 (2014)
2. Cheng, Y.L., Zhang, X.Q., Fang, C.Q., Chen, J., Wang, Z.: Discoloration mechanism, structures and recent applications of thermochromic materials via different methods: a review. J. Mater. Sci. Technol. **34**, 2225–2234 (2018)

Study on Hyperelasticity of Photosensitive Resin Plate

Xin Wang[1,2,3,4], Yingcai Yuan[1,2,3,4](✉), Zhenyu Fan[1,2,3,4], Junwei Qiao[5], Xuan Wang[1], and Chen Zhang[1]

[1] School of Mechanical and Electrical Engineering, Beijing Institute of Graphic Communication, Beijing, China
yuan_yc@bigc.edu.cn

[2] Beijing Key Laboratory of Digitalized Printing Equipments, Beijing Institute of Graphic Communication, Beijing, China

[3] Printing Equipment of Beijing Institute of Printing Engineering Research Center of Beijing Higher Education, Beijing Institute of Graphic Communication, Beijing, China

[4] Beijing Printing Institute Beijing Printing Electronic Engineering Technology R & D Center, Beijing Institute of Graphic Communication, Beijing, China

[5] Shanghai Publishing and Printing College, Shanghai 200093, China

Abstract. The work aims to study the hyperelasticity of photosensitive resin plates and analyze their mechanical properties. The compression experiments were carried out on different photosensitive resin plates, and the stress-strain data were obtained, which were fitted by Mooney-Rivlin constitutive model. The fitting curve, material parameters C_{10}, C_{01} and fitting curve residual value of Mooney-Rivlin constitutive model was obtained. Conclusion Mooney-Rivlin constitutive model can well fit the material's hyperelastic relationship, and the plate's deformation degree is directly proportional to the hardness, which provides a theoretical basis for the study of the deformation of the flexographic plate.

Keywords: Flexible resin plate · Compression experimental · Mooney-Rivlin constitutive model · Curve fitting

1 Introduction

Plate deformation is an important factor affecting the quality of flexographic printing. The commonly used solid photosensitive resin flexographic plate is a solid polyester composite material composed of protective film, photosensitive resin layer and supporting film. Composites have hyperelastic properties [1]. Plate characteristics are essential factors affecting printing quality. Based on the theory of hyperelasticity and the compression experiment of flexographic plate, the paper analyzes the hyperelasticity of various flexographic plates, studies the influence of hardness and model parameters on the mechanical properties of flexographic plates, summarizes the laws of the hyperelasticity of flexographic plates, and lays a foundation for more accurate analysis of the deformation of flexographic plates in the printing process.

M. Xu et al. (Eds.): CACPP 2022, LNEE 991, pp. 597–601, 2023.
https://doi.org/10.1007/978-981-19-9024-3_77

2 Research Status of Hyperelasticity Theory

For hyperelastic materials, stress may not be obtained only by strain. Many scholars at home and abroad have studied different Hyperelastic Materials and proposed constitutive models of various elastic potential energy functions.

C. Liu et al. Fitted the uniaxial tensile test data of two rubber materials with three different constitutive models, and obtained the optimal fitting model of the material [2]. Hassan Mansour Raheemaet et al. studied the hyperelastic constitutive model of hydrogel, and determined the best model for Ogden ($N = 1$) model for hydrogel mechanical behavior [3]. Huyi Wang et al. compounded the foam silicone rubber uniaxial compression and fitted the experimental data. When the nominal strain is less than 60%, it has an excellent fitting effect [4]. Hui Guo, et al. have introduced the constitutive model of foam rubber. The accuracy of the constitutive model is verified by experiments [5].

3 Mooney-Rivlin Hyperelastic Constitutive Model

3.1 Strain Energy Density Function of Hyperelastic Materials

The strain state equation of elastic material is:

$$\left(\lambda^2\right)^3 - I_1\left(\lambda^2\right)^2 + I_2\left(\lambda^2\right) - I_3 = 0 \tag{1}$$

I_1, I_2 and I_3 are the three invariants of strain tensor.

$$I_1 = \lambda_1^2 + \lambda_2^2 + \lambda_3^2, I_2 = \lambda_1^2\lambda_2^2 + \lambda_2^2\lambda_3^2 + \lambda_3^2\lambda_1^2, I_3 = \lambda_1^2\lambda_2^2\lambda_3^2 \tag{2}$$

λ_1, λ_2, λ_3 are the elongation ratio in each principal direction.

Since the resin material is incompressible, then:

$$I_3 = \lambda_1^2\lambda_2^2\lambda_3^2 \equiv 1 \tag{3}$$

For uniaxial compression:

$$\lambda_2 = \lambda_3 = \frac{1}{\sqrt{\lambda_1}} \tag{4}$$

Order $\lambda = \lambda_1$, the strain tensor invariants of uniaxial compression are:

$$I_1 = \lambda^2 + \frac{2}{\lambda}, I_2 = 2\lambda + \frac{1}{\lambda^2}, I_3 \equiv 1 \tag{5}$$

According to the series form of strain tensor invariant derived by Rivlin, the Mooney-Rivlin model of resin material can be expressed as [6]:

$$W = C_{10}(I_1 - 3) + C_{01}(I_2 - 3) \tag{6}$$

C_{10} and C_{01} are the constitutive model parameters of the material.

$$C_{10} = \frac{\partial W}{\partial I_1}, C_{01} = \frac{\partial W}{\partial I_2} \tag{7}$$

3.2 Stress-Strain Relationship of Resin Materials

The stress expression of uniaxial compression is:

$$\sigma = \frac{\partial W}{\partial \lambda} = \frac{\partial W}{\partial I_1}\frac{dI_1}{d\lambda} + \frac{\partial W}{\partial I_2}\frac{dI_2}{d\lambda} = 2\frac{\partial W}{\partial I_1}\left(\lambda - \frac{1}{\lambda^2}\right) + 2\frac{\partial W}{\partial I_2}\left(1 - \frac{1}{\lambda^3}\right) \quad (8)$$

λ can be used the principal stress in compression direction ε expressed as:

$$\lambda = 1 + \varepsilon \quad (9)$$

Bring Eqs. (7), (9) into Eq. (8):

$$\sigma = 2\left[C_{10}\left(1 + \varepsilon - \frac{1}{(1+\varepsilon)^2}\right) + C_{01}\left(1 - \frac{1}{(1+\varepsilon)^3}\right)\right] \quad (10)$$

4 Experimental Process and Data Analysis

Compression tests were carried out on six different types of flexible resin plates of Flint, and the plate parameters are shown in Table 1. The compression speed is 0.2 *mm/min*, each compression is 0.08 *mm*, and the compression range is 0–0.8 *mm*. The average value is taken for three measurements and the curve is drawn as shown in Fig. 1. The relationship between pressure and compression is nonlinear and proportional; The pressure compression curves of different types of plates do not intersect each other; Under the same force, the scale with high hardness has large deformation.

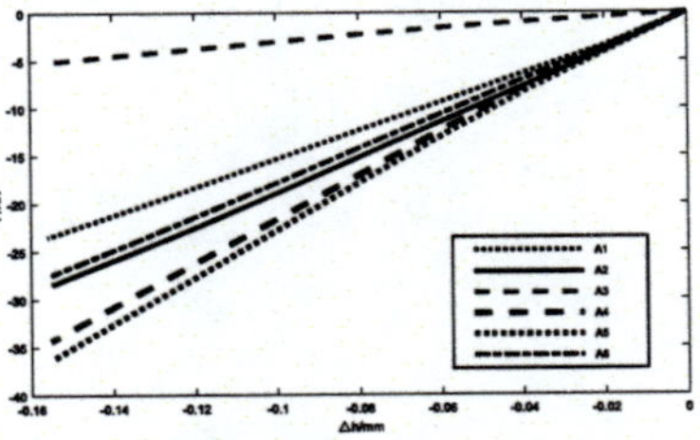

Fig. 1. Experimental data and measurement results

Since the resin material is incompressible, the stress and strain obtained by keeping its volume unchanged before and after compression is [7]:

$$\sigma = \frac{F}{A_1} = F\frac{(h + \Delta h)}{A_0 h} = \frac{F}{A_0}(1 + \varepsilon) \quad (11)$$

$$\varepsilon = \frac{\Delta h}{h} \quad (12)$$

The Mooney-Rivlin hyperelastic constitutive model is used to fit the data.

is parameters determined. The following conclusions can be drawn according to the experimental data and fitting curve.

1) The stress-strain curves of several types of plates used in the experiment are nonlinear. Therefore, it can be seen that the photosensitive resin flexible plate belongs to hyperelastic material.
2) According to the fitting results, it can be seen that within a specific range, the Mooney-Rivlin model has an excellent fitting effect on the flexographic plate, which can be used to predict the mechanical behavior of solid photosensitive resin flexographic plate.
3) Under the same pressure, the plate with high hardness has large deformation; When the strain is slight, the constitutive model parameters $C_{10} + C_{01}$ can reflect the trend of plate deformation. In the printing process, the plates with the appropriate constitutive model parameters can be selected according to the total pressure of the press.

Acknowledgements. Bidding project for the Key Laboratory of "intelligent and flexographic printing" of the State Press and Publication Administration (NO. ZBKT202103 and NO. ZBKT202006). "Research on flexographic printing ink transfer and innovative plate making technology" supported by the excellent Key Laboratory of the State Press and Publication Administration in 2019 (Z6E-0404-20-01-01y). National College Students' innovation and Entrepreneurship Program in 2021.

References

1. Lining, Y.: Flexible composites and their applications. Mech. Prog. **23**(3), 386–397 (1993)
2. Liu, C., Cady, C.M., Lovato, M.L., Orler, E.B.: Uniaxial tension of thin rubber liner sheets and hyperelastic model investigation. J. Mater. Sci. **50**(3), 1401–1411 (2014). https://doi.org/10.1007/s10853-014-8700-7
3. Hassan Mansour, R., Al-Mukhtar, A.M.: Experimental and analytical study of the hyperelastic behavior of the hydrogel under unconfined compression. Procedia Struct. Integr. **25**, 3–7 (2020)
4. Wang, H., Hu, W., Zhao, F.: Numerical simulation of quasi-static compression on a complex rubber foam. Acta Mech. Solida Sin. **30**(3), 285–290 (2017). https://doi.org/10.1016/j.camss.2017.03.009
5. Hui, G., Wenjun, H., Junlin, T.: Hyperelastic constitutive model of foam rubber material. J. Comput. Mech. **30**(4), 575–579 (2013)
6. Mingjun, Z., Wenjing, W., Zhengnan, C.: Determination of mechanical property constants of rubber Mooney Rivlin model. Rubber Ind. **50**(8), 462–465 (2003)
7. Xiaomin, C.: Study on data processing method of rubber uniaxial tensile test. World Rubber Ind. **44**(10), 34–38 (2017)

Study on the Influence of the Ratio of Solvent-Free Composite A and B on the Properties of Solvent-Free Composite Products

Hongwei Xu[1](✉), Wenbin Ye[1], Zhaohua Ma[1], Xiao Xu[1], Zhicheng Xue[2], and Darun Xi[2]

[1] Xi'an University of Technology, Xi'an, China
xuhongwei@xaut.edu.cn
[2] Shaanxi Beiren Printing Machinery Co., Ltd., Weinan, China

Abstract. The solvent-free composite products have some problems, such as low peel resistance and long curing time, which affect the development of solvent-free composite technology. Based on the analysis of the function of solvent-free compound mixer, this paper carried out an experimental study on the influence of the ratio of solvent-free compound A and B on the performance of solvent-free compound mixer. Through the experiment of the influence of the ratio of solvent-free compound A and B on the viscosity, curing speed and peel resistance of the solvent-free compound, the influence of the ratio of solvent-free compound A and B on the relevant properties of the solvent-free compound was revealed, and the importance of the ratio of solvent-free compound A and B in practical application was pointed out. It is concluded that the solvent-free compound A and B can show better performance in all aspects when the ratio is 100:70.

Keywords: Solvent-free composite · Ratio · Experiment · Peel resistance

1 Introduction

At present, solvent-free compounding has been widely used in the production of printing and packaging. However, due to the low peel resistance and long curing time of solvent-free compounding products, it cannot completely replace the dry compounding process. As we all know, solvent-free compounding has no VOCs release compared with dry compounding, and no oven drying is required. It is an environmental protection and energy saving trend in the future [1].

Solvent-free compounding equipment is mainly divided into solvent-free compounding machine and solvent-free compounding mixer. The function of solvent-free compounding mixer is to mix solvent-free compounding materials A and B in proportion, and inject the mixed glue into the glue tank of solvent-free compounding machine for compounding. The proportioning system is the key component of the solvent-free compound mixer, and the gear pump is its core [2]. The two gear pumps for conveying

M. Xu et al. (Eds.): CACPP 2022, LNEE 991, pp. 602–607, 2023.
https://doi.org/10.1007/978-981-19-9024-3_78

solvent-free compound A and B materials through the main pipe have different speeds to achieve their different proportions [3]. It can be seen that the ratio is a key factor affecting the function of solvent-free compound. In this paper, the effect of proportion on the function of solvent-free compound was analyzed.

On the basis of the previous research on solvent-free compounding, this paper makes an experimental analysis on the relevant performance parameters of solvent-free compounding compound such as viscosity, curing time, peel resistance, folding resistance, etc. according to the different proportions of solvent-free compounding materials A and B. The solvent-free composite materials used are solvent-free composite materials A and B produced by Shanghai Kangda Chemical Co., Ltd. and their performance parameters are shown in Table 1 [4].

Table 1. Performance parameters of solvent-free composite materials A and B

Item	Material A	Material B
Viscosity 23° (mPa·s)	1000–1200	600–700
Density (Kg/m^3)	1120	980

2 Effect of Solvent-Free Compound Mixing Ratio on Viscosity of Solvent-Free Compound

Viscosity is a measure of the viscosity of a liquid and a reflection of the flow force of a fluid on its internal friction phenomenon [5]. The viscosity of solvent-free compound has a certain influence on the coating uniformity during solvent-free compounding, so the viscosity of solvent-free compound is a parameter to characterize the performance of solvent-free compound.

The viscometer used in the experiment is SNB series viscometer, which is powered by motor and drives the rotor to rotate through hairspring and rotating shaft. When testing the sample, the rotor will be subject to the resistance brought to the rotor by the viscosity of the sample. At this time, the hairspring will generate torque. When the torque is balanced with the resistance brought by the viscosity of the sample, the liquid viscosity value (mPa•s) will be displayed on the night crystal screen with night vision function through the output signal of the photoelectric sensor. Since the ratio of solvent-free composite A and B produced by Shanghai Kangda company is required to be 100:70, the ratios used in this experiment are 100:60,100:65, 100:70, 100:75 and 100:80 respectively, which are weighed by electronic balance and fully mixed by stirring. After mixing, measure with a viscometer to obtain the corresponding viscosity, as shown in Table 2.

Table 2 shows that when the ratio of solvent-free compound A and B is 100:70, the maximum viscosity of the mixed glue solution is 1020 mPa•s. It is also proved that when the ratio of solvent-free compound A and B is 100:70, the effect of gluing and coating is also the best.

Table 2. Viscosity of solvent-free compound adhesive at different ratios

The ratios of material A/B	Rotation Speed (r/min)	Viscosity (mPa•s)
100/60	30	859.8
100/65	30	959.8
100/70	30	1020
100/75	30	879.8
100/80	30	779.8

3 Effect of the Proportion of Solvent-Free Compound on the Curing Rate of Solvent-Free Compound

Generally, the curing time of solvent-free composite adhesive is 24 h after coating. That is to say, after the printed matter is compounded with a layer of plastic film without solvent, it needs to wait 24 h before the next process can be carried out. This undoubtedly brings trouble to the production of printing and packaging, and is not conducive to the production efficiency of related products. For producers, the faster the curing, the more convenient it is to enter the next process. The ratio of solvent-free compound A and B is also a factor affecting the curing of solvent-free compound. This paper will study this through experiments.

In order to better quantify the curing time, this paper describes the curing speed by measuring the change of the viscosity of the mixed adhesive with time. The greater the viscosity changes with time, the faster the curing time. Similarly, the solvent-free composite materials A and B are proportioned at 100:60, 100:65, 100:70, 100:75 and 100:80, and fully stirred. Viscosity measurement V_1 shall be conducted immediately after mixing, and annual measurement V_2 shall be conducted 75 min later, and the curing rate shall be calculated as $(V_2-V_1)/75$. The corresponding data obtained after the experiment are shown in Table 3.

Table 3. Curing speed of solvent-free composite adhesive at different ratios

The ratios of materials A/B	Rotation speed (r/min)	Curing speed (mPa•s /min)
100:60	30	138.11
100:65	30	183.69
100:70	30	188.76
100:75	30	156.76
100:80	30	151.17

Table 3 shows that when the ratio of materials A, B is 100:70, the curing speed is biggest. Therefore, the proper ratio of solvent-free compound A and B can improve the curing speed of solvent-free compound.

4 Effect of the Proportion of Solvent-Free Compound Material A and B on the Peel Resistance of Solvent-Free Compound Glue

Peel resistance is a key index to evaluate the quality of solvent-free composite products. In the flexible packaging production process, the main reason why the solvent-free compounding process can not completely replace the dry compounding process is its low peel resistance. Here, the influence of the ratio of solvent-free compound A and B on the peel resistance of solvent-free compound is the focus of our research.

4.1 Cross-Cut Test

Generally, the experimental research on the peel resistance performance of the binder mainly uses the cross-cut test [6]. The core tool of the cross-cut test is the cross-cut tester (Fig. 1).

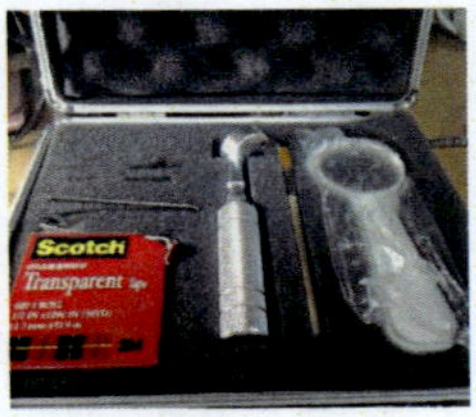

Fig. 1. cross-cut test tools

In the cross-cut test, using the cross-cut tester to cut all the coating layers with the test, and cut the base material. One knife is cut horizontally and longitudinally. Since the 100 grids knife has 11 blades, 100 small grids are formed. After cutting, paste the tape on the 100 small squares and press them tightly. Then quickly tear up the adhesive tape, observe whether the coating layer of the small square is lifted, and determine the peel resistance performance of the adhesive according to the lifting situation and the corresponding ISO grade standard.

Table 4 shows that when ISO grade is 0, the peel resistance performance is the highest, while when ISO grade is 5, the peel resistance performance is the lowest.

4.2 Experimental Analysis of the Influence of the Ratio on the Peel Resistance Performance

Generally, PET, BOPP and PE are often used as flexible packaging materials. These three materials are used as bottom materials in this experiment. Fully mix in the proportions of 100:60, 100:65, 100:70, 100:75 and 100:80 respectively, and evenly apply the mixed solvent-free composite adhesive on the bottom material. After 24 h of curing.

Performing the cross-cut test on the plastic film coated with different proportions of mixed adhesives, and recording the corresponding data, as shown in Table 5.

Table 4. ISO test standard for peel resistance of adhesives

ISO degree	Test standard
0	The edges of the cuts are smooth, and the lattice and the edges of the lattice are free of any peeling
1	There is a small scale of spalling at the intersection of the cuts, and the damage degree of the material in the delimited area is $\leq 5\%$
2	There is peeling at the edge of the notch or at the intersection of the lattice, and the peeling area is between 5 and 15%
3	There is partial peeling along the edge of the cut, or there is large peeling, or part of the lattice is peeled off by the whole piece. The spalling area is between 15 and 35%
4	The edge of the incision is largely peeled off, or some square parts are partially or completely peeled off, and the peeling area is between 35 and 65% of the cross-cut area
5	There are pieces of material falling off at the edges and intersections of the scribed lines, and the total area of peeling off is greater than 65%

Table 5. Peel resistance of solvent-free composite adhesives with different proportions

Material	Ratio of materials A, B	IOS grade
PET	100:60	2
PET	100:65	1
PET	100:70	1
PET	100:75	2
PET	100:80	2
BOPP	100:60	4
BOPP	100:65	4
BOPP	100:70	3
BOPP	100:75	3
BOPP	100:80	4
PE	100:60	5
PE	100:65	5
PE	100:70	4
PE	100:75	4
PE	100:80	5

Table 5 shows that the peel strength of mixed adhesives with different proportions is different. Not only the ratio but also the base material affects the peel strength of the solvent-free composite adhesive. Obviously, the bottom material is PET, and the solvent-free composite A and B materials with the ratio of 100:65 and 100:70 have the highest peel resistance. When the bottom material is PE and the ratio of solvent-free compound A and B is 100:60, 100:65 and 100:80, the peel resistance performance of the mixed adhesive is the lowest.

5 Conclusion

From the experimental analysis, which shows that the ratio of solvent-free compound A and B has an important influence on the properties of solvent-free compound. The proper ratio can ensure the proper viscosity and better curing speed of the mixed adhesive. At the same time, it can be seen from the experiment that the ratio of solvent-free compound A and B has a great influence on the key performance parameter of solvent-free compound adhesive - peel resistance, and the appropriate ratio can ensure better peel resistance performance, and the bottom material also has a great impact on the peel resistance performance of the solvent-free composite adhesive. Therefore, in order to make the solvent-free composite process better applied, the appropriate ratio of solvent-free composite A and B materials is the key factor to be paid attention to. Therefore, the high-precision ratio system is the key factor to ensure the quality of solvent-free composite products.

Acknowledgements. This research project is supported by the Key R&D Plan of Shaanxi Province (approve number: 2021GY-262).

References

1. Xu, H., Wang, X., Lei, R., et al.: Experimental analysis of the effect of vacuum degassing technology on the solventless laminating adhesive performance. In: 2016 China Academic Conference on Printing & Packaging and Media Technology (2016)
2. Wei, Y., Xu, H., Han, Y., Feng, S.: Tooth profile optimization for mixing proportional pump of solvent-free laminating mixer. In: 2019 China Academic Conference on Printing and Packaging (2019)
3. Xing, B., Xu, H., Chen, X., Li, X.: A control algorithm for improving the ratio precision of solventless laminating. In: Proceedings of 2019 IEEE 2nd International Conference on Automation, Electronics and Electrical Engineering, AUTEEE (2019)
4. Xu, H., Ma, Z, Liu, L., et al.: Study on influence of four various structure static mixers on mixing effect during solvent-less compound mixing. In: 2021 China Academic Conference on Printing and Packaging (2021)
5. Yusof, N., Green, J., Pitt, K., et al.: A novel method for the analysis of particle coating behavior via contact spreading in a tumbling drum: effect of coating liquid viscosity. Powder Technol. **351**, 102–114 (2019)
6. Sui, C., Luo, Q., He, X., et al.: A study of mechanical peeling behavior in a junction assembled by two individual carbon nanotubes. Carbon **107**, 651–657 (2016)

Research Status and Progress of Biomass-Based 3D Printing Materials

Hanyu Zhao, Ying Jia, Guangxue Chen(✉), Minghui He, Junfei Tian, and Qifeng Chen

State Key Laboratory of Pulp and Paper Engineering, South China University of Technology, Guangzhou, China
chengx@scut.edu.cn

Abstract. 3D printing technology, also known as "additive manufacturing", can be used to design highly complex structures to produce products quickly on demand and at low cost, and has great application prospects in rapid processing. The application of biomass materials in 3D printing can significantly reduce the negative impact of petroleum based materials on resource shortage and environment. Therefore, 3D printing of environmentally friendly materials has attracted extensive attention from researchers. This review introduces the preparation of biomass materials for 3D printing, fused deposition modeling (FDM), direct ink writing (DIW), selective laser sintering (SLS), laminated object manufacturing, and the characterization method of biomass-based 3D printing materials.

Keywords: 3D printing · Additive manufacturing · Rapid processing

1 Introduction

In recent years, 3D printing with its advantages of low cost and high efficiency has entered the attention of researchers. The commonly used 3D printing technologies include fused deposition modeling (FDM), direct ink writing (DIW), selective laser sintering (SLS), ink-jet printing, stereolithography (SLA) and electron beam melting [1–8]. With the rapid development of 3D printing technology, as an advanced technology, 3D printing has been applied in complex and high-precision intelligent manufacturing industry. Various materials and technologies are used in 3D printing. Conventional polymers and polymer composites can be used in 3D printing. With the shortage of petroleum resources and the increasingly serious problem of environmental pollution, environmental protection materials have attracted extensive attention. In the field of 3D printing, the commonly used biomass materials are cellulose, hemicellulose, lignin, protein, etc. The above materials have the advantages of wide range of sources, good biocompatibility and so on [9–11].

2 Application of Biomass Materials in 3D Printing

2.1 Wood materials

In order to protect the environment that human beings depend on, scientists are trying to use natural biomass materials to replace petroleum derived materials which cause

M. Xu et al. (Eds.): CACPP 2022, LNEE 991, pp. 608–615, 2023.
https://doi.org/10.1007/978-981-19-9024-3_79

serious environmental pollution. Cellulose, as the most abundant biomass material, has developed rapidly in the past decades. According to the current technological progress, cellulose mainly includes cellulose nanofibrils (CNF), cellulose nanocrystals (CNC), bacterial cellulose (BC) [12, 13]. Dong et al. [14] improved the dispersion of CNF in polylactic acid by using Pickering emulsion. The tensile modulus and tensile strength of the composite were increased by 28% and 63%, respectively. The composite filament was very suitable for 3D printing. For porous CNF grafted polyethyleneimine (PEI), the chelate was used as a persistent aerogel elastic matrix for the adsorption of heavy metal ions Cu^{2+}.The aerogel has good wet stability, shape memory and can be used for recycling. At the same time, the existence of CNF makes the aerogels have good porosity and mechanical stability. In the continuous column experiment, the aerogel sample adsorbed 88 bed volume of Cu^{2+}containing wastewater, which opened up a new way for industrial scale treatment of wastewater containing Cu^{2+} [15]. The researchers [16] used CNC combined with xanthan gum to create a bio-ink that could precisely control the 11-layer lattice structure. The bio ink has excellent viscoelasticity, and the printed sample has high fidelity and excellent resolution. After freeze-drying, the porosity of the sample is as high as 70%, and the swelling is 11 g/g, which provides an ideal condition for the preparation of tissue engineering scaffolds. This study highlights the potential of 3D printing products in soft tissue engineering applications. Shin et al. [17] hydrophobic modification of carboxyl CNF with methyl trimethoxysilane and used as a printing mechanism. Carboxymethyl CNF hydrogel was prepared by monochloroacetic acid. The carboxyl CNF hydrogel was printed on the surface of the modified hydrogel to form an immiscible and unique 3D structure. By removing the Vaseline ink from the CNF hydrogel, a channel was created, which showed excellent potential for drug delivery.

Lignin and cellulose as a variety of biomass resources on the earth, has a long history of research. As a complex aromatic polymer network. According to the current extraction technology of lignin, most of the lignin separated have dark color, heavy smell, wide molecular weight distribution and limited reactivity [18]. However, the excellent anti-aging, flame retardant and UV absorption properties of lignin can provide a variety of potential functions for printed matter, which provides convenience for the application of lignin in 3D printing research, which has attracted the attention of researchers [19]. Silica mineralized lignin aerogels were prepared based on lignin/siloxane colloids by water-induced self-assembly and in-situ mineralization. The aerogels have excellent moisture resistance, heat insulation, flame retardancy, no disintegration at 1200 °C, and good self-cleaning and superhydrophobic properties. These nanostructures have good thermal conductivity and low moisture resistance; SiO_2 ensures excellent fire resistance. The adjustable multi-scale microstructure of lignin and silica and the interaction of their chemical properties make the materials have excellent optical properties and self-cleaning properties. These excellent properties ensure that the aerogel can be used in extreme environments as an excellent thermal insulation material [20].

2.2 Other Biomass Materials

Other biomass mainly includes starch, protein, chitosan and so on. The application of 3D printing in the field of food has also received extensive attention. The researchers used potato, wheat and corn starch as raw materials to study the performance analysis of different types of starch on 3D printing samples. The experimental results show that the three kinds of starch gels can be used to prepare 3D printing model. Among them, because of its low viscosity and good extrudability, the size of 3D printing sample produced by wheat starch is the closest to CAD design model. Wheat starch gel. Compared with potato starch and corn starch, the microstructure of 3D printing samples treated with wheat starch showed more regular network structure and better storage performance. This study provides theoretical guidance and technical support for the application of 3D printing technology in food production [21]. Biomass materials such as starch and protein are also used in 3D printing. Liu et al. Studied the rheological properties and 3D printing behavior of mashed potato (MP) with starch (PS). The results of flow gradient experiment and dynamic oscillation frequency showed that the mixture appeared shear thinning, and the viscosity, τ_0, G' increased with the increase of starch content. Thus, it plays a positive role in supporting and maintaining the printability in 3D printing process [22].

3 3D Printing Process Classification

3.1 Fused Deposition Modeling

Fused deposition modeling (FDM) is the most widely used 3D printing technology. FDM has the advantages of simple operation, low cost and high ink utilization rate. By using nanocellulose (CNC) reinforced poly (3-hydroxybutyrate-co-3-hydroxycaproate) (PHBH), the acetylated CNCs content was further investigated between 5 and 20%, and the composite was prepared by melting method. In the comparison experiment, it was found that the thermal stability temperature of PHBH was 220 °C, and the thermal stability temperature of CNC-enhanced composite was significantly increased to 265 °C. Moreover, when the CNC content is increased to 20 wt%, the memory modulus is increased by 1.5 times. Based on the above excellent performance, the composite material is used in FDM 3D printing. The researchers printed a complex wearable finger with 170 layers and a 20° tilt on the Z-axis, which is ergonomic. The 3D-printed device could replace plaster in the medical industry, in biomedical engineering, surgical implants can be customized for patients, the parts of an automobile; electronic equipment and other materials are expected to replace traditional petroleum-based plastics [23]. Based on the introduction of biomass materials, their degradation properties have been favored by scholars, and the use of fiber reinforced polymers for the preparation of 3D printing products has been widely studied. The researchers used hydroxypropyl methylcellulose (HPMC) combined with polylactic acid (PLA) to 3D print parts with complex structures. Mini castles were successfully 3D printed by 5%HPMC/PLA, indicating that the composite has good 3D printing performance. However, the crystallization temperature of PLA decreased by 7 °C after HPMC composite. The addition of 5 or 7% HPMC has no significant effect on the mechanical properties of the material. However, with the increase of HPMC content, the tensile strength and impact strength of the composites decreased due to the increase of porosity. And the contact Angle is reduced by 30° [24]. The above literature shows that the biomass-reinforced polymer can be successfully used in 3D printing, but it will also cause some adverse effects on the performance of the composite.

3.2 Direct Ink Writing

Direct Ink Writing (DIW) is the most widely used 3D printing technology, and the appropriate rheological behavior of DIW precursor ink can be used to print any material. Furthermore, DIW simplifies overall manufacturing time, saves energy, reduces costs and maintains excellent material performance through single-step processing. Ethyl cellulose is widely used as a thickener, and inks with different viscosities can be prepared by dissolving it in α-terpenol solvents with different solid contents for application in DIW3D printing. However, the traditional DIW printer did not consider the functional requirements of cellulose materials, so scholars first developed the DIW printer. The traditional FDM printer is modified by replacing the printing head with a spiral screw. Introduce air ducts to reduce the air pressure to 2–3 psi. Use a step electrode to control the ink increment. Small tolerances are solved by perforating the housing. The effects of feed rate, nozzle height and extrusion rate on printed matter were studied by using the modified printer. The results show that the feed rate and nozzle height have no obvious effect on the product when the extrusion rate is constant. However, ethyl cellulose after DIW printing is still a gel with high viscosity, which needs to be heated by an infrared lamp to remove the solvent to complete the curing. It takes 30 min to complete the curing at 80 °C, and the curing time is different for samples with different thickness printed. Therefore, ethyl cellulose dissolved in α-terpenol solvent can be used as a good printing ink, but curing affects the promotion of the ink [25]. Researchers have studied the printability of printing ink made from a blend of two polysaccharides, aloe vera gel (AV) and nanocellulose (TOCNF). The ink with different blending ratios showed good fluidity, no needle clogging and liquid diffusion problems, and the printed samples showed good self-supporting performance. All printed samples also have excellent performance of high precision and high fidelity [26].

3.3 Selective Laser Sintering

Selective laser sintering (SLS) is one of the processes related to the use of final samples, which is currently in the stage of small batch production [27, 28]. Some scholars have studied the application of biomass materials in SLS inks. Polyamide is a common material in SLS. Lignin is added to reduce the cost and increase the processability of the product. Lignin was used to prepare ink for SLS printing at a volume concentration of 60 vol%. In the printing process, the temperature in the printing process has no effect on the composite ink. Compared with the pure polyamide, the porosity of the products containing lignin is increased by 10%, the tensile strength is slightly reduced, and the Young's modulus is increased by 16%. The different roughness and composition of the samples have a great impact on the contact tentacles, with a difference of about 71°. The experiment proved that the addition of lignin was beneficial to the stability and processability of SLS products [27].

4 Characterization Methods of 3D Printed Products

4.1 Mechanical Properties

Tensile properties and impact resistance are commonly used in 3D printing. Tensile strength, also known as ultimate tensile strength or ultimate strength, is the maximum stress of the specimen before fracture [29, 30]. Lignin was extracted from the fiber of oil palm hollow fruit bundle (EFB). Then, lignin and graphene nanosheets (GNP) were used as fillers and reinforcing materials of UV curable polyurethane (PU) for stereolithography 3D printing. The tensile strength of the composite lignin and graphene was increased by 27%, indicating that the composite had good compatibility [31].

The impact resistance is generally measured on the impact testing machine. The impact test is to measure the impact strength of the notched specimen, that is, the energy absorbed per unit area; the bending property can be obtained by bending test; bending modulus, also known as elastic bending modulus, refers to the ratio of stress and strain under the elastic limit. Bending strength is the biggest bending stress in bending test. The bending test is to fix the specimen on the bracket from both ends and apply pressure until the specimen breaks/breaks [30, 32, 33].

4.2 Stability

At present, the stability of most 3D printing products is an important performance index. Because most of the 3D printing inks are composite materials, the stability of composite materials determines whether the 3D printing ink can be used in practice. Among them, the stability of 3D printing products is mainly characterized by thermogravimetric analysis (TGA) and differential scanning calorimetry (DSC) [34–36]. TGA is a technology to characterize the thermal stability of materials in detail by real-time monitoring the change of material mass with temperature and time. The thermogravimetric analyzer is mainly composed of a thermobalance, a temperature controller, a data collection device, a container which is inert to the sample and a gas flow control device. The composition, thermal stability, thermal decomposition and products of the multicomponent system can be measured by analyzing the thermogravimetric curves [34, 37]. Differential scanning calorimetry (DSC) is a thermal analysis method different from TGA. It mainly characterizes the change of material and heat flow rate at program temperature. The typical DSC curves represent the endothermic and exothermic properties of the samples with temperature and heat flux. There are two kinds of DSC: heat flux DSC and power component DSC [38, 39].

4.3 Morphological Representation

Morphology analysis is also a common characterization method for 3D printing samples. 3D printing can improve the shape of raw materials and improve the quality of the products. The commonly used methods of morphology characterization are scanning electron microscope, optical microscope and so on. Scanning electron microscope (SEM) is a kind of electron microscope which can observe the surface morphology of samples by secondary electron signal imaging. It can be used to characterize the binding characteristics, morphology, interface compatibility, port and porosity of 3D printing samples [40, 41]. Optical microscope is the use of optical principles to enlarge and reduce the visual image of objects. Optical microscopy is used to characterize surface roughness, dispersion, shrinkage and crystal morphology [42].

4.4 Degradation Performance

The degradation criteria of biomass materials cannot be separated from time and environment. At present, the biodegradation methods mainly include soil burial degradation, compost degradation, water degradation and so on. Soil burial degradation generally needs to meet the requirements of complete biodegradation, degradation cycle of 0.5–2 years, and degradation products without pollution to the soil. The evaluation standard of compost degradation is ASTM D5338, compost is mainly the use of polymer macromolecules under the action of microorganisms to break and form small molecules to be degraded. And the final compost products are mainly CO_2, H_2O, inorganic salts and other products. Water degradation generally uses sea water and sewage water. Different water bodies are mainly used according to different degradation materials. CO_2 emissions and oxygen consumption during degradation were used to calculate biodegradation rates.

5 Conclusions

3D printing technology provides a promising manufacturing strategy for reducing waste generation, energy consumption and manufacturing complex samples. Cellulose, lignin, starch, whole biomass and their derivatives have been widely used in 3D printing technology, such as melt deposition modeling, stereolithography, adhesive jet, direct ink writing and so on. The application of biomass materials in 3D printing can significantly reduce the negative impact of petroleum-based materials on resource shortage and environment. And through the modification of biomass materials, in order to improve the performance of materials, and then engrave the poor liquidity of 3D printing materials, low product recovery rate, let 3D printing in-depth benefit our life and environment, and more applied in industrial technology and other fields.

Acknowledgements. This work has been supported by the National Natural Science Foundation(61973127); Science and Technology Program of Guangdong Province (2017B090901064); SCUT Liyan (Guangdong) New Material Technology Co., Ltd (High-tech industrialization entrepreneurship team project of Foshan High-tech Zone, FSBG2021021); Guangdong Basic and Applied Basic Research Foundation) (No.2021A1515010899).

References

1. Conner, B.P., et al.: Making sense of 3-D printing: creating a map of additive manufacturing products and services. Smart Mater. Struct. **1**, 1310–1311 (2015)

2. Mohamed, O.A., Masood, S.H., Bhowmik, J.L.: Optimization of fused deposition modeling process parameters: a review of current research and future prospects. Adv. Manuf. **3**(1), 42–53 (2015). https://doi.org/10.1007/s40436-014-0097-7
3. Sochol, R.D., et al.: 3D printed microfluidics and microelectronics. Microelectron. Eng. **189**, 52–68 (2018)
4. Haque, R.I., et al.: Self-powered triboelectric touch sensor made of 3D printed materials. Nano Energy **52**, 54–62 (2018)
5. Destino, J.F., et al.: 3D printed optical quality silica and silica–titania glasses from sol–gel feedstocks. Adv. Mater. Technol. **3**(6) (2018)
6. Fan, Z., Feng, H.: Study on selective laser melting and heat treatment of Ti-6Al-4V alloy. Results Phys. **10**, 660–664 (2018)
7. Ledford, C., et al.: Characteristics and processing of hydrogen-treated copper powders for EB-PBF additive manufacturing. Appl. Sci. (Switzerland) **9**(19) (2019)
8. Sochol, R.D., et al.: 3D printed microfluidic circuitry via multijet-based additive manufacturing. Lab Chip **16**(4), 668–678 (2016)
9. Ji, A., et al.: 3D printing of biomass-derived composites: application and characterization approaches. RSC Adv. **10**(37), 21698–21723 (2020)
10. Yoo, C.G., et al.: The critical role of lignin in lignocellulosic biomass conversion and recent pretreatment strategies: a comprehensive review. Bioresour. Technol. **301**, 122784 (2020)
11. Pérez, J., et al.: Biodegradation and biological treatments of cellulose, hemicellulose and lignin: an overview. Int. Microbiol. **5**(2), 53–63 (2002)
12. Julie, C., Thomas, B.: Cellulose: from biocompatible to bioactive material. J. Mater. Chem. B **2**(30), 4767–4788 (2014)
13. Siqueira, G., et al.: Cellulose nanocrystal inks for 3D printing of textured cellular architectures. Adv. Func. Mater. **27**(12), 1604619 (2017)
14. Dong, J., et al.: The influence of grafted cellulose nanofibers and postextrusion annealing treatment on selected properties of poly(lactic acid) filaments for 3D printing. J. Polym. Sci. Part B: Polym. Phys. **55**(11), 847–855 (2017)
15. Hong, H.-J., et al.: Fabrication of cylindrical 3D cellulose nanofibril (CNF) aerogel for continuous removal of copper (Cu2+) from wastewater. Chemosphere **278**, 130288 (2021)
16. Baniasadi, H., et al.: High-resolution 3D printing of xanthan gum/nanocellulose bio-inks. Int. J. Biol. Macromol. **209**, 2020–2031 (2022)
17. Shin, S., Kwak, H., Hyun, J.: Transparent cellulose nanofiber based open cell culture platform using matrix-assisted 3D printing. Carbohydr. Polym. **225**, 115235 (2019)
18. Ragauskas, A.J., et al.: Lignin valorization: improving lignin processing in the biorefinery. Science **344**(6185), 1246843 (2014)
19. Zhang, S., et al.: Stereolithography 3D printing of lignin-reinforced composites with enhanced mechanical properties. ACS Omega **4**(23), 20197–20204 (2019)
20. Fan, Q., et al.: Water-induced self-assembly and in situ mineralization within plant phenolic glycol-gel toward ultrastrong and multifunctional thermal insulating aerogels. ACS Nano **16**(6), 9062–9076 (2022)
21. Abdullah, T., et al.: Development of nanocoated filaments for 3d fused deposition modeling of antibacterial and antioxidant materials. Polymers **14**(13) (2022)
22. Liu, Z., et al.: Impact of rheological properties of mashed potatoes on 3D printing. J. Food Eng. **220**, 76–82 (2018)
23. Giubilini, A., et al.: 3D-Printing nanocellulose-poly(3-hydroxybutyrate-co-3-hydroxyhexanoate) biodegradable composites by fused deposition modeling. ACS Sustain. Chem. Eng. **8**(27), 10292–10302 (2020)
24. Jiang, G., et al.: Investigation into hydroxypropyl-methylcellulose-reinforced polylactide composites for fused deposition modelling. Ind. Crops Prod. **146**, 112174 (2020)

25. Adams, D., Ounaies, Z., Basak, A.: Printability assessment of ethyl cellulose biopolymer using direct ink writing. JOM **73**(12), 3761–3770 (2021). https://doi.org/10.1007/s11837-021-04911-8
26. Baniasadi, H., et al.: Direct ink writing of aloe vera/cellulose nanofibrils bio-hydrogels. Carbohydr. Polym. **266**, 118114 (2021)
27. Ajdary, R., et al.: Selective laser sintering of lignin-based composites. ACS Sustain. Chem. Eng. **9**(7), 2727–2735 (2021)
28. Wörz, A., Drummer, D.: Tribological anisotropy of selective laser sintered PA12 parts. Polym. Test. **70**, 117–126 (2018)
29. Ma, H., Suhling, J.C.: A review of mechanical properties of lead-free solders for electronic packaging. J. Mater. Sci. **44**(5), 1141–1158 (2009). https://doi.org/10.1007/s10853-008-3125-9
30. Long, H., et al.: Mechanical and thermal properties of bamboo fiber reinforced polypropylene/polylactic acid composites for 3D printing. Polym. Eng. Sci. **59**(s2), E247–E260 (2019)
31. Ibrahim, F., et al.: Evaluation of the compatibility of organosolv lignin-graphene nanoplatelets with photo-curable polyurethane in stereolithography 3D printing. Polymers **11**(10) (2019)
32. Mimini, V., et al.: Compatibility of kraft lignin, organosolv lignin and lignosulfonate with PLA in 3D printing. J. Wood Chem. Technol. **39**(1), 14–30 (2019)
33. Kuo, C.-C., et al.: Preparation of starch/acrylonitrile-butadiene-styrene copolymers (ABS) biomass alloys and their feasible evaluation for 3D printing applications. Compos. Part B: Eng. **86**, 36–39 (2016)
34. Tanase-Opedal, M., et al.: Lignin: a biopolymer from forestry biomass for biocomposites and 3D printing. Materials **12**(18) (2019)
35. Agnoli, E., et al.: Additive manufacturing of geopolymers modified with microalgal biomass biofiller from wastewater treatment plants. Materials **12**(7) (2019)
36. Nguyen, N.A., Bowland, C.C., Naskar, A.K.: Mechanical, thermal, morphological, and rheological characteristics of high performance 3D-printing lignin-based composites for additive manufacturing applications. Data Brief **19**, 936–950 (2018)
37. Wilkie, C.A.: TGA/FTIR: an extremely useful technique for studying polymer degradation. Polym. Degrad. Stab. **66**(3), 301–306 (1999)
38. Forrest, J.A., Dalnoki-Veress, K.: The glass transition in thin polymer films. Adv. Colloid Interface Sci. **94**(1), 167–195 (2001)
39. Wellen, R.M.R., Rabello, M.S.: The kinetics of isothermal cold crystallization and tensile properties of poly(ethylene terephthalate). J. Mater. Sci. **40**(23), 6099–6104 (2005)
40. Shariatnia, S., et al.: Atomization of cellulose nanocrystals aqueous suspensions in fused deposition modeling: a scalable technique to improve the strength of 3D printed polymers. Compos. Part B: Eng. **177**, 107291 (2019)
41. Tekinalp, H.L., et al.: High modulus biocomposites via additive manufacturing: cellulose nanofibril networks as "microsponges". Compos. Part B: Eng. **173**, 106817 (2019)
42. Wang, L., et al.: Spray-dried cellulose nanofibril-reinforced polypropylene composites for extrusion-based additive manufacturing: nonisothermal crystallization kinetics and thermal expansion. J. Compos. Sci. **2**(1) (2018)

Preparation of the Lignin-Based Carbon Fibers Reinforced Composite

Xiaojuan Shi(✉) and Yahui Tang

School of Light Industry and Chemical Engineering, Dalian Polytechnic University, Liaoning, China

shixj@dlpu.edu.cn

Abstract. In order to investigate the practical application effect of self-prepared lignin-based micro/nano carbon fibers as reinforcement, and improve the comprehensive mechanical properties of epoxy resin composites, the lignin-based carbon fibers reinforced composites were prepared with epoxy resin as matrix phase and ultra-short cut lignin-based carbon fibers as nano-dispersed reinforcing phase. With the aspect ratio of 11 and the additive amount of 0.4 wt%, nano-dispersed reinforcing phase dispersed well in the matrix, and the mechanical properties of the composite were the highest. The tensile strength, bending strength, impact strength and hardness of the composite enhanced by 61%, 77%, 35% and 23% respectively compared with those of pure epoxy resin.

Keywords: Lignin · Carbon fiber · Reinforced composite

1 Introduction

Due to the excellent chemical stability and preferable mechanical properties, epoxy resin is a common matrix material of composites. However, unsatisfactory impact resistance and toughness are the main reasons limiting its application in the preparation of composites. The carbon fiber (CF) reinforced composite provides the effective scheme for the manufacture of advanced epoxy resin matrix composites.

Since the 1990s [1, 2], researchers have dispersed a variety of nano-dispersed reinforced phases (nanoparticles, carbon nanotubes, etc.) into the resin matrix to improve its mechanical properties. Godara et al. [3] incorporated carbon nanotubes (0.5 wt%) in epoxy resin matrix, the prepared composites were obtained a substantial increase in fracture toughness by over 80%. Jiang et al. [4] found that the appropriate addition of fullerene nanoparticles (2 wt%), led to a significant improvement in fiber/matrix bond strength. Chen et al. [5] dispersed SiO_2 nanoparticles with a volume fraction of 8.7% into epoxy resin matrix. Yokozeki et al. [6, 7] found the aspect ratio (AR) of nano-dispersed phase had a great influence on the properties of reinforced matrix. A large number of research results prove that, carbon nanotubes are ideal nano-dispersed reinforced phase because of low amounts, controllable AR, and high strength. However, it is not widely used because of its expensive price.

M. Xu et al. (Eds.): CACPP 2022, LNEE 991, pp. 616–620, 2023.
https://doi.org/10.1007/978-981-19-9024-3_80

In this paper, self-prepared low-cost lignin-based micro/nano CF was used as nano-dispersed phase to prepare composites. The practical application effect and mechanism of lignin-based CFs reinforced epoxy resin matrix composites were investigated and studied.

2 Experimental

2.1 Materials

F_3-CF was prepared according to the method described in the literature [8]. Epoxy resin AB adhesive was purchased from Beijing Hongsheng Ruida building materials Co., Ltd..

2.2 Preparation of Lignin-Based Carbon Fibers Reinforced Composite

The F_3-CF was sheared and grinded into ultra-short cut carbon nano-CFs. Epoxy resin A glue and B glue were mixed with the mass ratio of 3:1. A certain mass percentage (0–1.5 wt%) of ultra-short cut nano-CFs (F_3-CF) was added in the resin paste, and then the mixture was stirred (1200 rpm) for 10 min. The mixture was injected into the silica gel mold (place horizontally) and placed 48 h until solidified. Then, the reinforced matrix was remolded and prepared into samples for testing.

2.3 Characterization

Micro morphology. The micro morphology of ultra-short cut carbon nano-CFs and samples fracture surface were observed using a scanning electron microscope (SEM, Jeol JSM-7800F, Japan). The average fiber length and diameter were analyzed by ImageJ software.

Mechanical properties test. Tensile strength and bending strength were tested by universal testing machine (Shimadzu, AGS-X, Japan). Tensile strength test rate was 10 mm/min. The bending strength test speed was 2 mm/min, the deflection was 30 mm, and the span was 60 mm. The impact strength was tested by cantilever combined impact tester (Chengdu Yuheng Electronics Co., Ltd, HY-JJ-5). The notch depth was 2 mm. The hardness was measured by shore hardness tester (HD), and the test temperature was 20 °C.

3 Results and Discussion

3.1 Effect of Nano-dispersed Reinforcing Phase on Mechanical Properties of Reinforced Composite

The aspect ratio (AR) and additive amount of dispersed phase are the main factors affecting its uniform distribution. The physical photos and electron microscope of nano-dispersed reinforcing phase were shown in Fig. 1. The average length of carbon fiber was 4.73 μm and AR was about 11.

Fig. 1. Photos (a) and SEM images (b) of ultra-short nano-CFs

Figure 2 showed the results of mechanical properties of the composite with different nano-dispersed reinforcing phase amounts. It could be seen from Fig. 2 (a) that the change trend of elongation at break of composite was consistent with that of tensile strength. When the additive amount of nano-dispersed reinforcing phase was 0.4 wt %, the elongation at break reached 3.2%, which was nearly twice that of pure epoxy resin (0 wt %), which showed that the additive of reinforcing phase greatly improved the toughness of epoxy resin matrix material. When the additive amount of nano-dispersed reinforcing phase was 0.4 wt%, the average tensile strength (65.2 MPa) of the composite was 61% higher than that of pure EP (40.4 MPa). When the additive amount continues to increase, the tensile strength decreased instead, and was lower than that of pure EP when the additive amount was 1.5 wt %. This was probably caused by the non-uniform distribution of nano-dispersed reinforcing phase in the matrix after the additive amount was greater than 0.4 wt %.

Similarly, the test results of bending strength, impact strength and hardness also showed similar changes. The average values of the properties reached the maximum when the additive amount of reinforcing phase was 0.4 wt % (bending strength 97.3 MPa, impact strength 8.4 kJ/m^2 and hardness 86.3 HD). Compared with the corresponding average values of pure EP (bending strength 54.9 MPa, impact strength 6.2 kJ/m^2, hardness 70.2 HD), they were increased by 77% (bending strength), 35% (impact strength) and 23% (hardness).

3.2 Enhancement Mechanism Analysis

Figure 3 was the fracture surface SEM images of the sample after tensile strength test. The fracture surface of pure EP (0 wt %) was very flat. The crack direction was consistent and arranged in parallel line, which indicated that when pure EP was cracked by external tensile force, the crack spread rapidly inside the material without obstruction and broke instantaneously. When the additive amount was 0.4 wt% (Fig. 3 (b)), the dispersion uniformity of carbon nano-fibers in EP matrix was preferable, and its blocking behavior to cracks significantly increased the fracture folds. The increase of fracture area and the fine lines through the carbon nano-fibers meant more energy consumption. This showed that due to the additive of reinforcing phase, when the matrix material was damaged by

Fig. 2. Effect of reinforcing phase amounts on the mechanical properties of the composite. (a) Representative stress-strain curves. (b) Representative stress-displacement curves. (c) Average impact strength (d) Average hardness

external force, the energy driving its deformation was weakened, which slowed down the trend of failure, which was directly reflected in the increase of elongation at break, that was, the improvement of toughness [9].

Fig. 3. Fracture surface SEM images of tensile strength samples before and after adding nano dispersed phase (a) Additive amount of 0 wt% (Pure EP) (b) Additive amount of 0.4 wt%

4 Conclusions

Self-prepared low-cost lignin-based micro/nano CF as reinforcing phase could significantly improve the mechanical properties of epoxy resin. When the aspect ratio (AR) of ultra-short cut F_3-CF was 11 and the addition amount was 0.4 wt%, the tensile strength, bending strength, impact strength and hardness of composite were 65.2 MPa, 97.3 MPa, 8.4 kJ/m^2 and 86.3 HD respectively. Compared with pure epoxy resin, the increase of strength (tensile strength 40.4 MPa, bending strength 54.9 MPa, impact strength 6.2 kJ/m^2 and hardness 70.2 HD) was 61%, 77%, 35% and 23% respectively, and all mechanical properties reach the maximum.

Acknowledgement. The authors would like to acknowledge the financial support from Doctoral Research Initial Foundation of Liaoning (No.2021-BS-224).

References

1. Hussain, M., Nakahira, A., Niihara, K.: Mechanical property improvement of carbon fiber reinforced epoxy composites by Al_2O_3 filler dispersion. Mater. Lett. **26**, 185–191 (1996)
2. Down, W.B., Baker, R.T.K.: Modification of the surface properties of carbon fibers via the catalytic growth of carbon nanofibers. J. Mater. Res. **10**(3), 625–633 (1995). https://doi.org/10.1557/JMR.1995.0625
3. Godara, A., Mezzo, L., Luizi, F., Warrier, A., Lomov, S.V., Vuure, A., et al.: Influence of carbon nanotube reinforcement on the processing and the mechanical behaviour of carbon fiber/epoxy composites. Carbon **47**, 2914–2923 (2009)
4. Jiang, Z., Hui, Z., Zhong, Z., Murayama, H., Okamoto, K.: Improved bonding between PAN-based carbon fibers and fullerene-modified epoxy matrix. Compos. A Appl. Sci. Manuf. **39**, 1762–1767 (2008)
5. Chen, W., Liu, Y., Jiang, Z., Tang, L., Liu, Z., Zhou, L.: Modeling of compressive strength for unidirectional fiber reinforced composites with nanoparticle modified epoxy matrix. Materials **12** (2019)
6. Yokozeki, T., Iwahori, Y., Ishiwata, S., Enomoto, K.: Mechanical properties of CFRP laminates manufactured from unidirectional prepregs using CSCNT-dispersed epoxy. Compos. A **38**, 2121–2130 (2007)
7. Yokozeki, T., Iwahori, Y., Ishibashi, M., Yanagisawa, T., Imai, K., Arai, M., et al.: Fracture toughness improvement of CFRP laminates by dispersion of cup-stacked carbon nanotubes. Compos. Sci. Technol. **69**, 2268–2273 (2009)
8. Shi, X., Dai, Z., Cao, Q., et al.: Stepwise fractionation extracted lignin for high strength lignin-based carbon fibers. New J. Chem. **43**, 18868 (2019)
9. Yokozeki, T., Iwahori, Y., Ishibashi, M., Yanagisawa, T.: Characterization of nonlinear behaviors of CSCNT/carbon fiber-reinforced epoxy laminates. Adv. Compos. Mater. **18**, 251–264 (2009)

Research Progress of Electromagnetic Shielding Performance of MXene ($Ti_3C_2T_x$) Composites

Yue Han, Ying Jia, and Guangxue Chen(✉)

State Key Laboratory of Pulp and Paper Engineering, South China University of Technology, Guangzhou, China
chengx@scut.edu.cn

Abstract. The global electromagnetic pollution problem is becoming increasingly serious as science and technology advance, affecting the normal use of electronic equipment. As a result, finding materials that are light, flexible, and have good electromagnetic shielding properties is becoming increasingly critical. In comparison to traditional electromagnetic interference shielding materials, the two-dimensional transition metal material MXene has low density, excellent flexibility, and superior conductivity, while waves of electromagnetic energy can dissipate in the form of heat energy after multiple reflections in its interior, resulting in good electromagnetic shielding effectiveness. The preparation method and electromagnetic shielding mechanism of MXene are described in this paper, as well as the MXene-based composite materials with various structures. And its application prospect in many fields such as aerospace, weapons and wearable electronic devices is prospected.

Keywords: MXene · Composite materials · Electromagnetic shielding

1 Introduction

The frequency and transmission power of electronic systems are increasing due to current communication technology's rapid development, particularly the miniaturization of electronic equipment and circuits. Signal crosstalk between electronic components, as well as electromagnetic radiation into the environment, has caused a slew of issues [1, 2]. Adverse electromagnetic interference (EMI) caused by these devices can impair human health by producing headaches, nausea, cancer, and other symptoms, in addition to degrading the operation of an electronic system [3]. Furthermore, in the alternating electromagnetic field, implants or devices (like hearing aids, insulin pumps and cardiac pacemakers) have a tendency to fail [4]. To adapt to the miniaturization, integration, and lightweight of electronic information systems, electromagnetic interference shielding materials must be developed in order to meet the demands of the future. The ongoing

M. Xu et al. (Eds.): CACPP 2022, LNEE 991, pp. 621–635, 2023.
https://doi.org/10.1007/978-981-19-9024-3_81

development of innovative materials such as graphene, ultra-fine magnetic nanoparticles, carbon nanotubes, and MXene has opened up new avenues for the study and implementation of lightweight and efficient electromagnetic shielding materials. MXene, a material with a unique two-dimensional layered structure that exhibits excellent conductive properties and controllable surface activity, is a typical representative of these materials [5]. There are currently numerous reports on the research and application of MXene materials in a variety of fields, including aerospace, weaponry, and wearable electronic equipment. To improve electromagnetic shielding performance of MXene, researchers have studied a variety of MXene composite materials, such as film materials, porous foams, and aerogels [6]. MXene-based electromagnetic shielding materials with a variety of structural forms are reviewed in this paper, including the structure, Special properties, and preparation process of MXene, and the next generation of MXene-based electromagnetic shielding materials is proposed and prospected.

2 Properties and Preparation of MXene

Carbide, carbonitride, or nitride of transition metals with a layered structure is 2D MXene. The standard formula is $M_{n+1}X_nT_x$, where M represents an early transition metal (such as Ti, Zr, V, Nb, Ta, or Mo), X denotes carbon or nitrogen, and T_x represents functional groups on the surface of MXene [7]. Compared with hydrophobic graphene materials and other carbon-based materials with small specific surface area, the two-dimensional layered material MXene-$Ti_3C_2T_x$ has excellent electrical conductivity, large specific surface area and contains a large number of hydrophilic groups, which exhibits excellent electrical conductivity and controllable surface activity. Compared with hydrophobic graphene materials and other carbon-based materials with smaller specific surface areas, there are many hydrophilic groups in the two-dimensional layered material Mxene-$Ti_3C_2T_x$, which contribute to its outstanding conductivity and tunable surface activity. The distinctive structure of MXene makes it have excellent optical and electrical characteristics and strong microwave attenuation ability, which meets the basic characteristics required by efficient EMI shielding materials. Therefore, MXene has been an ideal electromagnetic shielding and absorbing composite material [8].

The preparation method of MXene is somewhat homogeneous, which involves etching away the A-atom layer in the MAX phase by chemical etching to weaken the bonding between the layers, followed by intercalation peeling or ultrasonic-assisted phase peeling to peel off a single nanosheet and stabilize it by dispersing it into a specific solvent [9]. MAX is blocky as a whole, and its structure is dense and difficult to be modified. MXene obtained after etching is monolayer or less layered, which has more obvious advantages in material modification [7]. At present, the main etching method is the HF solution etching method, in addition, there are other preparation methods, such as molten fluoride etching, solution phase flocculation etching method are improved based on HF solution etching method.

2.1 Hydrofluoric Acid (HF) Etching Method

In 2011, Naguib [10] proposed for the first time to remove Al in Ti_3AlC_2 by using HF aqueous solution as etch agent to obtain binary crystal polymer Ti_3C_2 with graphene structure, and finally you get $Ti_3C_2F_2$, etching reaction equation is as follows [11]:

$$Ti_3AlC_2 + 3HF = Ti_3C_2 + AlF_3 + \frac{2}{3}H_2 \tag{1}$$

$$Ti_3C_2 + H_2O = Ti_3C_2(OH)_2 + H_2 \tag{2}$$

$$Ti_3C_2 + 2HF = Ti_3C_2F_2 + H_2 \tag{3}$$

In reaction (1), HF selectively etches the Al of Ti_3AlC_2, during which the Al atoms are replaced by O, OH, or F atoms. Compounds like $Ti_3C_2(OH)_2$ and $Ti_3C_2F_2$ are formed in the reaction medium when H_2O and HF come into contact with the exposed Ti atom layer. The synthesized $Ti_3C_2T_X$ sample is composed of multilayer thin slices. Due to the removal of Al atoms, the interaction between $M_{n+1}X_n$ layers is weak and the layers are easy to peel off. Therefore, layered MXene can be obtained by ultrasonic stripping.

2.2 Lithium Fluoride (Li) and Hydrochloric Acid (HCl) Etching Method

Due to the spontaneous intercalation of cationic (Li^+, Na^+, etc.) metal ions into the MXene layers in the salt solution, which increases the interlayer spacing and weakens the interlayer interaction, Lukatskaya [12] simplified the original etching procedure and designed a method that allows for one-step etching and intercalation: Ti_3AlC_2 is first added to HCl mixed with LiF and mixed with stirring at 40 °C for 45 h. The resulting precipitate is then washed to remove the reaction products and centrifuged to raise its pH. The method offers the possibility of applying the inserted electrodes to batteries and supercapacitors.

2.3 Ammonium Hydrogen Fluoride (NH_4HF_2) Etching Method

Halim [11] selectively etched the Al in the $TiAlC_2$ film by NH_4HF_2 aqueous solution with the equation shown below to obtain the $TiAlC_2T_x$ thin layer, whose resistivity increases with decreasing temperature at 100 K and shows negative values at low temperatures, ensuring that the obtained $TiAlC_2T_x$ conforms to the properties of 2D metal. And it can be concluded from the experimental results that the thin layers etched with NH_4HF_2 possess higher transparency and electrical conductivity. This method changed the etchant from the highly volatile HF to the more moderate NH_4HF_2, but also introduced some impurities such as $(NH_4)_3AlF_6$ into it, resulting in a less pure MXene.

$$Ti_3AlC_2 + 3NH_4HF_2 = (NH_4)_3AlF_6 + Ti_3C_2 + \frac{2}{3}H_2 \tag{4}$$

$$Ti_3C_2 + aNH_4HF_2 = (NH_3)_c(NH_4)_dTi_3C_2(OH)_xF_y \tag{5}$$

2.4 Polar Organic Solvent Anhydrous Etching Method

Most etching methods for synthesizing MXene use water as the primary solvent, limiting the use of MXene in water-sensitive applications. Natu et al. [13] demonstrated the use of an organic polar solvent and ammonium dihydrogen fluoride to etch and delaminate MXene in the absence of water, and when propargylene was tested as an electrode in a sodium ion battery (SIB) in an electrolyte containing propylene carbonate (PC), the electrode in PC $Ti_3C_2T_z$ synthesized in PC had almost twice the capacity of the same MXene etched in water. Besides, the use of low boiling solvents (e.g. acetonitrile) for etching allows easy recovery and purification of the solvent for reuse, which is of great interest when considering the industrial synthesis of MXene.

2.5 Molten Fluoride Salt Etching Method

MXene have a higher electrical conductivity than carbides, making them suitable candidates for electrodes in electrochemical capacitors. Although treatment in acidic aqueous solutions produces carbide and carbon-nitride Mxene well, they cannot etch A layers from the nitride-based MAX phase for reasons that are not yet clear. The synthesis of two-dimensional transition metal nitride Ti_4N_3-based MXene was first proposed by Urbankowski et al. [14] Under an argon atmosphere at 550 °C, Al was etched from Ti_4AlN_3 powder precursors using molten fluoride salts, and then the obtained MXene was layered to obtain $Ti_4N_3T_x$ few-layer nanosheets and monolayers. Although this method can obtain MXene, fluorine-containing impurities will be introduced, and if pure MXene is to be obtained, it needs to be purified with sulfuric acid, which will be more tedious in the process.

2.6 Alkali-Assisted Hydrothermal Etching Method

For the present, most of the preparation methods for synthesis in etching MXene will choose HF solution, however, HF is very corrosive, which may have damage to the performance of some electronic devices, such as lithium-ion batteries and supercapacitors, during the subsequent application. Inspired by the Bayer method, Xie et al. [15] reported the preparation of $Ti_3C_2T_x$ by NaOH-assisted hydrothermal etching of Ti_3AlC_2. The results showed that Al could be successfully removed selectively from Ti_3AlC_2 in NaOH solution at 270 °C, and $Ti_3C_2T_x$ powder with 92 wt % purity could be successfully prepared, and the whole process was completely fluorine-free. The $Ti_3C_2T_x$ thin film electrode obtained in H_2SO_4 without any F terminations has a weight capacitance of 314 Fg^{-1} and a volume capacitance of 511 Fcm^{-3} at 2 mV^{-1}, which is about 214% higher than that of $Ti_3C_2T_x$ obtained by using HF etching. This fluorine-free method provides an alkali etching strategy that can be used to explore novel MXene for the removal of interlayer amphiphilic acidic atoms from the pristine MAX phase, but the method itself is relatively time-consuming and labor-intensive, with poor etching results and a narrow range of applications.

2.7 Solution Phase Flocculation Etching Method

In order to solve the problem of stacking and aggregation between MXene nanosheet layers, Zhang [16] proposed a solution phase flocculation method, using the characteristics of negatively charged MXene colloidal sheet surface and mutual repulsion between nanosheet layers, adding NH_4^+ to MXene aqueous solution to reduce the electrostatic repulsion phenomenon, and after completing electrostatic flocculation, freeze-drying and annealing, we can obtain $Ti_3C_2T_x$/MXene nanosheet powder. To further promote the application of this method, it has been improved by adding a stepwise layering and assisted centrifugation step to the electrostatic flocculation process. The freeze-drying time can be further reduced in the subsequent centrifugation process. The proposed method can be used to prepare a variety of layer less MXene powders, broadening the scope of application of MXene.

3 Electromagnetic Shielding Mechanism

The shielding mechanism of electromagnetic shielding material is shown in Fig. 1. A portion of the incident power (P_I) is reflected both ways when electromagnetic waves interact with the shielding layer because of the mismatch in impedance between shielding and air layers (P_R). Due to attenuation and transmission (PT), the residual power is absorbed or transferred and dissipated as heat in the shield [6]. The protective effect of the shielding layer on an incident electromagnetic wave is defined as EMI SE, Schelkunoff theory [17]. According to the following equation, electromagnetic shielding efficacy is the total of reflection, absorption, and multiple reflections attenuation:

$$SE_T(dB) = SE_R + SE_A + SE_M \tag{6}$$

3.1 Reflection Loss (SE_R)

When two media with different impedances or refractive indexes come into contact, a reflection occurs, such as air and a shielding layer, and is the primary electromagnetic shielding mechanism. For highly conductive shielding, in order to attain a high reflection loss, one must use shielding material with a high conductivity, as evidenced by the fact that SER increases with conductivity. The reflection loss is not solely determined by conductivity, which the permeability of the shield and the frequency of the electromagnetic wave also plays an impact.

3.2 Absorption Loss (SE_A)

Electromagnetic waves in lossy media (that is, the shielding materials) of the transmission are absorbed, marvelous absorption loss requires high conductivity of ohmic losses, increases the interaction of high electron density and the incident electromagnetic wave, the larger dielectric constant and dielectric loss and magnetic hysteresis loss and eddy current loss of consumption magnetic conductivity of conductive shielding layer, the thickness of the material and electrical conductivity have a bigger influence on the absorption, What determines the absorption loss is the dielectric constant and permeability.

3.3 Multiple Reflections (SE_M)

In a membrane shielding material, due to multiple reflections, the ultimate transmission is influenced by reflections from the back surface, because the reflected radiation is reflected again on the front surface, contributing to a second transmission. The above process can be replicated until the wave energy is completely dissipated when $SE_T >$ 10 dB, the loss of multiple reflections is negligible.

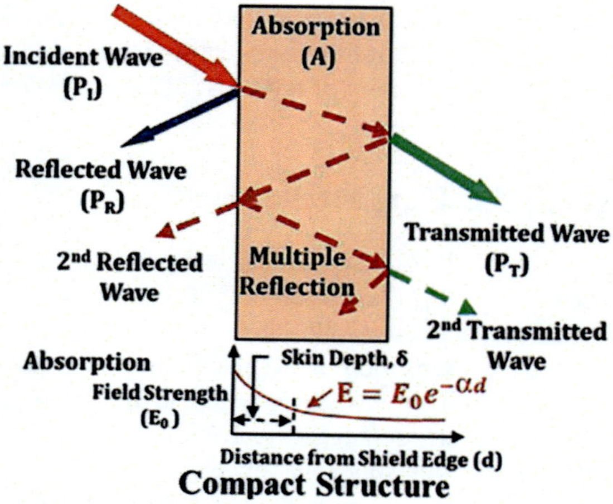

Fig. 1. Electromagnetic shielding mechanism. Reproduced with permission. [6] Copyright 2020, advanced functional materials.

4 Typical Structures of MXene/Composites in the Field of Electromagnetic Shielding

4.1 MXene Composite Membranes

$Ti_3C_2T_X$ films are usually prepared from $Ti_3C_2T_x$ suspensions by vacuum filtration or spraying, in terms of preparing flexible films by using MXene on a large scale, however, the effect is not ideal, because MXene films have relatively small sheet size ($\approx$ 200 nm) [18] and feeblish inter-sheet interplay [19], and mechanical strength is generally lacking. By the addition of polymers within the MXene layers, the mechanical characteristics of this film can be modified. Shahzad et al. [1] prepared $Ti_3C_2T_x$/SA composites using $Ti_3C_2T_x$ compounded with sodium alginate (SA). $Ti_3C_2T_x$/SA composite film (0.008 mm, 90 wt.% $Ti_3C_2T_x$) had electromagnetic shielding effectiveness of 92 dB and 57 dB, respectively, when compared to pure $Ti_3C_2T_x$ (0.045 mm) film. This is due to the excellent conductivity and two-dimensional layer structure of $Ti_3C_2T_x$ films, electromagnetic waves can generate multiple internal reflections between $Ti_3C_2T_x$ layers of $Ti_3C_2T_x$ /SA composite films, so $Ti_3C_2T_x$/SA composite films have high electromagnetic shielding effectiveness. Two-dimensional (2D) laminated MXene materials compounded with polymeric materials offer good process flexibility and can also effectively improve the mechanical strength of the composites. By introducing 30% poly(3,4-ethylenedioxythiophene)(PEDOT): poly(styrenesulfonate)(PSS) in $Ti_3C_2T_x$ film followed by acid treatment, EMI-SE levels of up to 40.5 dB were achieved by Wan et al. [20] in the composite film that they manufactured. The tensile strength of the film was 38.5 ± 2.9 MPa, which increased the tensile strength by 155% compared with pure $Ti_3C_2T_x$ film, thus achieving mechanical and shielding qualities in perfect harmon. Similarly, as shown in Fig. 2, Shi et al. [21] utilized the layered deposition encapsulation method to encapsulate MXene ($Ti_3C_2T_x$), silver nanowires, and graphene oxide-encapsulated hollow carbon fibers sequentially and prepared them into thin films. Due to effect of the continuous conductor network synergism and porous structure, with an 11-μm thickness, the EMI shielding efficacy of film is 73.2 dB. Moreover, the p-LMHA thin film further enhances the reliability of hydrophobicity and resistance to harsh environment, integrates the rapid response behavior of electrothermal/photothermal, and provides a simple solution for preparing devices with integrated characteristics.

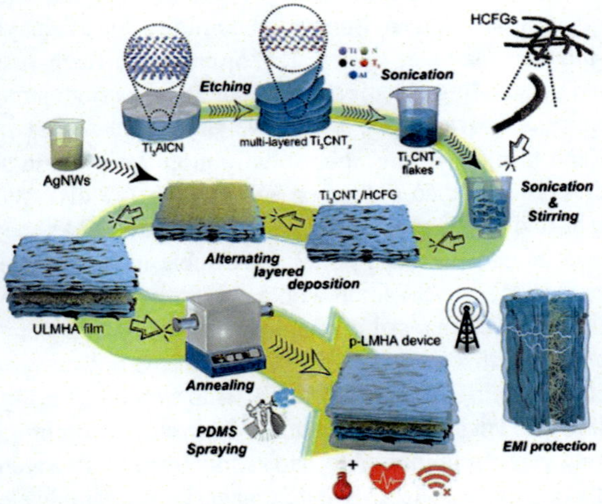

Fig. 2. Schematic diagram of PDMS encapsulated $Ti_3C_2T_x$/HCFG/AgNW (p-LMHA) composite film synthesis process. [21] Copyright 2022, ACS nano

MXene composite film can not only act as a barrier to electromagnetic scattering but also as a photothermal and energy conversion material. Zhou et al. [22] prepared an asymmetric sandwich (CNF@MXene@AgNW) film containing cellulose nanofiber (CNF) skin layer, self-supporting $Ti_3C_2T_x$ and silver nanowire (AgNW) layers, MXene and AgNW layers provide a directional highly conductive network for the MXene composite film, and the external CNF layer provides support to ensure the structural stability of the MXene composite film, so the CNF@MXene@AgNW film has outstanding tensile strength (118 MPa), excellent electrical conductivity (37378.2 S m^{-1}) and effective EMI shielding effectiveness (SE, 55.9 dB), high flexibility (minimum bending radius of ~ 85 μm) and marvellous specific SE (SSE/t, 10647.6 dB cm^2 g^{-1}) and high in-plane thermal conductivity (15.53 W m^{-1} K^{-1}), so this unique asymmetric sandwich structure design can effectively improve its electromagnetic shielding performance and thermal conductivity. Using a controlled self-assembly method, Xiang et al. [23] created multifunctional nano/microstructured Ti_3CNT_x/Ni@C composites for efficient electromagnetic attenuation and Joule energy conversion applications. The Ti_3CNT_x/Ni@C composite can produce a thin film with a thickness of just 1.5 mm (8 wt.% filler) and a strong electromagnetic wave absorption performance (65.7 dB), as well as an electromagnetic shielding rate of 66.7 dB. Furthermore, the effective Joule energy conversion, good flame retardancy, and high infrared shielding function of MXene composite films increase their application range and stimulate the development of MXene composite films in the field of energy conversion.

4.2 MXene-Based Porous Foam

Although two-dimensional transition metal carbide (MXene) has excellent metallic conductivity and electromagnetic shielding properties, the hydrophilicity of MXene films, can be a major drawback in humid conditions [24, 25]. In order that shielding properties of MXene do not deteriorate with environmental humidity due to its hydrophilic nature, using hydrazine-induced foaming, Liu et al. [26] created a foam structure from film of MXene that had outstanding electrical conductivity, ultra-high strength, and exceptional hydrophobicity, as shown in Fig. 3. Lightweight MXene foam with the same mass as unfoamed MXene film (70 dB) is substantially more effective in shielding against electromagnetic interference than unfoamed MXene film (53 dB). With hydrophobic, flexible, lightweight, and excellent EMI shielding performance, MXene foam has a clear application prospect in aerospace and portable wearable intelligent electronic devices. Wu et al. [27] prepared MXene/SA foam with a polydimethylsiloxane (PDMS) surface coating to impart excellent compressibility and durability to the 3D network. The electrical conductivity of MXene/SA mixed foam with PDMS covering may reach 2211 S m^{-1}, and it has a high EMI shielding efficiency (70.5 dB). Furthermore, the high EMI shielding effectiveness of 48.2 dB is maintained even after 500 cycles of compression and release. Because of this, lightweight, compressible, and conductive PDMS covered MXene foam is anticipated to find applications in EMI shielding gaskets, wearable electronic devices, sensors, and other industries. Lightweight absorbing electromagnetic interference (EMI) shielding materials are more attractive than traditional reflective shielding materials and can minimize secondary pollution caused by reflected electromagnetic waves (EM). By using a straightforward freeze-drying technique, Xu et al. [28] were able to produce a multi-MXene/PVA composite foam. This foam was made up of several layers of MXene and polyvinyl alcohol (PVA). Specific shielding efficiency (ESS) reached 5136 dB cm^2 g^{-1} at 0.15 vol%, while its reflection efficiency (SER) fell below 2 dB, exhibiting exceptional absorption-dominated performance in shielding. Fan et al. [29] introduced highly conductive 2D $Ti_3C_2T_x$-MXene nanosheets into graphene (GO), and prepared a lightweight MXene/graphene hybrid foam (MX-rGO) by freeze-drying and reduction heat treatment, with excellent EMI shielding effectiveness (50.7 dB). With a specific EMI shielding effectiveness (SSE) of 6217 cm^3 g^{-1}, MX-rGO has the ability to protect against electromagnetic interference (EMI). Because of its superior electromagnetic interference shielding capability, this lightweight conductive MXene/graphene hybrid foam may be used in aviation field and artificial intelligence applications.

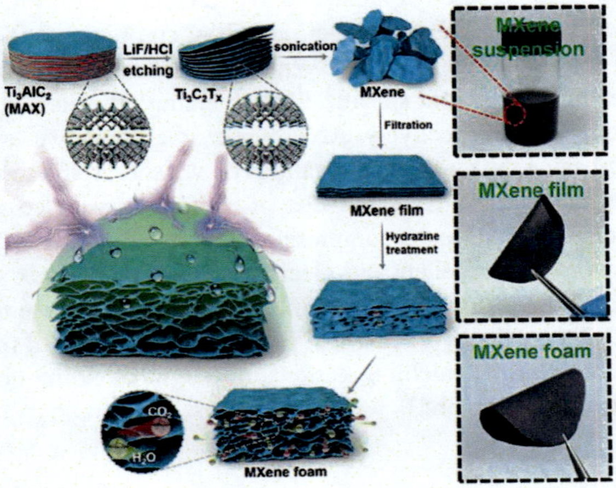

Fig. 3. MXene foam preparation schematic. [26] Copyright 2017, advanced materials

4.3 MXene-Based Composite Aerogel

3D aerogel has abundant pores, which can carry out more reflections and absorb EMI more effectively than planar composite materials. Moreover, abundant pores can reduce the density of MXene materials. MXene aerogel materials with the same volume have lighter mass, which is suitable for aerospace and military fields [6].

Without using external carriers, MXene aerogel with different structures can be obtained by changing the preparation method of MXene. The freezing casting process can be used to produce large-size MXene aerogels. The cold casting process by Bian et al. [30] can be used to produce large-size MXene aerogels that contain tiny pores determined by ice crystal morphology and exhibit excellent EMI shielding performance up to 75 dB, yielding specific shielding effectiveness up to 9904 dB cm^3 g^{-1}. Han et al. [31] assembled a series of MXene by a bi-directional freeze casting technique. As shown in Fig. 4, MXene aerogel has excellent compressibility, ultra-high mechanical strength, low specific gravity and excellent hydrophobicity because of its special interlayer gap and honeycomb-like structure. When the MXene density was about 11.0 mg cm^{-3}, the EMI SEs of $Ti_3C_2T_x$, Ti_2CT_x, and Ti_3CNT_x aerogels reached 70.5, 69.2, and 54.1 dB, respectively, and exhibited some resilience. Zhu et al. [32] used natural wood to prepare wood aerogels by delignification and assembled with highly conductive flexible multilayer $Ti_3C_2T_x$-MXene (f-$Ti_3C_2T_x$-MXene) nanosheets to obtain MXene@Wood (M@W) nanocomposite aerogels. The shielding performance of the prepared M@W aerogel is anisotropic, with excellent EMI shielding performance (72 dB) in the horizontal direction, and super strong EM absorption performance in the vertical direction, which can reach 12.4 GHz. After 400 cycles of compression release testing along the growing axis vertically, the original compressive strength is up to 93.3%, which suggests that structural alteration of wood aerogel confers remarkable compressibility upon it. By preparing MXene into aerogels, the original material may be less robust in the longitudinal direction due to the increased volume, and to overcome this drawback, Lu et al. [33] prepared MXene($Ti_3C_2T_x$)/ANF combination gel by combining MXene with aramid fiber (ANFs) via controlled freeze-drying. When the thickness is 1.9 mm in the X-band, the EMI shielding efficiency and specific EMI shielding efficiency approach 56.8 dB and 3645.7 dB $cm^2\ g^{-1}$, respectively. ANFs serve as a protective layer and strength carrier, hence increasing the mechanical qualities of MXene. ANFs aerogel retains 95% of its original characteristics after 120 cycles of compression under varying compressive strains, as demonstrated by the compression experiment. MXene ($Ti_3C_2T_x$)/ANFs hybrid aerogel not only has excellent anti-electromagnetic interference performance, but also has high compressive stress and reversible elasticity at the same time. As a result, the MXene ($Ti_3C_2T_x$)/ANFs hybrid aerogel is anticipated to be used for military applications such as durable EMI shielding gaskets, ultralight shielding design of portable and flexible electronic equipment.

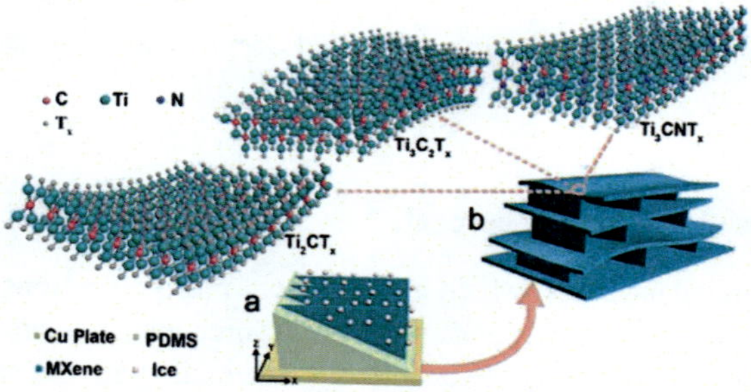

Fig. 4. (a) The mechanism of two-way freezing casting. Reproduced with permission, (b) the layered structure of MXene aerogel assembled from different MXene sheets. [31] Copyright 2019, advanced optical materials.

For MXene aerogel, its electrochemical properties can in principle be tuned by changing the terms of its constituent elements and surface groups. Since the end group of MXene is a mixture of O, OH, or F, the synthesized MXene is hydrophobic and conductive [9]. In the experiments based on the spontaneous redox reaction of MXene nanosheets on metal template substrates, Yun [34] explored the basic principles of the multidimensional architecture process based on pure MXene. It is precisely because part of the oxygen-containing functional groups is removed during the interface reduction process that the electrical conductivity of MXene is effectively increased by 60%, which greatly improves the electrical conductivity of MXene. When used as an aerogel electrode, the material shows ultra-high electron conductivity, and its weight capacitance reaches 298 F/g even at 2000 mV/s. Sambyal et al. fabricated $Ti_3C_2T_x$/CNT hybrid aerogel with strong conductivity and good EMI shielding performance at X-band frequencies using a bidirectional freezing process. Sambyal et al. [35] fabricated $Ti_3C_2T_x$/CNT hybrid aerogel with strong conductivity and good EMI shielding performance at X-band frequencies using a bidirectional freezing process. Carbon nanotubes were added to MXene as enhanced fillers, which improved the mechanical properties of porous $Ti_3C_2T_x$ while enhancing the electrical conductivity of MXene. The synthesized hybrid aerogel is lightweight, highly conductive, and has superior EMI shielding capability. To achieve a stable shape of polymer-free adhesive aerogel, Ding et al. [36] designed Mg^{2+}-Mxene aerogel with certain shielding ability as well as superior electrical conductivity up to 758.4 S m^{-1} by freeze-drying. The highly conductive MXene aerogel demonstrates a wide range of applications for electrodes and quasi-solid micro-supercapacitors (QMSCs). As a CDI electrode, the Mg^{2+}-Mxene aerogel exhibited a salt adsorption capacity of 33.3 mg g^{-1} and could operate for more than 30 cycles. In addition, the QMSC with staggered Mg^{2+}-Mxene aerogel electrodes showed an area capacitance of 409.3 mF cm^{-2} with higher power density and energy density. As an EMI shielding film and as a supercapacitor electrode with high capacitance and rate capability, MXene provides capacitance much in excess of that of carbon, which makes MXene become an ideal material. The applicability of MXene in electrochemical energy storage and green EMI shielding materials will be accelerated by this innovative notion (Table 1).

5 Summary and Prospect

Because of its excellent metallic conductivity, $Ti_3C_2T_x$ has the most advanced EMI shielding properties, outperforming other conductive materials and their composites at the same thickness value of the film material. This paper focuses on research progress of MXene composites in the field of electromagnetic shielding, evaluating three different structural forms of MXene composite films, porous foams, and aerogels. In wearable electronic devices, MXene composite membrane is an ideal ultra-thin shielding material with excellent mechanical properties. The MXene-based porous foam solves the problem that MXene films are unstable in humid environments due to their hydrophilic nature, further broadening the application range of MXene composites. MXene aerogels provide a trade-off between density and thickness for absolute shielding effectiveness (SSE/t). Some achievements have been made in the preparation of MXene composite membranes and porous materials in the existing research, but there are problems of difficulty in preparation and low yield, so preparing nanocomposites with excellent electromagnetic shielding properties while using small amounts of MXene remains a significant challenge. So far, Mxene composites research in the field of electromagnetic shielding is still in its early stages, with plenty of room for further investigation.

Table 1. Structure and application of typical MXene composites

Category	Composite materials	Properties	Applications	Refs.
MXene composite membranes	HCFG/AgNW	EMI value of 73.2 dB for electric and optical heating over a wide temperature range and in extreme environments	Healthcare monitoring sensor	[21]
	CNF/AgNW	EMI value of 55.9 dB, high tensile strength, ultra-high flexibility and solar thermal conversion capability	Electromagnetic shielding device in extremely cold environments	[22]
	Ni/C	EMI value of 65.7 dB, with good flame retardancy and high infrared shielding	Practical applications in complex and harsh environments	[23]
MXene-based porous foam	Hydrazine induced foaming	EMI value reaches 70.3 dB, hydrophobic, flexible, lightweight	Aerospace and portable wearable smart electronics	[26]
	PDMS/SA	Conductivity of 2211 S m^{-1}, EMI of 70.5 dB, lightweight, compressible	EMI shielding gaskets, wearable electronics, sensors	[27]
	GO	Shielding effectiveness EMI value of 50.7 dB and SSE/t value of 43690 dB cm^2 g^{-1}	Aerospace, next-generation intelligent shielding equipment	[29]
MXene-based composite aerogel	ANFs	EMI shielding efficiency of 56.8 dB, high compressive stress and reversible elasticity	Durable EMI shielding gaskets, ultra-lightweight shielding designs for portable and flexible electronics	[33]
	CNT	EMI SE of 103.9 dB, light weight, good mechanical strength and high conductivity	EMI shielding in the RF field with potential applications	[35]

(*continued*)

Table 1. (*continued*)

Category	Composite materials	Properties	Applications	Refs.
	Mg^{2+}	Electrical conductivity can reach 758.4 S m^{-1}, and has a certain electromagnetic shielding effectiveness	Quasi-solid-state micro-supercapacitors (QMSCs), as CDI electrodes	[36]

Acknowledgements. This work has been supported by the National Natural Science Foundation of China (Grant No. 61973127), Natural Science Foundation of Guangdong Province (Grant No. 2022A1515011416), and the Guangdong Provincial Science and technology project (Grant NO. 2017B09091064).

References

1. Shahzad, F., et al.: Electromagnetic interference shielding with 2D transition metal carbides (MXenes). Science **353**(6304), 1137–1140 (2016)
2. Yun, T., et al.: Electromagnetic shielding of monolayer MXene assemblies. Adv. Mater. **32**(9), 1906769 (2020)
3. Stam, R., Yamaguchi-Sekino, S.: Occupational exposure to electromagnetic fields from medical sources. Ind. Health **56**(2), 96–105 (2018)
4. Hocking, B., Mild, K.H.: Guidance note: risk management of workers with medical electronic devices and metallic implants in electromagnetic fields. Int. J. Occup. Saf. Ergon. **14**(2), 217–222 (2008)
5. Han, M., et al.: Ti_3C_2 MXenes with modified surface for high-performance electromagnetic absorption and shielding in the X-band. ACS Appl. Mater. Interfaces **8**(32), 21011–21019 (2016)
6. Iqbal, A., Sambyal, P., Koo, C.M.: 2D MXenes for electromagnetic shielding: a review. Adv. Func. Mater. **30**(47), 2000883 (2020)
7. Zhang, C.: Interfacial assembly of two-dimensional MXenes. J. Energy Chem. **60**, 417–434 (2021)
8. Li, X., et al.: 2D carbide MXene Ti_2CT_x as a novel high-performance electromagnetic interference shielding material. Carbon **146**, 210–217 (2019)
9. Naguib, M., et al.: 25th anniversary article: MXenes: a new family of two-dimensional materials. Adv. Mater. **26**(7), 992–1005 (2014)
10. Naguib, M., et al.: Two-dimensional nanocrystals produced by exfoliation of Ti_3 AlC_2. Adv. Mater. **23**(37), 4248–4253 (2011)
11. Halim, J., et al.: Transparent conductive two-dimensional titanium carbide epitaxial thin films. Chem. Mater. **26**(7), 2374–2381 (2014)
12. Lukatskaya, M.R., et al.: Cation intercalation and high volumetric capacitance of two-dimensional titanium carbide. Science **341**(6153), 1502–1505 (2013)
13. Natu, V., et al.: 2D $Ti_3C_2T_z$ MXene synthesized by water-free etching of Ti_3AlC_2 in polar organic solvents. Chem **6**(3), 616–630 (2020)
14. Urbankowski, P., et al.: Synthesis of two-dimensional titanium nitride Ti_4N_3 (MXene). Nanoscale **8**(22), 11385–11391 (2016)

15. Xie, X., et al.: Surface Al leached Ti_3AlC_2 as a substitute for carbon for use as a catalyst support in a harsh corrosive electrochemical system. Nanoscale **6**(19), 11035–11040 (2014)
16. Zhang, S., et al.: Fast and universal solution-phase flocculation strategy for scalable synthesis of various few-layered MXene powders. J. Phys. Chem. Lett. **11**(4), 1247–1254 (2020)
17. Schelkunoff, S.A.: Methods of electromagnetic field analysis. Bell Syst. Tech. J. **27**(3), 487–509 (1948)
18. Li, H., et al.: Flexible all-solid-state supercapacitors with high volumetric capacitances boosted by solution processable MXene and electrochemically exfoliated graphene. Adv. Energy Mater. **7**(4), 1601847 (2017)
19. Chang, T.H., et al.: Controlled crumpling of two-dimensional titanium carbide (MXene) for highly stretchable, bendable, efficient supercapacitors. ACS Nano **12**(8), 8048–8059 (2018)
20. Wan, Y.-J., et al.: Lightweight, flexible MXene/polymer film with simultaneously excellent mechanical property and high-performance electromagnetic interference shielding. Compos. Part A: Appl. Sci. Manuf. **130**, 105764 (2020)
21. Shi, Y., et al.: Multi-interface assembled N-doped MXene/HCFG/AgNW films for wearable electromagnetic shielding devices with multimodal energy conversion and healthcare monitoring performances. ACS Nano **16**(5), 7816–7833 (2022)
22. Zhou, B., et al.: An asymmetric sandwich structural cellulose-based film with self-supported MXene and AgNW layers for flexible electromagnetic interference shielding and thermal management. Nanoscale **13**(4), 2378–2388 (2021)
23. Zhen, X., et al.: Self-assembly of nano/microstructured 2D Ti_3CNT_x MXene-based composites for electromagnetic pollution elimination and Joule energy conversion application. Carbon **189**, 305–318 (2022)
24. Lipatov, A., et al.: Effect of synthesis on quality, electronic properties and environmental stability of individual monolayer Ti_3C_2 MXene flakes. Adv. Electron. Mater. **2**(12), 1600255 (2016)
25. Zhang, C.J., et al.: Oxidation stability of colloidal two-dimensional titanium carbides (MXenes). Chem. Mater. **29**(11), 4848–4856 (2017)
26. Liu, J., et al.: Hydrophobic, flexible, and lightweight MXene foams for high-performance electromagnetic-interference shielding. Adv. Mater. **29**(38), 1702367 (2017)
27. Wu, X., et al.: Compressible, durable and conductive polydimethylsiloxane-coated MXene foams for high-performance electromagnetic interference shielding. Chem. Eng. J. **381**, 122622 (2020)
28. Xu, H., et al.: Lightweight Ti_2CT_x MXene/poly(vinyl alcohol) composite foams for electromagnetic wave shielding with absorption-dominated feature. ACS Appl. Mater. Interfaces **11**(10), 10198–10207 (2019)
29. Fan, Z., et al.: A lightweight and conductive MXene/graphene hybrid foam for superior electromagnetic interference shielding. Chem. Eng. J. **381**, 122696 (2020)
30. Bian, R., et al.: Ultralight MXene-based aerogels with high electromagnetic interference shielding performance. J. Mater. Chem. C **7**(3), 474–478 (2019)
31. Han, M., et al.: Anisotropic MXene aerogels with a mechanically tunable ratio of electromagnetic wave reflection to absorption. Adv. Opt. Mater. **7**(10), 1900267 (2019)
32. Zhu, M., et al.: Ultralight, compressible, and anisotropic MXene@wood nanocomposite aerogel with excellent electromagnetic wave shielding and absorbing properties at different directions. Carbon **182**, 806–814 (2021)
33. Lu, Z., et al.: Micro-porous MXene/aramid nanofibers hybrid aerogel with reversible compression and efficient EMI shielding performance. Compos. Part B: Eng. **217**, 108853 (2021)
34. Yun, T., et al.: Multidimensional $Ti_3C_2T_x$ MXene architectures via interfacial electrochemical self-assembly. ACS Nano **15**(6), 10058–10066 (2021)

35. Sambyal, P., et al.: Ultralight and mechanically robust $Ti_3C_2T_x$ hybrid aerogel reinforced by carbon nanotubes for electromagnetic interference shielding. ACS Appl. Mater. Interfaces **11**(41), 38046–38054 (2019)
36. Ding, M., et al.: Metal ion-induced assembly of MXene aerogels via biomimetic microtextures for electromagnetic interference shielding, capacitive deionization, and microsupercapacitors. Adv. Energy Mater. **11**(35), 2101494 (2021)

Numerical Simulation of Femtosecond Laser Ablation of 304 Stainless Steel

Zhuang Liu(✉) and Junjie Tang

College of Light Industry, Harbin University of Commerce, Harbin 150028, China
hitprint@163.com

Abstract. The ablative effect of femtosecond laser on 304 stainless steel was investigated by using the double temperature equation. The distribution of the electron-lattice system temperature under the action of single-pulse and the relaxation time required to achieve equilibrium was numerically simulated, the ablation threshold of the material was calculated, and the variation of ablative depth and ablative radius under different energy densities were derived. The cumulative effect of energy during multiple pulses action under different periods was analyzed. The concept of overlap rate was introduced to compare the relationship between the ablative depth in the overlap area and the ablative depth in the centre of the spot at different overlap rates, which provided theoretical support for subsequent laser processing experiments.

Keywords: Double temperature equation · Relaxation time · Multi-pulse cumulative effect · Overlap rate

1 Introduction

Laser processing is currently the most common method of machining materials and manufacturing micro-nano structures on surface. Ultrashort pulse lasers have been regarded as an ideal light source due to their low thermal effect, high energy density, high processing accuracy and the wide range of processing materials [1]. Since the 1980s, with the rapid development of ultrashort pulsed laser technology, femtosecond pulsed lasers with pulse width in the range of several hundred femtoseconds had been successfully developed [2].

Femtosecond lasers have a unique advantage over normal long-pulse lasers due to their extremely small pulse width, resulting in virtually no heat-affected area during processing. The physical process of femtosecond laser ablation of solid materials is complex and involves the energy coupling between electrons and photons in the material, the transfer of laser energy to the lattice by electron-acoustic coupling processes [3]. About femtosecond laser processing, Wang et al. [4] investigated the ablative effect of ultrashort pulses on metallic materials and the number of pulses required to reach the ablation threshold at different laser energy densities. Zhang et al. [5] used an improved double temperature model to numerically simulate the temperature distribution of femtosecond laser three-pulse ablation of nitinol alloy. Kiran et al. [6] established a two-dimensional

M. Xu et al. (Eds.): CACPP 2022, LNEE 991, pp. 636–642, 2023.
https://doi.org/10.1007/978-981-19-9024-3_82

axisymmetric model to solve the interaction between the ultrashort pulse laser and the titanium alloy target using spatial and time-domain equations. The fact that successive laser pulses and the movement of the laser spot also has a crucial influence on laser processing, which makes it impossible to consider only single-pulse processes. In this paper, the femtosecond laser processing was numerically simulated under the premise of considering the actual processing conditions, which can form the basis for subsequent processing experiments.

2 Ablative Model of Single-Pulse

2.1 Numerical Simulation of Femtosecond Laser

In 1974, the control equations of electron-lattice temperature were established and coupled by the Soviet scholar S.I. Anisimov based on the unique mechanism of femtosecond laser interaction with metals. Double temperature equation has been elaborated by two partial differential equations for heat conduction.

$$C_e \frac{\partial T_e}{\partial t} = \frac{\partial}{\partial x}\left(K_e \frac{\partial T_e}{\partial x}\right) - G(T_e - T_l) + Q_{rzt} \tag{1}$$

$$C_l \frac{\partial T_l}{\partial t} = \frac{\partial}{\partial x}\left(K_l \frac{\partial T_l}{\partial x}\right) + G(T_e - T_l) \tag{2}$$

where T_e, T_l denote the temperature of the electron and the temperature of the lattice, C_e, C_l denote heat capacity, and K_e, K_l denote thermal conductivity, and G denotes the coupling coefficient of the electron-lattice. Q_{rzt} is the femtosecond laser heat source. On femtosecond timescales, the transfer of energy from lattice to lattice is negligible [7], i.e. K_l has been considered to be 0.

$$Q_{rzt} = (1 - R)\frac{F}{\tau}\alpha \exp\left(-2\frac{r^2}{r_0^2}\right) \exp\left(4 \ln 2\left(\frac{(t-\tau)^2}{\tau^2}\right)\right) Lambert(z) \tag{3}$$

where R denotes the reflectance of the material to the laser, F denotes the energy density of laser, τ is the laser pulse width, and r_0 is spot radius.

Lambert(z) has indicated that the absorption of laser by metallic materials follows the Beer-Lambert law, with the following equation:

$$Lambert(z) = \exp(-\alpha z) \tag{4}$$

where α is the absorption coefficient of the material.

2.2 Ablation Threshold for Materials at a Single-Pulse

The ablation threshold is the laser energy density required to remove a single layer of material when irreversible damage is caused to the target material in femtosecond laser processing [8]. The ablation threshold of femtosecond laser ablation 304 stainless steel

Table 1. Main parameters of 304 stainless steel [9]

Name	Numerical values	Unit
C_e	706.4	$J/(m^2k^2)$
C_l	3.66×10^6	$J/(m^2k)$
T_m	1700	K
T_b	3200	K
T_f	1.28×10^5	K
G	1.3×10^{18}	–
k_0	15	W/(mk)

under single-pulse conditions was calculated by using the double temperature equation (Table 1).

When the electron temperature is less than the Fermi temperature, the electronic thermal conductivity has be approximated as [10]:

$$K_e = k_0 \frac{T_e}{T_f} \tag{5}$$

where k_0 is the electronic thermal conductivity at room temperature.

When the electron temperature exceeds the Fermi temperature, the electronic thermal conductivity of the material has been expressed as [11]:

$$K_e = k \frac{\left(\theta_e^2 + 0.16\right)^{1.25}\left(\theta_e^2 + 0.44\right)\theta_e}{\left(\theta_e^2 + 0.092\right)^{0.5}\left(\theta_e^2 + \beta\theta_l\right)} \tag{6}$$

where θ_e, θ_l are the Fermi energy, and $\theta_e = T_e/T_f, \theta_l = T_l/T_f$, k and β are the constants associated with the material.

In this paper, the double temperature equation was numerically simulated by using COMSOL Multiphysics, which is an advanced numerical simulation software. It is widely used in various fields of scientific research and engineering calculations to simulate a wide range of physical processes in science and engineering. The double temperature equation was solved numerically using a backward differential method with the laser pulse width of 200 fs and the period of 5 ns. As shown in Fig. 1. When the energy density was 1.5 J/cm^2, the material ablation threshold was reached, as the energy density increased, the relaxation time for the electron-lattice system to reach equilibrium also increased.

2.3 Variation of Ablative Areas Under Different Energy Densities

The variation in the radius and depth of the ablation areas was simulated at different energy densities, and the results were shown in Fig. 2.

When the laser energy density reaches the ablation threshold of the material, the material had just been ablated, and the ablation area was very small. As the energy

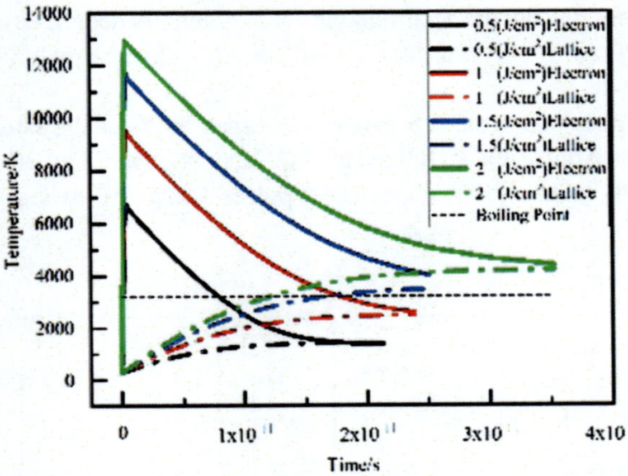

Fig. 1. Variation of electron-lattice temperature versus time at different energy densities

densities increased, the radius and the depth of ablation increased, the ablation rate in the radius direction was faster than in the depth direction and the radius of the heat affected areas gradually exceeded the radius of the spot.

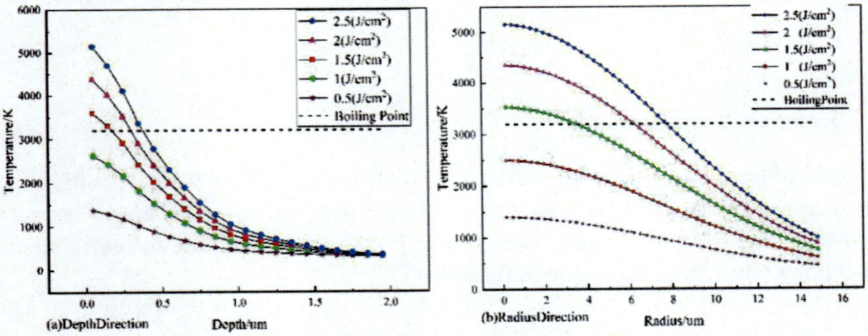

Fig. 2. The relationship of the ablation areas with the energy densities

3 Ablative Model of Multi-pulse

3.1 Energy Accumulative Effect of Multi-pulse

When multiple pulses act continuously on the material, the cumulative effect of heat is generated when the interval between two adjacent pulses is very short, that is, the residual heat of the previous pulse has not been completely diffused, and the next pulse acts on the surface of the material, so that its temperature has increased greatly compared with the maximum temperature generated by the previous pulse. Figure 3(a) showed that as the number of pulses increased, the lattice temperature gradually increased due to the accumulation of energy, but the accumulated value gradually decreases over time. Due to the Beer-Lambert law, energy has decayed with depth within the material and eventually tends to steady state.

When the period was 1 us, the maximum temperature of the lattice tended to be consistent after each pulse and stabilized at 1400 K, as shown in Fig. 3(b). This indicated that due to the long interval between adjacent pulses, there was no heat accumulation of energy.

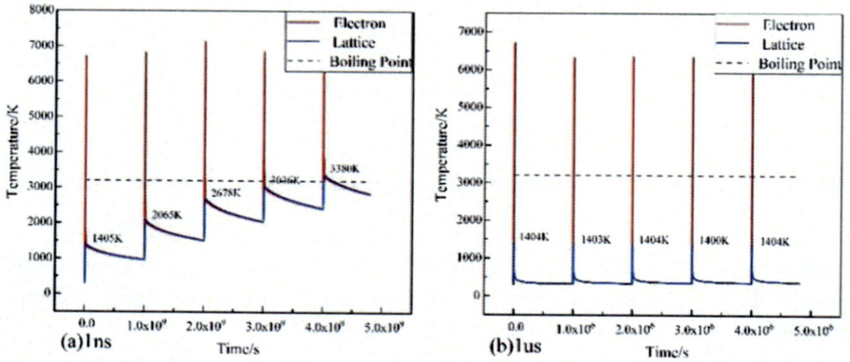

Fig. 3. Temperature accumulation of electron-lattice system at different periods

3.2 Laser Overlap Rate

In actual processing, there is an overlap phenomenon between two adjacent light spots due to the movement of the laser, the surface quality of the processed material was directly affected by the overlap of spots, as shown in Fig. 4(a). To facilitate the calculation of the overlap rate, it was expressed in formula (7).

$$\eta = \frac{W}{D} \times 100\% = \frac{D - L}{D} \times 100\% \tag{7}$$

where η is the overlap rate, and W is the width of the overlap area, and L is the laser moving distance, and D is the spot diameter.

The variation of ablation depth with overlap rate at a laser energy density of 2 J/cm^2 was shown in Fig. 4(b). As the overlap rate of the laser increased, the distance between the two spot centers was close, and the ablation depth of the center of the overlap area approached gradually the ablation depth of the center of the spot. At an overlap rate of 45%, the ablation depth of the center of the overlap area was almost the same as the ablation depth of the center of the spot, the small surface roughness after laser processing at this overlap rate can be indicated. When the overlap rate was 50%, the ablation depth of the overlap area was slightly deeper than the depth of the center of the spot, this meant that the material with an excessive laser overlap rate was over-ablated.

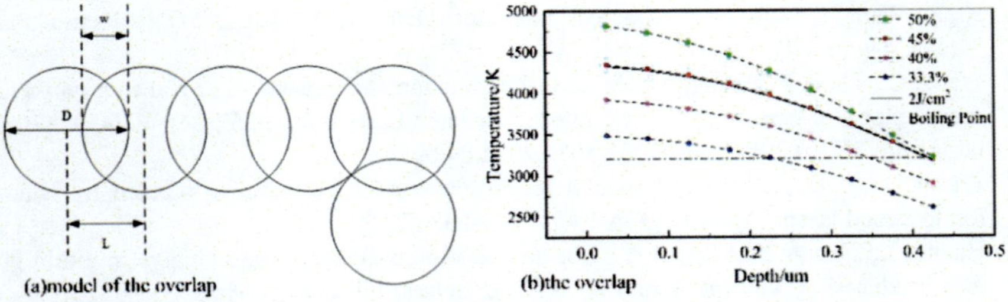

Fig. 4. The variation of the ablation depth under different overlap rates

4 Conclusion

Based on 304 stainless steel, the variation of the temperature of the electron-lattice system after the action of the femtosecond laser on the material was numerically simulated.

The ablation threshold of the material was calculated, the variation of the depth and radius of ablation under different energy densities were obtained. The model was applied to analyze the accumulation effect of laser energy at different periods when multiple pulses act, at lower energy densities, the ablation threshold of the material was reached by adjusting the pulse period. Finally, the relationship between the ablation depth in the centre of the spot and the ablation depth in the overlapped area was investigated at different overlap rates, the surface quality of the material can be effectively improved by adjusting the overlap rate between the spots.

References

1. Li, Z., Wang, X., Nie, J.: High frequency femtosecond laser induced periodic spatial structure on silicon surface. Infrared Laser Eng. **47**(1), 0106003 (2018). https://doi.org/10.3788/IRLA201847.0106003
2. Xiao, K., Li, M., Yang X.: Research status and prospect of femtosecond laser processing of materials. J. Univ. Sci. Technol. Liaoning **42**(3), 179–185 (2019). https://doi.org/10.13988/j.ustl.2019.03.004
3. Sundaram, S.K., Mazur, E.: Inducing and probing non-thermal transitions in semiconductors using femtosecond laser pulses. Nat. Mater. **1**, 217–224 (2002). https://doi.org/10.1038/nmat767
4. Wang, Z., Fu, W., Zhang, R.: Numerical simulation of femtosecond laser multi-pulse ablation of metal iron. Infrared Laser Eng. **48**(7), 0706002 (2019). https://doi.org/10.3788/IRLA201948.0706002
5. Zhang, Y., Wang, L., Gong, J.: Numerical simulation of femtosecond laser multi-pulse ablation of Ni-Ti alloy. Acta Photonica Sin. **45**(5), 13–18 (2016). https://doi.org/10.3788/gzxb20164505.0514002
6. Kumar, K.K., Samuel, G.L., Shunmugam, M.S.: Theoretical and experimental investigations of ultra-short pulse laser interaction on Ti6Al4V alloy. J. Mater. Process. Technol. **263**, 266–275 (2019)
7. Ming, X., Jin, L., Xiao, Y.: Femtosecond laser ablation characteristics of gear material 20CrMnTi. Acta Photonica Sin. **49**(12), 79–88 (2020)

8. Ni, X., Wang, Q.: Advances in femtosecond laser ablation research. Laser Optoelectron. Prog. **39**(12), 4–9 (2002)
9. Wang, C., Zhou, Z., Wang, Y.: Numerical simulation and experimental research on surface micro-nano structure prepared by femtosecond laser. Laser Infrared **52**(04), 497–502 (2022). https://doi.org/10.3969/j.issn.1001-5078.2022.04.005
10. Zhang, Y., Chen, J.K.: Melting and resolidification of gold film irradiated by nano-to femtosecond lasers. Appl. Phys. A **88**(2), 289–297 (2007)
11. Huang, J., Zhang, Y., Chen, J.K.: Ultrafast solid-liquid-vapor phase change in a thin gold film irradiated by multiple femtosecond laser pulses. Int. J. Heat Mass Transf. **52**(13–14), 3091–3100 (2009)

Simulation Analysis of Temperature Field of Ni60 Nickel-Based Alloy Femtosecond Laser Cladding High-Speed Steel Substrate

Zhuang Liu(✉) and Ruhai Yan

College of Light Industry, Harbin University of Commerce, Harbin 150028, China
hitprint@163.com

Abstract. Based on the finite element numerical simulation research method, the simulation process of the temperature field of Ni60 nickel-based alloy powder cladding $W_6Mo_5Cr_4V_2$ high-speed steel by femtosecond laser single-pass was studied. The effects of three process parameters of laser power-scanning speed-spot diameter on temperature field cladding were explored. The influence of different process parameters on the quality of the cladding layer is judged by factors such as the temperature field cloud map, the material temperature rise curve, and the material thermal physical property parameters. The temperature rise of the cladding material was solved by the iterative method, and the cladding state of the Ni60 nickel-based alloy powder after being irradiated by the laser source was judged. Objective: To screen out different process parameters to provide a theoretical basis for subsequent laser cladding experiments.

Keywords: Temperature field · $W_6Mo_5Cr_4V_2$ high-speed steel · Laser cladding

1 Introduction

In recent years, many scholars have made a series of studies on the temperature field of laser cladding. In 2014, Farahmand and Kovacevic [1] conducted numerical simulation of AISiH13 multi-channel temperature field. In 2021, Kai et al. [2] studied the changes in the temperature field of the Mo_2NiB_2 cladding layer under optimal process parameters. In 2005, Wei Lu and Nanhai Hao et al. [3, 4] carried out a numerical simulation of the powder feeding process using ANSYS unit birth and death technology. At present, the numerical simulation of the temperature field is mainly carried out from two aspects. The laser process parameters are predicted by establishing the simulation temperature field model. Through temperature field simulation, the molten pool's morphology, the cladding layer's quality, and the bonding force between cladding layers are speculated under different process parameters.

2 Cladding Material and Simulation Pretreatment Stage

2.1 Laser-Powder Interaction

During the laser cladding process, before the laser beam reaches the substrate, the cladding powder will absorb, reflect, and block part of the laser beam energy, and the

M. Xu et al. (Eds.): CACPP 2022, LNEE 991, pp. 643–647, 2023.
https://doi.org/10.1007/978-981-19-9024-3_83

loss ratio is about 10% [5]. The state of cladding powder is divided into three states: solid phase, liquid phase, and solid-liquid mixed phase.

The temperature rise of powder before it reaches the matrix [6]:

$$T_P - T_0 = \frac{3h}{4\pi r_p^3 \rho_p C_p V_0}(E_1 - E_2 - E_3) \tag{1}$$

The thermal physical parameters of Ni60 alloy [7] are $C_p = 464$ J/Kg.K, $\alpha = 0.1$, h = 10 mm, $r_p = 0.08$ mm, $\varepsilon = 0.53$, $\rho_p = 83780$ kg/m^3.

When the spot diameter is 1 mm, Under the laser power of 400 W, 500 W, 600 W, and 800 W, the temperature rise of Ni60 alloy are 1088°, 1334.8°, 1565.7°, and 1977°, respectively. Under the condition that the spot diameter is 1mm and the laser power is less than 400 W, the cladding material will be fused with the high-speed steel matrix in the form of solid powder. When the laser power is 800 W, the melting point of the M2 high-speed steel matrix is considered to be 1250 °C, which will cause the matrix melting pool melting depth is too large. Therefore, when the spot diameter is 1 mm, it is reasonable to regulate the range of laser power within the range of 400–600 W. When the spot diameter of Ni60 material is 1.5 mm, the laser power is 800 W, 1000 W, 1200 W, the temperature rise value is 974°, 1199°, 1413°; When the spot diameter is 2 mm, the laser power is 1200 W, 1500 W and 1800 W respectively, and the temperature rise is 827°, 1024°, and 1213°. Data shows that when the spot diameter is 1.5 mm, the laser power range is reasonable between 1000 and 1200 W; when the spot diameter is 2 mm, the laser power range is reasonable between 1500 and 1800 W. There is no linear relationship between spot diameter and material temperature rise.

2.2 Establishment of the Model

Matrix material: 20 mm long, 10 mm wide, 2 mm high; Laser cladding area: length 10 mm, width 3 mm, height 0.2 mm. Grid size: 2976 nodes, 444 cells, size 0.5 mm, **SOLID**186 cell division. The cladding material is Ni60 alloy, and the matrix material is high-speed steel. The initial reference temperature of femtosecond laser cladding was set as 22 °C, and the outer surface of the matrix and cladding layer was set as the convective heat transfer boundary.

3 Analysis of Temperature Field Results

3.1 Law of Temperature Field Changing with Location

The position of path 1 is from the starting point (5, 0, 0.2 mm) to the endpoint (15, 0, 0.2 mm), moving along the x direction; paths 2, 3 and 4 are separated by 0.1 mm along the z direction, respectively.

Figure 1 shows that the peak temperatures of paths 1, 2, and 3 are all up to the melting point of Ni60 alloy cladding powder. The depth of the matrix pool is 0.2 mm. The temperature variation of path 1 shows that the temperature curve shows transient nonlinear variation.

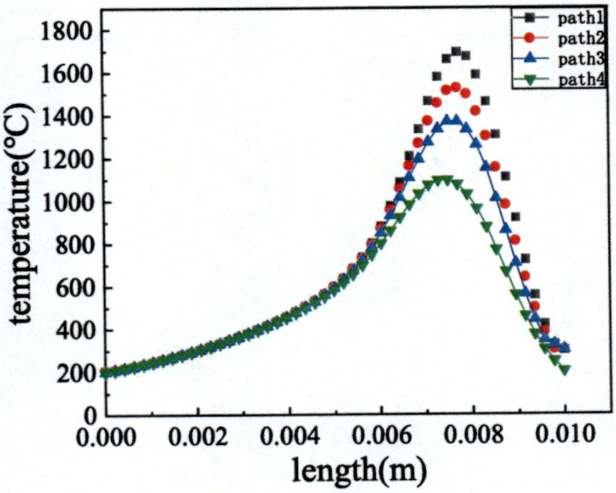

Fig. 1. Temperature field changes on different paths

3.2 Process Parameters - Temperature Field Results Influence

Seven groups of control experiments are designed in this paper, including three groups of variable laser power control experiments when the spot diameter is 1 mm, the scanning speed is 8 mm/s, and the laser power is 600, 800, and 1000 W. In the control group, the spot diameter was 1, 1.5, and 2 mm, and the scanning speed was 5, 8, and 10 mm/s. At the initial position coordinate origin x = 7 mm, y = 0, along the thickness z = 0.2 mm, z = 0, path A is set on the cladding layer, and path B; path C and path D are set at the substrate surface z = 0 and the substrate z = − 1 mm. When the laser power is 600, 800, and 1000 W, and the spot diameter is 1, 1.5, and 2 mm, the influence of the variable spot diameter and laser power on the temperature field is explored. Figure 2(a) shows that the temperature change curves corresponding to the three powers are distributed in the form of left and right symmetry. The peak temperature increases with the increase of laser power. Figure 2(b)shows that the peak temperature decreases with the increase of spot diameter. The temperature trend of the curve in Fig. 2(c)is the same. According to the formula of laser power density, the scanning speed determines the absorption rate of laser cladding energy. Figure 2(d) shows that when V = 5 mm/s, the peak temperature is 1606 °C, reaching the melting point of cladding material Ni60.

Figure 3(a)(b)(c) shows that the overall temperature field presents a semicircle shape, with the peak temperature appearing at the center. When the laser power is 600 W, the peak temperature of the matrix is lower than the melting point of the matrix. When the laser power is 800 W, the temperature gradient of the cladding layer ranges from 2373.2 to 2708 °C, and the peak temperature of the matrix is 2038 °C, forming an excellent metallurgical combination. When the laser power is 1000 W, the peak temperature of the cladding layer reaches 3692.4 °C. At this time, the cladding powder will be evaporated and disappear, and the matrix material will also cause edge collapse due to the high temperature of the laser beam [8]. Figure 3(d)(e)(f) shows that when the laser power and moving speed are the same, the comet-like tail gradually becomes more extended, and the range of the heat-affected zone increases. Figure 3(g)(h) shows that with the

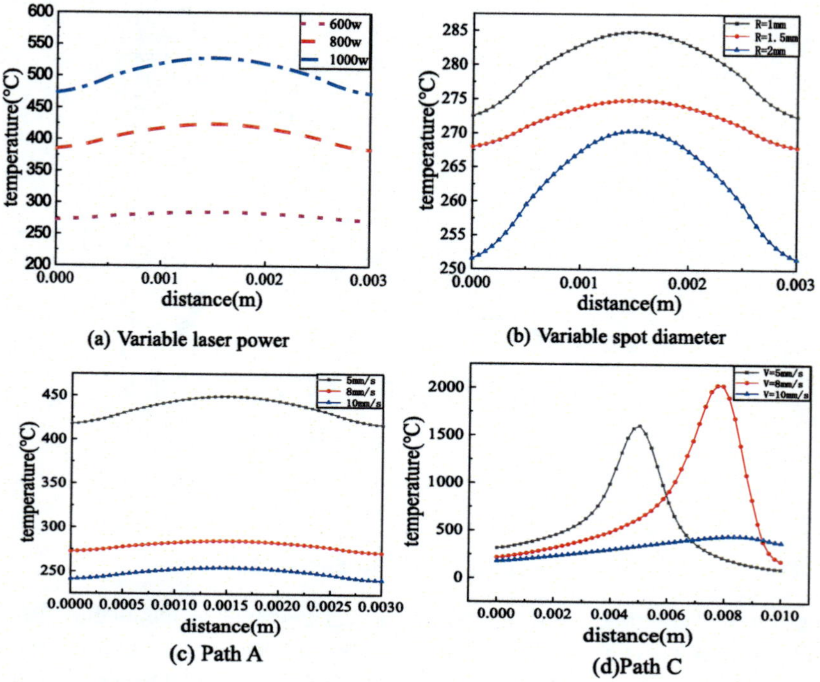

Fig. 2. Effect of process parameters on temperature field

increase of the spot moving speed, the heat affected zone value and temperature gradient of the heat source in the thickness direction gradually decrease. The contact surface temperature between the matrix and the cladding material is 964.9–1342 °C.

4 Conclusion

The relationship between the temperature rise of cladding material and laser power and spot diameter is studied in this paper. When the spot diameter is 1 mm, the laser power is adjusted to 400–600 W. When the spot diameter is 1.5 mm, the laser power is 1000–1200 W; When the spot diameter is 2 mm, the laser power range can be adjusted from 1500 to 1800 W. The peak temperature field, the heat-affected zone range, and the molten pool depth were compared with different process parameters. Screening process parameters P = 800 W, D = 1 mm, V = 8 mm/s; P = 600 W, D = 1 mm, V = 5 mm/s.

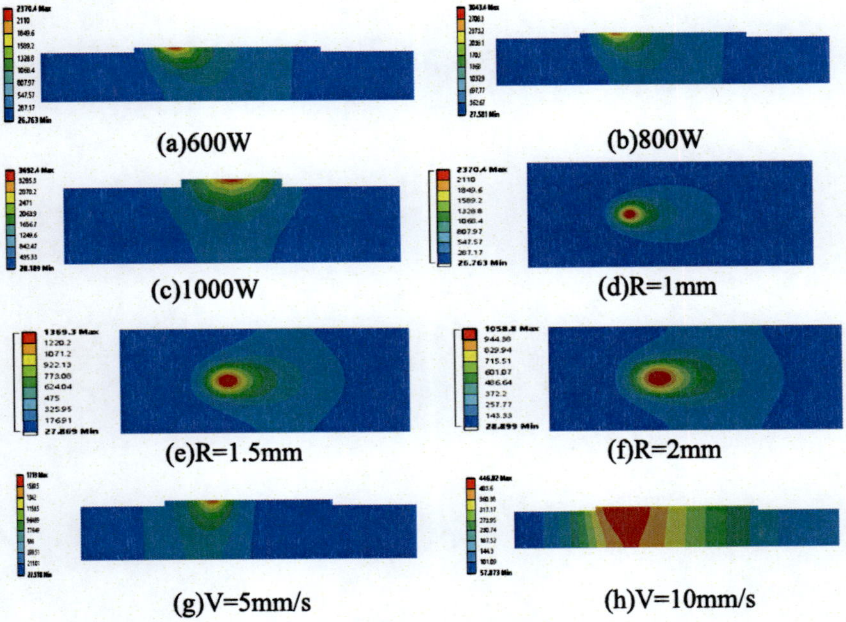

Fig. 3. Temperature field cloud image

References

1. Farahmand, P., Kovacevic, R.: An experimental-numerical investigation of heat distribution and stress field in single- and multi-track laser cladding by a high-power direct diode laser. J. Opt. Laser Technol. **63**, 154–168 (2014)
2. Kai, J.W., Yin, L., Hu, Z.W.: A simulation study on temperature field and stress field of laser cladding Mo2NiB2 layer. J. Acta Ceram. Sin. **42**(06), 1064–1071 (2021)
3. Hao, N., Lu, W., Zuo, T.: Analysis of thermal-mechanical coupled finite element stress field in laser cladding process. J. Chin. Table Surf. Eng. **1**(70), 20–23 (2005)
4. Hao, N., Lu, W., Zuo, T.: Thermomechanical coupled finite element temperature field analysis of laser cladding process. J. Chin. Table Surf. Eng. **6**(69), 10–14 (2004)
5. Yunchang, F., Loredo, A., Martin, B.: A theoretical model for laser and powder particles interaction during laser cladding. J. Mater. Process. Technol. **128**, 106–112 (2002)
6. Lin, J.: A simple model of powder catchment in coaxial laser cladding. J. Opt. Laser Technol. **31**(3), 233–238(1999)
7. Yang, Y., Song, Y.: Interaction of a laser beam and alloy powders in powder-feed laser cladding. Chin. J. Lasers **A25**(3), 280–284(1998)
8. Liu, F., Liu, D.: Study on grain refinement and parameter optimization of high-speed steel by laser quenching. D. Yanshan University (2021)

Author Index

M. Xu et al. (Eds.): CACPP 2022, LNEE 991, pp. 649–651, 2023.
https://doi.org/10.1007/978-981-19-9024-3